Cooperative Radio Communications for Green Smart Environments

RIVER PUBLISHERS SERIES IN COMMUNICATIONS

Volume 47

The "River Publishers Series in Communications" is a series of comprehensive academic and professional books which focus on communication and network systems. The series focuses on topics ranging from the theory and use of systems involving all terminals, computers, and information processors; wired and wireless networks; and network layouts, protocols, architectures, and implementations. Furthermore, developments toward new market demands in systems, products, and technologies such as personal communications services, multimedia systems, enterprise networks, and optical communications systems are also covered.

Books published in the series include research monographs, edited volumes, handbooks and textbooks. The books provide professionals, researchers, educators, and advanced students in the field with an invaluable insight into the latest research and developments.

Topics covered in the series include, but are by no means restricted to the following:

- Wireless Communications
- Networks
- Security
- Antennas & Propagation
- Microwaves
- Software Defined Radio

For a list of other books in this series, visit www.riverpublishers.com

Cooperative Radio Communications for Green Smart Environments

Editor

Narcis Cardona

Chairman of EU COST IC1004

Universitat Politecnica de Valencia

Spain

Published 2016 by River Publishers
River Publishers
Alsbjergvej 10, 9260 Gistrup, Denmark
www.riverpublishers.com

Distributed exclusively by Routledge
4 Park Square, Milton Park, Abingdon, Oxon OX14 4RN
605 Third Avenue, New York, NY 10158

First published in paperback 2024

Cooperative Radio Communications for Green Smart Environments / by Narcis Cardona.

Routledge is an imprint of the Taylor & Francis Group, an informa business

ISBN: 978-87-93379-15-2 (hbk)
ISBN: 978-87-7004-458-5 (pbk)
ISBN: 978-1-003-33772-0 (ebk)

DOI: 10.1201/9781003337720

Contents

Preface

This book presents the scientific results achieved within the framework of the European Cooperation in Science and Technology (COST) Action IC1004 "Cooperative Communications for Green and Smart Environments" (see www.ic1004.org for details).

COST Actions have succeeded in providing one of the best networking instruments of FP6 and FP7, and currently in H2020. In COST the research excellence of universities meets the technological leadership and applied view of the Industry, increasing the synergies between both, and getting mutual benefit from joint discussions. COST Actions base their success in the networking effects produced within its framework, which are not relying on any partnership contract or specific funded competitive objectives, but on the bottom-up approach of sharing results, and the technical discussions among its participants. COST actions are better understood as a networking instrument, where results from projects funded by many different sources are discussed, and a natural coordination between the participating researchers' interests is reached. A total of 600 individuals from more than 140 institutions have participated to IC1004, both from Academia and Industries, coming from 34 countries in Europe, America and Asia.

The scientific focus of COST IC1004 has been the evolution of Mobile and Wireless technologies towards cooperative networking strategies, in all the new scenarios of application and challenges that came up after the exponential increase of broadband data service demand. COST IC1004 run in parallel with the deployment of the 4G (LTE) networks and its evolution, as well as with the initial steps in the definition of 5G concepts, being many of the green and smart scenarios identified at the beginning of the Action those currently included in the ITU and 3GPP roadmaps. The topics discussed and reported during the Action period (May 2011–May 2015) produced 980 technical documents, which results are described in this book and contain, among many other aspects, radio propagation models, testing methods, networking algorithms, and contributions to future standards on Wireless Communications.

Being this book the result of the work of many colleagues, I would like to acknowledge all COST IC1004 participants, to its Management Committee for their continued commitment and contributions, and in particular to the Steering Committee, who helped me in the direct management of the Action: Prof. Roberto Verdone, Prof. Luis Correia, Prof. Silvia Ruiz, Prof. Alain Sibille, Prof. Claude Oestges, Prof. Alister Burr, Prof. Christoph Mecklenbräuker, and all the working group chairs, who served also as chapter editors, Prof. Buon Kiong Lau, Prof. Sana Salous, Dr. Denis Rose, Dr. Raffaele D'Errico, Dr. Kamya Y. Yazdandoost, Dr. Levent Ekiz, Prof. Erik Strom, Prof. Gert F. Pedersen, Prof. Wim Kotterman, Prof. Pawel Kulakowski, Prof. Klaus Witrisal, Prof. Katsuyuki Haneda, Dr. Tommi Jämsä, Prof. Jan Sykora. It is also of outmost importance the constant support we received from the COST Office, from our Officer Dr. Ralph Stübner, and the excellent administrative management done by the Secretary of this Action Ms. Silvia Zampese.

Finally, I would like to mention that the "spirit" of COST Actions is to become inclusive research networks, in which any researcher from any institution can hold scientific discussions based on a bottom-up approach, and not constrained to specific a-priori technological objectives. The technologies of the future are based on the science of today, and today's science needs fresh and new ideas that can only come from new generations of researchers. That is why COST is a perfect framework for Early Stage Researchers (ESRs) in the European Research Area to meet experienced colleagues from Academia and Industry, and that is the reason why this book is dedicated to all of them, and in memory of Michal.

Narcís Cardona
(COST IC1004 Chairman)
Valencia

Acknowledgments

This publication is supported by COST.

COST (European Cooperation in Science and Technology) is a pan-European intergovernmental framework. Its mission is to enable break-through scientific and technological developments leading to new concepts and products and thereby contribute to strengthening Europe's research and innovation capacities. www.cost.eu

This article is based upon work from COST Action IC1004, Cooperative Radio Communications for Green Smart Environments, supported by COST (European Cooperation in Science and Technology).

COST is supported
by the EU Framework
Programme Horizon 2020

List of Contributors

Taimoor Abbas, *Volvo Car AB, Sweden*

Sławomir J. Ambroziak, *Faculty of Electronics, Telecommunications and Informatics, Gdańsk University of Technology, Poland*

Pevand Bahramzy, *Intel Corporation, Denmark*

Aliou Bamba, *Ghent University, Belgium*

Sandra Caporal Del Barrio, *Aalborg University, Denmark*

Vasile Bota, *Technical University Cluj-Napoca, Romania*

Conor Brennan, *School of Electronic Engineering, Dublin City University, Ireland*

Alister Burr, *University of York, York, UK*

Narcis Cardona, *iTEAM – Universitat Politècnica de Valencia, Spain*

Symeon Chatzinotas, *University of Luxembourg, Luxembourg*

Laurent Clavier, *Télécom Lille, Univ. Lille, CNRS, UMR 8520 – IEMN – Institut d'Électronique, de micro Électronique et de Nanotechnologie, F-59000 Lille, France*

Luis Correia, *Instituto Superior Técnico – University of Lisbon, Portugal*

Simon Cotton, *The Institute of Electronics, Communications and Information Technology (ECIT), Queen's University Belfast, Belfast, UK*

Raffaele D'Errico, *Systems and Solutions Integration Department, Antenna and Propagation Laboratory, Commissariat à l'énergieatomiqueet aux énergies alternatives, MINATEC Campus | Grenoble Cedex, France*

Ghassan Dahman, *Lund University, Sweden*

Margot Deruyck, *University of Ghent, Belgium*

Goran Dimić, *University of Belgrade, Institute Mihajlo Pupin, Volgina 15, 11060 Belgrade, Serbia*

Levent Ekiz, *BMW, Germany*

Wei Fan, *Institut for Elektroniske Systemer, Aalborg Universitet, Denmark*

Lucio Studer Ferreira, *Instituto Superior Tecnico Lisboa, Portugal*

Jose Flordelis, *Lund University, Sweden*

Franco Fuschini, *Department of Electrical, Electronic, and Information Engineering "G. Marconi", Alma Mater Studiorum University of Bologna, Italy*

Christoph Gagern, *Rhode & Schwartz, Germany*

Concepción García Pardo, *Universitat Politecnica de Valencia, Spain*

Mario García-Lozano, *Universitat Politècnica de Catalunya, Spain*

Philip K. Gentner, *KATHREIN-Werke KG, Germany*

David Gonzalez, *Aalto University, Finland*

Paolo Grazioso, *Fundazione Ugo Bordoni, Italy*

Carl Gustafson, *Lund University, Sweden*

Katsuyuki Haneda, *Aalto University, Finland*

Ruisi He, *Beijing Jiaotong University, China*

Sooyoung Hur, *Samsung, Korea*

Morten Lomholt Jakobsen, *Department of Electronic Systems, Aalborg University, Denmark*

Tommi Jamsa, *Huawei Technologies, Sweden*

Tomaz Javornik, *Jožef Stefan Institute, Jamovacesta 39, Slovenia*

Wout Joseph, *Ghent University, Belgium*

Wim Kotterman, *Institut für Informationstechnik, Technische Universität Ilmenau, Germany*

Pawel Kulakowski, *AGH University of Science and Technology, Poland*

Timo Kumpuniemi, *Centre for Wireless Communications (CWC), Department of Communications Engineering (DCE), University of Oulu, Finland*

Pekka Kyösti, *Anite Telecoms Oy, Finland*

Gregor Lasser, *University of Colorado, Boulder, USA*

Buon Kiong Lau, *Lund University, Sweden*

Per-Hjalmar Lehne, *Telenor ASA Research, Norway*

Erik Leitinger, *Graz University of Technology, Austria*

Maria Lema, *King's College London, UK*

Hui Li, *Dalian University of Technology, China*

Michal Mackowiak (deceased before printing), *Instituto Superior Tecnico – University of Lisbon, Portugal*

Zachary Miers, *Lund University, Sweden*

Jose Maria Molina-Garcia-Pardo, *Universidad Politécnica de Cartagena, Spain*

Paul Meissner, *Graz University of Technology, Austria*

Mohamad Mostafa, *Deutsches Zentrumfür Luft- und Raumfahrt (DLR), Institutfür Kommunikation und Navigation Nachrichtensysteme, Oberpfaffenhofen-Wessling, Germany*

Andrés Navarro, *Universidad ICESI Cali, Colombia*

Jörg Nuckelt, *Siemens AG, Germany*

Claude Oestges, *ICTEAM Universitécatholique de Louvain, Louvain-la-neuve, Belgium*

Joan Olmos, *Barcelona TECH-UPC, Universitat Politècnica de Catalunya, Spain*

Jan Papaj, *Technical University of Kosice, Slovakia*

Jeongho. Jh. Park, *Samsung, Korea*

Gert F. Pedersen, *Department of Electronic Systems, Aalborg University, Denmark*

Michael Peter, *HHI, Berlin, Germany*

Robert Rehammar, *Bluetest AB, Sweden*

Denis Rose, *Technical University of Braunschweig, Germany*

Ramona Rosini, *Lepidaspa, Viale Aldo Moro, 64-40127 Bologna, Italy*

Silvia Ruiz, *Universitat Politècnica de Catalunya, BarcelonaTech, Spain*

Moray Rumney, *Keysight Technologies, United Kingdom*

Sana Salous, *School of Engineering and Computing Sciences, Durham University, UK*

Kamran Sayrafian, *Information Technology Laboratory, National Institute of Standards and Technology, MD, USA*

Christian Schneider, *Institute for Information Technology, Technische Universität Ilmenau, Germany*

Werner L. Schroeder, *RheinMain University of Applied Sciences, Germany*

Veronika Shivaldova, *Kathrein Group, Germany*

Alain Sibille, *LTCI, CNRS, Télécom ParisTech, Université Paris-Saclay, Paris, France*

Erik G. Ström, *Chalmers Univ. of Technology, Sweden*

Jan Sykora, *Czech Technical University in Prague, Czech Republic*

Istvan Szini, *Antenna Innovation Lab Research, Motorola Mobility, USA*

Emmeric Tanghe, *Ghent University, Belgium*

Alexandru Tamiromescu, *Aalborg University, Denmark*

Werner Teich, *Ulm University, Institute of Communications Engineering, Ulm, Germany*

Fredrik Tufvesson, *Lund University, Sweden*

Fernando J. Velez, *Instituto de Telecomunicações – DEM, Universidade da Beira Interior, Faculdade de Engenharia 6201-001 Covilhã, Portugal*

Roberto Verdone, *Università di Bologna, Italy*

Usman Tahir Virk, *Aalto University, Finland*

Enrico Maria Vitucci, *University of Bologna, Italy*

Christoph von Gagern, *Rohde & Schwarz GmbH & Co. KG, Germany*

Thomas Werthmann, *University of Stuttgart, Germany*

Klaus Witrisal, *Graz University of Technology, Austria*

Kamya Yekeh Yazdandoost, *Dependable Wireless Laboratory, Wireless Network Research Institute, National Institute of Information and Communications Technology, Yokosuka, Japan*

Yan Zhang, *Simula Research Laboratory and University of Oslo, Norway*

List of Figures

List of Tables

List of Abbreviations

μTVWS	Micro TV White Space
2-D	Two-Dimensional
2DP	2 Dominant Paths
2G	Second Generation
2-WRC	2-Way Relay Channel
3-D	Three-Dimensional
3G	Third Generation
3GPP	3rd Generation Mobile Group
3GPP	3rd Generation Partnership Project
4G	Fourth Generation
5G	Fifth Generation
5G PPP	5G infrastructure Public Private Partnership
AADA	Adaptive Alamouti Decoding Algorithm
AADS	Alamouti Adaptive Distributed Scheme
AAS	Advanced Antenna System
AF	Amplify and Forward
AFR	Aggregation with Fragment Repetition
AICc	Second Order Akaike Information Criterion
ANC	Analogue Network Coding
ANT	Antenna
AoA	Angle-of-Arrival
AoD	Angle of Departure
AAoD	Azimuth of Departure
AP	Access Point
APP	*A Posteriori* Probability
ARQ	Automatic Retransmission Request
ASU	Arbitrary Strength Unit
ATF	Antenna Test Function
ATPC	Adaptive Transmission Power Control
AUT	Antenna Under Test
AWGN	Additive White Gaussian Noise

B2B	Body-to-Body
BAN	Body Area Network
BBU	Base Band Unit
BCJR	Bahl, Cocke, Jelinek, Raviv
BER	Bit Error Rate
BICM	Bit-Interleaved Coded Modulation
BICM-ID	Block-Interleaved Coded Modulation-Iterative Detection
BLER	Block Error Rate
BPSK	Binary Phase Shift Keying
BS	Base Station
BTB	Best-To-Best
BT-LE	BlueTooth-Low Energy
BTW	Best-To-Worst
C/I	Carrier-to-Interference ratio
C2C	Car-to-Car
C2i	Car to infrastructure
CAGP	Clustering Algorithm for Gateway Positioning
CCDF	Complementary Cumulative Distribution Function
CCMSim	Car2X Channel Model Simulation
CDF	Cumulative Distribution Function
CDMA	Code Division Multiple Access
CE	Channel Emulator
Cell ID	Cell IDentity
CEO	Chief Executive Officer
CEPT	European Conference of Postal and Telecommunications Administrations
CFC	Carbon-Fibre Composite
CIFA	Circular Inverted-F Antenna
CIN	Cooperative Interference Neutralisation
CIR	Carrier to Interference Ratio
CIR	Channel Impulse Response
C-ITSs	Cooperative Intelligent Transport Systems
CL-SM	Closed-Loop Spatial Multiplexing
CM	Channel Model
CMD	Correlation Matrix Distance
CoF	Compute and Forward
COMP	Cooperative Multi-Point
CoMP	Coordinated Multi-Point
COST	European Cooperation in Science and Technology

COTS	Commercial Off The Shelf
CPE	Customer Premises Equipment
CPU	Central Processing Unit
CQI	Channel Quality Indicator
CR	Cognitive Radio
C-RAN	Cloud Radio Access Network
CRC	Cyclic Redundancy Check
CRLB	Cramér Rao Lower Bound
CRRM	Common Radio Resource Management
CSI	Channel State Information
CSMA	Carrier Sense Multiple Access
CSMA/CA	Carrier Sense Multiple Access with Collision Avoidance
CSO	Cell Switch Off
CTC	Convolutional Turbo Codes
CTIA	International Association for the Wireless Telecommunications Industry
CTS	Clear-To-Send
CW	Continuous Wave
D	Destination
D2D	Device-to-Device
D-ACC	Doped Accumulator
DAS	Distributed Antenna System
DCC	Decentralised Congestion Control
DCRS	Distributed Cooperative Relay System
DD	Digital Dividend
DD2	Second Digital Dividend
DF	Decode and Forward
DFT	Discrete Fourier Transform
Dir	Direct
DJSCC	Distributed Joint Source-Channel Coding
DL	Downlink
DMC	Dense Multipath Components
DNF	DeNoise and Forward
dNLOS	dominant Non-Line-Of-Sight
DoA	Direction-of-Arrival
DoF	Degree of Freedom
DOF	Degrees-Of-Freedom
DPC	Dirty-Paper Coding
DPSK	Differential Phase Shift Keying

DP-UMCS	Dual-Polarised Ultra-wideband Multi-Channel Sounder
DS	Delay Spread
DS	Diffuse Scattering
DSR	Dynamic Source Routing protocol
DSRC	Dedicated Short-Range Communications
DTN	Delay Tolerant Network
DTT	Digital Terrestrial Television
DuT	Device under Test
DVB	Digital Video Broadcasting
DVB-H	Digital Video Broadcasting-Handheld
DVB-NGF	Digital Video Broadcasting-Next Generation Handheld
DVB-T	Digital Video Broadcasting-Terrestrial
DVB-T2	Digital Video Broadcasting-Terrestrial 2nd Generation
EADF	Effective Aperture Distribution Function
EC	Energy Capture
ECC	Envelope Correlation Coefficient
ECC	Electronic Communications Committee
ECDF	Empirical Distribution Function
EDOF	Effective Degree-Of-Freedom
EE	Energy Efficiency
EIRP	Equivalent Isotropic Radiated Power
EKF	Extended Kalman Filter
EM	Expectation Maximisation
eNB	enhanced Node-B
eNB	evolved Node-B
EoD	Elevation of Departure
EPC	Evolved Packet Core
ER	Effective Roughness
ESNR	Effective Signal-to-Noise Ratio
ESPRIT	Estimation of Signal Parameters via Rotational Invariance Techniques
ETSI	European Telecommunications Standards Institute
EuQ	Equipment under Test
EVA	Extended Vehicular A
EXIT	EXtrinsic-Information Transfer
FBMC	Filter Bank Multi-Carrier
FC	Fusion Centre
FDD	Frequency Division Duplex
FDTD	Finite Difference Time Domain

FEC	Forward Error Correction
FER	Frame Error Rate
FFR	Fractional Frequency Reuse
FFT	Fast Fourier Transform
FG	Fixed Gain
FLC	Fuzzy Logic Controller
FMCW	Frequency-Modulated Continuous Wave
FoM	Figure-of-Merit
FRC	Fixed Reference Channel
FSA	Fixed Spectrum Assignment
FSK	Frequency Shift Keying
FSS	Frequency Selective Surface
GBSC	Geometry-Based Stochastic Channel
GDoF	Generalised Degree of Freedom
GEV	Generalised Extreme Value Distribution
GMT	Generalised Multipole Technique
GO	Geometrical Optics
GPS	Global Positioning System
GSCM	Geometry-based Stochastic Channel Model
GSE	Green Smart Environment
GSM	Global System for Mobile Communication
GTP	Geometrical Theory of Propagation
H-ARQ	Hybrid Automat Repeat Request
HD	High Dynamic
HDF	Hierarchical Decode and Forward
HetNet	Heterogeneous Network
HF	High Frequency
H-IFC	Hierarchical Interference Cancellation
HNC	Hierarchical Network Code
HND	Hard Network Decoding
HO	HandOver
HOF	HandOver Failure
HOM	HandOver Margin
HSI	Hierarchical Side Information
HSPA	High-Speed Packet Access
HSR	High-Speed Railway
HU	Hub Unit
i.i.d.	independent and identically distributed
IA	Interference Alignment

IC	Interference Cancellation
IC	Interference Channel
ICI	Inter Carrier Interference
ICIC	InterCell Interference Coordination
ICNIRP	International Commission on Non-Ionising Radiation Protection
IDMA	Interleave Division Multiple Access
IEEE	Institute of Electrical and Electronics Engineers
IF	Intermediate Frequency
IFA	Inverted-F Antenna
IIC	Iterative Interference Cancellation
IIC-RNN	Iterative Interference Cancellation-based Recurrent Neural Network
IID	Independent and Identically Distributed
ILA	Iterative Local Approximation
IMT	International Mobile Telecommunications
IN	Impulsive Noise
INR	Interference-to-Noise Ratio
IO	Interacting Object
IoT	Internet of Things
IR	Incremental Redundancy
IRC	Interference Rejection Combining
IRC	Interference Relay Channel
ISI	Inter Symbol Interference
ISM	Industrial, Scientific and Medical
ISD	Iterative Spatial Demapping
ITS	Intelligent Transportation Systems
ITU	International Telecommunications Union
ITU-R	ITU Radiocommunication sector
KLT	Karhunen–Loeve Transform
KPIs	Key Performance Indicators
K–S	Kolmogorov–Smirnov
L1	Layer 1
L2	Layer 2
LA	Link Adaptation
LAC	Location Area Code
LCP	Liquid Crystal Polymer
LCR	Level Crossing Rate
LCX	Leaky Coaxial Cable

LD	Low Dynamic
LDHC	Long Delay High Correlation
LDLC	Low-Density Lattice Code
LDPC	Low-Density Parity Check
LER	Largest Eigenmode Relaying
LLR	Log-Likelihood Ratio
LoS	Line-of-Sight
LSF	Large-Scale Fading
LSP	Large-Scale Parameter
LTE	Long-Term Evolution
LTE-A	Long Term Evolution-Advanced
LUT	Look-Up-Table
M2i	Machine-to-infrastructure
M2M	Machine-to-Machine
MAC	Medium Access Control
MAC	Multiple Access Channel
MAI	Multiple-Access Interference
MANET	Mobile *Ad hoc* Network
MAP	Maximum *A Posteriori*
MAP	Multi-radio mesh Access Points
MAT	Maddah-Ali-Tse
MBS	Macrocell Base Station
MC-CDMA	MultiCarrier Code Division Multiple Access
MCS	Modulation and Code Scheme
MCT	Mobile Container Terminal
MD	Medium Dynamic
ME	Mean Error
MF	Matched Filtering
MFSK	M-ary Frequency Shift Keying
MI	Mutual Information
MICS	Medical Implant Communication Service
MID	Molded Interconnect Device
MIESM	Mutual Information Effective Signal-to-noise Mapping
MIMO	Multiple-Input Multiple-Output
MISO	Multiple-Input Single-Output
ML	Maximum Likelihood
MMIB	Mean Mutual Information per coded Bit
MMSE	Minimum Mean-Squared Error
mmW	millimetre Wave

MoM	Method of Moments
MOS	Mean Opinion Score
MOSG	MIMO OTA Sub Group (of CTIA)
MPAC	Multi-Probe Anechoic Chamber
MPC	MultiPath Component
MPP	Mesh Point Portal
MPR	Maximum Power Reduction
MRC	Maximum Ratio Combining
MRE	Maximum Radial Extend
MS	Mobile Station
MSD	Multi Stage Detector
MSE	Mean Square Error
MSR	Maximum Sum-Rate
MT	Mobile Terminal
MTC	Machine-Type Communication
MU	Measurement Uncertainty
MU-MIMO	Multi-User MIMO
MWM	Multi-Wall Model
NB	Nash Bargaining
NCM	Network Coded Modulation
NE	Nash Equilibrium
NFV	Network Functions Virtualisation
NGMN	Next Generation Mobile Networks
NIG	Normal Inverse Gaussian
NIST	National Institute of Standards and Technology
NLOS	Non-Line-Of-Sight
nLoS	non-Line-of-Sight
NOMA	Non-Orthogonal Multiple Access
NOSS	Non-Orthogonal Spectrum Sharing
OFDM	Orthogonal Frequency Division Multiplex
OFDMA	Orthogonal Frequency Division Multiple Access
OLOS	Obstructed Line-Of-Sight
OLPC	Open Loop Power Control
OL-SM	Open-Loop Spatial Multiplexing
OP	Optimistic
OP	Outage Probability
OppNet	Opportunistic Networks
OR	Operating Region
OSI	Open System Interconnection

OSS	Orthogonal Spectrum Sharing
OSTBC	Orthogonal STBC
OTA	Over-The-Air (Testing)
PA	Power Amplifier
PAPR	Peak-to-Average Power Ratio
PAS	Power Angular Spectrum
PAS	Power Azimuth Spectrum
PCB	Printed Circuit Board
PCI	Physical Cell Identifier
P-CSMA	Priority-based Carrier-Sense Multiple Access
PDCCH	Physical Downlink Control CHannel
PDF	Probability Density Function
pdf	probability density function
PDP	Power Delay Profile
PDR	Packet Delivery Ratio
PE	Pessimistic
PEB	Position Error Bound
PEEC	Partial Element Equivalent Circuit
PER	Packet Error Rate
PF	Proportional Fair
PFS	Pre-Faded Synthesis
PHY	Physical Layer
PIFA	Planar Inverted-F Antenna
PL	Path Loss
PLNC	Physical Layer Network Coding
PLR	Packet Loss Rate
PM	Planar Monopole
PMI	Precoding Matrix Indicator
PN	Phase Noise
PO	Physical Optics
PP	Pilot Pattern
PPP	Poisson Point Process
PR	Protection Ratio
PRB	Physical Resource Block
PRBS	Pseudo-Random Binary Sequences
PSD	Power Spectral Density
PSK	Phase Shift Keying
PSS	Primary Synchronisation Signal
PWS	Plane Wave Synthesis

QAM	Quadrature Amplitude Modulation
Q-D	Quasi-Deterministic
QoE	Quality of Experience
QoS	Quality of Service
QOSTBC	Quasi-Orthogonal STBC
RA	Rate Adaptation
RA	Relay-Assisted
RAN	Radio Access Network
RANaaS	Radio Access Network as a Service
RAR	Repeat Accumulate Repeat
RAT	Radio Access Technology
RAU	Remote Antenna Unit
RBIR	Received Bit Information Rate
RC	Reverberation Chamber
RCS	Relay Control Station
REM	Room ElectroMagnetics
RF	Radio Frequency
RFID	Radio-Frequency IDentification
RI	Rank Indicator
RIMP	RIch MultiPath
RL	Realistic
RMS	Root Mean Square
RMSE	Root Mean Square Error
RN	Relay Node
RNN	Recurrent Neural Network
RoF	Radio-over-Fibre
RRC	Radio Resource Control
RRH	Remote Radio Head
RRHs	Remote Radio Heads
RRM	Radio Resource Management
RRU	Radio Resources Unit
RS	Reference Signals
RS EPRE	Reference Signal Energy Per Resource Element
RSAP	Reference Signal Antenna Power
RSARP	Reference Signal Antenna Relative Phase
RSC	Recursive Systematic Convolutional
RSRP	Reference Signal Received Power
RSRQ	Reference Signal Received Quality
RSS	Received Signal Strength
RSSI	Received Signal Strength Indicator

RSU	RoadSide Unit
RT	Ray Tracing
RTS/CTS	Request-To-Send/Clear-To-Send
RU	Radio Unit
RV	Redundancy Version
RX	Receiver
RXQUAL	Rx Quality
S	Source
SAGE	Space-Alternating Generalised Expectation Maximisation
SAR	Specific Absorption Rate
SBACK-MAC	Sensor Block Acknowledgement
SC	Single Carrier
SC	Superposition Coding
SCC	Spatial Cross Correlation
SC-FDMA	Single Carrier-Frequency Division Multiple Access
SCM	SubCarrier Mapping
SCME	3GPP Spatial Channel Model Extended
SCN	Small Cell Network
SCO	Semi-Closed Obstacles
SD	Spectral Divergence
SDLC	Short Delay Low Correlation
SDMA	Space-Division Multiple Access
SDN	Software Defined Network
SDoF	Spatial Degrees-of-Freedom
SE	Stackelberg Equilibrium
SE	Smart Environment
SEE	Standard Error of Estimate
SER	Symbol Error Rate
SF	Super Frame
SFR	Soft Frequency Reuse
SG	Stackelberg Game
SI	Study Item
SIC	Successive Interference Cancellation
SIMO	Single-Input Multiple-Output
SINR	Signal-to-Interference plus Noise Ratio
SIR	Signal-to-Interference Ratio
SISO	Single-Input Single-Output
SLA	Service Level Agreement
SM	Spatial Multiplexing

SNIR Signal to Noise plus Interference Ratio
SNR Signal-to-Noise Ratio
SoC System-on-a-Chip
SON Self-Organising Network
SOR Successive Over Relaxation
SOSF Second-Order Scattering Fading
SoTDMA Self-organising Time Division Multiple Access
SPCC Spatial and Polarisation Cross Correlation
SpE Spectral Efficiency
SR Selective Relaying
SRS Sounding Reference Signal
SS HO Small cell-to-Small cell HandOver
SSF Small-Scale Fading
ST(F)BC Space–Time (Frequency) Block Coding
STA Station
STB Set Top Box
STIA Space–Time Interference Alignment
TAC Traction Area Code
TAPAS Travel and Activity PAtterns Simulation
TASPS Time Averaged Simultaneous Peak SAR
TB Transport Block
TCM Theory of Characteristic Modes
TCP Transport Control Protocol
TD Transmit Diversity
TDMΛ Timc Division Multiplc Acccss
TDOA Time-Difference-Of-Arrival
TG Task Group
TIS Total Isotropic Sensitivity
TLM Top Loaded Monopole
TM2 Transmission Mode 2
TOA Time-Of-Arrival
TP ThroughPut
TPC Transmission Power Control
TRx Transceiver
TSB Truncated Shannon Bound
TTI Transmission Time Interval
TTT Time-To-Trigger
TVWS TV White Spaces
TWG Topical Working Group

TX	Transmitter
Tx	Transmission
UE	User Equipment
UHF	Ultra High Frequency
UL	Uplink
ULA	Uniform Linear Array
UMa	Urban-Macro
UMi	Urban-Micro
UMTS	Universal Mobile Telecommunications System
US	Uncorrelated Scattering
USB	Universal Serial Bus
USP	User Social Pattern
UTD	Uniform Theory of Diffraction
UTRAN	UMTS Terrestrial Radio Access Network
UWB	Ultra-WideBand
V2i	Vehicular to infrastructure
V2V	Vehicle-to-Vehicle
V2X	Vehicular to any (Communications)
VA	Virtual Anchor
VAA	Virtual Antenna Array
VANET	Vehicular *Ad Hoc* Network
VBS	Virtual Base Station
VEE	Virtual Electromagnetic Environment
VLSI	Very Large-Scale Integration
VNA	Vector Network Analyzer
VNet	Virtual Network
VNO	Virtual Network Operator
VoIP	Voice over IP
VPL	Vehicle Penetration Loss
VR	Visibility Region
VRRA	Virtual Network Radio Resource Allocation
VRRM	Virtual Radio Resource Management
VS	Virtual Source
VSimRTI	V2X Simulation RunTime Infrastructure
VTX	Virtual Transmitter
WAVE	Wireless Access in Vehicular Environments
WBE	Wireless Body Environment
WBEN	Wireless Body Environment Network
WCDMA	Wideband Code Division Multiple Access

WCPE	Wireless Channel Parameter Estimation
WGN	White Gaussian Noise
WI	Walfisch–Ikegami model
WiFi	Wireless Fidelity
WiMAX	Worldwide interoperability for Microwave Access
WINNER	Wireless World Initiative New Radio
WIoTs	Wireless Internet of Things
WLAN	Wireless Local Area Network
WMN	Wireless Mesh Network
WNC	Wireless Network Coding
WPAN	Wireless Personal Area Network
WPNC	Wireless Physical layer Network Coding
WRAN	Wireless Regional Area Network
WRC	World Radio communication Conference
WRC-07	2007 World Radiocommunication Conference
WRC-12	2012 World Radiocommunication Conference
WRC-15	2015 World Radiocommunication Conference
WSD	White Space Devices
WSN	Wireless Sensor Network
WSS	Wide-Sense Stationary
WSS-US	Wide-Sense Stationary-Uncorrelated Scattering
WTPC	Wireless Transceiver Power Consumption
WWRF	Wireless World Research Forum
XPD	Cross-Polarisation Discrimination
XPR	Cross-Polarisation Ratio
ZF	Zero Forcing
ZF-BLE	Zero-Forcing Block Linear Equaliser

1

Introduction

Narcis Cardona, Luis Correia and Roberto Verdone

1.1 Technology Trends

1.1.1 Mobile Networks and the Wireless Internet of *Everything*

Mobile Communications are exponentially evolving to allow any electronic device wirelessly to connect to the Internet. The first decade of the 20th century saw how the computer and the phone have converged into a single concept of user terminal, making mobile telephony and nomadic computing facilities coexist in a single device: the smartphone. As smartphones have become a revolution in the wireless user's experience, a second revolution at the terminal side may be expected, and the future traffic growth rate will be huge, due to the extension of mobile data communications to machines, vehicles, sensors, and smart objects. All these heterogeneous wireless devices must be connected to each other in a massive moving data scenario that is already being called wireless internet of things (WIoTs).

Every WIoT scenario is essentially an smart environment (SE), i.e., a physical (layer) (PHY) space populated by sensors, actuators, embedded systems, user terminals, and any other type of communicating device, which cooperatively pursue given tasks by exchanging information and share all types of resources, such as radio spectrum or energy. Some examples of SEs are the human body, vehicles on a road, and a smart building or a smart city. Any of the SEs in mobile communications was seen as the raising of a new generation of wireless sensor networks (WSNs), either local or wide area, with coordination and communication entities and group mobility. Nowadays all those SEs are intended to be merged with the current and future mobile radio access networks (RANs). In some cases, like connected vehicles and body area networks (BANs), a coordinated group of sensors is intending to communicate with an infrastructure network or other mobile devices. In other scenarios, such as Smart City and Connected Home, a moving terminal is intended to get information from neighbouring wireless devices and sensors.

Cooperative Radio Communications for Green Smart Environments, 1–16.

It is then necessary to find ways to efficiently merge those types of SEs with the current and future mobile RANs.

SEs are delimited in space, but not in scope. For example, wireless devices on or inside the human body can improve everyday life for patients needing continuous health care (Smart Health), but can also help the professional sports-man/woman to improve his/her performance; inter-vehicular and vehicular-to-infrastructure (V2i) communications can assist the driver and increase road safety (Smart Cars) or enable the provision of Internet services in the car; wireless sensors in buildings can provide information helpful for energy control purposes (Smart Energy) or support rescue teams in emergency situations.

SEs can provide better safety and lifestyle, and can help reduce global energy consumption by contributing to an intelligent distributed management of the energy resource. SEs may be green smart environments (GSEs) because of their application: especially, in enabling energy efficient lifestyles, for example in enabling home working and teleconferencing to reduce the need for travel. However, they need to become green themselves. The impact on the environment and global energy consumption of information and communication technologies (ICT) must be minimised; SEs must be green also in the sense that their deployment and use must follow energy efficient paradigms, regardless of their goals and application areas. Progress will come from better hardware design, but also improved transmission techniques and protocols.

Moreover, GSEs need to be efficient overall, with better use of energy (Joule/bit), radio spectrum (bits/s/Hz/m^2), computing resources, etc. To achieve this goal, GSEs have to make proper use of the concept of cooperation among network nodes, both at link level to maximise end-to-end throughput between nodes, through relaying, network coding, or other forms of cooperation, and at network level to maximise network capacity in a dense heterogeneous environment. Cooperation also involves the proper detection, mitigation, and management of inter-network interference, which arises from the possibility of autonomous networks merging and splitting.

1.1.2 Mobile Communication Scenarios

Cars have been the first focus in those types of scenarios, with the development of some standards for V2i and vehicular to vehicular (V2V) communications along the last decade. Nowadays, the automotive industry is increasingly adapting wireless sensors and communication systems to improve the connectivity of their vehicles. Four main aspects benefit from radio communications to cars: infotainment services to passengers, vehicular cloud services, traffic safety, and traffic efficiency. In beyond 4G wireless communications, vehicles will be integrated parts of the system, not just end-nodes. In essence, vehicles can act as mobile base stations, which will be beneficial with respect to filling in coverage holes, supporting local capacity needs that appear unpredictably in time and space, and providing good quality of experience for passengers. In addition to serving as a mobile base station, a vehicle can sense its environment to support a multitude of applications. For instance, real-time traffic and environmental monitoring is truly enabled, which in turn enables real-time traffic management to increase the efficiency of the transport system and reduce its environmental impact. Connecting vehicles to the cloud with a stable, high-rate, low-latency wireless link will bring many benefits. Heavy calculations and storage can be off-loaded to the cloud, vehicle maintenance will be facilitated, and novel services can be delivered to vehicle customers with a very short time-to-market. Traffic safety and traffic efficiency applications require vehicles and road infrastructure-like road signs, traffic lights, toll booths, etc., to exchange information to make transport safer and more efficient, reducing accidents, traffic jams, fuel consumption, and emissions. Hence, wireless communications is a crucial enabling technology for these applications.

Another raising SE is the wireless body environment (WBE), which is supposed to revolutionise health monitoring, with its huge number of possible applications in the home and hospital, for elderly care and emergency cases. WBE communications will enable a new generation of services and applications that give the user an enhanced and intuitive interaction with surrounding technologies. This interaction is boosted by a network of body implanted or wearable devices, operating in the immediate environment around and inside the human body, that can exchange important data, health parameters, in real time. Moreover wireless body environment networks (WBENs) are expected to provide new functionalities for applications in people's every-day lives, such as sport, leisure, gaming, and social networks. As this service permits remote monitoring of several patients simultaneously, it could also potentially decrease health care costs. Health systems will also benefit from the integration of wireless on- and in-body sensors, with applications to healthcare, remote monitoring, wellness and assisted surgery. WBENs are developing as a more advanced and separate addition to wireless communications in general. While its basic operating characteristics are the same as all radio systems, there are many features and specific problems that justify dealing with it separately from other forms of wireless communication, such as battery life, low power, low cost, small size, body dynamics, and in-body propagation. Body environment applications are designed for short-range applications with relatively low power and are regulated by the telecommunication authorities. Also, the devices are operated in or close to a human body, which affects communication performance. Therefore, it is needed to find models to predict a reliable operating range for these systems, based on propagation characteristics, human posture, and movements.

The *Smart Cities* concept has been developed during recent years mainly, but not only, on the basis of urban sensor networks, which provide information either to a centralised management system or to the moving devices around them. The Smart City sensors are intended to help the control of lighting, temperature, pollution, flow of traffic, gas, water, and electricity parameters, as well as monitoring streets, pipes and bridges, emergency responses and public transportation, among others. This concept is an evolution of the intelligent room and intelligent building ideas, and is addressed to both people living and working in the city and people organising and administering the urban infrastructures. Current ongoing applications to provide the bases of the Smart City deployment are: optimised transportation, personalised services, community services, and management of big data. From the radio access deployment perspective, a huge set of advanced technologies is required for a full implementation of Smart Cities, and substantial effort to integrate them is essential. The same happens with manufacturers, operators, service providers, and city administrators, since close collaboration is envisaged among them. All the existing wireless networks (cellular 2G/3G/4G and beyond, point-to-point, wireless local area network (WLAN), wireless personal area network (WPAN), wireless metropolitan area network (WMAN), and mesh networks), and also user terminals, mostly smartphones, which are today's main moving transceivers, will play an important role to facilitate Smart City deployment. Their role as personal standalone network access devices is changing in future networks, when the handheld will be active as a cooperative networking node, and then be able to measure environmental parameters through its own sensors, as well as to geo-reference them, thus contributing to Smart City monitoring procedures.

On the basis mentioned above that every electronic device is to be wirelessly connected, future scenarios in radio communications will give rise to situations where a huge number of devices are located in PHY proximity (in the space domain, relative to radio coverage ranges), while generating independent traffic with different patterns and needing to share the same pool of radio resources to create some type of network topology. These situations come from static and low mobility devices, as in machine to machine (M2M) and machine to infrastructure (M2I) communications, which are to improve industrial automation; or the connected home multimedia systems and sensors, where revenues are expected because of the increase of wireless data traffic, while improving user experience within the concept of connected living.

Also high-dense traffic occurs in high mobility scenarios, such as public events, either planned or unexpected, and in others like the smart cities mentioned above, offices, smart grids, and smart metering. In any of these situations, the number of contenders for the radio resources can potentially be much higher than those manageable by traditional wireless architectures, protocols, and procedures. Such extreme situations do not occur often, but may happen in some cases in regular operation, for example in stadiums, in clothing stores (where thousands of radio frequency identification (RFID) tags may exist), or in a huge depot of some company. In all cases, the possibility of maintaining highly efficient connectivity using wireless communications can provide significant added value as it may solve many problems. In fact the huge number of moving connected devices in certain areas, either accessing centralised services or getting information from sensors in their surroundings, increases the interest in developing selfish *ad hoc* dynamic network configuration strategies. The popular term to describe these situations is "*Moving Networks*", where many new concepts in network coding, relaying, self-organisation, or opportunistic caching, among others, are to be developed.

1.2 RANs-Enabling Technologies

1.2.1 Small Cells in Very High-Dense Deployment

There is already a natural trend in the infrastructures of mobile networks to reduce the range, and hence the size and complexity of the base stations, while increasing the number and bandwidth of the PHY connections in between smaller cell sites. The reason for this is the continuously increasing data traffic demand, which generates new opportunities for the provision of new services, which at the end give rise to a further increase of traffic capacity and throughput requirements. The wide deployment of optical communications networks, with fibre connections closer to end users, makes sense also for those wideband connections in between small cells, changing the current basic concept of traffic-scaled cellular deployment to a modern view of opportunistic spectrum-access-based cooperative networking.

In RAN deployment, it is by now obvious that installing smaller cells in areas where the infrastructure can provide the required backhaul connectivity is the current natural evolution of the RAN infrastructure. The cost of its deployment versus the amount of data it can handle is beginning to be competitive compared to other solutions. But these small cells will also require new techniques for configuration, management and optimisation. Self-configuration of such elements of access networks is expected, as well as coordinated behaviour among groups of small cells, together with neighbouring network sites.

The deployment of new radio networks is based on facilitating access to devices by installing small cells or access nodes, which are interconnected in groups by high-speed backhaul networks, and cooperate within the group to manage the resources in a joint cooperative area. Small cells change the classical concept of the cellular network, from the traditional geometric approaches for coverage and service area analysis, and change also resource management concepts, such as "neighbour" to "partner" cell, the whole concept of "cellular" being progressively replaced by "*cooperative*", as the major infrastructure embodied in base stations evolves to a connected sub-network of small cells.

1.2.2 Moving and Relaying Nodes

In addition, terminals have ceased to be the edge of mobile and wireless communications networks, and are also becoming a local area communications node entity for many of the current scenarios, so the terminal acts as a local manager of radio communications, not only for the user but also for the surrounding smart devices. This concept, together with the reduction of cells size and the changes to the cellular based deployment concept mentioned above, leads to a view of a future convergence between a mobile device and a small base station, the terminal as a *local access enabler*, either via a fixed connection to the fibre loop or by a wireless connection to another access point or mobile device.

This scenario of wireless access enablers, both fixed and mobile, requires an inclusive approach of the current technologies on mobile opportunistic relaying, cooperative networking, distributed antenna systems (DASs), dynamic spectrum resources allocation, cognitive radio (CR), and massive and distributed multiple-input multiple-output (MIMO).

The evolution towards very dense small cells infrastructure brings radio network architectures to consider the roaming user device (on the bus, in the street, inside the car, at home, etc.) as a *relaying node* able to provide coverage extension and to act as an access point to the Internet for the "things" equipped with IP address; a similar service will be provided by urban radio backhauls, deployed using non-cellular low-energy and low-cost radio interfaces. This can make the IoT paradigm become true through a network of mobile and fixed gateways, interconnected according to the random mobility behaviour of humans.

1.2.3 Virtualisation, Cloud, and Ultra-Flexible RANs

Networks architectures are becoming more and more flexible, developing towards the Cloud RAN concept, based on technologies such as DASs, and evolving beyond it towards the so-called "Ultra-flexible RANs". The operation of a mobile network, which traditionally was based on a single owner of the Core and Radio Access infrastructures, who is at the same time the service provider for its customers, tends to become diffused in the virtual operation of shared RAN facilities, opening the door for Virtual Operators as service providers, Radio Resource Providers who own the spectrum and manage its access, RAN providers owning the infrastructure, etc. This scenario for Infrastructure Networks is in principle the basis for the future ultra-flexible RAN, in both technology and operation.

A first step in the evolution of network infrastructures has been the use of DASs for mobile access in small areas, mainly – but not only – indoors, which largely eliminates the concept of "cell". This occurs at least in the sense that cells will not be the stable projection of the service area created by some radiating elements from a single site, but a dynamic set of positions to which a service connection is provided by a combination of signals generated in a cooperative manner from several distributed antennas.

The approach that makes RANs even more efficient is referred to as cloud radio access network (C-RAN), a centralised processing, collaborative radio, real-time cloud computing, and clean RAN systems. Based on real-time virtualisation technology, C-RAN minimises CAPEX and OPEX costs. It enables the fast, flexible, and optimised deployment and upgrade of RANs, supporting pay-per-use models. It also eases the flexible and on-demand adaptation of resources to non-uniform traffic. Besides this, the centralised processing of a large cluster of remote radio units (RUs) also enables the efficient operation of inter-cell interference reduction, and coordinated multi-point (CoMP) transmission and reception mechanisms, and eases mobility between RUs.

To make the future mobile networks sustainable from an economic viewpoint, it should provide more and cost less, both metrics being equally important. Flexible architecture concepts like Cloud RAN address these challenges, although in many cases the research effort highlights the performance enhancement aspect only. However, beyond 4G mobile services provision in a 2020 time horizon also requires cost-effective technologies for ubiquitous coverage, together with the flexible management of centralised resources. With this dual *performance–cost* goal in mind, the Cloud RAN concept is being expanded to include not only the RAN but also many of the evolved packet core (EPC) functionalities, in the so-called *Network Virtualisation*.

Network Virtualisation refers to the capability of partitioning and/or pooling underlying PHY resources (e.g., sites, racks, and base band cards) or logical elements (e.g., RAN and EPC nodes) in a network, and it is usually associated with the concepts of software-defined networking (SDN) and cloud services. Some operators consider Network Virtualisation as a fundamental tool for making the network manageable, and a lever for modifying (opening) the mobile network infrastructure ecosystem, without precluding the incorporation of performance-enhancement technologies, such as CoMP or interference management, which is not necessarily a primary goal.

The role of Operators is also evolving, from the current RAN sharing approaches to a full virtual operation concept. In fact, the operation of mobile networks, which traditionally was based on a single owner of the Core and Radio Access infrastructures, who is at the same time the service provider for its customers, tends to become diffused in virtual operation of shared RAN facilities, opening the door for a combined architecture with Virtual Network Operators as service providers, Radio Resource Providers who own the spectrum and manage its access, RAN providers owning the infrastructure, etc. This has a large impact on the development of novel integrated and flexible services based on RAN-as-a-service (RANaaS). The scenario of Infrastructure Networks is in principle the basis for a future ultra-flexible RAN, in both technology and operation.

1.2.4 Energy- and Spectrum-Efficient Networking

From the mobile communications perspective, and looking also to other related disciplines, the user interface revolution is crucial for the required evolution of terminals and networks. A factor that changed dramatically with the launch of touch-screen terminals was the "user latency". If the "time-to-type" a command in a mobile keyboard in previous years is compared to the "time-to touch" in today's tablets, a factor of 10 is easy to reach. Hence, the time between uplink packet transmissions from these new generation terminals has greatly decreased, and previous traffic models for uplink load have become obsolete. The activity of the user when accessing wireless services might give rise to a new revolution in the coming decade, since other interfaces are under development, such as gestural (in glasses or lenses), muscular (using on-skin sensors), or even brain activity sensors in the long term. Any such human interfaces will boost a new era of applications and services, started to be referred as "Tactile Internet" which, of course, give rise to new requirements for wireless connectivity and mobile networking. The only way to make the coming services and applications efficient at the RAN side is to reduce the signalling load, and latency, to enrich user experience, and to maximise spectrum efficiency.

To date, the improvement of user's experience has been achieved by widening the network, the inclusion of new infrastructures, new frequency bands, and additional transmission systems. Three generations of Radio Access already coexist in 3GPP mobile networks, and little effort has been expended to improve the overall energy and spectrum efficiency. Thus, the metric of Joules per Bit per user is so far useless in RAN deployment. Moreover, even with the deployment of 4G technology, RANs continue to offer broadband access to a limited percentage of locations. This is true even in densely populated areas, where broadband radio access is available, but where interference limits capacity to well below its theoretical maximum. This means that overall the deployed RAN is still making sub-optimal use of the spectrum and energy (bits/Joule), mainly because it is based on fixed spectrum allocation and hierarchical network infrastructure. In the current decade, and beyond 2020, new terminals, devices, applications and services will continuously surge onto the market, forcing RANs to offer extreme levels of throughput, capacity, coverage, and ubiquity. This boosts a fundamental review of the basis of the current RANs, leading to the above mentioned Ultra-flexible architectures, combined with extremely efficient RATs and new approaches for a smarter spectrum management and sharing strategies.

The above-mentioned scenarios of WIoT require dramatically reduced energy consumption, to enable long-endurance self-powered nodes, or nodes powered by energy harvesting, while radio spectrum and overall energy resources remain strictly limited. Especially for IoT, total energy requirements in the order of 1 pJ/bit/node, including all contributions to node energy consumption, will probably be necessary.

To achieve such goals, the implementation technology has to be based on low-power hardware architecture and energy-efficient signal processing. *Multihop cooperative networks* have the capability to greatly increase capacity density and reduce energy consumption, by bringing the access network closer to the end-user. Anyway, the existing "layered" protocols, with their requirements for retransmissions, and multiple acknowledgements, may be highly inefficient, and in multihop networks result in bottlenecks that prevent them scaling as required both for high capacity density access networks and large scale IoT networks. wireless network coding (WNC) – a.k.a. physical layer network coding (PLNC), is a technique that has potential to become a "disruptive" technology for such networks. It has the capability of naturally solving problems related to dense, cloud-like, massively-interacting networks of nodes. It can also be regarded as an example of the more general concept of the "network-aware physical layer", in which functions conventionally performed at high layers of the protocol, such as routing, are more efficiently carried out at the PHY layer, which alone has the capability of processing signals directly, without loss of information. These networks will nevertheless need to be self-managing to optimise their efficiency and adapting to varying demands and resource availability.

1.2.5 New Spectrum Bands for Mobile Broadband

Spectrum availability is always crucial for the evolution of Radio Communications. There is no doubt that for mobile operation in open areas the UHF band is the best possible allocation, but these bands are becoming overloaded under the current access technologies, RAN infrastructures and services provision schemes. Some spectrum bands were already identified by international telecommunications union (ITU) during the World Radio communication Conference of 2012 in Geneva [ITU2012] as potential allocations for spectrum sharing technologies. More recently, the 2015 edition [ITU2015] has confirmed additional spectrum allocations to mobile services on a primary basis, and identified additional frequency bands above 1 GHz for international mobile telecommunications (IMTs) to facilitate the development of future terrestrial mobile broadband applications.

Assuming that the amount of data traffic is to grow by two orders of magnitude in the coming decade, no matter what will be the type of traffic and the "killer" applications or services in these future networks, the need for more resources to deal with even more than 100 times more mobile data traffic in the 2020s is apparent. These "resources" could of course come from additional spectrum, although this is already scarce below 3 GHz and less feasible after the second digital dividend, but obviously still available in higher frequency bands for short-range broadband connections.

While UHF remains being the main allocation for mobile services, with many thousands of moving terminals per square kilometre, new concepts for efficient spectrum usage will have to appear. Among other approaches, opportunistic access to certain frequency bands, CR, co-primary usage in shared bands, and massive and distributed MIMO are already under deployment. *Sharing spectrum* requires consideration not only of technologies but also of regulatory policies, and the effect on applications and services for the end users derived from such approaches.

In addition, exploitation the mm waves band for the next generation of mobile communication standards (5G) has started to gain considerable traction within the wireless industry, EU's Horizon 2020 such as the 5G PPP initiative [5GPPP], regulators and the ITU. Study Groups of the ITU keep working on potential IMT at higher frequencies, above 6 GHz, as part of the preparations for next World Radio Conference (WRC'19). Propagation measurements and channel modelling in the higher frequency bands are also part of the objectives of the ITU SGs after the WRC'15 and for the coming 4 years.

Finally, systems at 60 GHz and above, even up to 400 GHz, are currently under consideration as the natural way of providing wireless Gigabit links, but this is limited to short-range radio access to infrastructure, focussing the wireless "networking" problem on managing the location of terminals and the distribution of services though access points over wired infrastructure.

The inclusion of several new bands in simultaneous operation for the same network access tends to increase very much the complexity at the terminal side. In fact, the heterogeneity of the RAN spectrum and access modes has been enforced by the rapid increase in demand for mobile data, but it is nowadays becoming a drawback for the design of radio terminals, and its energy efficiency. If the future of mobile communications evolves through the inclusion of new spectrum bands on current cellular infrastructures, terminals and antenna systems may reach their limit of reliability.

1.2.6 Radio Channels and Propagation Modelling

The radio channel is central to the paradigm of the current and future mobile communications scenarios: multiple antenna systems, interference recognition and the high degree of cooperation among separate network nodes, require a multi-dimensional description of the radio channel characteristics, jointly modelling of space, time, frequency, polarisation, etc. Moreover, the deployment contexts include the human body, the vehicular environment, dense urban areas, indoors, and many other where characterisation of radio propagation is complex. All this motivates the need for accurate models describing radio propagation in a multi-dimensional fashion, and proper evaluation of how the radio channel characteristics affect the link and network performance, and vice-versa.

Radio propagation for wireless networks has been extensively investigated for over 20 years, mainly towards networks planning for 2G, 3G, and 4G mobile communications. Simple and sophisticated models have been developed in COST IC1004, and some of them incorporated into standards, following the activities of COST 207, 231, 259, 273, 2100, WINNER, and METIS in Europe, as well as of bodies such as 3GPP, ETSI, and IEEE 802 among others. There was a strong body of knowledge for outdoor, rural, and urban channels, covering "classical" use cases of communication between, e.g., a mobile terminal and a base station. However emerging cases, such as BANs, vehicle-to-vehicle links or wireless sensor networks, were much less known from the radio channel point of view, and have been studied in IC1004. The same case applies to cooperation and relaying, being one of the major mitigation techniques to combat channel attenuation and consequent power consumption, and in which IC1004 has assessed multi-link radio channels. There was also a significant lack of knowledge in radio propagation for the more unusual environments considered for the above mentioned SEs, about transitions from one kind of SE to another, models for heterogeneous communication systems, models for the behaviour of real world terminals in real environmental conditions and an interference model for the crowded spectrum conditions determined by a very large number of co-existing wireless objects. Many of these lacks have been covered in COST IC1004.

Also in the *development of RANs*, the supporting knowledge required is that of the radio channel. The requirement for robust and resilient wireless access requires that the characteristics of the propagation channel are addressed in a more complete manner than before, when link outage could be accepted by the user to some extent. Site specific *network planning* will tend to be used much more than at present, in order to provide adequate access in cases that would previously be treated as poorly-performing outliers, not specifically addressed. Thanks to the widespread availability of computing power, PHY propagation tools that make use of digital terrain and building maps will be used within operational networks in order to better optimise the allocation of resources and ensure correct access in the worst cases and in quasi real time. Obviously, further progress in the validity and accuracy of such tools are required, and drives research in this area so as to find a suitable trade-off between precision and computation time. In addition, the possibly distributed character of the nodes challenges the traditional cellular model, with concepts such as stationarity and shadowing becoming less well defined than before. This calls radio channel modelling research to move away from base-station centric approaches and take a whole new range of concepts into account.

Finally, according to the current trend in the Wireless Industry, future wireless networks are expected to operate not only in the allocated frequency bands below 6 GHz, but also in the higher frequency bands between 30 and 90 GHz (the millimetre wave (mmW) frequency range), where the availability of large contiguous blocks of spectrum could be exploited, enabling the possibility of very significant increase in bandwidth. In these higher frequency bands, the wavelength and, consequently, antenna elements are smaller, which facilitate the implementation of large array antennas for beamforming to compensate for propagation losses, and achieve significant system capacity and throughput gains. Although initial results and trials on the use of mmW for mobile broadband access look extremely promising, a number of important challenges need to be overcome, and radio channel modelling in such new bands is one of extreme importance.

1.3 Scope of the Book

This book is structured into twelve chapters.

Chapter 2 provides an understanding of the propagation phenomena in the diverse urban environments, to enable the design of efficient wireless networks in future bands and scenarios. The chapter includes results of studies related to 4G and future 5G radio systems for both outdoor and outdoor to indoor, while scenarios include rural and highway, base station to pedestrian users, vehicular-to-vehicular, V2i, container terminals, vegetation and high-speed mobility such as trains.

Chapter 3 addresses aspects of radio link and system design in indoor scenarios for supporting ubiquitous indoor data connectivity and location and tracking services. In particular, the chapter deals with modelling of the complex indoor radio channels for effective antenna deployment, evaluation of mmW radios for supporting higher data rates, and indoor localisation and tracking techniques.

Chapter 4 focuses on key issues in vehicular to any (communications) (V2X), including propagation, antennas, and Medium access control (MAC) and MAC layer algorithms. The chapter includes reports on measurements, characterisation, and modelling of vehicular radio channel for road, railway, and special environments.

Chapter 5 provides an overview of the last advances in research on WBANs, carried out in the framework of COST IC1004, and proposes a radio channel model for these particular types of scenarios. Given the wide range of applications of WBAN communications, the chapter is organised in the analysis of both antennas and propagation aspects for In-Body, On-Body and Off-Body cases, as well as some insights on the Body-to-Body communications scenario.

Chapter 6 deals with the evolution of RANs, and summarises advances done by COST IC1004 researchers on advanced resource management ecosystem, considering resource scheduling, interference, power, and mobility management. The chapter also describes the recent energy efficiency new strategies in 3G and 4G RANs, as well as on Spectrum Management in Cognitive Networks, specifically for TV white spaces (TVWS). Virtualised and Cloud architectures for realistic scenarios are proposed and analysed in this chapter, as well as reconfigurable radio for Heterogeneous Networks (HetNets).

Chapter 7 addresses PLNC, machine-to-machine (M2M), Relaying and CR networks, which will be key enabling technologies for 5G and beyond. Among other techniques, this chapter covers the virtual MIMO, RA distributed processing, interference analysis and cancellation algorithms in relay-assisted (RA) wireless transmissions, network coded modulation (NCM), hierarchical network code (HNC) maps, and relay/destination decoding techniques.

Chapter 8 discusses the progress on the PHY layer for next generation wireless systems, covering the range of "SEs". The topics in chapter 8 include interference characterisation and alignment, models for the energy consumption of wireless terminals, energy efficient relaying, MIMO precoding for adaptation on realistic channels, beam-space MIMO and spatial modulation, iterative methods in modulation and coding and models of the PHY layer to incorporate into system-level simulations.

Chapter 9 is dedicated to radio channel measurement and modelling techniques for beyond 4G Networks. In particular, this chapter includes new measurement techniques not only targeting radio channels, but also material properties in new frequency bands, improved PHY models, covering full-wave as well as ray-based methods, with a specific sub-section dealing with diffuse scattering and complex surfaces, progress in analytical models, new channel estimation tools for model development based on channel sounding data. The COST2100 channel models are updated and presented here, enabling to include new features, such as massive and distributed MIMO aspects.

Chapter 10 deals with recent advances and innovations in antenna systems for communications. The contents include the design of multiple antenna systems, the optimisation of antenna performance using decoupling techniques, smart reconfigurable antennas, RFID and sensor antenna innovations, holistic characterisation and measurement of antenna patterns and performance, including mmW frequencies.

Chapter 11 gives first an insight on general underlying concepts of the state of the art of MIMO over-the-air (testing) (OTA) technologies, and then describes four specific methods for MIMO terminals testing proposed by COST IC1004 to standardisation bodies: multi-probe anechoic, reverberation, two-stage, and decomposition.

Finally, Chapter 12 attempts to risk a vision on the future challenges and developments in mobile and wireless communications, on a time scale of about 5 years. This is based on the results obtained by the COST IC1004 network of scientific experts, and on the three main "pillars" of the success story of follow-up Radio communication COST Actions, which are antennas and channels, signal processing, and device networking.

2

Urban Radio Access Networks

Sana Salous, Thomas Werthmann, Ghassan Dahman, Jose Flordelis, Michael Peter, Sooyoung Hur, Jeongho. Jh. Park, Denis Rose and Andrés Navarro

The increase in demand for high-data rates on the move in the complex urban environment requires either the allocation of new spectrum such as available contiguous spectrum in the mm-wave band or the use of novel configurations such as the application of massive multiple-input multiple-output (MIMO) technology. To enable the design of efficient wireless networks, an understanding of the propagation phenomena in the diverse urban environments is fundamental. In this chapter, we present results of studies related to fourth generation (4G) and future 5G radio systems both outdoor and outdoor-to-indoor. Classifications include rural and highway, BS to pedestrian users, vehicular-to-vehicular, vehicular-to-infrastructure, container terminals, vegetation, and high-speed mobility such as trains. Results for path loss (PL) and shadow fading are presented from various studies of stochastic and deterministic channel models based in outdoor, indoor-to-outdoor, hotspots, vehicular, and train environments. Relay stations and the impact of antenna placement in vehicles, antenna terminal height, and the presence of pedestrians are discussed. Results of angular spread and *rms* delay spread of wideband channels are presented for the frequency bands below 6 GHz allocated for 4G networks, and preliminary results in the mm-wave band, envisaged for 5G networks, including prediction of the impact of rain. To facilitate the simulation of radio networks in urban environments the Hannover Scenario is proposed to give a common simulation environment.

2.1 Radio Propagation in Urban Scenarios

2.1.1 Radio Propagation Measurement and Stochastic Modelling

In this section, updated channel parameters for various urban environments are reported in Section 2.1.1.1. Section 2.1.1.2 reports measurement results

Cooperative Radio Communications for Green Smart Environments, 17–70.

in challenging conditions such as container terminals, rain, vegetation and trains. In Section 2.1.1.3, results on the effect of the user mobility, existence of pedestrians, and transmit/receive antenna height on the channel parameters are reported.

2.1.1.1 Channel characterisation in various urban scenarios

Relay

Extensive measurements were performed to study the statistics of shadow fading in relay links and the correlation of shadow fading between a BS-mobile station (MS) link and a RS-MS link or between two RS-MS links [CM12, CCC11]. In the work by Chu et al. [CCC11], shadow fading is modelled as a zero-mean log-normal distribution with 5–11 dB standard deviation (STD). For each RS location, the correlation of shadow fading with the same RS-MS link was calculated for three different RS antenna heights: (4.7, 8.8, and 12.7 m). The correlation of shadow fading of the same RS-MS link for the different RS antenna height was found to be 0.75–0.96, indicating high correlation. However, the correlation of the shadow fading between the BS-MS, and the RS-MS links (and also between two RS-MS links), varied from –0.04 to 0.57. The correlation values were found to be inversely proportional to the difference angle between the two considered links. For an angular difference less than 10°, high-correlation values from 0.4 to 0.6 were found. When the difference angle increased to about 60°–80°, the correlation coefficients rapidly decreased towards zero as seen in Figure 2.1. The correlation also decreases when the distance between transmitters (TXs) increases. Moreover, the highest coefficient of about 0.5 was observed between TX pairs having a small separation as illustrated in Figure 2.2.

The PL model was studied in the work by Conrat and Maaz [CF12, CM13, CM14], where the relay antenna height was found to have an impact on the model in non-line-of-sight (NLoS) propagation. In the paper by Conrat and Maaz [CM13], the line-of-sight (LOS) PL was found to consist of two distance zones called the near zone (20–30 m from the RS) and the far zone. For the near zone, the PL depends on the antenna gains of the RS and the MS but generally remains flat and thus the log-distance model is not applicable as seen in Figure 2.3. For the far zone, the free-space log-distance model with an additional term representing a shift of about ±6 dB is suggested to be added in order to incorporate the effect of the RS position Equation (2.1).

$$\mathrm{PL_{dB}} = 20\log_{10}(d) + 39 + \mathrm{EF} + \mathrm{SF}, \tag{2.1}$$

where, SF is a Gaussian variable with a STD of 5.8 dB, and EF is uniformly distributed random variable ±6 dB.

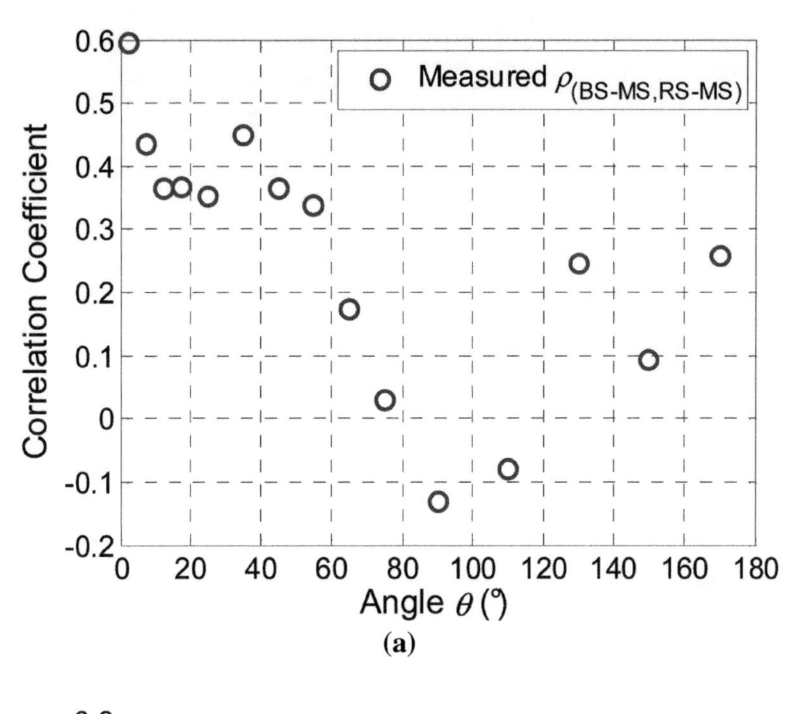

(a)

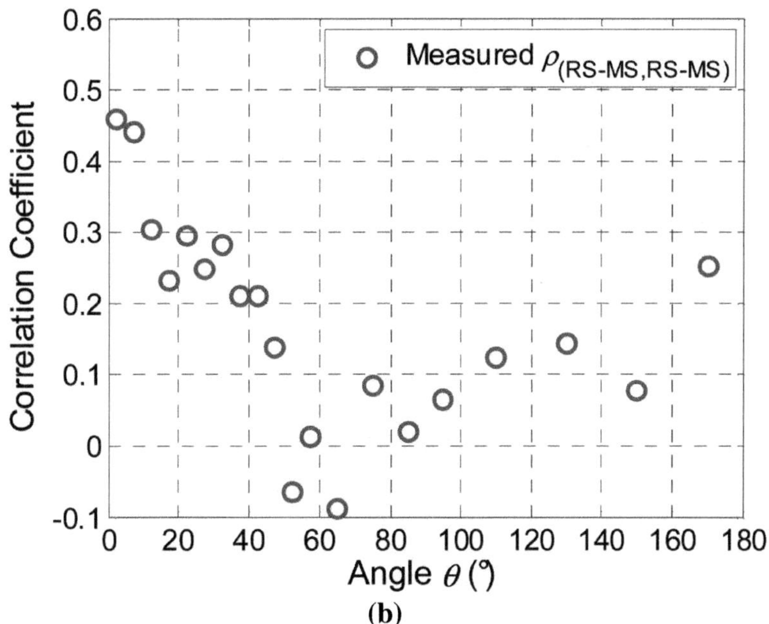

(b)

Figure 2.1 Measured shadow fading correlation versus difference angle: (a) $\rho_{(\mathrm{BS\text{-}MS,RS\text{-}MS})}$, and (b) $\rho_{(\mathrm{RS\text{-}MS,RS\text{-}MS})}$.

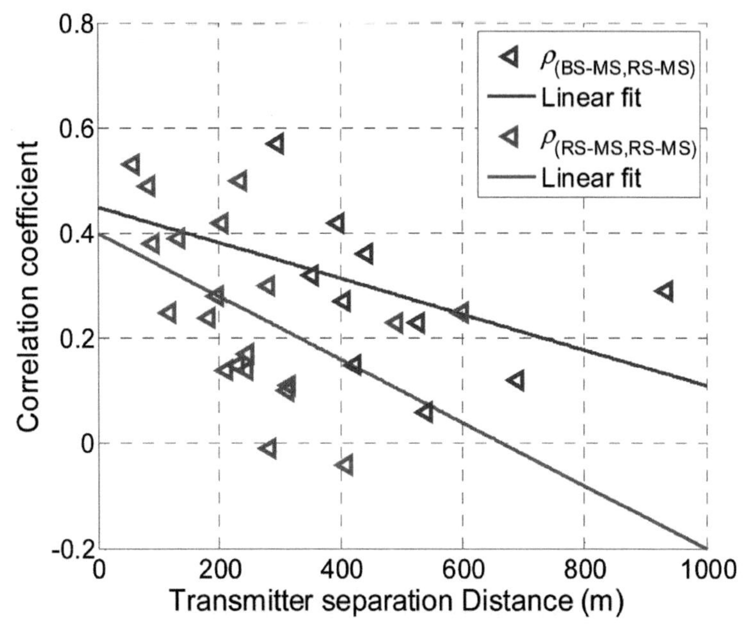

Figure 2.2 Dependence of shadow fading correlation on the TX separation distance.

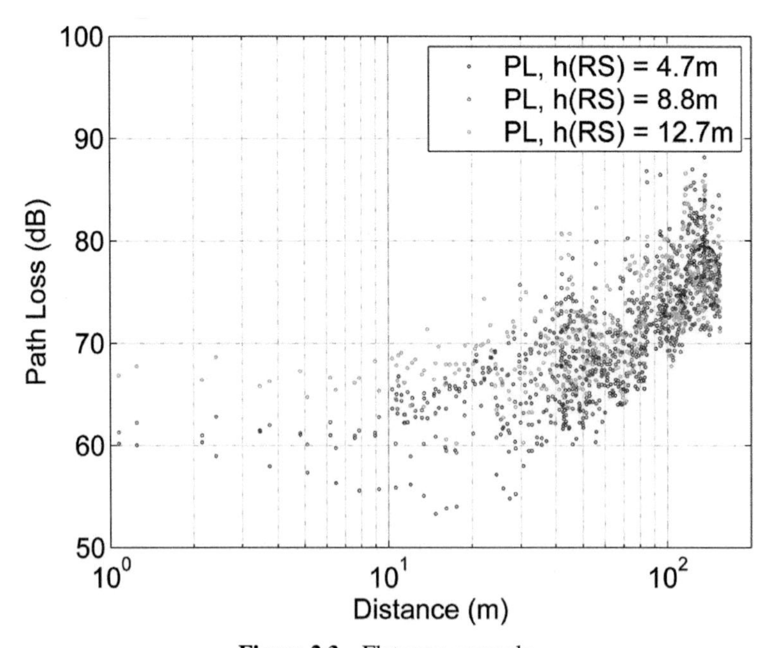

Figure 2.3 Flat zone example.

In the NLoS case, the PL is affected by the RS antenna height and the PL model can be presented as in Equation (2.2).

$$\mathrm{PL_{dB}} = 46\log_{10}(d) + 14.2 - 0.6\,h_{\mathrm{RS}} + \mathrm{SF}, \tag{2.2}$$

where, SF is a Gaussian variable with a STD of 9.6 dB, and h_{RS} is the RS antenna height.

In Conrat and Maaz [CF12], a simple refined PL model is introduced based on existing PL models with an antenna height correction factor. A measurement-based comparison was performed in the work by Conrat and Maaz [CM15] and it showed that the COST231 WI and wireless world initiative new radio (WINNER)+PL models overestimate the PL prediction. However, models proposed by 3rd generation partnership project (3GPP) and the ITU give the best performance.

In Yin et al. [YPKC12], the cross-correlation of small-scale fading (SSF) was evaluated for different locations of the base station (BS), relay station and mobile station in a three-node cooperative relay system. SSF was found to be more correlated when (i) the difference angle between the direct links from the BS to the MS and from the RS to the MS decreases; (ii) when the MS has similar distance to the BS and to the RS; and (iii) when the MS moves away from the region where the BS and RS are located.

Outdoor to/from indoor

The PL from outdoor-to-indoor at three different frequency bands 700, 450, and 150 MHz were compared in the work Bultitude et al. [BSCZ12]. It was found that the absolute and the excess outdoor-to-indoor propagation losses were highest in the 700-MHz band. Figure 2.4 shows the CCDF for propagation loss where the medians at 700 and 450 MHz are seen to be 13 and 7 dB higher than the median at 150 MHz, respectively.

The CDF of static RMS delay spread (calculated from a single impulse response) shown in Figure 2.5, indicate median values of 88 ns, at 150 and 450 MHz and 95 ns at 700 MHz. The probability of very large delay spreads, greater than 1200 ns is almost zero at 450 and 700 MHz, whereas it is non-zero at 150 MHz.

When comparing the propagation loss in outdoor-to-indoor environments at 1.8 GHz with those at 900 MHz, an extra 3 dB loss was found [RK12]. Furthermore, the distance from the building to the BS and the LoS conditions between them play an important role. When there is no direct LoS path, the power levels inside and around the buildings are more homogeneous and thus the fluctuations inside the buildings are less abrupt. This leads to the conclusion

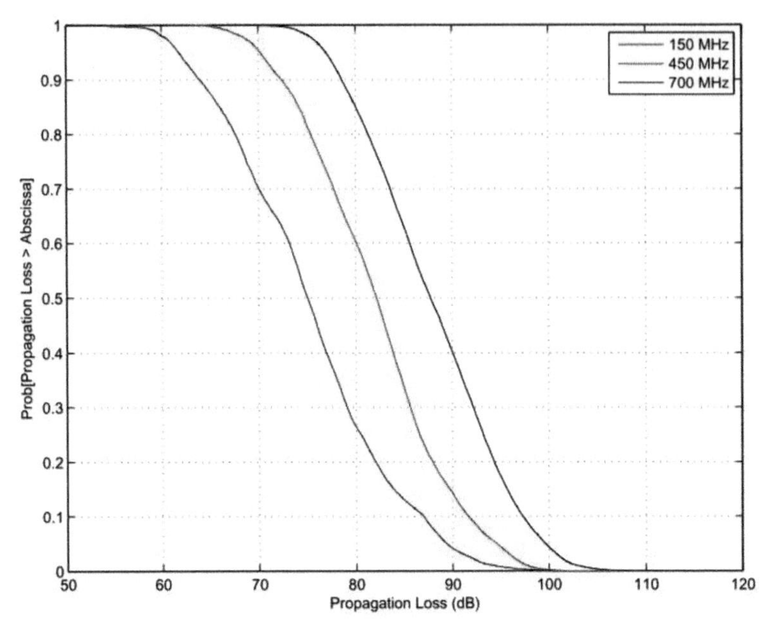

Figure 2.4 CCDFs for outdoor-to-indoor propagation measured in building 2 at CRC.

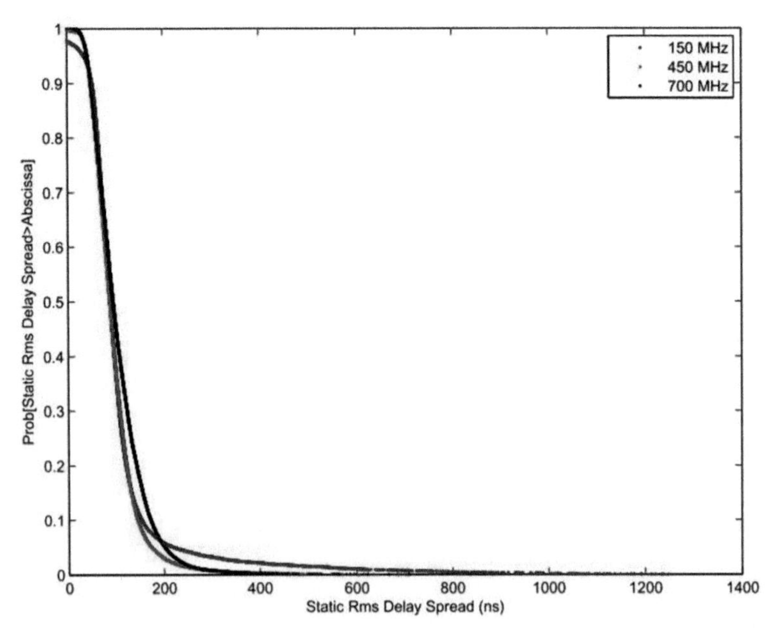

Figure 2.5 CCDF for static RMS delay spreads, estimated from all data measured in building 2 at CRC.

that even simple building models and less advanced indoor prediction models provide realistic assumptions on the coverage inside by means of empirical modelling. In the case when the BS is nearby and under LoS condition, a single path that follows the direct penetration line often has a dominant influence and thus results in a more inhomogeneous indoor coverage, which suggests the usefulness of using a semi-empirical model.

The indoor-to-outdoor propagation can be modelled using a hybrid model by combining two models: the extended angle-dependent multi-wall model, and the vertical knife-edge diffraction model. This can be done by modelling the effects on the direct propagation line (i.e., the penetration loss) between the femto-cell and the mobile terminal with a multi-wall model approach [COST231]. When the material parameters for the building are unavailable, empirical attenuation for different classes of walls can be used. Second, the diffraction loss can be modelled using the knife-edge diffraction model based on the elevation angles between the home-eNodeB, the window frame and the mobile terminal on the street. A hybrid-model that combines the dominant propagation effects from both models can be used by choosing the smaller PL of each. Thus the maximum received power from both predictions is achieved. This model provides good accuracy in the areas in front of the windows, but since only vertical diffraction is considered, the prediction is somewhat pessimistic in the areas on the left and on the right of the window [RJK12].

Vehicular-to-X

Different studies have been performed in order to evaluate the properties of vehicular-to-X channels and how these properties are affected by system parameters such as the position of antennas. These include vehicular-to-vehicular (V2V) and vehicular-to-infrastructure (V2I) in different propagation scenarios such as highway, urban, and rural.

The impact of antenna placement on V2V communications is analyzed in the work by Abbas et al. [AKT11]. Three propagation environments were chosen: highway, urban, and rural, where two Volvo V70 cars were used, each of which was provided with four omni-directional antennas, in azimuth, mounted at four different positions. The measurements were performed at a centre frequency of 5.6 GHz with 200 MHz bandwidth, where the TX and RX cars were moving in a convoy and in opposite directions. It was suggested that a pair of antennas with complementary properties, e.g., roof or left-side mirror mounted antenna together with a bumper antenna would reduce the shadowing on the RX antenna when vehicles are moving in opposite directions as shown in Figure 2.6.

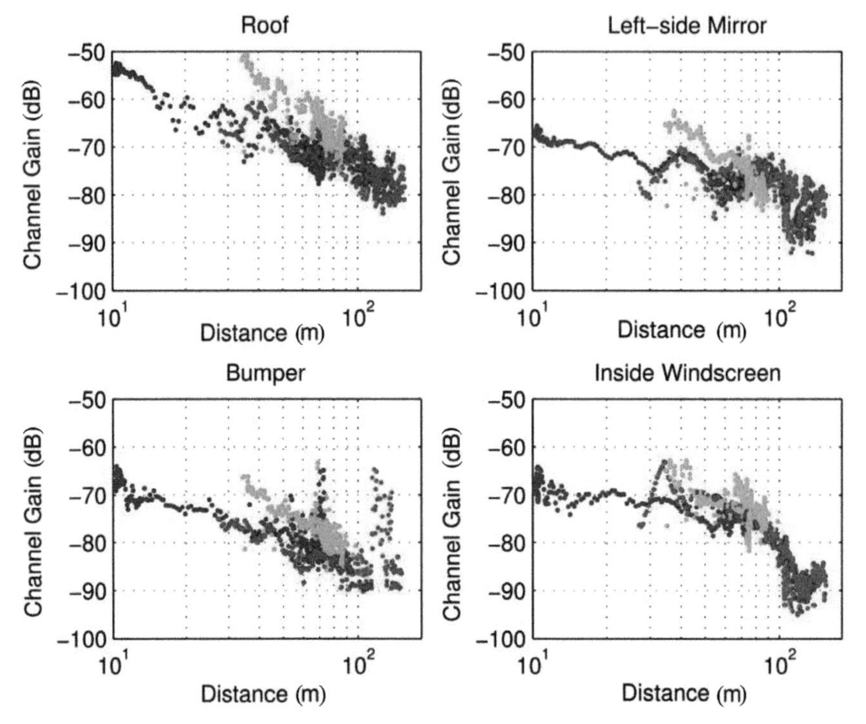

Figure 2.6 Measured channel gain for urban scenario when both vehicles were moving in a convoy. Results from four different measurements are shown with four different colours.

The influence of common shadowing objects like trucks on the wireless channel properties was addressed in Vlastaras et al. [Dim14], where a measurement campaign with a TX car incorporating six antennas and a RX car incorporating a single roof antenna was conducted. Two scenarios were considered; rural and highway. In both cases, a convoy formation was considered, where the TX was in the front, the RX was in the back and the truck was in the middle whenever present. See the different positions of the antennas in Figure 2.7.

The crucial rule of the antenna design, especially in regards of the radiation pattern of the MIMO antenna system for V2I communications was confirmed by Ekiz et al. [EKM12], where shadowing between the antennas found to cause a diminishing performance in areas close to the BS. It was pointed out that such effects can be circumvented by misalignment of the feed point of both antennas to obtain a radiation pattern being more congruent and closer to an omni-directional characteristic.

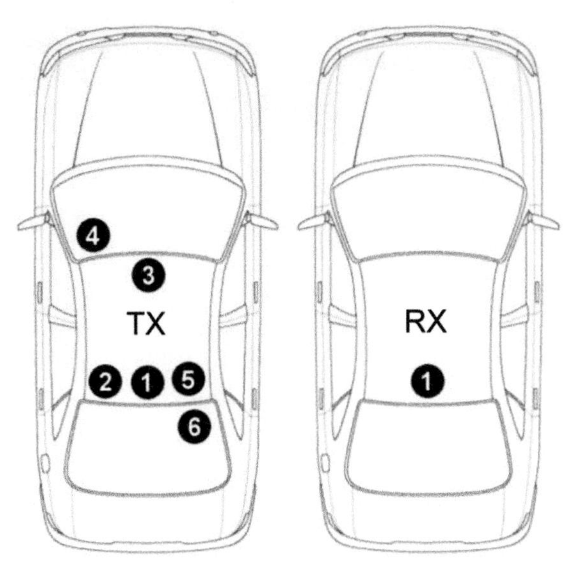

Figure 2.7 Antennas placement on the TX and RX cars. TX: Antennas 1, 2, 3, and 5 were omni-directional shark-n antennas mounted on the roof of the car. Antennas 4 and 6 were omni-directional antennas mounted inside of the front and rear windshields respectively. RX: A single omni-directional antenna was mounted on the roof of the car.

Hot-spots

A measurement campaign was performed in order to model radio channels with either LoS or NLoS connection between a UE and a NodeB in an operating universal mobile telecommunications system (UMTS). The statistics of the distance between the NodeB and the UE in LoS/NLoS scenarios, the life-distance of LoS channel, the LoS existence probability per location, and per NodeB, the power variation from LoS to NLoS transition and vice versa, as well as the transition duration were extracted [CYT$^+$14].

Figure 2.8 depicts the empirical and fitted CDFs of the channel gains in the LoS and NLoS scenarios. It can be observed that the channel gains for LoS are concentrated in the region more to the right of the abscissa than the NLoS scenario. The results show that both CDFs are well fitted by normal distributions with parameters ($\mu = -90.68$ dB, $\sigma = 8.27$ dB) and ($\mu = -96.80$ dB, $\sigma = 12.92$ dB) for the LoS and the NLoS scenarios, respectively. It can be observed that the average of path loss PL for the LoS scenario is 6.12 dB higher than that in the NLoS scenario, and the STD of the channel gain in the LoS scenario is 4.65 dB less than that in the NLoS scenario. These statistical differences indicate that the existence of the LoS path in the channel

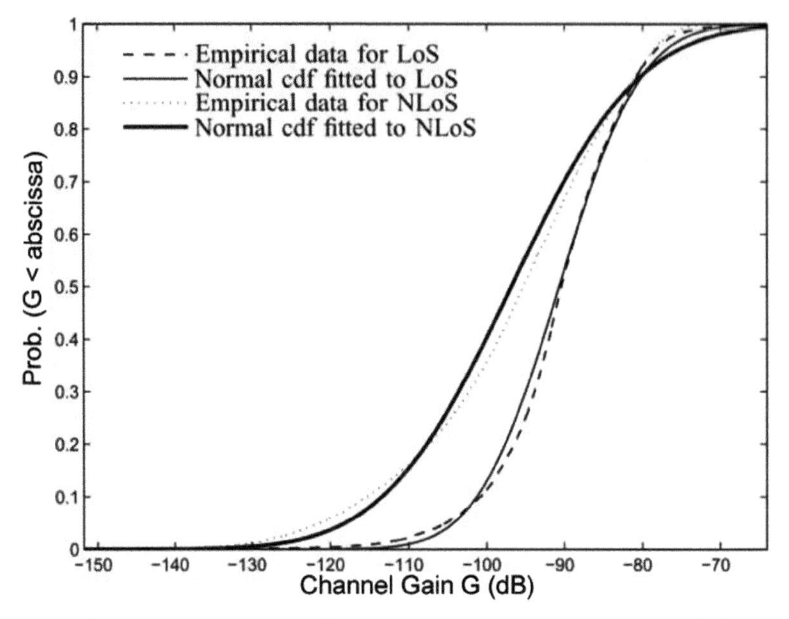

Figure 2.8 The empirical and fitted CDFs of the channel gains in the LoS and NLoS scenarios.

can not only bring 6 dB gain on average for the received signal, but also reduce the variation of the channel gain significantly.

Figure 2.9 illustrates the resultant probability density functions (PDFs) of direct distance from the NodeB to UE in LoS and NLoS scenarios, respectively. It can be observed that the NLoS users are at a maximum of 2.5 km, whereas the majority of the LoS users are located within 200 m from the NodeB. The NLoS users are widely spread within the range from 0 to 4 km, and the majority are within 300 m from the NodeBs. The non-zero probability observed for large range up to 4 km might be due to the existence of macro-cellular NodeBs. It was also found that the LoS and NLoS life-distances are usually less than 10 m, and the LoS-to-NLoS transition can be achieved in less than 10 ms.

2.1.1.2 Updated models in challenging conditions

Container terminals

In the work by Ambroziak and Katulski [AK13], the propagation loss based on the mobile, container, and terminal (MCT) model is compared to almost 290,000 measurements in a real container terminal environment. The obtained standard error of estimate (SEE) is 4.45 dB, which proves the accuracy and usefulness of the MCT model which may be used for frequencies from

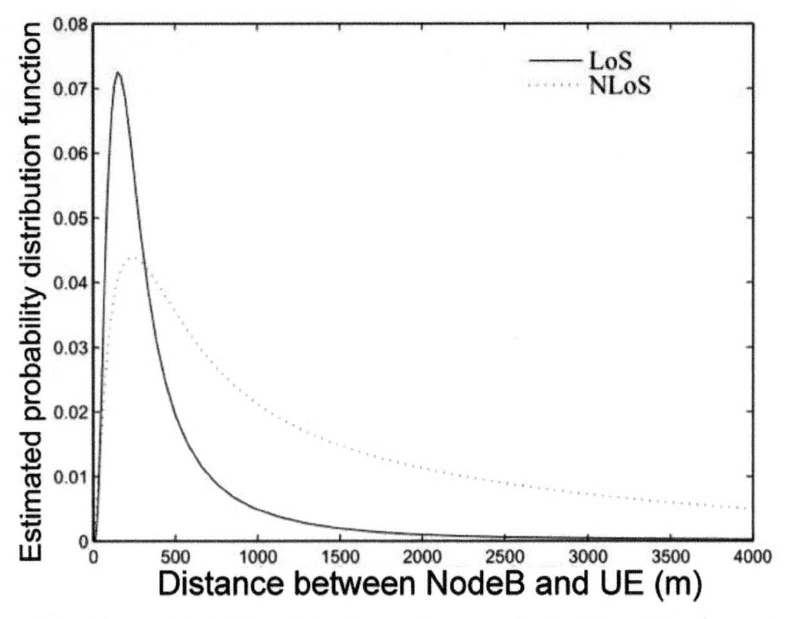

Figure 2.9 The empirical PDFs of the distance between the NodeB and the user equipment (UE) in LoS and NLoS scenarios.

500 MHz to 4 GHz, and path lengths from 50 to 620 m and BS antenna height between 12 and 36 m. The mean error (ME), and the SEE are listed in Table 2.1, where L_{LoS}, L_{Cont}, and L_{offT} are the PL for LoS, container area, and off-terminal area, respectively.

Rain

In order to investigate the effect of rain on attenuating the communication link in a wireless network, transformation of measured attenuation data available for long communication links to a hypothetical short link of 200 m length was applied in the work by Kantor et al. [KDB14]. Rain fading measurement data covering a span of 5 years were used assuming a carrier frequency of 38 GHz and horizontal polarisation. The attenuation on the hypothetical short link can then be calculated from Equation (2.3), where A_{r} (dB) is the

Table 2.1 Verification results of the MCT propagation model

Parameter	L_{LoS}	L_{cont}	L_{offT}	L_{mct}
ME (dB)	0.85	1.03	–0.26	0.72
SEE (dB)	4.40	4.53	4.30	4.45

attenuation on the real link that is exceeded with a probability $p = 0.01\%$ of the time, A_h is the attenuation on the hypothetical link, k_r, α_r (k_h, α_h) are frequency and polarisation-dependent empirical coefficients pertaining to the real (and hypothetical) links [ITU12], and r_r (r_h) is the distance factor of the real (hypothetical) link [ITU12].

$$A_h = k_h \cdot \left[\frac{A_r}{k_r d_r r_r}\right]^{\left(\frac{\alpha_h}{\alpha_r}\right)} d_h r_h. \tag{2.3}$$

The proposed transformation is adequate on $A_{0.01}$ (i.e., with $p = 0.01\%$); however, on different A_p values, the applicability of the proposed method has to be verified. Comparison of rain attenuation CCDFs of the reference link and of various hypothetical transformed links with the ITU theoretical curves, based on the $A_{0.01}$ values, is depicted in Figure 2.10. The transformed CCDFs and the ITU theoretical A_p curves for reduced path lengths are also nearly identical, except between exceedance probabilities of $p = 10^{-4}$ to $p = 10^{-6}$; however, this dissimilarity comes from the shape of the CCDF of the rain attenuation on the reference link. Therefore, it can be concluded that the proposed link transformation method is applicable on both short and long links since it takes the distribution of the rain intensity along the link path into account by considering the r distance factor as a constant for short (<250 m) path lengths.

In the work by Kntor et al. [KCD+14], the ratio of the disconnected nodes in a 5G microwave mesh network was investigated at high rain intensities in

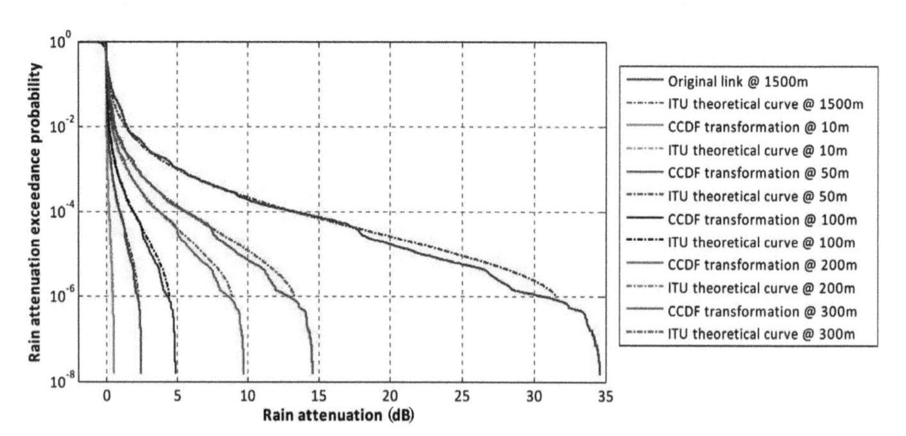

Figure 2.10 CCDF of the rain attenuation the transformed links according to the proposed method on different link lengths compared to the corresponding ITU theoretical curves based on $A_{0.01}$ (f = 38 GHz, pol = h).

a scenario when only rerouting is applied as a technique to improve network resilience. It was found that at high rain intensities (i.e., at low rain attenuation exceeding probabilities) the average ratio of the disconnected nodes can be considerable. Moreover, in a disadvantageous node deployment the effect of rain attenuation can be excessively harmful.

Vegetation

In the work by Torrico et al. [TCK12] and Chee et al. [CTK12], an analytical approach to compute the propagation loss for a typical mobile radio system in a vegetated residential area is discussed. In the described scenario, the transmitting antenna is elevated or around the average rooftop height, whereas the RX is located at street level and possesses no direct LoS to the TX. Using this approach, the total propagation loss Equation (2.4) is broken in three components, namely the free space loss, $\mathrm{FSL}_{\mathrm{dB}}$, multi-screen diffraction loss, from the elevated transmitting antenna to the last screen adjacent to the mobile RX $L_{\mathrm{msd,dB}}$ and the rooftop-to-street diffraction loss combined with tree scaterring $L_{\mathrm{rts,dB}}$ [CTK12].

$$\mathrm{PL}_{\mathrm{dB}} = \mathrm{FSL}_{\mathrm{dB}} + L_{\mathrm{msd,dB}} + L_{\mathrm{rts,dB}}, \tag{2.4}$$

Train

Measurement campaigns between a high-speed train and a BS (i.e., a cellular deployment) were performed at 800 MHz, 930 MHz, and 2.6 GHz [Ke13, Flo14]. In the work by Guan et al. [GZAK13a], a semi-deterministic model for the propagation of high-speed railway was proposed, and it was found that the extended Hata model and multi-edge diffraction models can be conjunctively utilised to predict the PL in the viaduct and the cutting scenarios, which on average make-up more than 70% of the whole railway line. It was also found that the average estimated PL component is 3.2 and 3.5 dB at 800 MHz and 2.6 GHz, respectively which is in line with established PL models for rural areas [KBA+14].

2.1.1.3 Effects of user mobility, existence of pedestrians, and Tx/Rx antenna height

Effect of user mobility

The effect of arranging virtual sources of traced rays that arrive from the BS to the user in urban scenarios is addressed in Zentner et al. [ZMD12]. It was found that the common practice of placing virtual sources at the edges of the buildings where diffraction occurs yields correct results only for static cases. When mobile users are considered, it results in the wrong estimate of Doppler shift. To overcome this problem, instead of locating the virtual sources at diffraction interaction points, it is suggested to set the virtual sources exclusively at a location that appears as a source to the mobile user [ZMD12].

To generate channel instances that take realistic user's movements inside a building into consideration, a two-step motion simulator was introduced in Rose et al. [RJHK13]. First, the motion between the rooms and the duration of time a user stays inside these rooms is calculated using a Markov jump process, which requires transition rates for each room to be stated in advance. These rates vary strongly depending on the time of the day and the type of building. Second, some spontaneous motion within the rooms is added to the simulation, such as: walking around, change position in the same room, walk back and forth between two positions, and visit another room (office) and come back to the original position. The last case represents a special case, since the non-spontaneous movement between rooms is determined in the first step.

These types of motion also vary highly with respect to daytime and the type of the building. Consequently, their associated probabilities are defined differently.

In the work done by Hahn et al. [HRSK15], a mobility model for simulating user mobility from indoor to outdoor and vice verse is introduced. The model simulates users moving from an entry point of a building to another based on real geographical data. The proposed model was compared with the well-known random walk mobility model. The comparison was conducted in the context of two prediction models: the extended Hata model and a complex 3D ray-optical propagation model. The results show that if a more advanced prediction model, e.g., the used ray-optical PL model, is applied, the use of a traditional random mobility model tends to produce results with more outliers towards lower and upper bounds.

Effect of existence of pedestrians

The effect of pedestrians on the properties of MIMO channels in small-cell setup was studied in Saito et al. [SIO14] based on measurements in the 2 GHz band. The measurements were carried out in a plaza of a typical urban railway station in Tokyo. To elucidate the influence of pedestrians on the channel properties, the measurements were carried out both in the daytime and at midnight. The height of the MS was fixed at 1.45 m, while two heights were considered for the BS, 1.45 m (L-scenario), and 2.9 m (H-scenario). As detailed in Table 2.2, the results indicate that in the day time, the received power and

Table 2.2 Large-scale parameters (LSPs) extracted from measurements for two BS heights 1.45 m (L-scenario), and 2.9 m (H-scenario), versus the ITU-RM2135 UMi (LoS)

Scenario		ITU-RM2135 UMi (LoS)	Measurement (H-scenario, Daytime)	Measurement (H-scenario, Midnight)	Measurement (L-scenario, Daytime)	Measurement (L-scenario, Midnight)
Delay spread (DS)	μ	–7.19 (65 ns)	–7.42 (38 ns)	–7.30 (50 ns)	–7.59 (26 ns)	–7.50 (32 ns)
Iog10 (s)	ρ	0.4	0.18	0.17	0.21	0.19
AoA azimuth spread	μ	1.75 (56°)	1.70 (50°)	1.73 (54°)	1.61 (41°)	1.62 (42°)
(ASAazm) log10 (°)	ρ	0.19	0.16	0.16	0.17	0.16
AoA elevation spread	μ		1.00 (10°)	0.91 (8°)	1.16 (14°)	1.07 (12°)
(ASAelv) log10 (°)	ρ		0.31	0.32	0.31	0.31
AoD azimuth spread	μ	1.2 (16°)	1.86 (72°)	1.85 (71°)	1.61 (41°)	1.62 (42°)
(ASDazm) log10 (°)	ρ	0.43	0.08	0.08	0.17	0.16
AoD elevation spread	μ		0.90 (8°)	0.88 (8°)	1.16 (14°)	1.07 (12°)
(ASDelv) log10 (°)	ρ		0.30	0.25	0.31	0.31
Cluster ASDazm (°)		3	5.6	6.0	5.3	5.2
Cluster ASDelv (°)			3.6	3.4	6.1	4.9
Cluster ASAazm (°)		17	5.9	6.0	5.3	5.2
Cluster ASAelv (°)			4.5	3.7	6.1	4.9

the delay spreads decreased, while the azimuth and angular spreads increased due to the change of propagation environment caused by the mobility and shadowing of pedestrians.

Effect of transmit/receive antenna height

The effect of BS height on LSPs was studied in the work by Böttcher et al. [Ann11] and Sommerkorn et al. [SKST12, SMS+14] based on several measurement campaigns at 2.53 GHz. In Böttcher et al. [BSVT11], antenna heights at 25 and 15 m at the BS were measured. It was found that under NLoS conditions, the delay-spread, K-factor and XPR significantly depend on the considered BS height and on the absolute distance to the mobile station. The DS was found to increase with increasing distance, while the K-factor and the XPR decrease as detailed in Tables 2.3–2.5.

In [SKST12] measurement with extremely elevated BS (approximately 118 m) were performed and the dependency of the LSPs on the elevation angle was confirmed, especially for the rms delay spread under NLoS conditions, where the average delay spread ranged from 72 ns to 151 ns (high to low elevation angle). Furthermore, the dependency of the LSPs on the elevation angle leads to the usefulness of sectorisation based on elevation beamforming as confirmed in [SMS+14].

For indoor environments, the effect of the height of the transmit and receive antenna on the delay spread was reported in the work by Salous et al. [SCR14], where the transmit antenna was placed either at 1.5 m or at the ceiling height and the receive antenna was either mounted on a trolley at 1.5 m or at the ceiling level at 2.6 m. Measurements were performed both in corridors and in offices. Table 2.6 summarises the results of the delay spreads in the different setups.

2.1.2 RT Techniques

For the 4G wireless communication system, the conventional semi-empirical or stochastic propagation prediction models are insufficient for network planning. Time dispersion and angular dispersion in a radio channel are important for the performance of the 4G network. The bandwidth of sub-carriers of OFDM system is determined by the knowledge of time dispersion in the radio channel. Moreover, smart antennas, consisting of adaptive array and MIMO antennas, are used for the 4G networks. Angular dispersion due to multipath propagation affects the spatial filter characteristics of the smart antennas and the effect is different for different kinds of smart antennas.

Table 2.3 Values for the mean and STD for the delay spread for different antenna heights and distance ranges

	NLoS		LoS	
	Mean (µs)	STD (µs)	Mean (µs)	STD (µs)
60–200m/25m	0.09	0.04	0.06	0.03
200–400m/25m	0.16	0.09	0.11	0.03
400–640m/25m	0.14	0.14	0.12	0.06
60–200m/15m	n.a.	n.a.	n.a.	n.a.
200–400m/15m	0.12	0.09	n.a.	n.a.
400–640m/15m	0.23	0.29	n.a.	n.a.

Table 2.4 Values for the mean and STD for the *K*-factor for different antenna heights and distance ranges

	NLoS		LoS	
	Mean (dB)	STD (dB)	Mean (dB)	STD (dB)
60–200m/25m	7.51	8.30	10.40	6.87
200–400m/25m	6.13	6.74	10.32	6.81
400–640m/25m	5.24	6.62	5.86	5.22
60–200m/15m	n.a.	n.a.	n.a.	n.a.
200–400m/15m	6.06 dB	7.47 dB	n.a.	n.a.
400–640m/15m	4.35 dB	6.55 dB	n.a.	n.a.

Table 2.5 Values for the mean and STD for the horizontal XPR for different antenna heights and distance ranges

	NLoS		LoS	
	Mean (dB)	STD (dB)	Mean (dB)	STD (dB)
60–200m/25m	8.09	5.33	7.63	4.66
200–400m/25m	7.40	3.62	11.63	3.67
400–640m/25m	3.87	2.88	6.96	2.52
60–200m/15m	n.a.	n.a.	n.a.	n.a.
200–400m/15m	6.10	2.83	n.a.	n.a.
400–640m/15m	2.79	2.77	n.a.	n.a.

Table 2.6 Summary of RMS delay spread obtained with different transmit and receive antenna height in indoor environment

	TX in Office with Antennas at Ceiling Height 2.6 m	TX in Corridor with Antennas at 1.5 m	TX in Office with Antennas at 1.5 m
Median value (ns)	11	18.53	13.74
90% value (ns)	12.5	25.16	20.15
10% value (ns)	8	8.49	10.74

For adaptive antenna arrays, angular dispersion degrades the performance of adaptive beamforming. However, for MIMO a wide angular spread of the multipath waves produces a large decorrelation of the spatial channels and hence increases diversity performance. The information of spread in angle domain and time domain cannot be predicted with the conventional empirical propagation models. Instead, deterministic prediction models become more interesting to predict the propagation channels for the 4G networks. The RT model is one of the popular deterministic models nowadays. It uses physical models of radio propagation mechanisms, such as reflection and diffraction, and detailed information of the environment to provide deep insight into the propagation channels. Most RT models with a detailed building database in urban scenarios results in excessive computational complexity, which limits the use by the mobile system operators. Most of the current research in the area of deterministic propagation modelling deals with reducing the computational complexity without losing the prediction accuracy, modelling effects like diffuse scattering or the use of efficient computational techniques to reduce time.

In this section, different uses of ray based techniques are shown, as a result of the work of different institutions in COST IC1004. From traditional approaches to improve computation time, to channel modelling or comparison with FDTD techniques, as well as just model simplifications for the 5G networks.

2.1.2.1 Diffuse scattering

The diffuse scattering phenomenon has been studied for years because of its relevance in the field of wave propagation and in many other fields of application as well (remote sensing, optics, physics, etc.), but its understanding is still far from being complete. It is well known that in the presence of smooth and homogeneous walls the basic interaction mechanisms can be analyzed using the geometrical optics approximation, thus all interactions are regarded as reflections, transmissions, and diffractions. Nevertheless, in a real propagation scenario the case of a perfectly smooth slab is rarely present, especially in dense urban areas where buildings have highly irregular structures and volume inhomogeneities: therefore, modelling only specular and transmitted paths might not be sufficient because of the presence of the so-called dense multi-path component (DMC), which have a significant impact on the radio link, especially in the case of NLoS propagation [LMV14].

In the work by Vitucci et al. [LTY14] a semi-deterministic propagation graph modelling approach is proposed based on the framework of graph theory. Unlike in the original stochastic propagation graph modelling, the graph in our method is set up by the aid of digital maps: walls and buildings are discretised into scatterers whose spatial distribution over planes or volumes reflect the shape of the actual obstacle they represent. Scatterer distribution and corresponding propagation coefficients are jointly defined in order to satisfy realistic propagation conditions, such as the power-distance decay law, power balance, transmission attenuation, etc. The proposed model is very suitable for highly diffuse scattering conditions, such as propagation at mm-wave and terahertz frequencies.

Some works in the diffuse scattering issue, by Vitucci et al. [EVC11] and Mani and Oestges [MO12], were evolving from the effect of smooth walls to the introduction of the effect of trees, trying to demonstrate the accuracy of the model, where simulations taking into account LoS propagation, reflection, diffraction, and diffuse scattering were tested.

In the work by Oesteges [Oes14], a measurement campaign to verify diffuse scattering in 12.5 and 30 GHz is described. This paper analyzes some limitations of the diffuse approach.

2.1.2.2 RT in vehicular networks

The evolution of vehicular networks led to the development of improved RT-based models to vehicular environments. Several works around the use of ray-based techniques in vehicular environments were proposed along the IC1004. Mainly, works by [JNK12], Werthmann et al. [TJT12], and Rose and Kürner [RK14] were devoted to the use of ray-based in vehicular environments.

RT for vehicular networks includes three major wave propagation mechanisms: (i) LoS, (ii) specular components, as well as (iii) diffuse scattering. It relies on the calculation of all propagation paths connecting the TX and the RX location for a given propagation mechanism. The RT channel model enables the calculation of the electric field in amplitude, phase and polarisation at the mobile terminal position.

As a deterministic channel model in the work by Werthmann et al. [TJT12], authors propose the use of a 3D ray-optical model that has initially been developed for car-to-car communication purposes. The model has been extended by a diffraction model for the development of the Hanover 3D reference scenario. The following types of rays are considered in the model: direct path, specular reflections up to *n*th order (2nd order in this case), diffuse scattering, and diffraction. The direct path between the TX and RX is identified by a LoS check using 3D polygons of the buildings of Hanover. The specular reflections are calculated based on the image method (RT). Due to the high computation time only 2nd order reflections are considered. On surfaces seen by both the TX and RX diffuse scattering is taken into account. Diffraction is considered using the knife edge model shows an example of the identified rays between the TX on a rooftop of a building and RX located on the street level.

The underlying simulation-based channel model that is used in the work by Nuckelt et al. [JNK12] belongs to the class of deterministic channel modelling approaches using 3D ray-optical algorithms. In order to characterise the channel between TX and RX, the direct path, specular reflections, as well as diffuse scattering in terms of non-specular reflections are taken into account. Specular reflections are calculated recursively up to a desired order, but depending on the complexity and level of detail of the environment only reflections up to order three or four are practical regarding computational effort. Faces of buildings or obstacles that can be seen by both the TX and the RX are treated as sources of non-specular reflections of first order, modelled by means of Lambertian emitters. Furthermore, the channel model is able to include the full-polarimetric antenna patterns of TX and RX, respectively.

2.1.2.3 Channel modelling for 5G networks

During the last years, the use of Ray Based techniques for channel modelling has been evolving, and during the COST IC1004, many works related with this issue were presented and different approaches were proposed. In the work by Zenter et al. [ZK13, ZH14, RZD12], the concept of a virtual source for deterministic reference channel models (RCMs) is introduced and developed. The feasible option for geometry-based RCM would be a set of RT-simulated environments. RT allows high-resolution simulations, thus providing a very detailed description of the radio environment and the propagation phenomena. The concept of ray entity reduces the computational complexity for RCMs in complex scenarios.

Kaltenberger et al. [MT13] propose the use of RT based on a spheroidal subspace, as a deterministic technique which is currently employed to predict wireless channel parameters such as delay spread, Doppler spread, and angular spread in a variety of environments and a wide range of frequency. The multipath channel model is the representation of complex phenomena involving several mechanisms of interaction between the radio wave and the environment. Accordingly, through spatial and time characterisation it is possible to design and theoretically evaluate the wireless communication systems. Time variant channel impulse response can be seen as a superposition of all propagation paths' contributions, which can be calculated as:

$$H_{\mathrm{RT}}(f, x(t)) = n = 1]N \sum \eta_{\mathrm{n}}(x(t))e^{j2\pi f T_{\mathrm{n}}(x(t))} \tag{2.5}$$

where n is the propagation path index, $\eta(x(t))$ is the complex-valued weighting coefficient of the n_{th} path, $T_n(x(t))$ is the delay, and N is the total number of paths.

In the work by Vitucci et al. [EVE14], the authors show the use of ray-launching techniques for channel modelling in large indoor environments, showing the restrictions presented when detailed information about the scenario is not present. This work compares simulating results with measurements in order to identify the elements required for an accurate modelling of the scenario.

Garcia-Pardo et al. [JPGLL] discuss a PDP estimation using a ray-launching tool implemented in Matlab, and compare results with measurements in 60 GHz band. In this work the authors implement diffuse scattering, based on the works described above.

In the work by Kitao et al. [NO15], the authors propose alternative model for buildings at the intersection to improve the accuracy of RT calculation. Specifically, the conventional model assumes box-shape as a building while the proposed model considers more detailed shape of building. The proposed and conventional models are used in RT calculation, respectively, and the accuracy is evaluated.

The approach of Baek et al. [SHP14a, SHP14b] is based on a simplification of a propagation model and channel estimation procedures using ray-based simulations to obtain slope parameters and cluster information, in order to accelerate simulations results for real implementations in the 5G networks.

For wideband channels, like those used in long-term evolution (LTE)-A and ultra wideband (UWB) networks, Gan et al. [PW13] uses the concept of simulating the propagation channels at multiple frequency points, which are the centre frequencies of the corresponding subbands. The accuracy of the subband divided RT is related to the number of the sub-bands: the larger this number the better the accuracy, but at the cost of a higher computational effort. Subband divided RT can be summarised with the following steps:

- The whole UWB bandwidth is divided into several subbands. In each subband, constant frequency characteristics can be assumed for all materials and mechanisms.

- Conventional RT is used to obtain the channel impulse response (CIR) at each sub-band centre frequency.
- The subband frequency responses are calculated by Fourier transforms. Afterwards all frequency responses over different subbands are combined into a complete frequency response over the whole UWB bandwidth.
- Finally, the CIR over the entire UWB bandwidth can be obtained by an inverse Fourier transform.

The complete frequency response can be expressed as:

$$H(f) = [i = 1]N \sum F\left\{h_i(T) \cdot R_i(f)\right\}, \tag{2.6}$$

where i is the sub-band index, N is the total number of subbands, $F\{\cdot\}$ is the Fourier transform, $h_i()$ is the CIR at the ith sub-band, and $R_i(f)$ is the rectangular window function associated with the ith sub-band.

2.1.2.4 Improvements of ray launching

It is well known that RT models are deterministic and, therefore, it is expected a very precise prediction for radio propagation. However, such models require information of material's constitutive parameters, increasing the difficulty of use because of the absence of such parameters. This is especially true in outdoor environments, because the diversity and quantity of building blocks typically found in outdoor environments. Although some works have been done trying to obtain constitutive parameters for different materials and building blocks, the diversity is such that it is almost impossible to characterise all possible environments. This is a major constraint related with the accuracy of RT methods, that is usually solved using some typical values for constitutive parameters, according to the environment. In Navarro et al. [DGC13], the problem of parameters calibration is discussed and some techniques to improve computational time are proposed.

In the work by Pascual-Garcia [MMIJL12], a 3D RT technique which has been fully programmed in Matlab was applied to calculate most important channel parameters. The mentioned method is based on image theory. Images of the TX are computed for each wall, and then reflections are considered if the ray hits the corresponding wall. This software is programmed to compute any order of reflection. The diffraction phenomena is also considered, and the software searches all wedges, and it also finds the images of the TX and RX (diffraction + reflection). The position of each image depends on the scenario element position where the reflection takes place. Therefore, the accuracy of the RT technique relies on the precision of the 3D model of the real scenario. In this work, special care has been taken in the definition of the 3D model.

In Andersen et al. [SFF12] simulations of free-space propagation and rooftop diffraction using the FDTD method and RT method are described and compared. Concerning the free-space propagation simulations, the two methods show similar results. However, the FDTD computation time is much larger than the RT one, demonstrate the importance of improving the ray-based algorithms for better propagation and channel estimation.

Vitucci et al. [EVB15] discusses the precision of RT models for traditional propagation signal losses estimation using a set of measurements in cities like San Francisco and the sensitivity of ray based models to the simulated scenario versus the real scenario.

Finally, Brem et al. [MME13] proposes a modification of the wave front to improve accuracy of RT in conditions where far field is not guaranteed. This approach simplifies the mathematical model and reduces the computation time.

2.1.3 Massive and Distributed MIMO

In this section we present recent advances in the topic of channel modelling for massive and distributed MIMO.

2.1.3.1 Massive MIMO

Massive MIMO is an emerging communication technology promising order-of-magnitude improvements in data TP, link reliability and transmit energy efficiency [Mar10, RPL+13, LTEM14, NLM13]. This approach involves multi-user MIMO (MU-MIMO) operation with an arbitrarily large number of BS antennas in a multi-cell environment. It is shown that, under these operating conditions, the effects of uncorrelated noise and fast fading disappear, as does the intra-cell interference, and the only remaining impediment is the inter-cell interference due to pilot contamination. In this regime, simple matched filtering (MF) linear pre-coding becomes optimal and capacity can be achieved with relatively inexpensive signal processing.

The fundamental idea of massive MIMO is that, as the number of BS antennas grows large, the channel vectors between users at the BS become pair-wise orthogonal. This ideal situation is often referred to as "favourable" propagation conditions. Experimental work is, therefore, of great importance to investigate the range of validity of this assumption: what benefits can we obtain at very large, but limited, number of BS antennas in a realistic scenario?

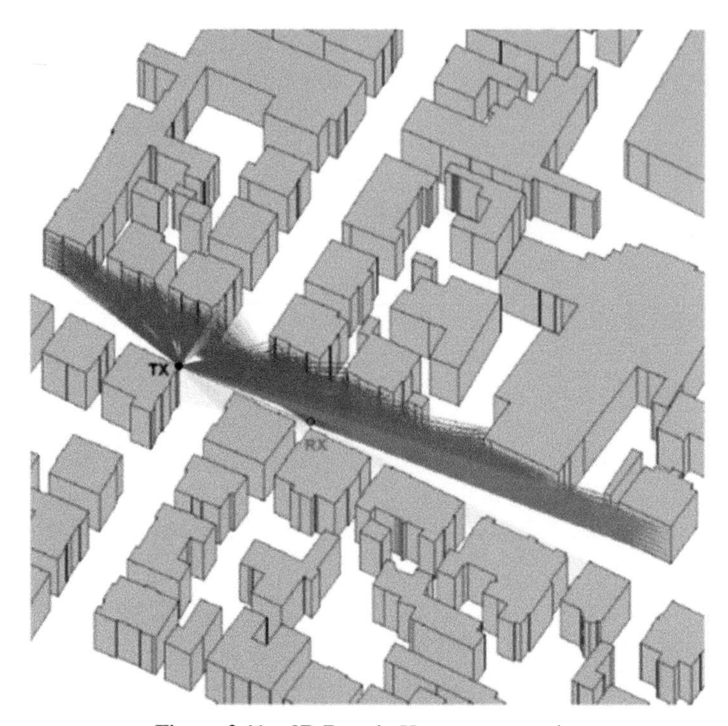

Figure 2.11 3D Rays in Hannover scenario.

In an effort to answer this question, measurements were conducted in a residential area north of Lund, Sweden, representative of a suburban environment [GERT11]. The BS array was placed indoor, while the user terminals were located at different outdoor positions with similar LoS directions to the BS. The array is a cylindrical antenna array with 64 dual-polarised antenna elements, giving in total 128 antenna ports (Table 2.7). The measurement data were recorded with the RUSK LUND channel sounder [THR+00] using a centre frequency of 2.6 GHz and a measurement bandwidth of 50 MHz. Some results on the sum-rates achievable by zero forcing (ZF) and minimum mean-squared error (MMSE) linear pre-coding schemes with random antenna selection are shown in Figure 2.11, for the case of two users being served. Clearly, the average correlation between the users' channels decreases as the number of antennas[1] at the BS increases. This suggests that practical very-large arrays can decorrelate multi-user channels

[1]Whenever there is no risk of confusion, we will use the shorter term "antenna" in place of "antenna port".

Table 2.7 Comparison of two large antenna arrays at the BS side

Cylindrical	Linear
128 antenna ports	128 antenna ports
Compact array (30 cm × 30 cm)	Physically-large array (7.3 m)
Two-dimensional	One-dimensional
Multiplexed	Virtual
Dual polarised	Single polarised
Directional elements	Omni-directional elements
Resolution in 2D	Superior (1D) angular resolution

in realistic propagation scenarios. Indeed, with as few as 20 BS antennas, the linear pre-coding sum-rates already reach 98% of the dirty-paper coding (DPC) capacity (Figure 2.12b). Similar conclusions are reached in Flordelis et al. [FGD+15], where results on the ability of massive MIMO systems to spatially separate eight users located close to each other in an outdoor environment with LoS to the BS are reported. These capacity/sum-rate results reinforce previous findings in the related literature that a large fraction of the predicted capacity gains of very-large MIMO are possible at sizes of the BS array as low as 10 times the number of users served.

The impact of the shape and size of the BS antenna array on the performance of massive MIMO systems is discussed in [GTER13]. Several considerations apply when selecting the form factor of very-large antenna arrays. For instance, from a practical point of view, it is preferable to have a physically-compact array with a large number of antennas at the BS. On the other hand, making the arrays smaller in size brings about some drawbacks such as higher antenna correlations. In [GTER13], the performances achievable using a 128-element cylindrical array and a 128-element linear array in realistic propagation environments are compared. The main properties of the BS antenna arrays used in this study are summarised in Table 2.7. Channel measurements were performed in a suburban environment, at the Faculty of Engineering (LTH) at Lund University, Lund, Sweden. The two BS antenna arrays were placed on the same roof. More precisely, the cylindrical array was positioned on the same line as the linear array, near its beginning. Both measurement data sets were recorded using a centre frequency of 2.6 GHz and a measurement bandwidth of 50 MHz. An omni-directional antenna was used at the user side.

It is observed that, under random antenna selection, the variations incapacity/sum-rate experienced by the cylindrical array are larger than those experienced by the linear array. In the same way, the average capacity/sum-rate offered by the linear array tends to be higher than the average capacity/sum-rate

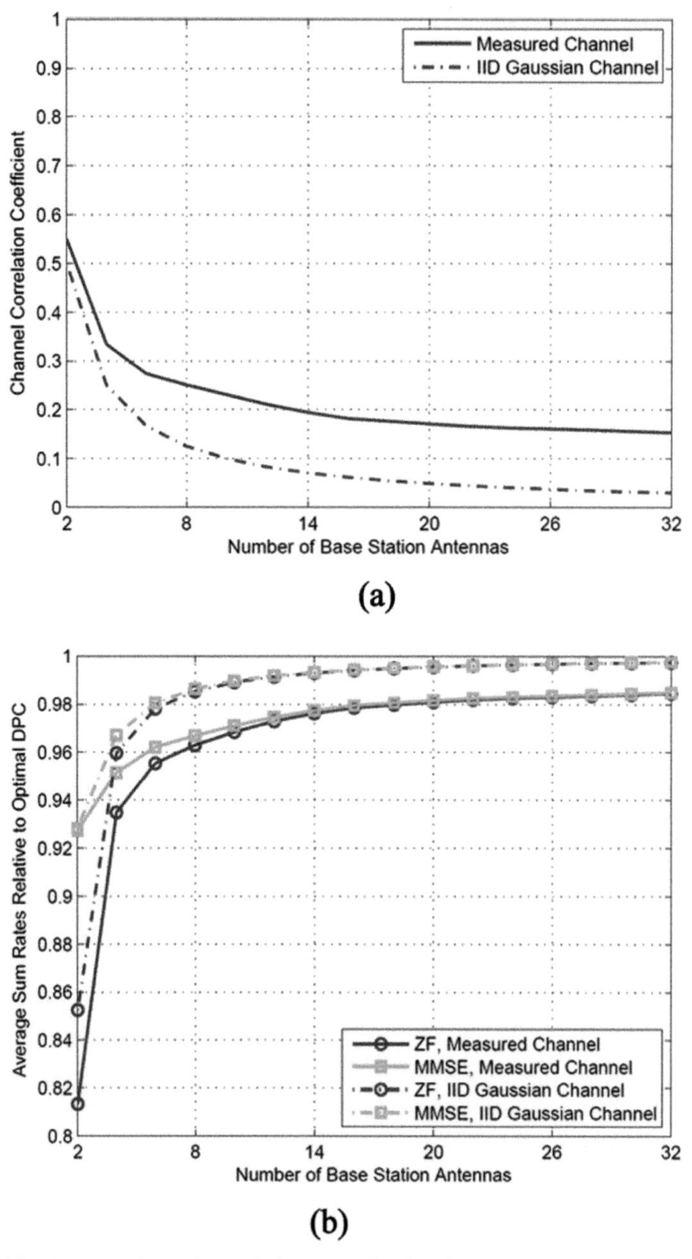

Figure 2.12 Average channel correlation (a) and ratio of average linear pre-coding sum-rate and DPC capacity (b) as a function of the number of BS antennas. The transmit signal-to-noise-ratio (SNR) is set to 20 dB and the total transmit power is kept unchanged.

at the cylindrical array. One possible explanation is that the variations induced by the polarisation and directionality characteristics of the antenna elements of the cylindrical array (the antenna elements of the cylindrical array are directional, dual-polarised patches, while the linear array is virtually formed by an omni-directional, single-polarised antenna element, (see Table 2.7) seem to have a larger impact on capacity/sum-rate than the large-scale fading (LSF) along the linear array caused by the environment where the measurements are performed (and assumed negligible for physically compact arrays).

The presence of LSF across the antenna elements of a physically large antenna array was noted in the work by Payami and Tufvesson [PT12]. The most important observation is that the channel cannot be seen as wide-sense stationary over the physically-large array. This situation is illustrated in Figure 2.13, which shows the angular power spectrum over a BS physically large antenna array for a user location with NLoS [GTER12, GZT+15]. Cluster parameters according to the COST 2100 channel model have been extracted following Czink et al. [CCS+06] and Czink [Czi07], and are also shown in Figure 2.13b. We note that some scatterers are not visible over the whole array and, for scatterers being visible over the whole array, their power contribution varies considerably. This power variation may be critical to algorithm design and performance evaluation of very-large MIMO systems. Therefore, it is important to model the LSF process over physically large arrays. The modelling approach attempted can be seen as an extension of the COST 2100 channel model [VZ12], in which only small and compact multiple antenna arrays are considered. In the work by Gao et al. [GTER12, GZT+15], the concept of cluster visibility region (VR) in the COST 2100 channel model is proposed to be used at the BS side, as well, to account for LSF over physically-large antenna arrays. At the mobile side, the VR concept is used without changes. This concept is illustrated in Figure 2.14. Now, for the propagation link between a BS antenna element and a MS, a cluster is active when the antenna element is inside the cluster's BS-VR *and* the MS is inside the corresponding MS-VR. Both conditions should be satisfied to declare a cluster being active.

Some new parameters are added to the COST 2100 channel model to account for the BS-VR model extension. These are:

- The total number of clusters that are visible over a physically large array.
- As can be seen from Figure 2.14, more clusters are visible for a physically-large array as compared to a compact array and so, the total number of clusters in the conventional model is not suitable any more.

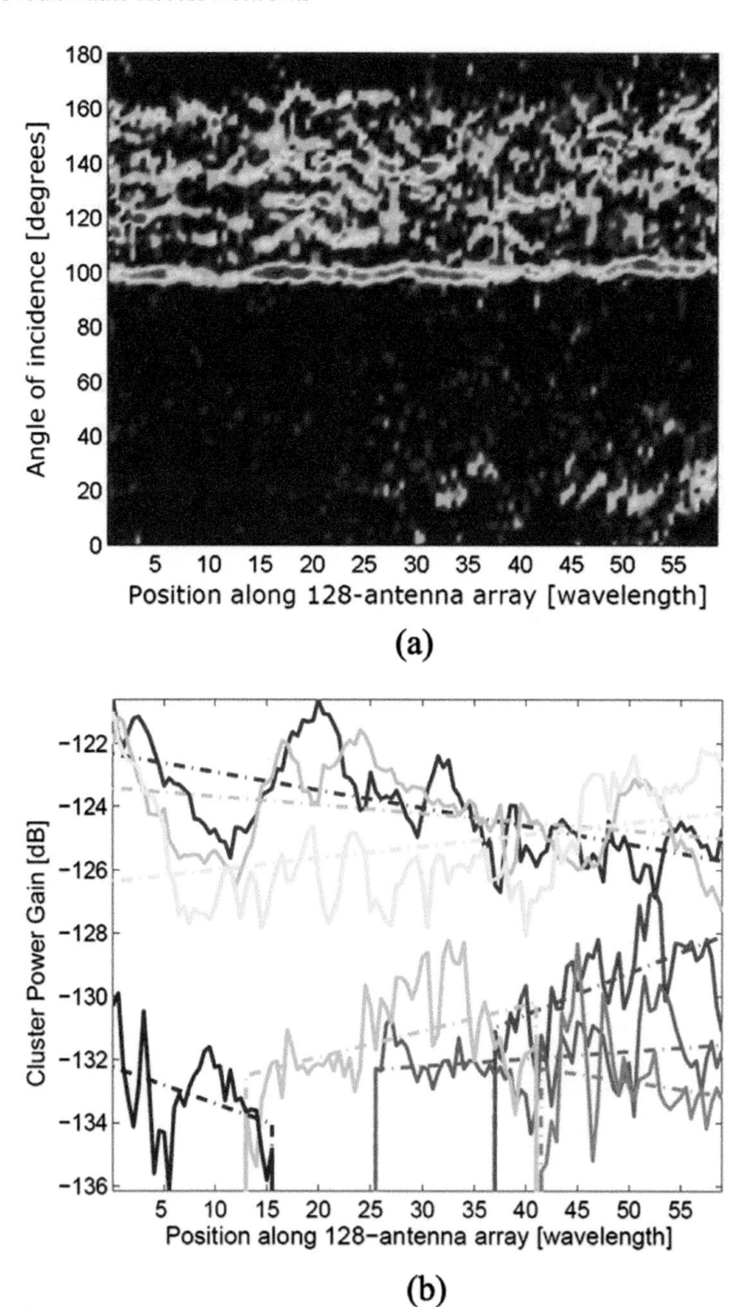

Figure 2.13 Angular power spectrum over the BS antenna linear array (a) and cluster power variation with least square linear fitting (b). The user is located at a position with NLoS.

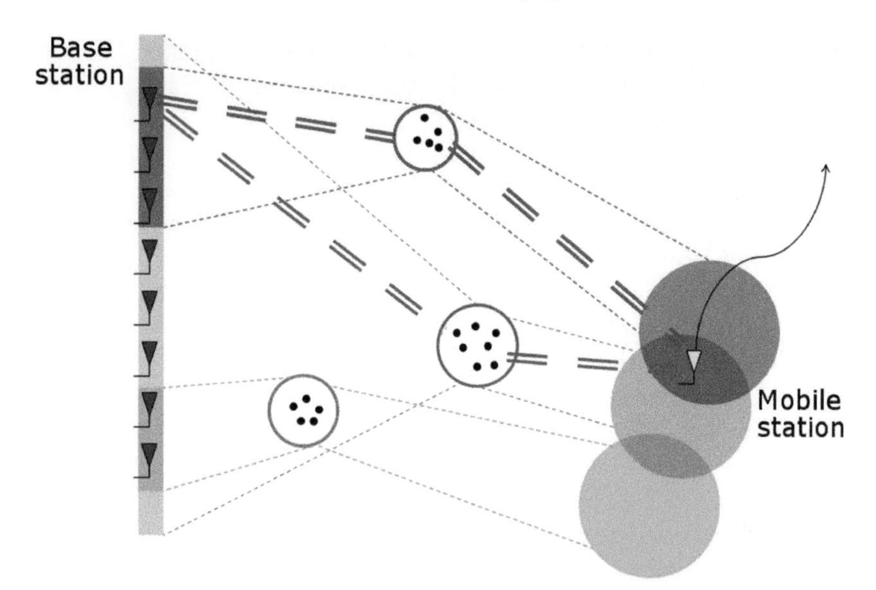

Figure 2.14 Extension of the concept of cluster VR to the BS side. The clusters in *red* and *grey* colours are active.

- The shape and size of BS-VR. For example, the BS-VRs of physically large linear arrays are modelled in the work by Gao et al. [GZT+15] as intervals on a line, rather than circular regions as those used for modelling the MS-VRs.
- Cluster power variations within BS-VR. It is found appropriate to replace the cluster visibility *gain function,* which describes the time evolution of clusters at the user side, by a slope—the cluster visibility *gain,* which characterises the power variation of those clusters visible at the BS side across the antenna array elements.

A detailed description of the modelling and parameterisation of BS-VRs for a physically-large antenna array in NLoS propagation conditions can be found in the work by Gao et al. [GTER12, GZT+15]. The modelling of MS-VRs in the conventional model cannot be directly applied to BS-VRs since MS and BS usually have very different propagation environments in their vicinity.

2.1.3.2 Distributed MIMO

Spatially distributed MIMO antenna systems, also known as coordinated multi-point (CoMP) or distributed MIMO, is a novel communication technology that aims at improving spectral efficiency and fairness for wireless mobile communication networks [MF11]. The main idea behind distributed MIMO systems is that deploying the antennas of the BSs over a large geographical area should increase the probability of a MS of experiencing high signal strength conditions on one or more *links,* simultaneously. Of course, this idea rests on the assumption that different links undergo independent LSF and, hence, the achievable gains of a distributed antenna system will ultimately depend on the correlation properties of the so-called LSPs of the links, such as LSF (F), delay spread (τ_{rms}), azimuth angular spread (ϕ_{rms}) and elevation angular spread (θ_{rms}). Correlation between LSPs belonging to different links is often explained by the existence of interacting objects (IOs) common to those links and, in this sense, it is environment specific.

In Dahman et al. [DFT13, DFT14], multi-site, fully-coherent measurements conducted with the RUSK LUND [THR$^+$00] channel sounder using a centre frequency of 2.6 GHz and with a measurement bandwidth of 40 MHz are reported. The measurement campaign took place at the Faculty of Engineering (LTH) at Lund University, Lund, Sweden, in a suburban environment with low buildings, rich vegetation, and a small pond at the centre. The setup chosen consists of four single-antenna BSs communicating with a MS provided with a 128 antenna ports array as described in Section 1.1.1. The BS antennas were placed outside the windows at the second and third floors of four different buildings, which corresponds to 5–12 m above the ground level and 10–20 m below the surrounding buildings. The distance between antenna sites varies between 60 and 200 m. BS antennas are connected to the channel sounder device by means of RoF links. A photo of the measurement area with the positions of the BSs can be seen in Figure 2.15. During the measurements, the MS circulated the pond counter-clockwise at pedestrian speed (less than 0.5 m/s) along a predefined route with a length of 490 m. The propagation conditions over the whole route can be described as obstructed line-of-sight (OLoS) due to the presence of high, leafy trees in the measurement area blocking the LoS from the MS to the BSs. The intensity of the tree blockage varies from one MS position to another. It must be noted, however, that the probability of having LoS between the MS and one of the BSs is significant when the MS is close to that BS.

Details about the extraction of the LSPs can be found in [DFT13, DFT14]. Following a methodology similar to that introduced in the work by Kyösti et al. [KMH$^+$08], we make use of the Box–Cox transformation in order to transform non-normally distributed LSPs to another domain where they have approximately normal distributions. Then, in the transformed domain, we estimate the parameters of the corresponding Gaussian distributions of the

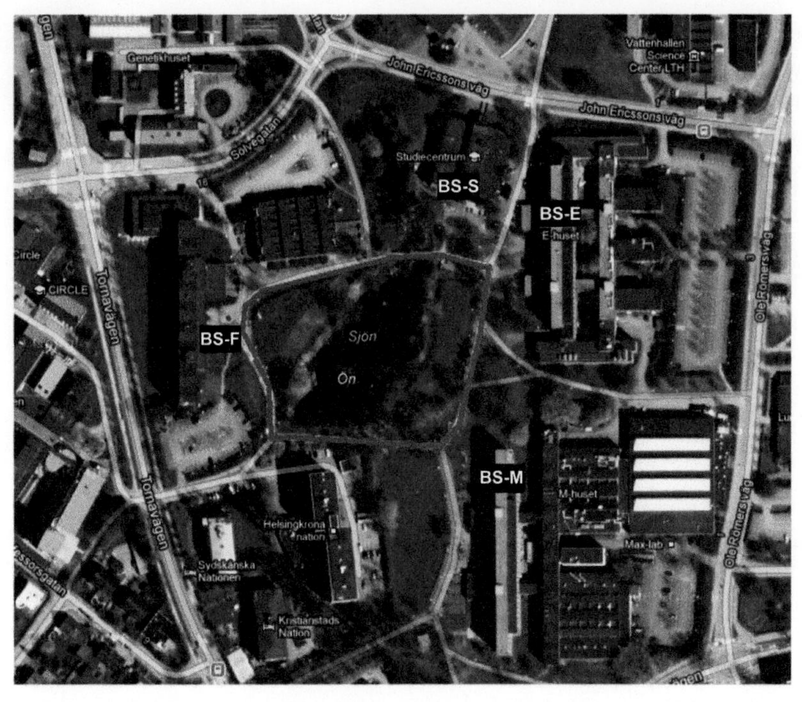

Figure 2.15 Aerial photograph of the measurement area. BSs locations are indicated by labels BS-E, BS-S, BS-F and BS-M. The measurement route is plotted in *blue* colour.

LSPs and the correlation coefficients between LSPs. The Box–Cox transform with *power parameter* is defined as

$$g(x)^{(\lambda)} = \begin{cases} \frac{x^{\lambda}-1}{\lambda}, & \lambda \neq 0 \\ \log(x) & \lambda = 0. \end{cases} \tag{2.7}$$

For each LSP and each BS, we choose the value leading to the distribution in the transformed domain that is closest to normality. The transformed delay spread, angular azimuth spread and angular elevation spread are denoted by $g(\tau_{\mathrm{rms}})$, $g(\phi_{\mathrm{rms}})$, and $g(\theta_{\mathrm{rms}})$, respectively, and their corresponding Box-Cox power parameters by λ_{τ}, λ_{ϕ}, and λ_{θ}. We note that $\lambda_{\phi} = 0$ and, therefore, it follows that $g(F) = F$. Tables with the fitted Box–Cox power parameters can be found in the work by Dahman et al. [DFT14a]. From the estimates of the LSPs obtained as described above, we then estimate the auto-correlation functions of the LSPs and the correlation coefficients of LSPs belonging to the same link (*intra-link* correlation coefficients) and to different links (*inter-link* correlation coefficients).

We first consider the LSPs belonging to the whole measurement route at once for estimating the auto-correlation functions and intra- and interlink correlation coefficients. We refer to this case as *global-scale* LSPs. The following remarks, which are in good agreement with reports available in the existing literature, can be made:

- The value of the auto-correlation distance[2] of the same LSP varies significantly from one link to another. See the work by Dahman et al. [DFT14a] for further details.
- For the intra-link case, there is a significant negative correlation between F and $g(\tau_{\mathrm{rms}})$, for all links. In general, F is negatively correlated with the rest of the (transformed) LSPs.
- For the intra-link case, there is a positive correlation between $g(\phi_{\mathrm{rms}})$ and both $g(\tau_{\mathrm{rms}})$ and $g(\theta_{\mathrm{rms}})$.
- Values of the correlation coefficients for global-scale, inter-link LSPs fall in the range ± 0.44. In general, inter-link LSPs display lower values of the correlation coefficient than corresponding intra-link LSPs.

The low values of the correlation coefficients of the inter-link LSPs found at the global-scale may suggest the viability of ignoring inter-link correlations when simulating multi-link systems. However, one should remember that these small values of the estimated global-scale inter-link correlation coefficients are obtained by considering the whole measurement route at once, a length of 490 m, which might exceed the extension of the joint stationarity region of the multi-link LSPs. Simulations of wireless channels are often performed as a series of *simulation drops,* each of which is between few meters to few tens of meters long, and each simulation drop is assumed to have constant LSPs. Therefore, it will be valuable to utilise our measurements in order to extract a model that describes the correlation coefficients among the LSPs of the different links at a *local-scale*, i.e., within short parts of the route of ten to few tens of metres length.

The use of the truncated Gaussian distribution, TG(μ, σ^2), to model both the intra- and inter-link correlation coefficients of the LSPs on the local-scale is proposed in the work by Dahman et al. [DFT14a]. The model can be used as follows:

1. Depending on the LSPs considered and on the intra-link or inter-link case, use Tables 2.8, 2.9, or 2.10 to generate values for $\mu \sim u(\mu_{\min}, \mu_{\max})$ and

[2]The *auto-correlation distance* is defined as the distance at which the magnitude of the auto-correlation function has decreased to e^{-1} of its peak value.

Table 2.8 Local-scale inter-link correlation coefficients (same LSPs)

	F	$g(\tau_{rms})$	$g(\phi_{rms})$	$g(\theta_{rms})$
μ_{min}	–0.25	–0.18	–0.16	–0.10
μ_{max}	0.28	0.33	0.23	0.28
σ_{min}	0.23	0.23	0.28	0.14
σ_{max}	0.52	0.47	0.39	0.39

Table 2.9 Local-scale inter-link correlation coefficients (different LSPs)

	F, $g(\tau_{rms})$	F, $g(\phi_{rms})$	F, $g(\theta_{rms})$	$g(\tau_{rms})$, $g(\tau_{rms})$	$g(\tau_{rms})$, $g(\phi_{rms})$	$g(\phi_{rms})$, $g(\theta_{rms})$
μ_{min}	–0.31	–0.28	–0.26	–0.17	–0.17	–0.11
μ_{max}	0.21	0.18	0.19	0.33	0.25	0.29
σ_{min}	0.24	0.20	0.20	0.20	0.18	0.17
σ_{max}	0.61	0.55	0.52	0.50	0.50	0.42

Table 2.10 Local-scale intra-link correlation coefficients

	F, $g(\tau_{rms})$	F, $g(\phi_{rms})$	F, $g(\theta_{rms})$	$g(\tau_{rms})$, $g(\tau_{rms})$	$g(\tau_{rms})$, $g(\phi_{rms})$	$g(\phi_{rms})$, $g(\theta_{rms})$
μ_{min}	–2.50	–2.50	–2.24	0.68	0.32	0.54
μ_{max}	–0.82	–0.58	–0.41	2.50	1.67	2.50
σ_{min}	0.09	0.10	0.26	0.08	0.26	0.19
σ_{max}	0.89	1.00	0.80	0.91	0.80	0.80

$\sigma \sim u(\sigma_{min}, \sigma_{max})$, where $u(a, b)$ represents a uniform distribution in the interval [a, b].

2. Generate $r \sim G(\mu, \sigma^2)$, where $G(\mu, \sigma^2)$ represents a Gaussian distribution with mean μ, and variance σ^2.
3. Repeat step 2 while $|\rho| > 1$.
4. Repeat steps 2 and 3 for each simulation drop.

The proposed method can indeed reproduce the large variability exhibited by the correlation coefficients on the local-scale. For example, if $\mu = -0.31$ and $\sigma = 0.14$ are used to model the inter-link correlation coefficient of one link pair, then the corresponding probability of having $\rho > 0.5$ is almost zero. However, for another simulated link pair, if $\mu = 0.33$ and $\sigma = 0.61$ are assumed, then the probability of having $\rho > 0.5$ will be 0.30. Performing step 1 only once guarantees that the values of the correlation coefficients are consistently generated over a whole batch of simulation drops since, for each LSP, its mean and variance remain unchanged. A further simplification of the model is possible if one observes that, for any pair of LSPs, the results reported in Tables 2.8 and 2.9 can be approximated by drawing μ and σ from uniform distributions in the intervals [–0.31, 0.33] and [0.14, 0.61], respectively (step 1).

It should be noted that, despite the fact that the parameters in Tables 2.8, 2.9, and 2.10 correspond to a Gaussian model, depending on the selected pair of parameters (i.e, μ and σ), the actual resulting distributions after truncation might have shapes not resembling that of a Gaussian distribution. For example, selecting ($\mu = -2.5, \sigma = 1$), ($\mu = 2.5, \sigma = 1$), and ($\mu = 0, \sigma = 1$) will result in PDFs that look like exponential PDF, truncated bell shape PDF, and reverse exponential PDF, respectively.

On a related topic, fully coherent measurements of a multi-link setup in which the antenna elements of the BS are separated up to several metres are reported in the work by Dahman et al. [DFT15], with a discussion of further extensions to the COST 2100 channel model to accurately describe radio propagation in this scenario.

2.1.4 Cellular mm-Wave

Mobile data traffic is projected to grow 5000-fold by the year 2030. This drastic increase can only be met through increase in performance, spectrum availability and massive network densification (small cells). The availability of large contiguous blocks of spectrum in the mm-wave band (30–90 GHz) could be exploited to enable very significant (1 GHz or more) increase in bandwidth in cellular communication systems. Also, at these higher frequencies, antenna elements are smaller which enables the implementation of large array antennas for beam forming to compensate for propagation losses, and to achieve significant system capacity and TP gains [RSP+14, PK11, RSM+13]. Having access to such large blocks of spectrum also makes possible in early deployments to trade-off spectral efficiency for bandwidth, i.e., high-data rates can be achieved even with low-order modulation schemes requiring lower powers and lower complexity and cost. To enable the development of next generation mobile systems, it is of utmost importance to characterise mm-wave propagation in small cell scenarios and to develop accurate channel models. Significant challenges arise with regard to channel sounder hardware development, channel measurement, deterministic channel simulation, and modelling itself.

2.1.4.1 General propagation characteristics

When considering propagation in the mm-wave band, one of the aspects to consider in the design of such radio links is the attenuation due to gases in the lower atmosphere and rain. Figure 2.16 displays the specific attenuation in dB/km due to oxygen absorption and water vapour for frequencies between 1 to ~300 GHz. In the frequency range up to 100 GHz, two peaks are visible, one due to oxygen at 60 GHz and the second peak at about 22 GHz due to water vapour. Though the high loss of up to 15 dB/km at 60 GHz prevents from transmission distances in excess of around 1 km, it can be used to advantage for frequency reuse and for secure communication. Losses due to rain can be on the order of 10 dB/km at 25 mm/h increasing to 30 dB/km at 100 mm/h rainfall. They can be crucial at longer transmission distances. However, if only short links below, e.g., 200–300 m are targeted, the losses are less significant. Another factor to consider is the penetration loss due to building material. Whereas coverage in buildings can usually be achieved easily in the classical mobile bands, solid walls become increasingly impenetrable at mm-wave frequencies. These additional losses will either require compensation through higher effective radiated powers or indoor coverage needs to be accomplished by other means.

In principle, a greater amount of rough-surface scattering is to be expected due to the smaller wavelength compared to the STD of surface-roughness, especially for walls made of rough concrete, bricks and other construction materials. Although a stronger DMC at mm-wave frequencies have been hypothesised compared to ultra high frequency (UHF), recent studies have shown that the actual ratio of the DMC to the specular component is similar to what have been observed at UHF [DFM+14] or even lower [HJK+14].

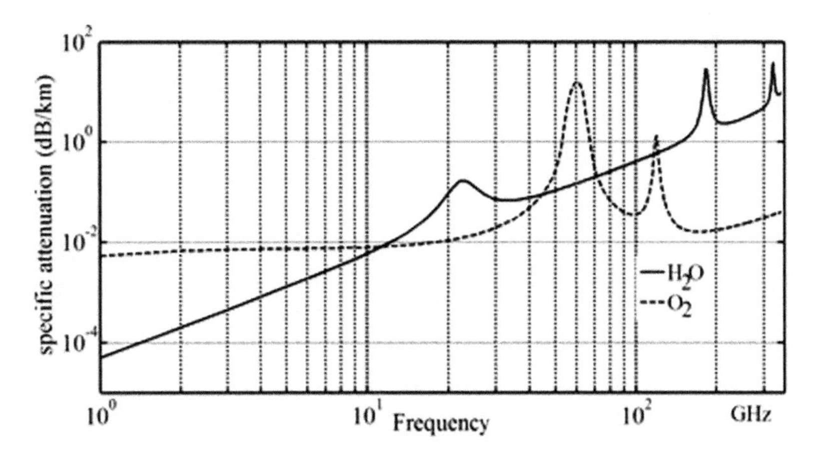

Figure 2.16 Specific attenuation versus frequency due to atmospheric gases: oxygen and water vapour.

This could be due to the lower level of multipath-richness. This aspect is very important to properly design and assess the performance of beamforming techniques recently proposed to enhance the SINR and to implement spatial multiplexing in future multi-gigabit transmission systems.

2.1.4.2 Channel measurement campaigns and results

Due to the large propagation loss and the low-power budget at mm-wave bands, the fundamental concern is the feasibility of NLoS links or of LoS links exceeding 200–300 m range. Therefore reliable PL and shadowing models are of vital importance for the design of mm-wave radio links. In addition, detailed information on the spatiotemporal characteristics of the channel are required for the development of appropriate channel models.

Two types of PL models have been proposed: Hata-like PL models based on measurements [PAC12, RSM+13, HCKP14, RJSJ14, SRC14a, KWPW14] and a dual-slope PL model for 28 GHz outdoor links based on RT simulations in downtown Ottawa, Canada [HCB+14].

In the work by Rappaport et al. [RJSJ14], wideband 28 and 73.5 GHz propagation measurements were performed at 75 distinct RX locations for three separate TX locations in Manhattan, for distances ranging from 30 to 200 m. For large distances, horn antennas with 24.5 dBi (10.9° beamwidth) at 28 GHz and 27 dBi (7° beamwidth) at 73.5 GHz were used at both the TX and at the RX. In NYU campus, the PL exponent in the NLoS condition was estimated as 3.4 at both 28 and 73.5 GHz.

Channel measurements at 28 GHz were conducted using the synchronised rotating channel sounder in different environments which included in building, outdoor campus, and urban environments for distances from 14 to 200 m [HCKP14]. Directional horn antennas with 24.5 dBi gain were used at both the TX and RX. The measurement angle in azimuth was set to cover all directions by rotating the transmit and receive antennas, from 0–360° while the elevation angle range of the RX was selected within –60–60° according to the measurement environment. The omni-directional PL model was then derived after synthesising the omni-directional channel. Figure 2.17 shows the aerial view of the urban environment. The PL exponent was then estimated assuming a reference of 1 m free space loss. The PL exponents and shadowing factors are summarised in Table 2.11.

A new multiband frequency-modulated continuous wave (FMCW) channel sounder with 4.4 GHz bandwidth and 2×2 MIMO capability which uses intermediate frequency (IF) at the TX and fully parallel reception with high dynamic range after back-to-back as well as over-the-air calibration was used to perform measurements in a reflective office environment at 60 GHz

Figure 2.17 PL measurement urban area at 28 GHz.

Table 2.11 The PL parameters

	d (m)	PLE, n	SF, σ (dB)
In-building LoS	$10 < d < 60$	1.91	3.29
In-building NLoS	$10 < d < 60$	3.08	8.92
Campus NLoS	$100 < d < 300$	2.68	7.54
Downtown NLoS	$40 < d < 300$	3.00	6.30

with a wireless local area network (WLAN)-oriented antenna arrangement for distances between 9.4–35 m [SRC14b]. The measurements yielded a PL exponent of 1.87. The median RMS delay spread revealed to be highly dependent on the applied relative threshold and values of 0.42, 5.13, and 8.46 ns were derived for 20, 30, and 40 dB threshold levels, respectively. This shows that only the most dominant components are captured if a 20-dB threshold is applied.

The same channel sounder was used to carry out 2 × 2 MIMO outdoor measurements at 60 GHz in a representative small cell LoS scenario at the science site of Durham University [SRC14a] covering ranges up to 200 m, with directional antennas with a nominal 20.7 dBi gain and 15.4° azimuthal angular width. A PL exponent of 1.93 was derived. The delay spreads calculated over all four MIMO channels are given in Table 2.12

Outdoor measurements at 70 GHz in the courtyard at the campus of TU Ilmenau for distances up to 45 m have been performed using a novel

Table 2.12 RMS delay spread for small cell measurements at Durham University (60 GHz) and TU Ilmenau (70 GHz) with different threshold (Th) levels

	60 GHz [SRC14a]		70 GHz [MDS+14]	
	20 dB Th		30 dB Th	
	LoS (ns)	NLoS (ns)	LoS (ns)	NLoS (ns)
95% CDF value	2.34	106.9	94.30	
50% CDF value	1.17s	34.64	65.34	

UWB dual-polarised ultra-wideband multi-channel sounder (DP-UMCS) [MDS+14] based on a 12 stage M-sequence radar chipset [SHK13]. Using dual polarimetric antennas, the vertical and horizontal characteristics are measured in parallel at the RX with fast switching between horizontal and vertical polarisation at the TX.

The measurements were performed with circular dual-polarised horn antennas with a 3-dB aperture angle of 14° and a maximum gain of 21 dBi. In the work by Müller et al. [MDS+14], results for vertical polarisation are presented for delay spread. For LoS and with (30 dB threshold) the delay spread varied between 18 and 107 ns. The lowest delay spread was observed for the highest receive powers mainly in the direction of the maximum antenna gain. The delay spread for the NLoS measurement shows values from around 36–94 ns. The 95% CDF values and the 50% CDF values are also given in Table 2.12. In the power Azimuth spectrum (PAS) first and second order reflections were observed. The average attenuation per reflection in the investigated scenario was around 20 dB. The results indicate that transmission ranges of several hundreds of metres are feasible at 70 GHz and even NLoS conditions might be supported by exploiting strong reflections from surrounding buildings, where double reflections can also contribute energy to the link budget.

Several 60 GHz, outdoor measurement campaigns were carried out in the framework of the MiWEBA project [MiW14]. Omni-directional as well as directional antennas were used in combination with two different wideband channel sounders providing 250 and 800 MHz bandwidth, respectively. The omni-directional setup allowed to capture the time-variant characteristics of the busy urban environment, including human body shadowing, as well as the effects of a mobile RX. Several million channel impulse responses were captured to provide a reliable data base for statistical analysis. The directional measurements focused on the investigation of the ground reflection. Overall, the measurements confirm that the ground reflection contributes to the channel in small cell scenarios and that the channel can by highly time-variant, even for a stationary user. On the basis of these findings and the acquired data a quasi-deterministic (Q-D) modelling approach was derived [MPK+14]. The estimated PL exponent for the LoS-dominant street canyon scenario was 2.13 [KWPW14].

The data from the 28-GHz measurement campaign in the work by Hur et al. [HCL+14] were analysed to extract the cluster parameters such as the number of clusters, *N*, the number of sub-paths, *M* and angular information. These parameters can be used for modelling the channel and the extracted parameters are summarised in Table 2.13.

2.1.4.3 RT investigations

Considering the propagation characteristics at mm-wave frequencies, deterministic ray models such as RT or ray launching seem to represent the natural base to model mm-wave propagation [SDNH14]. Being founded on a sound, albeit approximate, theory such as ray optics, ray-based models can be considered fairly reliable when the wavelength is small compared to the size of the obstacles and when the propagation environment is limited and, therefore, a detailed environment description is possible. Recent investigations show that diffuse scattering can be observed in the mm-wave band, but specular propagation mechanisms clearly dominate over DMC. In the work by Haneda et al. [HJK+14] 75–90% of the total power could be attributed to specular components, whereby the dominance was found to be more pronounced in larger environments. The limited degree of DMC could be due to the presence of smooth surfaces in the considered indoor environments. However, outdoor measurements indicate that also the urban mm-wave channel is mainly constituted by several strong components related to specular reflections [MPK+14, MiW14, RSM+13]. With respect to UHF frequencies, comparably small objects such as lamp posts, trash cans and metal window frames seem to have a significant, but more specular-like effect. Depending on the targeted accuracy, it can therefore be important to include them in the model. This especially applies to environments without smooth, planar surfaces and NLoS scenarios [MAB15]. However, in many cases, a deterministic model might not be reasonable anymore due to the unpredictable shape and position of the objects, so related reflections are to be treated as scattering and modelled in a statistical manner [DFM+14, MAB15].

Overall, RT approaches have to be further evaluated for centimetric and millimetric wavelengths. This also includes appropriate modelling of vegetation. In the work by Oesteges [Oes14], a RT tool accounting for vegetation effects is compared with continuous wave (CW) measurements at 12 and 30 GHz in a street canyon carried out in late summer and winter to

Table 2.13 Cluster-based channel model parameters in indoor building-scale environment and outdoor campus environment

Environment	In-Building				Outdoor Campus			
	LoS		NLoS		LoS		NLoS	
Parameters	Mean	Std	Mean	STD	Mean	STD	Mean	STD
Mean Excess Delay (ns)	6.81 ns	10.95 ns	20.17 ns	19.40 ns	45.95 ns	33.34 ns	48.09 ns	65.74 ns
RMS Delay Spread (ns)	5.94 ns	13.10 ns	18.73 ns	19.43 ns	28.68 ns	26.04 ns	34.98 ns	47.25 ns
Number of Clusters, N	1.26	0.63	2.24	1.01	1.77	0.52	1.81	0.80
Number of Sub-paths in Cluster, M	5.17	2.72	10.91	10.48	6.76	4.37	8.59	17.18
Angle Spread - Departure (ASD) (°)	8.80	11.61	10.45	10.27	4.39	0.47	6.60	3.36
Angle Spread - Arrival (ASA) (°)	5.84	5.32	21.88	17.49	4.39	0.94	9.54°	12.55
Intra-cluster ASD (°)	7.71	10.60	11.52	12.70	4.34	0.73	5.54°	2.59
Intra-cluster ASA (°)	4.72	2.69	11.84	13.35	4.19	1.28	5.87°	7.62

investigate the impact of two deciduous trees obstructing the line of sight. Corrugated circular horns with vertical polarisation were used with 20° beamwidth at the TX and 30° at the RX. Measurements were taken when the antennas were aligned on a 60-m TX track in the street with 1 m spacing, and the RX was fixed on a building.

Using the well-known Fresnel theory and the uniform theory of diffraction (UTD), the RT tool accounted for single-reflected rays by walls and the ground, single diffraction by building wedges and attenuation by trees. Diffuse scattering and transmission through buildings were neglected. Regarding vegetation, only the tree foliage was modelled by means of parallelepiped blocks applying the ITU-R 236-6 recommendation to calculate the attenuation [CCI86]. Despite the simple simulation model, good agreement between the simulations and the measurements of the average received power could be achieved for both frequencies.

According to Oesteges [Oes14] slightly rough surfaces at small wavelengths can be accounted for by introducing a reduction factor with respect to free-space propagation and Fresnel reflection for the specular reflections. Given the roughness of typical walls, e.g., brick walls, the diffuse component contribution at mm-wave frequencies is often negligible compared to the specular one. The diffuse power can be assessed by applying the Kirchhoff approximation or the first order pertubation solution, which better accounts for polarisation effects [Oes14, COS01]. However, diffuse scattering can become important if very directive antennas are used in NLoS scenarios.

Parameters estimated from measurements are limited by the resolution of the radio measurement equipment. RT on the other hand models only specular reflection, edge diffraction, and penetration. However, according to Takada et al. [MGI06], scattering cannot be ignored comparing the measurements with the RT simulation results in visualised photos which can be helpful to understand the propagation mechanisms. In Tsuji et al. [TKT14], visualisation of the propagation channel is compared with the radiophotos of the measurement and the simulation results. In Figure 2.18, the direct wave is appropriately reproduced in the simulation. However, the possibility remains that some reflections are underestimated in the simulation and scattering might not be considered.

2.1.4.4 Stochastic and semi-stochastic mm-wave channel models

Since the application of mm-wave transmission to mobile communication is closely linked to the exploitation of beam forming and MIMO techniques, multidimensional propagation models accounting for time and angle

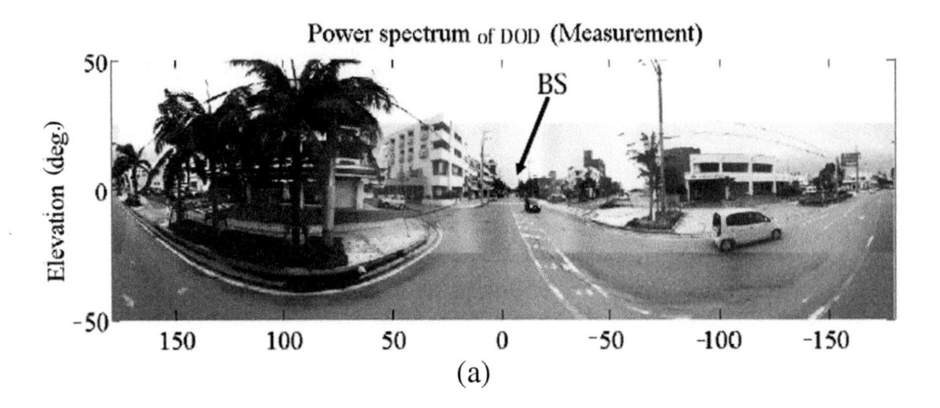

(a)

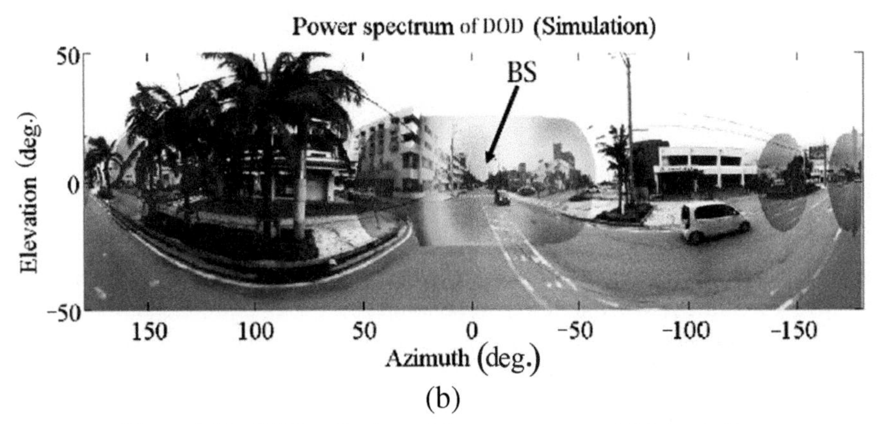

(b)

Figure 2.18 DOD radiophoto: (a) measurement and (b) simulation.

dispersion of the signal are required. Several models for 60 GHz indoor applications have been developed within standardisation bodies such as the statistical channel model IEEE 802.15 TG3c [Yon07] and the IEEE 802.11ad geometric-stochastic channel model [MMS$^+$10]. The latter is based on a mix of deterministic RT and measurement-based statistical modelling techniques and on the concept of multipath cluster [HJK$^+$14].

However, for cellular mm-wave transmission there is no channel model for outdoor long-distance range. Due to the long-distance range and the directional measurements, only limited measurement campaigns were conducted to extract angular spread. An extension of the standard channel model used in the ITU and 3GPP, to the mm-wave channel modelling in cellular system is proposed in the work by Hur et al. [HCKP14]. From the measurements, the spatio-temporal channel characteristics such as multipath delay, angular statistics, and PL are analyzed including clustering analysis. The corresponding channel models based on the 3GPP spatial channel model (SCM) are proposed with a set of mm-wave radio propagation parameters. Procedures similar to 3GPP SCM [3GPP10] and WINNER II model are applied to extend the model to cover mm-wave systems up to 250 MHz bandwidth with the proposed subpath delay/AoD/AoA distributions in the work by Hur et al. [HCKP14].

There is also a trend towards the introduction of deterministic elements into geometry stochastic channel model (GSCMs) to derive "map-based GSCM" models [MBH+14]. In the framework of the FP7 MiWEBA project a novel Q-D approach for modelling outdoor and indoor mm-wave channels is proposed [MPK+14, MiW14]. The Q-D modelling methodology is based on the representation of the channel impulse response as the superposition of a few Q-D strong rays (D-rays) and a number of relatively weak random rays (R-rays) and flashing rays (F-rays). D-rays can, e.g., be related to the direct and the ground-reflected ray. In the model they are explicitly calculated from geometry. R-rays and F-rays are attributed to faraway or moving ("flashing") reflectors and defined by random variables. Following this methodology, models for open area (university campus), street canyon and hotel lobby scenarios have been developed. They have been verified and parametrised by two independent 60 GHz channel measurement campaigns carried out by Intel and Fraunhofer HHI. The explicit introduction of deterministic and random rays enables the modelling of real dynamic outdoor environments, taking into account mobility and blockage effects. Moreover, the versatility of the Q-D methodology allows extending the developed model to other usage cases with the same environment geometries like device-to-device (D2D) and street-level backhaul links.

2.2 Urban Reference Scenarios

For the development and the evaluation of mechanisms for mobile networks, an adequate system model is an important foundation. To define a system model, different modelling and design decisions have to be considered depending on the evaluated effects. However, deviating assumptions and implementations of such models render it difficult to compare and reproduce research results. Common simulation scenarios are therefore important for the efficiency and quality of research, Section 2.2.1 gives an overview over different modelling approaches and summarises efforts to align the results of independent simulators.

Widespread simulation scenarios, like those proposed by the 3GPP, rely on regular geometries and uniform user distributions. Thus, they cannot represent the fine shades of real or realistic, complex, heterogeneous environments. In order to provide actual means for the research community to shift from simple simulation assumptions to a far more reasonable degree of detail – at lowest costs and efforts – the *Hannover Scenario* is designed to provide a common simulation environment in the city of Hannover, Germany, as described in Section 2.2.2. It ensures the comparability of simulation results, which are based on this scenario. Furthermore, it allows researchers bringing their focus back to the actual development of smart algorithms instead of reinventing simulation scenarios again and again. For this purpose, a complete LTE network has been modelled for the entire city area of Hannover.

2.2.1 Simulators for Urban Environments

For the evaluation of urban environments, typically system level simulators are used to grasp the complexity of interactions between multiple cells. Simulations can either be dynamic, i.e., chronologically modelling the reaction of the system to events, or static, i.e., analyzing a fixed environment.

Static simulations are often used for site planning or other long-running optimisations. In del Apio et al. [dAMC+11], the authors compare the energy efficiency of different cell deployments in the same static scenario. They show that adding femto cells to a traditional network layout leads to a significant improvement of system performance and energy efficiency. To plan the placement of femto cells, Deruyck et al. [DTJM12] presents a network planning tool which is based on abstract and efficient channel models and heuristics. As more cells are added to a system, interference becomes an even more relevant topic. To preserve sufficient system performance also for cell border users, the optimisation of power and bandwidth allocation is performed by Krasniqi [KMM11] based on simplified static scenarios.

Dynamic simulations serve to incorporate dynamic effects of user behaviour, e.g., mobility and traffic patterns. They have to be applied whenever algorithms or control loops operate iteratively over a range of time. The temporal granularity of the simulations has to be chosen to match the evaluated algorithms and effects. The comparison of self-organising network (SON) algorithms to tune handover parameters and transmit power for load balancing is simulated with a 100-ms interval in the work by Ruiz-Avilés et al. [RALRT+11, RLTR12, HRK14]. In contrast, to evaluate the quickly reacting uplink power control algorithms of LTE, the temporal granularity of the simulation has to be increased to 1 ms [LGOR11, LGRO+11]. The same fine temporal granularity is required for the investigation of scheduling algorithms as presented in the work by Robalo et al. [RVPP15].

For static as well as dynamic simulations, one can apply either synthetic or realistic scenarios. Typical example for a synthetic scenario are the simulation guidelines published by 3GPP [3GPP10]. BSs are placed in a regular hexagonal grid and user mobility is modelled randomly. In contrast, realistic models strive to use scenarios from the real world as prototypes.

In Uppoor et al. [UF12, UTFB14], the authors have built a realistic model for urban road traffic in the city of Cologne, Germany. The model is based on freely available street data and on the TAPAS Cologne project, which derives traffic demands from a large survey. To get realistic traffic density, some manual corrections to the data were required. The authors show that, by using this model in simulations, relations between cars are significantly different from those evaluated with coarser mobility models. The risk of overly simplified models is also pointed out by the publication [HRK14]. There, the authors show that the performance of SON algorithms for load balancing significantly differs between synthetic and realistic scenarios.

The trend to use realistic models is also seen in Hahn et al. [HRK14], Deruyck et al. [DTJM12], del Apio [dAMC+11], and Lema et al. [LGRO+11]. However, evaluations based on realistic models often lack comparability. One of the reasons is that data sources are often not freely accessible, as they are, e.g., based on confidential information of network operators or collected commercially and their usage is subject to a fee. To ease the sharing of geographical data, a data format for exchange is proposed in the work by Navarro and Londoño [NL14]. Also, in Subsection 2.2.2 a publicly available scenario is presented, which can be used for cellular network simulations.

Institutions do typically use different simulation programs for their evaluations. These are often self-implemented [RALRT+11, LGOR11, CIJW13], but there are also commercial [XS12] and open source simulators [MCIŠ+11, RVPP15] available. For the scientific community, it would be good to have comparable and reproducible simulation results. However, it is often difficult to reproduce simulations from the model specification given in a publication.

In Colom Ikuno et al. [CIJW13], the authors describe their effort to align simulation results of three simulators. The compared implementations are the LTE System Level Simulator implemented at the Vienna University of Technology [MCIŠ+11], the SiMoNe Simulator implemented at the Technische Universität Braunschweig [RBHK15], and the IKR RadioLib implemented at the Universität Stuttgart [IKR15] (based on the work by Sommer and Scharf

[SS10]). The focus of the Vienna simulator lies on good mapping of link level results and realistic TP prediction. In contrast, the SiMoNe simulator implements realistic cell layouts and user mobility, which is required to evaluate SON algorithms. The IKR RadioLib is typically used with synthetic scenarios to evaluate cross-layer effects and interactions of cells in Cloud-RAN systems. The latter two simulators employ rather abstract models of the physical layer to achieve higher simulation performance. This is required to evaluate SON algorithms and effects from data traffic, which operate on longer time frames and thereby require longer simulated times.

Two scenarios have been chosen for comparison. In the first scenario, a single cell with isotropic antenna is used to align the calculation of PL and the mapping of SNR to spectral efficiency. The evaluations show that the results match well. The second scenario consists of multiple interfering hexagonal cells and follows the definitions in [3GPP10]. Evaluations in this scenario show a good match for the RSRP of the mobiles, but only if shadowing is disabled. Although the compared simulators implement similar shadowing models, the different configuration of correlations between adjacent sectors and the interaction between shadowing and placement of users into cells results in different RSRP distributions. More deviations occur at the evaluation of spectral efficiency. These are probably caused by differences in the abstraction of the physical layer, i.e., the mapping of wide-band SNR to TP. In addition, the scheduling algorithm and granularity also influences these results. It proved to be difficult to achieve an exact match of simulation results, because the used simulators were designed for different applications. It remains an open task for the research community to improve comparability and reproducibility of simulation results.

Simulation guidelines and scenarios (like published by 3GPP) and widely accepted models (like the Winner II channel model) are a good starting point to facilitate comparability, but they do still leave aspects and implementation details open. In addition, studies of effects on different layers and time scales require problem-specific abstractions. If in the common guidelines relevant effects are not modelled with the required accuracy, it is inevitable to deviate from these guidelines. In addition, many small design and parameter decisions, which are not covered by published scenario descriptions, can have influence on the results. One way to ensure a comprehensive specification of the simulation model and to ease reproducibility is to publish the whole source code used for the simulations [MCIŠ+11]. However, retracing foreign source code can be a time-consuming and error-prone task. As an alternative, the usage of predefined scenario data might help to mitigate this problem. The following subsection describes a common scenario based on publicly available trace files. By using these trace files in simulations, less effects have to be modelled in the simulator, which eases comparison and reproduction.

2.2.2 Hannover Scenario

As stated above, the *Hannover Scenario* is designed to provide a common simulation environment in the city of Hannover, Germany. For this purpose, a complete LTE network has been modelled for the entire city area of Hannover. By using real 3D building data, it allows for realistic PL predictions and realistic network planning. In addition, sophisticated modelling approaches for macroscopic traffic and individual user mobility have been employed, in order to get a realistic baseline for the simulation of subscribers in the network. The described scenario provides the starting point for the development and implementation of the mentioned SON algorithms. They might also be used in context with other advanced network algorithms, which take the timely varying channel conditions of a plethora of different users into account.

Nevertheless, the current simulation scenario is subject to change. Most probably, this process will reveal simulation and/or environmental aspects that need to be optimised in future versions of the simulation scenarios, in order to cope with the special requirements of the individual cases. In order to ease the initial work, the scenario definitions of the *Hannover Scenario* are provided as traces in a common input format, which is explained in detail in the work by Rose et al. [RJT^{+}13].

2.2.2.1 LTE network in the Hannover scenario

The current evolution of this scenario consists of a macro cellular network, which covers mainly the suburban and urban regions of the city of Hannover. The area of the *Hannover Scenario,* for which the network is planned, spreads over 20×24 km^2. In order to minimise border effects in actual simulations, an area of only 3×5 km^2 is considered for simulation purposes and will be used for the collection of user and/or cell statistics. All mobile and static users reside in the inner area, only. Besides, all PL predictions for this scenario are currently made for a height of 1.5 m above ground. In the future development of this scenario, PL prediction layers for different heights are planned. In particular, to cope with the effects of moving users in buildings

Table 2.14 States the relevant aspects of this scenario and the provided models

Scenario	LTE 1800
Coordinates of the planned area (m)	E: 00000 ... 20000 (20 km) N: 00000 ... 24000 (24 km) E_{gk4}: 4336000 ... 4356000 (20 km) N_{gk4}: 5794000 ... 5818000 (24 km) 480 km^2
Coordinates of the scenario area (m)	E: 08500 ... 11500 (3 km) N: 11000 ... 16000 (5 km) E_{gk4}: 4344500 ... 4347500 (3 km) N_{gk4}: 5805000 ... 5810000 (5 km) 15 km^2
Macro cells	LTE 1800: 195 TX power: 46 dBm
Smaller cells	None
Mobility patterns	Static users (10000) Vehicular users (4620) Pedestrian users (5247)

[RJHK13] and indoor deployments of pico and femto cells on different storeys [RJK11].

The BSs used in this scenario are realistic, which means *close to the reality,* but nevertheless artificial. As for the modelled area no information of real BS configurations is publicly available, the model is based on data from the regulatory authority and on publicly reported sites. Based on this data, a single-layer network for LTE 1800 has been planned, using an iterative process of optimisation. Briefly explained this means, after each step of network geometry planning, the network's coverage has been predicted and analysed for coverage holes. Finally, this process resulted in the network presented herein: In total, 195 sectors for the 1800 MHz band have been planned.

Figure 2.19 shows the geometry of the macro cellular network at 1800 MHz for the whole planned area of 20×24 km^2.

2.2.2.2 User mobility models

One very important aspect in system-level simulations is the change of the environment over time especially in terms of varying spatial traffic. This means, for the simulations of individual user movements, the users' positions are generated using different approaches, ranging from static indoor users to highly mobile users, as briefly described in the following subsections.

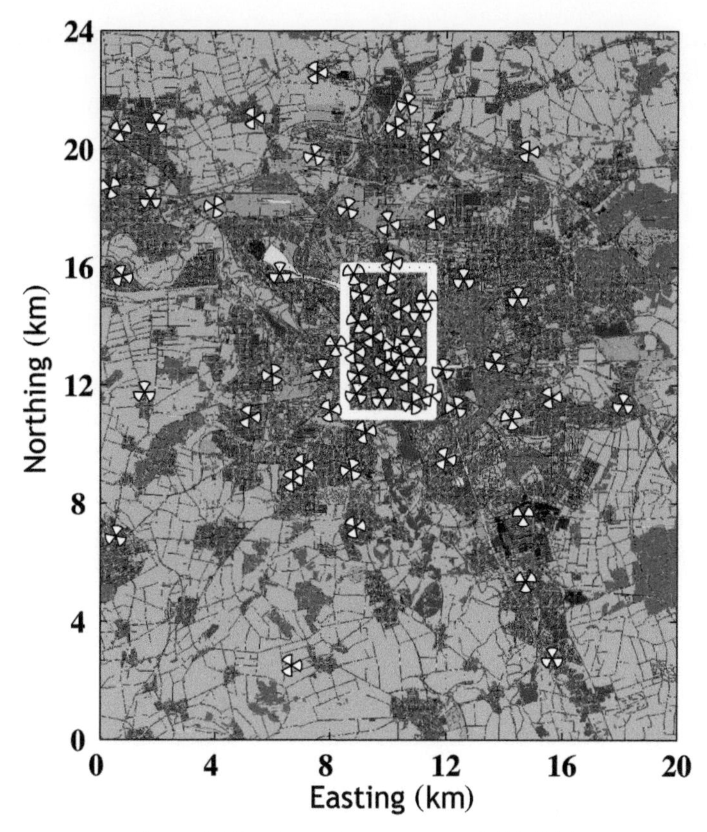

Figure 2.19 Network geometry for LTE 1800.

Static users

The simplest form of user *mobility* is actually where the position remains constant over time. These static users are useful in order to generate network traffic, which is associated with certain positions but no actual movements. All static users are located indoors.

Vehicular users

Vehicular users are generated using a highly realistic simulator, named SUMO (Simulation of Urban MObility). SUMO is a freely available microscopic road traffic simulator [DLR14], which simulates cars travelling along the streets. The actual mobile users are inside their cars. The simulation capabilities of SUMO include lane changes, overtaking of cars and acceleration and deceleration. Especially in situations close to junctions or crossings a queue in front of the traffic lights can be observed or the following of right-of-way rules. In other words, the simulated users are interacting with each other, which lead to very realistic user positions and movements.

Pedestrian users

Pedestrian users are generated by using a simulation approach stated in the study by Hahn et al. [HRSK15]. The resulting traces are representing individual users, which follow an aim-oriented behaviour. Besides paths that are provided as sidewalks in the geographical database, automatically generated short cuts a pedestrian might also take when leaving the path are considered. The contained traces simulate users on their way moving from an entry point of a building to another one, the route in between is determined by a routing algorithm.

2.2.2.3 Propagation models

In this subsection the principle process to derive PL predictions is described. The propagation models are sketched along with their general settings and considerations, which had been used for the creation of the *Hannover Scenario*. Since three-dimensional building information has been accessible for the city of Hannover, a ray optical model is best suitable for PL predictions in this area. More details on ray optical modelling, in particular on RT and ray-launching can be found in the work by Lostanlen and Kürner [LK12].

Ray optical outdoor predictions

A RT model (called the ray tracer), has been used to predict the outdoor coverage of the cells in this scenario. This predictor looks for the ray optical propagation paths between the TX and the RX antennas. The ray tracer follows the concept of sub-dividing the model into sub-models as described in Kürner [Kür99] and Kürner and Meier [KM02]. Here, a vertical plane model (VPM) and a multi-path model (MPM) have been used. The MPM model is based on the approach described by Kürner and Schack [KS10] and recognises the following types of paths:

- Reflected paths are considered up to the second reflection. Reflections can have a maximum image source distance [LK12] of up to 1000 m [Kür99].
- Scattered paths are considered up to a maximum distance of 500 m between the TX and the RX. Scatterers are searched up to a distance

of 550 m around the BS according to the findings in the work by Kürner [Kür99].

The VPM uses the following approach:

- In case of LoS the PL is calculated based on a distance dependent selection, which uses the free-space propagation loss (up to 200 m), the Okumura–Hata model with sub-urban correction (beyond 500 m) and a transition model in between.
- In case of NLoS the paths over the roof-tops are considered. The total PL is the sum of the PL (as stated for the LoS case) and a diffraction loss calculated according to Deygout [Dey66].

Outdoor-to-indoor predictions

The prediction of areas covered by buildings is done in a subsequent step, based on a *Ground Outdoor-to-Indoor* approach [GB94, RK12]. *Ground Outdoor* refers to the outdoor coverage at ground level, which is usually predicted for mobile terminal heights of 1.5 m. The signal level on the inside at ground level is directly related to the coverage on the outside at the same height. The prediction model is described in the work by Rose and Küner [RK12]. According to this publication, outer walls are contributing with very individual attenuations between 5 and 20 dB. So a general attenuation of 10 dB has been chosen for the predictions, which has also been used in other publications [OKI09, HWL99]. Based on the distance between the respective indoor position and the outer wall, a linear attenuation of 0.8 dB/m is added to the PL, according to the findings reported in the work by Rose and Küner [RK12].

2.2.2.4 Data traffic model

Besides the user mobility, the load of the cells in a cellular system is mainly influenced by the data traffic the users generate or request from the Internet. Simple models like full buffer or fixed rate do often fail to capture the dynamics of real Internet traffic. Therefore, we provide a model which is based on measurements in mobile and fixed networks. The dynamics of Internet traffic are mainly influenced by two effects: on a large time scale, the user or an autonomously acting application decides to send or request a certain amount of data. On a smaller time scale the transport protocol (typically TCP) fragments this data into packets and controls the rate at which these packets are sent. While the start times and the size of requests and responses can be pre-calculated and stored in trace files, the behaviour of TCP directly depends on the actions of the mobile network, especially the scheduling. Therefore, the behaviour of TCP has to be either modelled in the simulator or these effects have to be approximated.

2.2.2.5 Conclusion and future work

The presented IC 1004 Urban Hannover Scenario delivers a variety of scenario data in a realistic simulation environment. The currently available data set is a first step towards more complex and modular simulation scenarios and system simulations. It is hoped that the utilisation of a freely available, ready-to-use and easy to implement simulation scenario will ease the comparison of simulation results in the future. Feedback from the scientific community on the usability and completeness of the present scenario data will help to improve the data set for new research activities.

2.3 Summary and Future Directions

Prediction of the performance of radio networks in the diverse urban environment requires a variety of techniques. Deterministic models dictate the use of scenarios with known propagation characteristics, while empirical models are limited by the number of measured scenarios. To this end we propose the use of the Hannover scenario for common simulations. Results of studies in a variety of urban scenarios below the 6 GHz band including the challenging environments of trains and containers and the emerging massive MIMO technologies have been presented in this chapter as well as vehicular scenarios. Preliminary results of propagation measurements in the mm-wave band in line of sight scenarios indicate PL exponents close to free-space propagation on the order of 2 with the use of directional horn antennas. Higher coefficients of 3 have been estimated for non-line of sight scenarios. Delay spread values of LoS and NLoS in environments in the 60 and 70 GHz bands for two outdoor scenarios with single and dual antenna configurations give values varying from 2.3 to 94.3 ns for 95% of the measured locations. However, the number of measured scenarios is currently very limited to a few sets of data to draw comprehensive conclusions and the results are system dependent that further extensive measurement campaigns are needed to extract and validate the relevant channel parameters. A major challenge in the mm-wave band studies is to de-embed the antenna pattern from the measured channel since currently due to limited transmit powers, propagation studies are performed using high gain directional antennas. The deployment of an antenna array that enables such studies is a major challenge where fast switching to cover the high Doppler shifts in urban mobile environments would be needed. Due to the complexity of such measurements, RT approaches or a combination of RT and stochastic models are being proposed. Detailed information regarding the environment and the electrical properties of the materials are needed in RT modelling which can be a challenge. Further research to investigate the feasibility of beamforming, tracking, and interference between links in the mm-wave band is still at an early stage and requires further investigation.

3

Indoor Wireless Communications and Applications

Chapter Editor: K. Haneda, Section Editors: W. Joseph, E. Tanghe, A. Bamba, U.-T. Virk, E.-M. Vitucci, C. Gustafson, J.-M. Molina-Garcia-Pardo, K. Witrisal, P. Kulakowski, P. Meissner and E. Leitinger

Chapter 3 addresses challenges in radio link and system design in indoor scenarios. Given the fact that most human activities take place in indoor environments, the need for supporting ubiquitous indoor data connectivity and location/tracking service becomes even more important than in the previous decades. Specific technical challenges addressed in this section are (i), modelling complex indoor radio channels for effective antenna deployment, (ii), potential of millimeter-wave (mm-wave) radios for supporting higher data rates, and (iii), feasible indoor localisation and tracking techniques, which are summarised in three dedicated sections of this chapter.

3.1 Advances in Short-Range Radio System Design

In the forthcoming years, due to the increasing popularity of multiple-input multiple-output (MIMO), massive MIMO and wide- and ultra wideband (UWB) transmission schemes along with the increasing use of small cells, the conventional propagation models will no longer suffice. This section introduces methods for accurate characterisation of channels and advanced channel models, detailing about dense multipath components (DMCs), different channel simulation methods and modelling techniques for MIMO and UWB. Finally, new antenna deployment approaches and their performance are discussed.

Cooperative Radio Communications for Green Smart Environments, 71–120.

3.1.1 Characterisation of the Indoor Channel Using Room Electromagnetics (REM) and Diffuse Multipath

In this subsection, the diffuse components of the indoor channel are characterised using two approaches, namely, high-resolution algorithms and the REM. State-of-the-art REM models to determine the reverberation times will be presented and the contribution of the diffuse multipath components (DMCs) in industry will be determined using RiMAX.

3.1.1.1 Introduction: REM and diffuse multipath components (DMCs)

The radio channel not only consists of specular paths with well-defined *discrete* locations in the different radio channel dimensions (e.g., space, frequency, time, etc.) but also of DMC, which are *continuous* across these dimensions [Ric05, OCD+12]. These not only originate from distributed diffuse scattering on electrically small objects but also include all radio SMCs channel energy that cannot be associated with the due to the inconsistency of the specular multipath model with reality [PSH+11, MQO12]. The dense and SMCs are often also referred to as the incoherent and coherent parts of the radio channel, respectively [ETJ14]. It has been shown that the DMC power density may contribute significantly – up to 95% – to the total power density in an indoor environment [PSH+11]. Two approaches have been proposed to tackle the modelling of DMC in real indoor channels.

The first approach is based on the use of the high-resolution algorithms such as RiMAX, i.e., an iterative maximum-likelihood multipath search algorithm [Ric05]. The specular and DMCs can be estimated from the channel sounding data by means of RiMAX because it is built on a data model that allows for both specular and DMCs. The second modelling approach is based upon the REM [ANP+07]. Basically, the theory states that the PDP of indoor environments is comprised of two parts: the line-of-sight (LoS) signal if any and the diffuse fields. The LoS component (black line in Figure 3.1(a)) is due to the direct propagation between the transceivers. The diffuse fields stem from multiple scattering of the electromagnetic waves on rough surfaces (walls, ceiling, floor, furniture, etc.) and diffractions at corners.

3.1.1.2 REM models

Room electromagnetics modelling is used in [SPF+12, SPF+13a, GSML+14], to characterise the PDP and proposed for multi-link radio channels, UWB channels, the estimation of the absorption cross-section of a person, and hybrid raytracing.

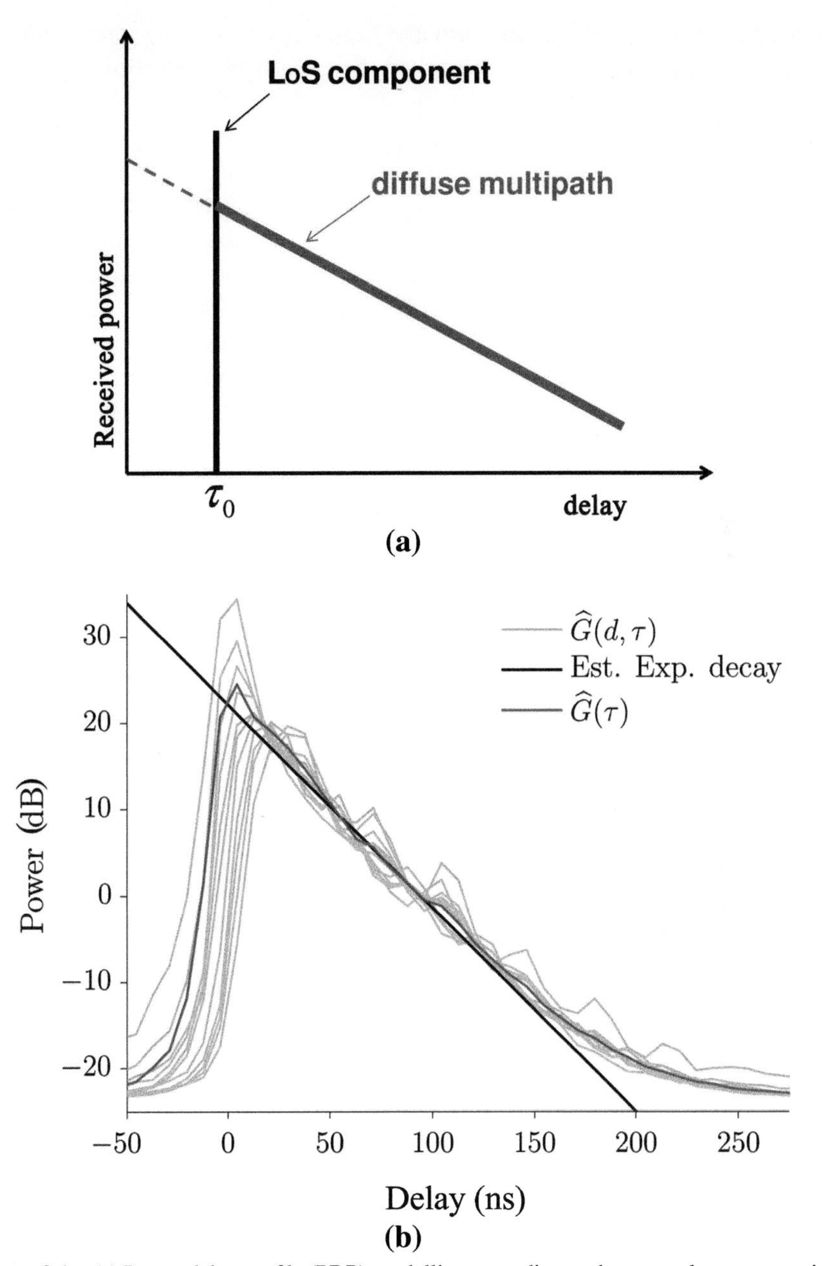

Figure 3.1 (a) Power delay profile (PDP) modelling according to the room electromagnetics (REM) theory, (b) spatially averaged PDPs obtained for transmitter–receiver distance intervals of 4 λ for receiver positions [SPF+12, SPF+13b].

Steinböck et al. [SPF+13a] validates experimentally the prediction accuracy of reverberation models applied to electromagnetics in typical indoor environments (consisting of different scenarios A–F). The predictions obtained with Sabine's [Sab92, Kut00] and Eyring's [Eyr30, Kut00] models are then compared with the experimental results. The experimental data of Figure 3.1(b) is used to perform the validation. Figure 3.2(a) shows that Sabine's model generally seems to predict too large reverberation times and the prediction error increases when the average absorption coefficient increases. The predictions obtained with Eyring's model shown in Figure 3.2(a) are close to the experimental results and their respective confidence intervals overlap except in Scenario E. The obtained results indicate that the reverberation models are valid for REM and that Eyring's model provides a better prediction. In Steinböck et al. [SPF15], Eyring's model is proposed to estimate the absorption cross-section of a person. The obtained values can be used to predict the change of reverberation time with persons in the propagation environment. This allows prediction of channel characteristics of communication systems in presence of humans.

Bamba et al. [BJT+12, BJT+13] aim to introduce an electrical circuit model based on the REM theory [ANP+07], while Bamba et al. [BMG+14] investigates the electromagnetic reverberation time characteristics as a function of frequency and for UWB. The measurements have been carried out in two laboratories from 2 to 10 GHz. The results demonstrate that, for a given frequency, the reverberation time is constant over a large bandwidth up to 900 MHz. Figure 3.2(b) shows the experimental (*black circles*) and predicted (*blue curve*) reverberation time values from 2 to 10 GHz. Maximum (resp. average) relative errors between the predicted and the measured values are about 22.30% (resp. 8.80%) for the investigated frequency range. These low deviations show that the model accurately predicts the reverberation time values. Figure 3.2(b) also shows the decrease of the reverberation time as the frequency increases. This indicates that the energy is fading faster away at higher frequencies compared to lower frequencies. In addition, the reverberation time decreases smoothly as the frequency increases, indicating that the diffuse fields fade at a faster rate at higher frequencies. This is attributed to the frequency dependence of the absorption coefficient of the building materials.

3.1.1.3 Diffuse multipath components (DMCs) in industry

Lienard et al. [ETJ13, ETJ14] present an analysis of DMC in an industrial workshop for shipping container restoration. Radio channel sounding experiments with a vector network analyser and virtual antenna arrays are carried

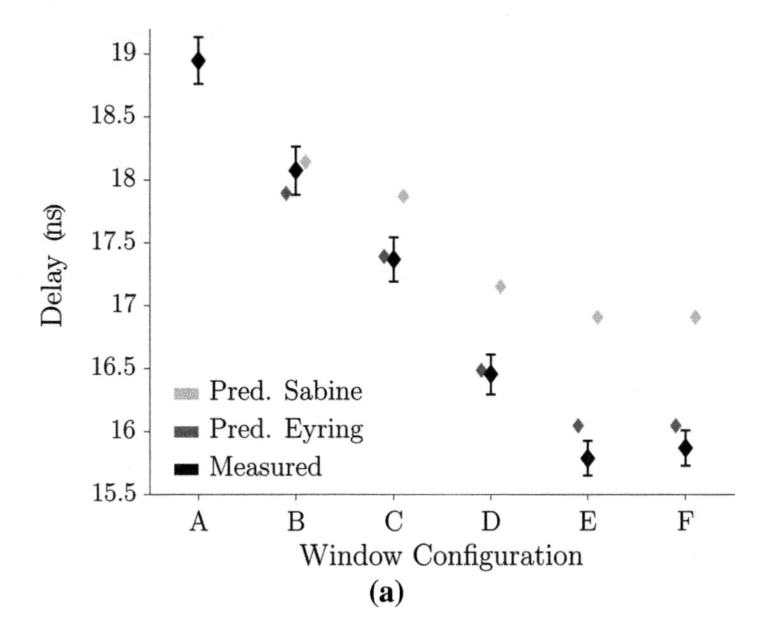

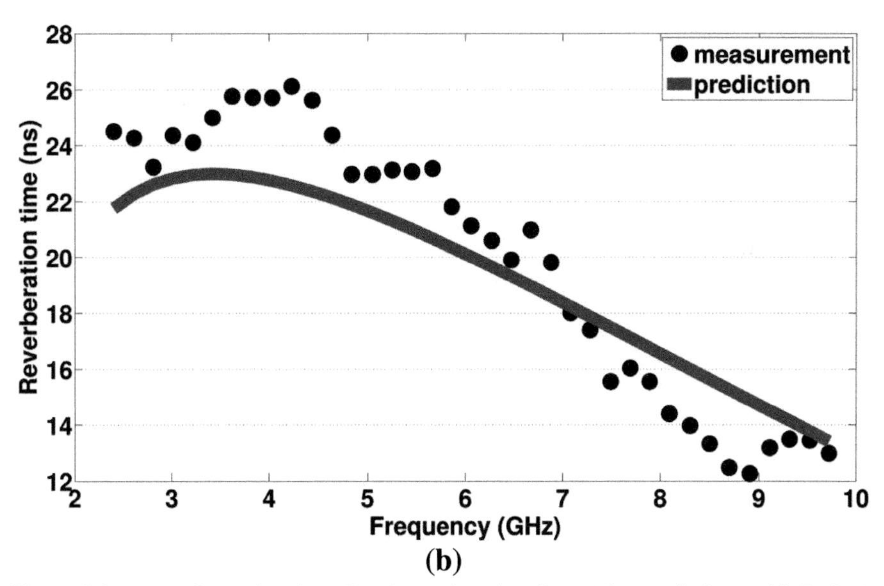

Figure 3.2 (a) Estimated and predicted reverberation times. The predictions with Eyring's and Sabine's models for the different scenarios are shown including the 95% confidence interval [SPF+13a], (b) experimental (*black circles*) and model (*blue line*) of τ from 2 and 10, i.e., 2 to 10 GHz.

out. Specular and DMCs are estimated from channel sounding data by means of RiMAX.

In this study of Lienard et al. [ETJ13, ETJ14], the DMC power is found to account for 23–70% of the total channel power. Significant difference between DMC powers in LoS and non-line-of-sight (NLoS) situations is discovered, and this difference can be largely attributed to the power of the LoS multipath component (MPC). Overall, DMC in the industrial workshop appears to be more important than in office environments, i.e., a larger fraction of channel power and longer reverberation times are observed: this is explained by the highly cluttered and metallic nature of the industrial environment. Table 3.1 presents a comparison of the fractional DMC power f_{DMC} with related work in office environments. The work of Poutanen et al. [PSH+11] also found a strong link between f_{DMC} and link shadowing category. The larger f_{DMC} in the industrial environment may be explained by presence of many clutters in the environment leading to diffuse scattering. Additionally, the measurements in Lienard et al. [ETJ13, ETJ14] are performed at a lower centre frequency than in Poutanen et al. [PSH+11].

A comprehensive analysis of the polarisation characteristics of SMCs and DMCs in a large industrial hall based on frequency-domain channel sounding experiments at 1.3 GHz with 22 MHz bandwidth is presented in Tanghe et al. [IV14]. In summary, strong (antenna de-embedded) SMC depolarisation is obtained for the horizontal (H) polarisation in obstructed-line-of-sight (OLoS) scenarios. On the other hand, DMC depolarisation is observed to be weaker than previously reported for indoor environments but constant across LoS/OLoS, polarisation, and distance. Finally, a full-polarimetric distance-dependent model of the PDP for large hall scenarios is proposed in Cheng et al. [CGT+15]. A two-step method is proposed to compute the path loss (PL) exponent of the estimated components from the measured data. This approach provides a deeper understanding of the indoor radio channel when including DMC.

Table 3.1 Fractional DMC power f_{DMC} for industrial and office environments

		Centre Frequency	Link Shadowing (%)		
Reference	Environment	(GHz)	LoS	OLoS	NLoS
Poutanen et al. [PSH+11]	Open hall in office building	5.3	10–25	35–65	60–90
Quitin et al. [QOHD10]	Office building	3.6	n/a	20–60	
Lienard et al. [ETJ13, ETJ14]	Industrial workshop	3.0	23–38	27–70	57–64

3.1.2 Characterisation of the Indoor Channel Using Simulations: Heuristic, Ray-based and Full Wave

This subsection introduces various examples of indoor radio channel modelling with different complexity and accuracy. The complexity and accuracy depends on the channel modelling method, such as full-wave methods, ray-tracing and a heuristic method. In case of site-specific channel modelling, the level of modelling accuracy depends on the quality of the environment description. Prospects and limitations of different indoor radio channel modelling approaches are illustrated through concrete example results.

3.1.2.1 Heuristic channel modelling

The work of Plets et al. [PJA+14] uses a heuristic electromagnetic pathloss prediction tool, called a WiCa Heuristic Indoor Propagation Prediction (WHIPP) tool [PJV+12], to evaluate the whole-body exposure due to indoor base station antennas PL or access points in downlink, and localised exposure due to mobile devices in uplink indoor channels. The tool calculates the PL by considering the effect of the environment on the radio channel and by determining a DP between transmitter and receiver, i.e., the path along which the signal encounters the lowest obstruction. The PL and subsequently the link budget estimates are calibrated by an RSS measurements of a commercial mobile phone. The calibration led to 3 dB accuracy in mobile transmit (Tx) power estimates as illustrated in Figure 3.3(a). Using the calibrated power estimates, three phone-call scenarios, i.e., universal mobile telecommunications system (UMTS) macrocell, UMTS femtocell, and WiFi voice-over-IP, are compared with respect to the localised specific absorption rate (SAR) distribution. The benefit of a low localised SAR_{10g} is illustrated in the UMTS uplink scenario due to the power control mechanism.

3.1.2.2 Full-wave channel modelling

The work of Virk et al. [VHK+13] reported characterisation of electromagnetic field propagation inside a room using the finite difference time domain (FDTD) method. Thanks to the advancement of computational power and availability of strong computational unit such as a graphical processing unit, it is possible to run full-wave simulation in a feasible time for a propagation environment with a limited electric volume. The predicted PL using the SEMCAD-X and the corresponding measurement are compared as shown in Figure 3.3(b). It was found that the distance-decaying factor of the PL and the small-scale fading were estimated well, but the absolute power level showed about 5 dB constant offset. Possible reasons behind the offset

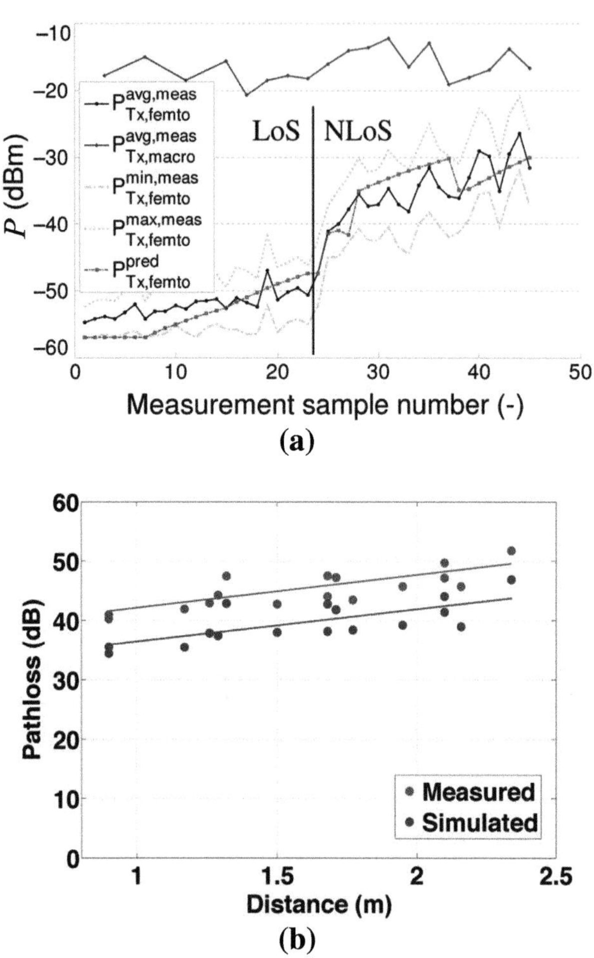

Figure 3.3 (a) Measured and predicted uplink power at a mobile, radiated to a femtocell base station at various mobile locations. The greater measurement sample number of the *x*-axis corresponds to greater distance from the base station [PJA+14]. (b) Comparison of path loss (PL) between measurements and finite difference time domain (FDTD) simulation in an office room.

are numerical wave dispersion, conversion of the electric field values to a PL estimate and imperfect absorbing boundary conditions.

The works of Kavanagh and Brennan [BK14] and Kavanagh and Brennan [KB15] report a two-dimensional full-wave solver of electromagnetic wave propagation in indoor environments based on the volume electric field integral equation (VEFIE) formulation. The formulation is in frequency domain and hence produces a matrix equation when discretised by the method of moments. The matrix equation can be iteratively solved and accelerated by using the fast fourier transform (FFT). The unknown fields are derived only on the scatterers, and those in the free-space were calculated through a simple post-processing to reduce the number of iterations required. Numerical results of electric field distribution in a hypothetical two-dimensional room are compared with the results of the exact solution, showing validity of the VEFIE formulation.

3.1.2.3 Advanced ray-based modelling

One of the key challenges in ray-based indoor channel modelling is acquisition of accurate description of the environment. Available blueprints of the indoor environments may not be sufficient for accurate channel modelling since they often miss necessary details such as fixtures and metallic structures that may reflect or block radio waves.

The work of Vitucci et al. [VFDE14] revealed that a proper description of the propagation environment with an accurate choice of material parameters is a key factor to predict radio channels accurately. A full three-dimensional ray-tracing tool [DEGd+04] is used to predict radio propagation inside a large exhibition hall, with the aim of reproducing root mean square (RMS) delay spread (DS) and angular spread accurately as reality. The peculiarity of the building was highly reflective floors and ceilings due to embedded metal structures and meshes. Figure 3.4(a) shows that inclusion of the reflective floors and ceilings and surrounding buildings is required to predict DSs accurately.

The works of Järveläinen and Haneda [JH1], Virk et al. [VHW14], and Wagen et al. [WVH15] proposed a ray-based channel modelling based on an accurate description of the physical environment, which is obtained from laser scanning, called a point cloud. The challenge in channel modelling based on the point cloud is the fact that there is no predefined surface for calculating specular reflections. The early works of Järveläinen and Haneda [JH1] and Virk et al. [VHW14], therefore, used only diffuse scattering, assuming that each point in the point cloud produces electromagnetic scattering. However, it was demonstrated in Wagen et al. [WVH15] that specular reflections can also be reproduced in addition to the diffuse scattering. The environmental description in the form of the point cloud allows us to analyse the level of structural details that is essential for accurate channel modelling. For example, Virk et al. [VHW14] showed that a simple empty room model suffices to predict PDPs of a small office room. The work compares the PDPs based on the point cloud to those with an empty room model having the same dimension as illustrated

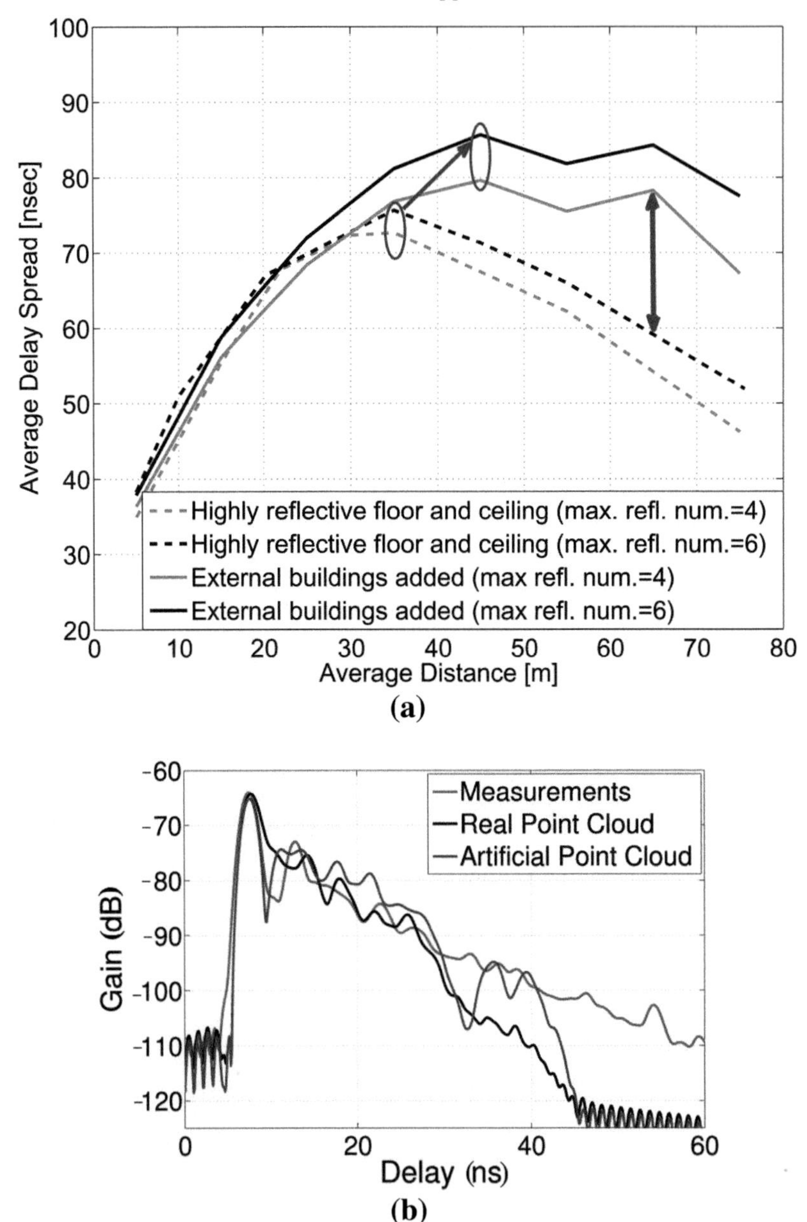

Figure 3.4 (a) Comparison of delay spreads (DSs) from ray-tracing simulations using two environmental settings; one with highly conductive floor and ceiling and another with surrounding buildings [VFDE14]. (b) Comparison of PDPs between measurements and ray-based channel modelling with accurate environmental description and an empty room model at 10 GHz radio frequency [VHW14].

in Figure 3.4(b). The figure shows a decent agreement of both modelling results to the measurement, because of the dominance of electromagnetic field reverberation inside the room.

3.1.3 Advanced Indoor Propagation Modelling

New advanced techniques for indoor propagation characterisation and modelling are presented in this subsection. Such techniques deal with: (i) spatial, temporal, and polarisation characteristics of multi-antenna and multi-link indoor channels; (ii) time and frequency dependence in UWB indoor channels; and (iii) accurate models for deployment of next-generation mobile networks.

3.1.3.1 Characterisation of multi-antenna and multi-link indoor channels

In Haneda et al. [HTKD11, HKD$^+$13], a method to identify an upper bound of spatial diversity gain in multipath channels is presented. The spatial diversity gain is derived from the measured propagation channels by applying the spherical wavemode expansion method [Han88]. The dependence of the attainable spatial diversity gain for a given array volume is shown in Figure 3.5 for several measurement locations at 6 GHz in an industrial room scenario: the gain increases with the size of the antenna array, and three-times larger diversity gains are achieved in OLoS scenarios compared to LoS scenarios. The attainable diversity gain is less than 3 dB indicating that *any* two-branch antenna array realised in the volume originates correlated signals in the tested propagation environments.

Distributed MIMO, i.e., MIMO with large separation between the elements of the antenna arrays, has been proposed as a solution to exploit the beneficial effect of spatial diversity gains. Actually, its performance is closely related to the low cross-correlation of the large-scale fading between multiple links. In Tian et al. [TZY12], the fading correlations for different Tx antenna configurations are analysed based on measurements data collected in an indoor office environment. It is shown that for some particular configurations, e.g., when the elements of the Tx array are distributed symmetrically with respect to both the receivers (or Rxs) and the propagation environment, the spatial cross correlation (SCC) is non-negligible (i.e., larger than 0.3). In such cases, an additional polarisation diversity can effectively reduce the correlation and enhance the capacity gain, since the joint spatial and polarisation cross correlation (SPCC) is shown to assume negligible values (around 0.1) in most of the considered cases.

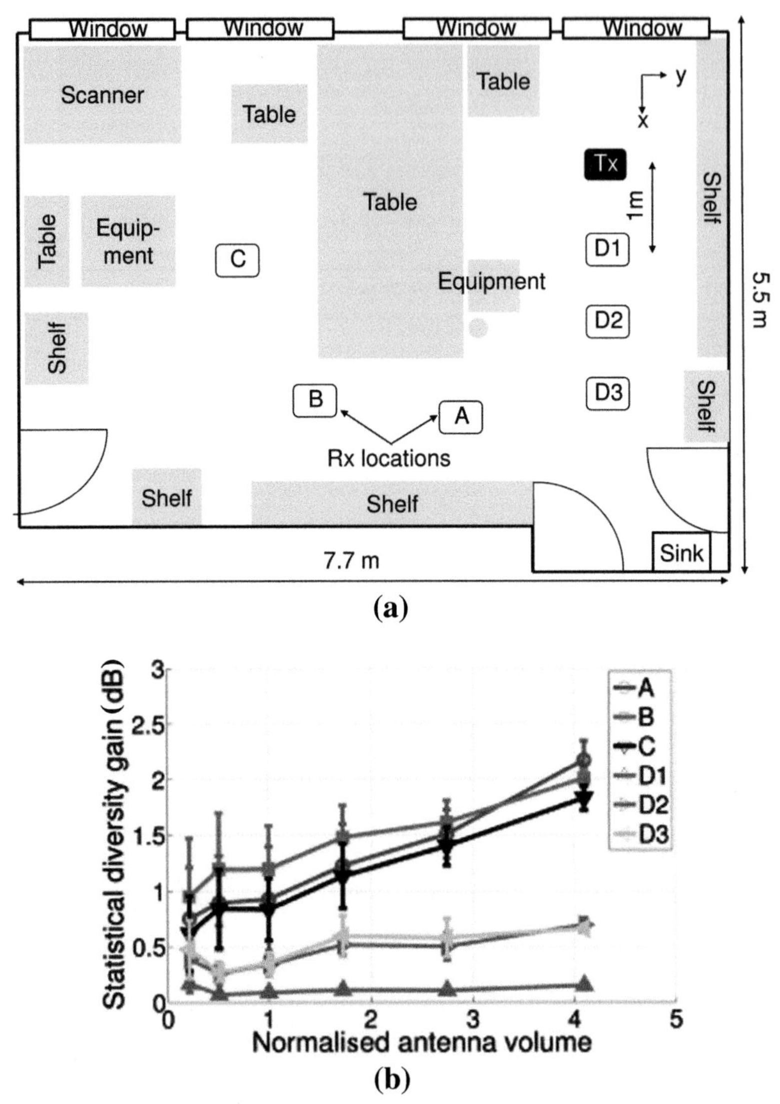

Figure 3.5 Dependence of the attainable spatial diversity gain against the volume of a cubical hollow array at 6 GHz, in an industrial room scenario [HTKD11].

In the work of Kim et al. [KKCT12] and Chang et al. [CKKT12], extensive measurement campaigns for wideband MIMO systems at 11 GHz with large arrays are presented. In particular, in Kim et al. [KKCT12], the influence of the polarisation on PL, DS and MIMO capacity is investigated for a 12×4

dual-polarised MIMO system. It is shown that the DS does not significantly depend on the polarisation, while PL is highly degraded in some NLoS locations when the horizontally polarised transmission is used, because of high losses originated by penetration, sidewall reflections, and diffraction on door edges. This leads to a significant degradation of the MIMO capacity, due to the well-known keyhole effect [CO13].

In Chang et al. [CKKT12], measurements with a 8×12 MIMO system in a dynamic scenario with randomly moving people are used to develop a new stochastic model, called 2 dominant paths (2DP) model. In addition to the LoS and stochastic-NLoS components of the well-known Rician MIMO channel model [CO13], an additional term is introduced, the so-called dominant non-line-of-sight (dNLoS) component. According to this model, the MIMO channel matrix for the i-th subcarrier ($\mathbf{H}_i{}^{\mathrm{2DP}}$) can be then expressed as:

$$\mathbf{H}_i^{2DP} = \sqrt{\frac{A}{1+A+B}}\mathbf{H}_i^{\mathrm{LoS}} + \sqrt{\frac{B}{1+A+B}}\mathbf{H}_i^{\mathrm{dNLoS}} + \sqrt{\frac{1}{1+A+B}}\mathbf{H}_i^{\mathrm{NLoS}}, \tag{3.1}$$

where the subscript i refers to the i-th subcarrier, $\mathbf{H}_i^{\mathrm{LoS}}$, $\mathbf{H}_i^{\mathrm{dNLoS}}$ and $\mathbf{H}_i^{\mathrm{NLoS}}$, are the channel matrices related the LoS, dNLoS, and NLoS components, and A, B are the power ratios of the LoS and dNLoS to the stochastic NLoS components. The 2DP model better agrees the measurements shown in Chang et al. [CKKT12] compared to the Rician model, while the LoS and dNLoS components are highly correlated to the first and second eigenvalue of the MIMO channel.

A statistical characterisation of fast-fading in indoor peer-to-peer networks based on channel measurements is provided in Vinogradov and Oestges [VO13]. The small-scale fading statistics are evaluated using the second-order scattering fading (SOSF) distribution which reflects any combination of Rician, Rayleigh, and double-Rayleigh fading [CO13]. It is shown that in double mobile scenarios the predominant fading mechanism is a combination of Rayleigh and double-Rayleigh fading, while Rician fading or combinations of LoS and double Rayleigh components are occasionally observed. In single mobile scenarios, fading is either Rician or Rayleigh distributed. These different fading behaviors mainly depend on the mobility of the nodes, and the transitions between them can be described by a hidden Markov model [EM02].

3.1.3.2 Modelling of time and frequency dependence in UWB indoor channels

In UWB wireless propagation channels, each MPC can exhibit delay dispersion, i.e., frequency dependence. This effect is investigated in Haneda et al. [HRM12], where a method to extract the significant MPCs based on the image principle is presented. Such a method identifies locations and intensity of both original radio sources and image sources. A map of image source distribution is experimentally constructed (Figure 3.6) based on UWB propagation measurements in a small office scenario, with the frequency varying from 3.1 to 10.6 GHz. A detection method is then applied to the map to identify the significant image sources, which are then tracked over different sub-bands to get the frequency dependence of their intensity and location. A MPC-wise frequency dependence model is finally proposed, which can be combined with existing channel models. The effectiveness of this approach is evident in Figure 3.6(c): the inclusion of the frequency dependence characteristics of MPCs in the IEEE802.15.4a channel model [MFP03] reveals noticeable differences in the shape of the channel impulse responses (CIRs), with a considerable impact on UWB radio system design.

In Hanssens [HTM+14], the frequency dependence of Doppler Spread for UWB indoor communication in a time-varying office environment is investigated. A measurement campaign ranging from 3.1 to 10.6 GHz was performed for several days, using a network analyser. Measurements show a clear frequency-dependent behavior: in particular, a heavy decrease of Doppler Spread is observed from 3.1 to 5.1 GHz, followed by a steady behavior up until 8 GHz, and a slight decrease toward 10.6 GHz. Results also confirm that the well-known Jakes' Doppler spectrum no longer applies in such a scenario, in agreement with theoretical predictions shown in Thoen et al. [TdPE02].

3.1.3.3 Accurate propagation models for mobile networks deployment

The increasing traffic demand from inoor users in future mobile communication networks will require the deployment of femto cells to cover areas inside buildings. For this reason, accurate PL models including building penetration losses are needed. An extensive measurement campaign in the ultra high frequency (UHF) band aiming at modelling the indoor-to-outdoor and indoor-to-indoor propagation for long term evolution (LTE) femto cells is described in Giménez et al. [GJRC12] and Ballester et al. [BGJ+13]. In Giménez et al. [GJRC12] measurement results are analysed

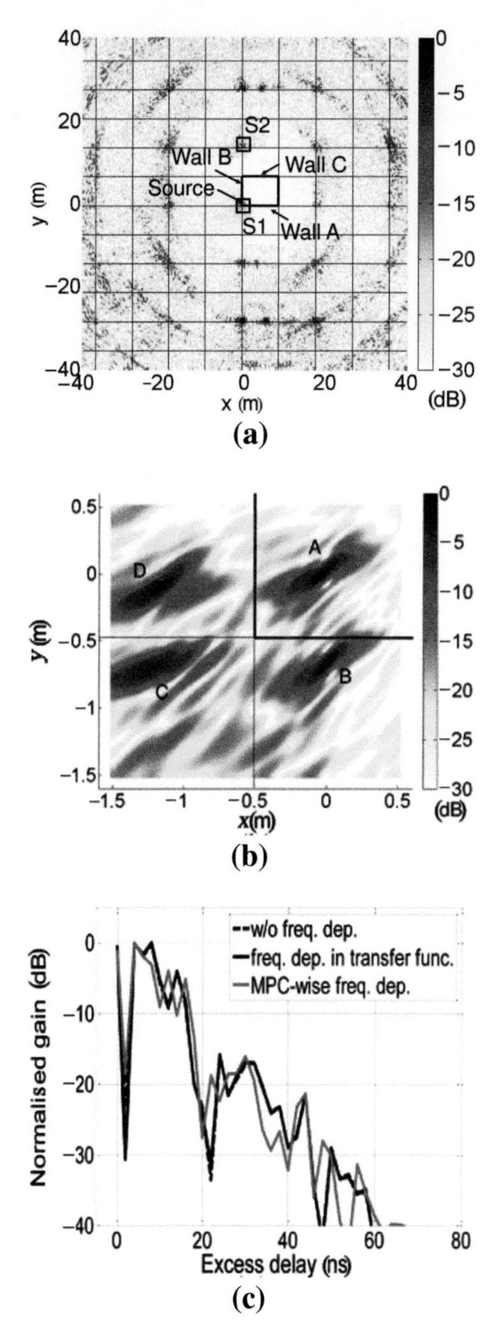

Figure 3.6 (a) Overall map of image source distribution and (b) enlarged map for the source group S1; and (c) effect of the multipath component (MPC)-wise frequency dependence [HRM12].

with respect to indoor-to-outdoor propagation characteristics, using the model described in Rose et al. [RJK11] as a reference. It is shown in particular that specular reflections from metal structures (e.g., the metal shield of elevators) can give strong contributions, increasing the total RSS. An analysis of the indoor-to-indoor propagation losses in the same measurement scenario is reported in Ballester et al. [BGJ+13]. The COST 231 multi-wall model (MWM) is taken as a reference and new values of the empirical parameters are derived in order to minimise the error relative to measurements, using an optimisation algorithm based on the Nelder–Mead Simplex Method [NM65]. The resultant expression for the model that fits the measurements is the following:

$$L = L_{\mathrm{FS}}(1.83, d) + 41.53 + \sum_{i=1}^{N} 3.98 \cdot K_{\mathrm{wi}} + 15.42 \cdot K_{\mathrm{f}}^{\{\frac{K_{\mathrm{f}}+2}{K_{\mathrm{f}}+1} - 0.48\}} \tag{3.2}$$

where L_{FS} is the free-space attenuation corrected with a proper PL exponent, d is the distance, K_{wi} is number of penetrated walls of type i, and K_{f} is number of penetrated floors. Using this formula, an average estimation error of 0.85 dB and a standard deviation of 4.63 dB are achieved for the measurements shown in Ballester et al. [BGJ+13].

3.1.4 Advanced MIMO Techniques: Leaky Coaxial Cables (LCXs), Distributed Antenna System (DAS) and Massive MIMO

This subsection briefly introduces two techniques for achieving a uniform wireless coverage, namely LCXs and DAS. It is discussed how these techniques can be used in conjunction with MIMO. The implication of massive MIMO on multipath estimation is also discussed in this subsection.

3.1.4.1 LCXs

The use of LCXs as antennas in MIMO wireless communication is discussed in Medbo and Nilsson [MN12]. The main advantage of LCXs in single-antenna systems is a uniform distribution of the signal strength due to the LCX's distributed nature [Mor99]. However, the performance of LCXs in a MIMO setup was largely unknown. A measurement-based analysis of the MIMO mutual information of a 2×2 system in an 80 MHz bandwidth centered at 2.44 GHz is presented in Medbo and Nilsson [MN12]. The receiving antennas are two LCXs of length 20 m and the transmitting antennas are regular electrical dipole and magnetic loop antennas. When the two LCXs are mounted about 1 m apart, the measurements show that the MIMO capacity is slightly better than that of an independent and identically distributed (i.i.d) MIMO channel. Interestingly, when the two LCXs are taped together, the MIMO capacity is only slightly worse than that of the i.i.d. channel.

3.1.4.2 DAS

Apart from leaky cables, DAS is another way of achieving a more uniform wireless coverage [SRR87]. In a traditional DAS, the same signal is distributed to a number of remote antenna units (RAUs) to achieve a better and more capillary radio coverage. The use of MIMO in conjunction with DAS is discussed in Vitucci et al. [VTF+14]. A straightforward implementation distributes the entire MIMO signal to RAUs equipped with multiple antennas and is called co-located MIMO DAS (c-MIMO DAS). In another implementation called interleaved MIMO DAS (i-MIMO DAS), the different MIMO branches are split up and sent to different RAUs. The advantage of i-MIMO DAS over c-MIMO DAS is that older SISO DAS deployments with single-antenna RAUs can be re-used to implement MIMO. The downside of i-MIMO DAS is that the MIMO branches become spatially well-separated, leading to power imbalances of these branches at the receiver side and thus to capacity losses.

Tian et al. [TNB14a] discuss the uplink error performance of a DAS system where the RAUs are LTE eNodeBs. Maximum-ratio combining (MRC) techniques are applied to the eNodeB signals before sending the combined signal to the LTE base station. Tian et al. [TNB14a] investigate two MRC strategies. The first one is a "space-only" MRC technique that applies different MRC weights to the different eNodeB signals. The second, more complex, "space-frequency" MRC technique is the same but also applies different weights to different frequency carriers. Simulations with the WINNER II channel model [KMH+08] show an expected performance improvement of the space-frequency MRC technique in highly frequency-selective channels. However, the authors argue that complexity of the space-frequency technique may outweigh the performance benefits in practical LTE DAS systems.

3.1.4.3 Massive MIMO

One of the main challenges in radio channel modelling for massive MIMO scenarios is how to ensure the spatial consistency of the model [MBH+14]. Conventional MIMO arrays are compact: it can be readily assumed that each MPC behaves as a plane wave. For massive MIMO arrays, however, the plane wave assumption is difficult to maintain and a spherical wavefront assumption is more appropriate. In the spherical wavefront assumption, the directions of arrival and departure change for each element of the array. Yin et al. [YWZ14] propose a modification of the space-alternating generalised expectation maximisation (SAGE) MPC estimator [FTH+99] to account for spherical wavefronts. The traditional parameters of each MPC are estimated: the directions of arrival and departure (for a reference antenna), the time-delay, the Doppler frequency, and the full-polarimetric complex amplitudes. To consider spherical wave fronts, two additional parameter dimensions are added: the two distances from Tx to the first bounce point and from the last bounce point to the Rx.

3.2 Towards Higher Frequencies: Millimeter-Wave Radios

With the growing number of wireless devices and the ever increasing demand for higher wireless data rates, next generation wireless systems for cellular and WiFi technology may have to utilise frequency bands that are well above the frequency bands below 6 GHz that are typically used today. Recent developments in RF technology have made it feasible to produce cost-effective radios that operate in different mm-wave frequency bands in the range from 30–300 GHz. With the large amount of underutilised spectrum available in the mm-wave range, this could potentially unlock huge swaths of bandwidth for cellular and WiFi technology.

For indoor short-range high data rate wireless communications, the 60 GHz band has attracted the most attention due to the large amount of unlicensed bandwidth (about 5–9 GHz) available worldwide. For this band, the IEEE802.11ad and IEEE802.15.3c standards has been defined, and commercial products are already available. Besides the unlicensed 60 GHz band, the potential of other frequency bands, such as the 28, 38, 70, and 80 GHz bands, are also being investigated. The frequency bands around 28 and 38 GHz have been pointed out as feasible candidates for next generation cellular technologies, whereas the 60, 70, and 80 GHz bands have been identified as viable candidates for indoor short range wireless communications, such as WiFi and device-to-device technologies.

3.2.1 Comparison with Lower Frequency Bands

Wireless propagation in the mm-wave range is inherently different from propagation in the lower frequency bands commonly used today. The attenuation associated with propagation in the mm-wave range is generally higher compared to lower frequency bands for the following reasons: (i) The free-space

PL is proportional to the square of the carrier frequency (assuming constant-gain antennas), and thus much higher than for the 2 and 5 GHz bands. (ii) The dimensions of physical objects in a room are typically large in relation to the wavelength of about 1–10 mm in the mm-wave range, resulting in sharp shadow zones. (iii) The transmission through obstacles such as walls is much lower compared to lower frequencies.

In Martinez-Ingles et al. [MMP+14], radio wave propagation in the 2–10 and 57–66 GHz bands are compared based on measurements in an indoor laboratory environment under LoS conditions. Metrics such as the PL, DS, coherence bandwidth and Rician K-factor for these two bands are compared. Typical values for the RMS DS was found to be about 9 and 5 ns for the 2–10 and 57–66 GHz bands, respectively. The Rician K-factor was found to be larger for the 60 GHz band, indicating that the ratio between the power in the direct path and the remaining scattered paths is larger at higher frequencies. In the paper, it is also argued that diffuse scattering plays a more important role in contributing to the overall received power in the 2–10 GHz band as compared to that at the 60 GHz band.

A comparison of the indoor propagation properties at 2.9 and 29 GHz was investigated in Koymen et al. [KPSL14]. Based on indoor measurement for two separate office floors, where one floor represents an office cubicle environment and the other floor represents office and hallway environments, estimated parameters for PL, excess delay, RMS DS and power profiles of the received paths are presented. The estimated PL parameters for the PL exponent and standard deviation of the log-normal shadowing distribution were found to be quite similar for the two bands. The RMS DS for 2.9 and 29 GHz was also evaluated, based on the measurements from each floor. In general, the RMS DS at 29 GHz is smaller. However, there were also occasions when the RMS DS was greater at 29 GHz, which the authors attribute to a better waveguide effect at 29 GHz in the hallways compared to at 2.9 GHz.

3.2.2 Characterisation of mm-Wave Radio Channels

3.2.2.1 Wideband characterisation

There are some contributions dealing with wideband characterisation at mm-wave frequency bands, including parameters such as PL and RMS DS. In Salous [Sal14], using a newly developed multi-band frequency-modulated continuous-wave (FMCW) channel sounder, 2×2 MIMO measurements where performed in an indoor environment typical of a large office working area. Results for the RMS DS give similar values for the four MIMO channels, and show a significant increase as the threshold level is increased from 20 to 40 dB (0.42, 5.13, and 8.46 ns for the 20, 30, and 40 dB threshold levels, respectively). A PL exponent of 1.8 was also estimated from the measurements. In Barratt et al. [BNB15], an experimental measurement campaign using 60 GHz transceivers with on-chip antenna arrays is described. Measurement-based estimates of the PL exponents for a laboratory, office, corridor and an atrium environment were found to be 1.55, 1.55, 1.44, and 2.05, respectively.

The wireless mm-wave channel in a hospital environment, for real-time video streaming for angiography and ultrasonic imaging applications, was studied in Kyrö et al. [KHS+12]. From the measurements, it was found that the shape of the power delay profile (PDP) was significantly different for the two scenarios, and seems attributed to the volume of the room and the degree of reverberation of radio waves in the room, as shown in Figure 3.7. The authors point out that PDPs with shapes similar to that of the ultrasonic inspection room also have been observed in an industrial environment where the reverberation of radio waves was significant due to the presence of a lot of metallic equipment in the environment. In the angiography room, the PDPs consist of a set of discrete specular components and the diffuse component having continuous power distribution.

In Haneda et al. [HJK+14], measurement results for large indoor environments at 60 and 70 GHz are presented. Figure 3.8 shows two typical PDPs that were measured at 60 and 70 GHz. Specular paths accounted for at least 75% of the received power in the office scenarios and more than 90% in the shopping mall and station scenarios. The results suggest a possibility to use a single structural framework to model 60 and 70 GHz channels.

Finally, in Hafner et al. [HDM+15], a fully polarimetric, wideband and directional radio channel sounding campaign conducted at 70 GHz is presented. In this contribution, propagation effects and characteristics that have to be considered in future Modelling were identified. It was found that more specular than diffuse components occur and that the channel is spatially sparse.

3.2.2.2 Spatial characterisation

In Martinez-Ingles et al. [MPR+13], 10×10 MIMO measurements were performed at 60 GHz in an indoor laboratory environment at University of Cartagena, Spain. This laboratory consists of a room of dimensions $4.5 \times 7 \times 3$ m^3 and is furnished with several closets, desktops, and computers. Apart from wideband characterisation, the high-resolution parameter

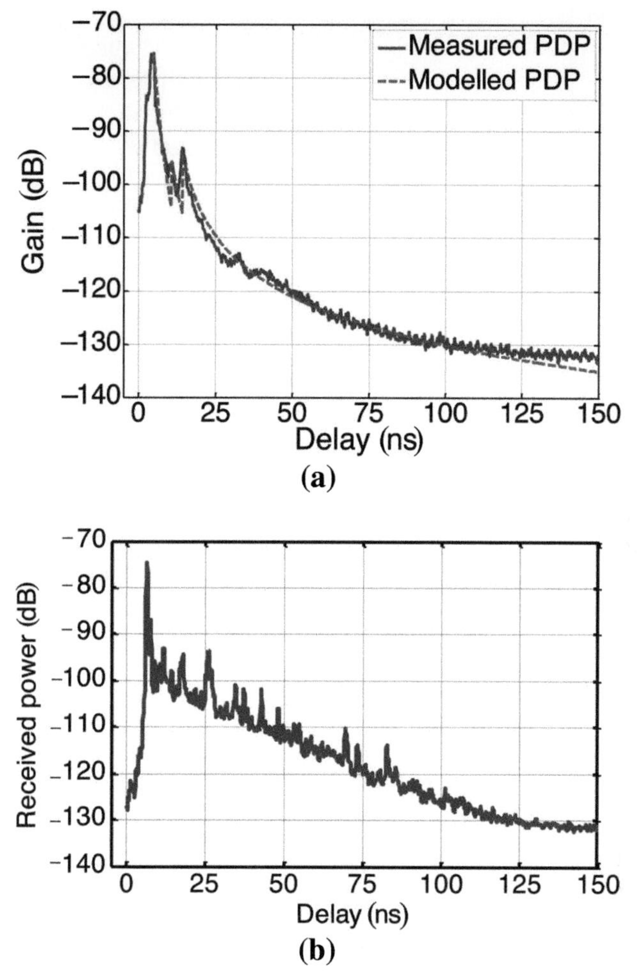

Figure 3.7 Typical PDPs of (a) the angiography room and (b) the ultrasonic inspection room [KHS+12].

estimator RiMAX was used to extract directional parameters, and compared to a 3D ray-tracing tool. The results show that the angular and DSs obtained from RiMAX are quite similar to the ones based on the 3D ray tracing tool.

Another contribution to angular characterisation can be found in Zhang et al. [ZYL+14], where channel measurement campaigns at centre frequency of 72 GHz with 2 GHz bandwidth were presented. The statistics of the composite and cluster-level spreads of channels in azimuth and elevation of

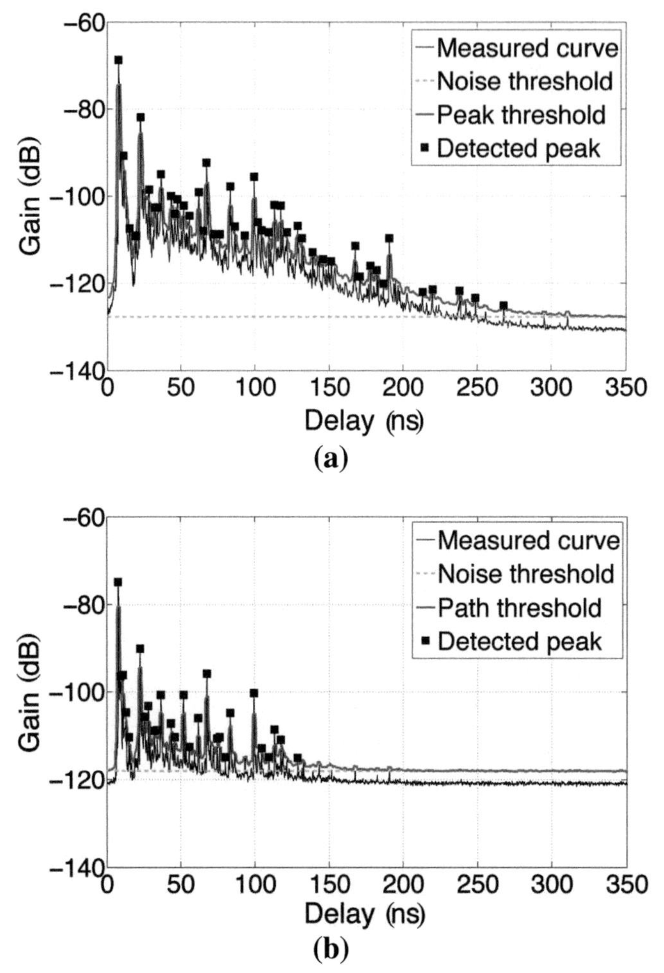

Figure 3.8 Exemplary PDPs in a large office: (a) 60 GHz and (b) 70 GHz. The PDPs were measured at the same Tx and Rx location [HJK+14].

arrival domains were calculated based on the angular power spectra. Less than four clusters can be identified in the direction of arrival domain. Laplacian distributions were found to fit well with the empirical occurrence of most of the angular-domain spread parameters considered. The estimated mean cluster azimuth spreads were found to be 6 and 9° for LoS and NLoS scenarios, respectively, and the mean cluster elevation spread is 6° in both cases.

In [Med15], 60 GHz measurement data based on an extreme size virtual antenna array ($25 \times 25 \times 25 = 15625$ elements), is used to provide highly

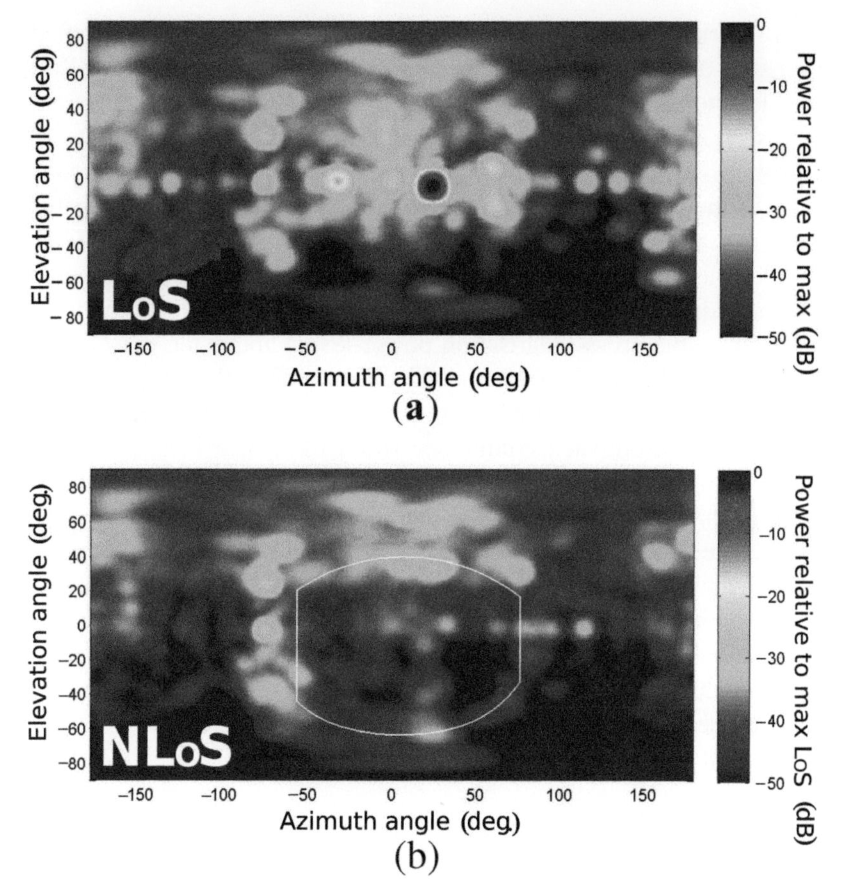

Figure 3.9 60 GHz directional power spectra for indoor line of sight (LoS) and non line of sight (NLoS) measurements. In the NLoS figure, the blockage by a white-board is indicated by a *white line*.

resolved directional properties of the radio propagation channel. The results indicated that the distinct spikes observed in the PDPs were caused mainly by specular reflections. There were however also a significant diffuse contribution due to scattering caused by the many smaller objects of the environment. Furthermore, corresponding LoS and NLoS channels were found to be very similar in directions where no blockage occurs, as seen in Figure 3.9.

3.2.2.3 Polarisation characterisation

The polarisation properties of mm-wave channels have been studied sparsely. In Dupleich et al. [DSF$^+$14], quad-polarised 60 GHz channel measurements have been carried out in a small-office environment. The identified strongest paths clearly indicate the need of using beamforming in NLoS scenarios. Furthermore, the presence of different paths with different polarisations shows the advantage of considering polarisation diversity in the beamformer.

A study of the cross-polarisation ratios (XPRs) of the propagation paths is presented in Karttunen et al. [KHJP15], based on measured indoor 70 GHz channels. Despite very extensive measurement campaigns with 23 channel sounding campaigns and 518 propagation paths detected in total, only 17 paths have detectable cross-polarisation power levels above the noise floor. A range of the XPRs between 10 and 30 dB is observed in Figure 3.10. Polarimetric radio channel measurements at 60 GHz in a small meeting room and in an empty, unfurnished conference room have also been presented in Gustafson and Tufvesson [GT15]. The results showed that the specular XPRs were in the range of 5–30 dB, which is consistent with the previous contribution. The XPRs can also be modeled as normally distributed random variables with a mean and standard deviation of 17.1 and 5.2 dB, respectively. Furthermore, in Degli-Esposti et al. [DEFV$^+$15], directional polarimetric 60 GHz indoor measurements were presented. These measurements were used to modify and calibrate a 3D ray-tracing model, and to tune the embedded effective-roughness diffuse scattering model.

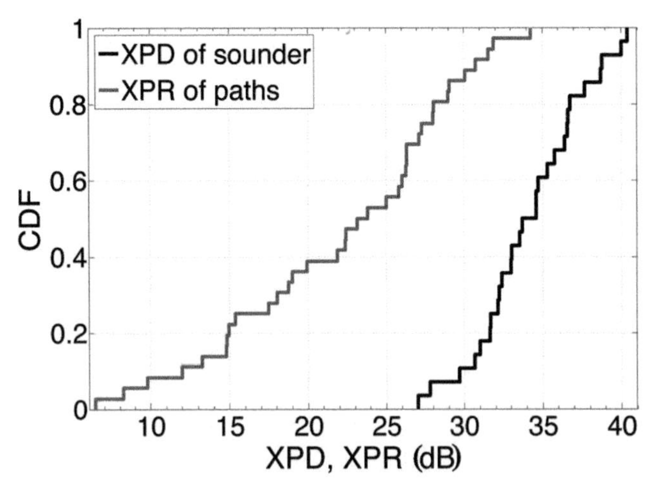

Figure 3.10 Cumulative distribution functions (CDFs) of the cross-polarisation discriminations (XPDs) and cross-polarisation power ratios (XPRs) for 70 GHz channels in large indoor environments [KHJP15].

3.2.3 Channels Models for mm-Wave Systems

Wireless channel modelling is always subject to a trade-off between the complexity and accuracy. Cluster-based, double-directional channel modelling is a popular compromise of the trade-off, as it supports wideband directional channel modelling, which is vital for beamforming, and can be used to derive a propagation channel model that is valid for arbitrary antenna elements and array configurations, while maintaining the complexity to a manageable extent.

3.2.3.1 Stochastic channel models

In Gustafson et al. [GHWT14], an indoor 60 GHz stochastic channel model was derived based on results from a measurement campaign in a conference room environment. The model is based on an extended Saleh–Valenzuela model, where the impulse response, h, is given by

$$h(t, \Theta_{\text{rx}}, \Theta_{\text{tx}}) = \sum_{l=0}^{L} \sum_{k=0}^{K_l} \beta_{k,l} e^{j\chi_{kl}} \delta(t - T_l - \tau_{k,l}) \delta(\Theta_{\text{rx}} - \Omega_l - \omega_{k,l}) \delta(\Theta_{\text{tx}} - \Psi_l - \psi_{k,l}). \tag{3.3}$$

Here, $\beta_{k,l}$ is the complex amplitude of the kth ray (i.e., MPC) in the lth cluster and T_l, Ω_l, and Ψ_l are the delay, direction of arrival (DOA) and direction of departure (DOD) of the lth cluster, respectively. Similarly $\tau_{k,l}$, $\omega_{k,l}$, and $\Psi_{k,l}$ are the delay, DOA and DOD of the kth ray in the lth cluster, respectively. Finally, $\delta(\cdot)$ is the Dirac delta function and the phase of each ray, $X_{k,l}$, is assumed to be described by statistically independent random variables uniformly distributed over $[0, 2\pi)$. The statistical properties of the inter-cluster parameters T_l, Ω_l, and Ψ_l, as well as the intra-cluster parameters $\tau_{k,l}$, $\omega_{k,l}$ and $\Psi_{k,l}$ were investigated. It was found that the ray and cluster fading were both appropriately modeled using a log-normal distribution, whereas the intra-cluster angles are modeled as Laplacian-distributed. The estimated channel model parameters can be found in Gustafson et al. [GHWT14].

In the work of Kyrö et al. [KHS+12], a different type of stochastic channel model was suggested for three different indoor hospital environments. This channel model does not include angular information, but is able to reproduce the typical shapes of the PDPs in the different environments. For the angiography room environment, a typical PDP consists of several narrow peaks and an exponentially decaying power distribution. These PDPs are modeled as a set of specular components and a so-called diffuse part caused by distributed diffuse scattering:

$$\mathrm{PDP}(\tau) = \sum_{n=1}^{N_s} P_n \delta(\tau - \tau_n) + P_{0\mathrm{dif}} \exp\{-(\tau - \tau_0)/\beta\}. \quad (3.4)$$

Here N_s is the number of identifiable specular components, P_n and τ_n is the power level and the excess delay of the nth specular component, respectively. $P_{0\mathrm{dif}}$ and β is the initial power level and the decay constant of the diffuse component, respectively, and τ_0 is the delay of the first arriving component.

Unlike the angiography room, in the ultrasonic imaging room, a PDP consists of one or several clusters, which start from a distinguishable peak followed by a decaying tail of power. For this reason, the PDP for the ultrasonic inspection room is instead modeled using a power law model for the power of the nth cluster, as

$$P_n(\tau) = P_{n0}(\tau + \tau_{n0})^{-\alpha_n}, \quad \tau \geq \lambda_n, \quad (3.5)$$

where P_{n0}, τ_{n0}, and α_n are the initial power value, offset delay, and decay constant for the nth cluster, respectively, and λ_n is the delay time from which the nth cluster begins. The modelling parameters, and the associated distributions, for the angiography and ultrasonic inspection room can be found in Kyrö et al. [KHS+12].

3.2.3.2 Site-specific channel models

As it has been observed that the diffuse spectrum seems to be less significant for mm-wave channels compared to lower frequencies, and the major part of the received power in many mm-wave scenarios stem from specular reflections, it has been argued that ray tracing [MGG+12] or other deterministic approaches [JHK+12] might be viable at mm-wave frequencies. In Järveläinen et al. [JHK+12], a site-specific model is presented, where the room dimensions of an ultrasonic inspection room have been measured with a laser scanner in order to produce an accurate, so-called point cloud model of the room. In the ray tracing simulations, each point in the point cloud generates one scattering component using a single-lobe directive scattering model. Results show that the model is able to predict the measured PDP reliably within a 20-dB dynamic range relative to the LoS component, using a single-bounce scattering assumption. Below this range, higher order interactions need to be considered, as evident from Figure 3.11.

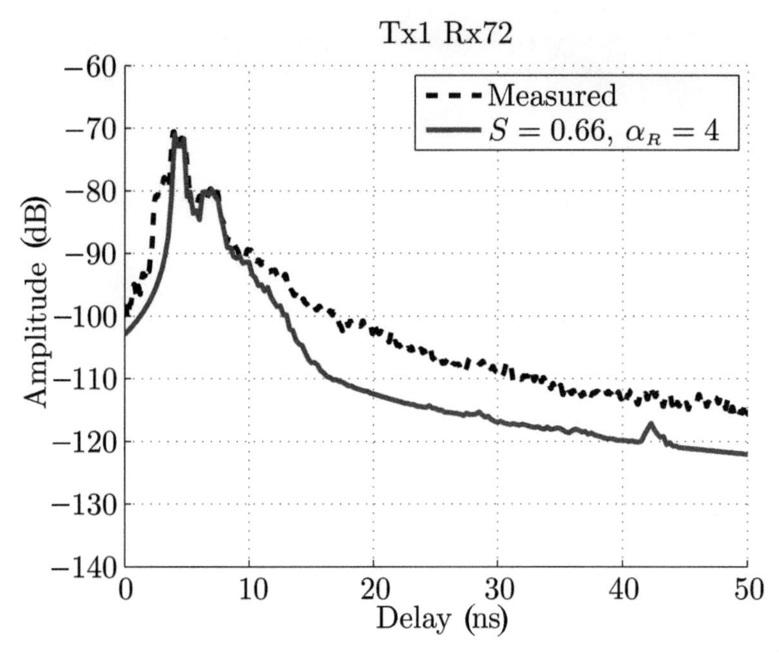

Figure 3.11 Comparison between measured and simulated PDP in a hospital room [HJK+14].

In Martinez-Ingles et al. [MGG+12], results from a measurement campaign in a laboratory environment is compared with the results from ray tracing simulations of the same environment. The 3D environment model in the ray tracing routine also includes important objects such closets, tables and windows, and appropriate values of the permittivity of different materials are assigned to different materials, based on values presented in the literature. By comparing the measured PDPs with the simulated ones based on ray tracing, it could be concluded that the ray tracing results agree fairly well with the measured PDPs in terms of the received powers and RMS DSs. However, when visually comparing the measured and simulated PDPs, it is clear that the ray tracing simulation is unable to predict some of the strong specular paths that are present in the measured PDPs.

3.2.3.3 Ray tracing and stochastic modelling

For mm-wave indoor channels, another popular approach is hybrid modelling with both deterministic and stochastic parts in combination. An example of this is the IEEE802.11ad channel model, where the inter-cluster angles are determined based on ray tracing and the gain of the LoS component is

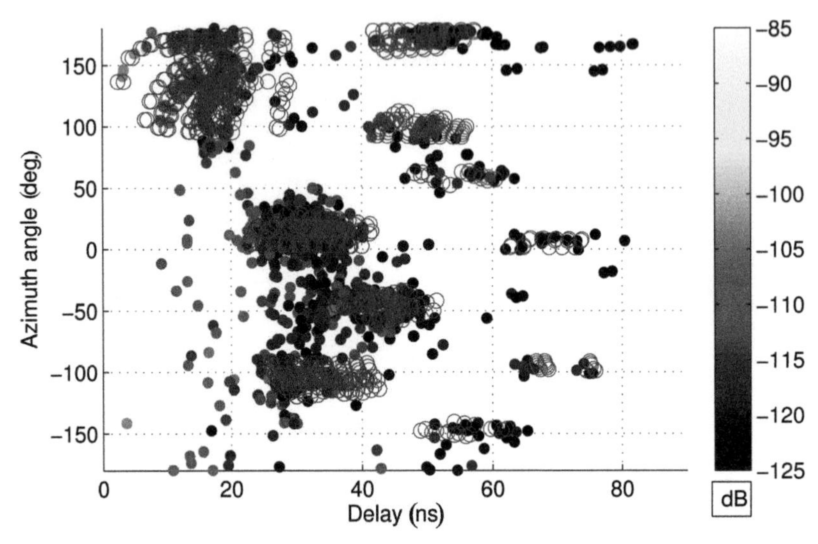

Figure 3.12 Delay and azimuth angles of estimated multi-path components (*dots*) and ray-tracing results for first, second and third order reflections (*circles*).

modeled using a deterministic expression. In Gustafson et al. [GHWT14], a purely stochastic channel model for a 60-GHz indoor scenario is compared with a hybrid channel model where the inter-cluster angles are determined using ray tracing, similar to the IEEE802.11ad model. The results in the paper show that both the stochastic and hybrid models offers a good agreement with the measured channels. In Figure 3.12, the delay and azimuth angles of the MPCs estimated with the high-resolution SAGE algorithm are compared with the estimated delays and angles from the ray tracing simulation, for all the measured scenarios in Gustafson et al. [GHWT14]. Similar results were obtained for the elevation angles. As seen in the figure, there is a there is a considerable overall agreement, but there are also many multi-path components with significant power that are not explained by the ray tracing simulations. This is a direct result of an over-simplified environment model in the ray tracing routine, indicating the fact that it is necessary to include finer structures in the model for the room environment.

3.2.4 Capacity Evaluation of mm-Wave Systems

3.2.4.1 OFDM and single carrier (SC)

Implementation of IEEE physical layer (PHY), such as IEEE802.15.3c and IEEE 802.11ad, have been one of the investigation areas in COST IC-1004.

The orthogonal frequency division multiplexing (OFDM) or SC specifications have been used in Takizawa et al. [TKH+12]. The evaluation of 60 GHz high-speed radio systems has been conducted through calculation of SIR, channel capacities, and achievable TPs by using measured PDPs in hospital environments. The results reveal that OFDM systems provide the required TP for the application scenario in all the measured PDPs. On the other hand, the SC systems without any equalisation are limited to 75% of the measured PDPs where the required TP is achieved. Figure 3.13 shows the TPs for OFDM and SC using both standards.

Extending this approach to MIMO, in Martinez-Ingles et al. [MSM+13], the performance of IEEE 802.15.3c MIMO–OFDM systems for four antenna configurations has been experimentally studied. The correlation was highly influenced by the antenna configurations, since the arrangement of the antenna has a large impact on the performance, especially over short distances. In addition, MIMO seems a reasonable solution for both increasing the TP and the communication distance, as it was shown that a 4×4 QSTBC-MIMO system could increase the data bit rate by a factor of 3.7 and also increase the maximum achievable distance by 1 m.

3.2.4.2 Spatial multiplexing

The potential of spatial multiplexing at mm-wave frequency band is analysed in Haneda et al. [HKGW13]. The paper provides a measurement-based investigation of the spatial degrees-of-freedom (SDoF) of 60 GHz indoor channels. The SDoF are defined in a way so that they only depend on the propagation conditions and the sizes of the Tx and Rx antenna apertures. This way, the SDoF provides an upper bound on the number of eigenchannels that can be realised with a given antenna aperture size under the given propagation conditions. For an antenna aperture size of 9 λ^2 at 60 GHz, as seen in Figure 3.14, the SDoF is more than one for a Tx–Rx antenna separation of more than 2 m. Furthermore, at most ten antennas on each side of the Tx and Rx are sufficient to perform effective spatial multiplexing in the considered propagation scenario. Even five antennas can work efficiently to capture eigenchannels down to –15 dB relative magnitude.

3.2.4.3 Beamforming

At mm-wave frequencies, physically small antenna apertures can be electrically large. Therefore, mm-wave radios are inherently much more capable of focusing beams and obtaining large array gains than microwave radios. In Haneda et al. [HGW13], the authors compare the capacity improvement

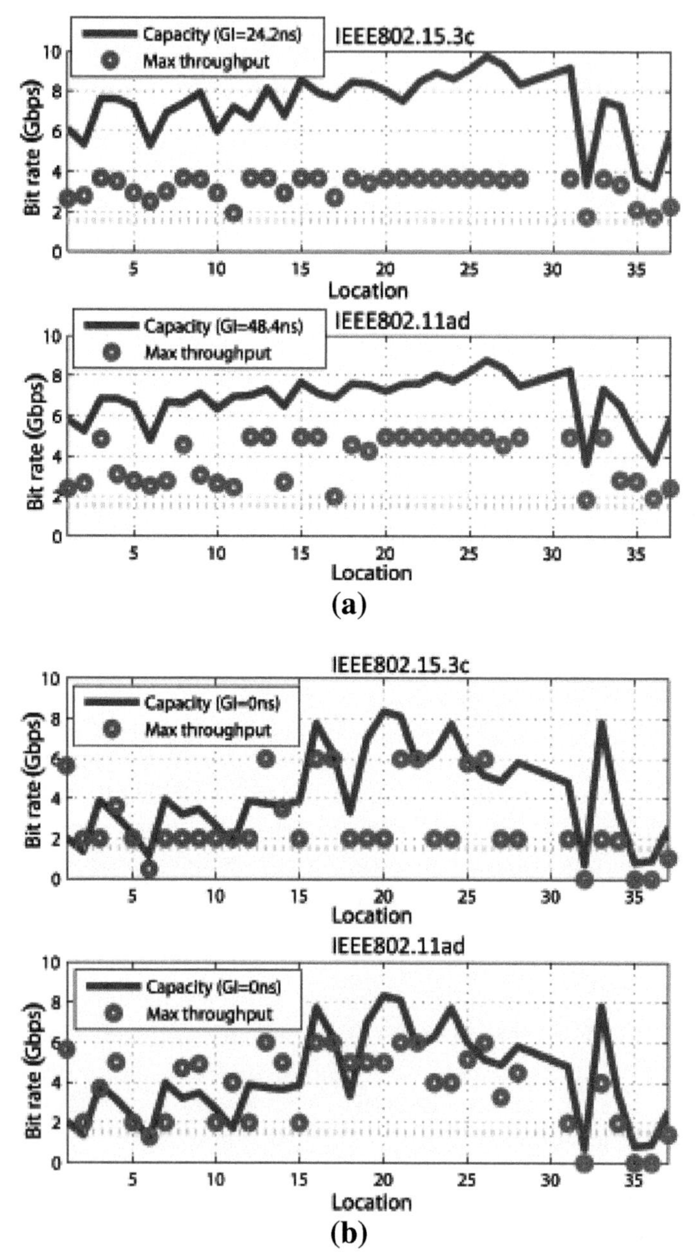

Figure 3.13 A comparison between the TPs (simulation results) and channel capacities calculated from measured PDPs (dotted lines show required TP of 1.53 Gbps); (a) Orthogonal frequency division multiplexing (OFDM), and (b) Single carrier (SC) [TKH+12].

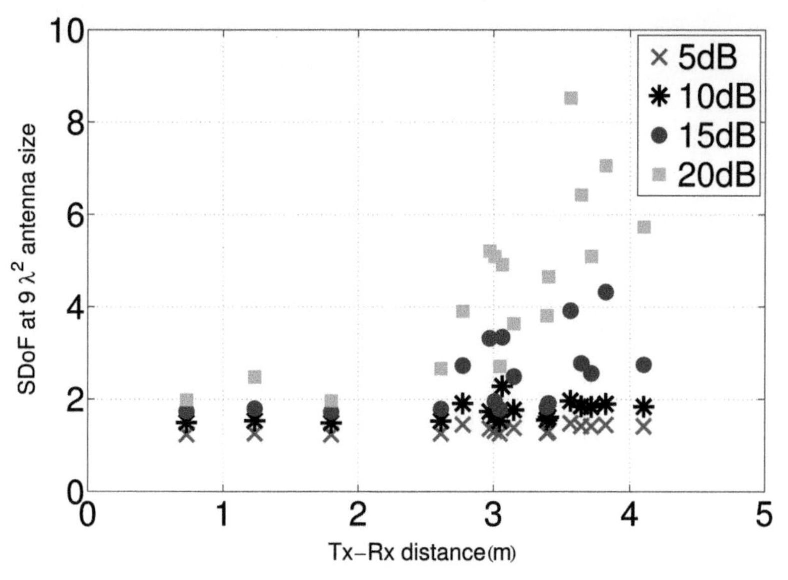

Figure 3.14 Distance dependency of the spatial degrees-of-freedom (SDoF) for an antenna aperture size of 9 λ^2. The SDoF is plotted for different threshold levels, i.e., t = 5, 10, 15, and 20 dB [HKGW13].

capability of spatial multiplexing and beamforming techniques for 60 GHz spatial transmissions in a multi-carrier radio system, as shown in Figure 3.15. It is concluded that beamforming is, in general, as robust as spatial multiplexing in providing capacity enhancement in the 60 GHz band. Spatial multiplexing would be a worthwhile option when the Rx SNR is favorable and a higher peak data rate is required. A hybrid approach of beamforming and spatial multiplexing can be a sound compromise to take advantage of both techniques.

3.3 Techniques and Models for Localisation

Location awareness is a key component of many future wireless applications and a key enabler for next-generation wireless systems. The core challenge is achieving the required level of localisation accuracy *robustly,* especially in indoor environments which are characterised by harsh multipath conditions. Promising candidate systems thus either provide remedies against multipath or they fuse information from complementary sensing technologies [SMW12, CDG$^+$14].

This section starts with the modelling and characterisation of "deterministic" channel properties that are relevant for positioning (Section 3.3.1).

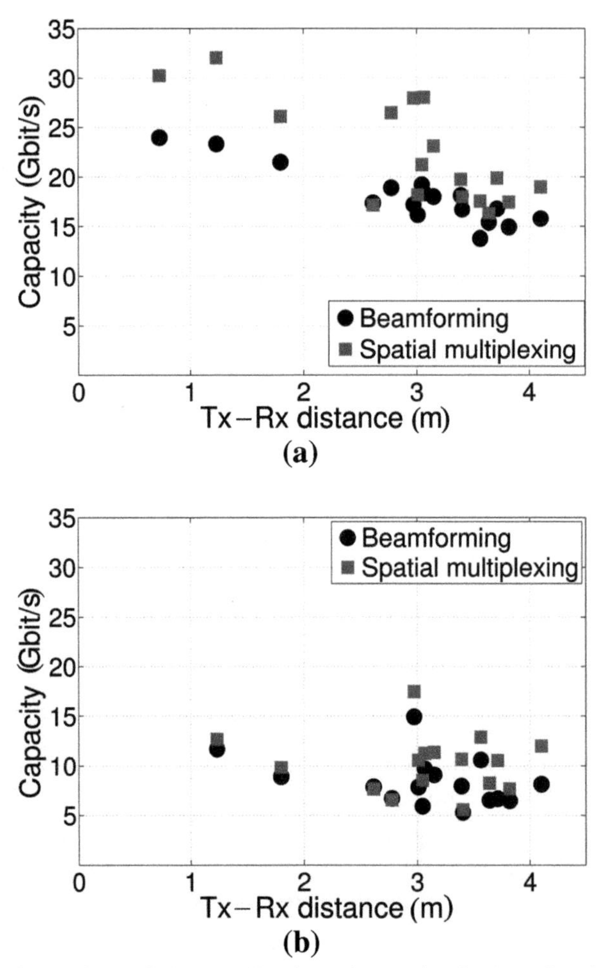

Figure 3.15 Comparison of the ergodic channel capacity for beamforming and spatial multiplexing. (a) 9 λ^2 antenna aperture size, and (b) NLoS scenario with antenna aperture size of 9 λ^2 [HGW13].

The acquisition of position measurements from received signal is then discussed in Section 3.3.2. Different system architectures that can be used for indoor positioning are addressed in Section 3.3.3, while Section 3.3.4 summarises algorithms that *exploit* position-related information from the multipath channel beyond the LoS component, as suggested by the discussion in Section 3.3.1.

3.3.1 Radio Channel Characterisation for Positioning

Geometric information related to the position of the agent (the radio device to be localised) is found not only in the LoS component, but also in "deterministic" reflected MPCs. This motivates to exploit such components for positioning and also to model such components in a geometrically consistent way. The analysis of performance bounds quantifies the amount of information obtained from these components.

3.3.1.1 Performance bounds motivating channel modelling

In Witrisal and Meissner [WM12], the (baseband) signal received at an agent has been described as

$$r(t) = \sum_{k=1}^{K} \alpha_k s(t - \tau_k) + s(t) * v(t) + w(t), \tag{3.6}$$

where, $s(t)$ is a signal that is transmitted from/to an anchor, a reference node at known location $\mathbf{p_1}$. The *deterministic* part of the channel is given by the sum over K MPCs, scaled and shifted by the complex amplitudes α_k and delays τ_k, respectively. The delays $\tau_k = \frac{1}{c} \|\mathbf{p} - \mathbf{p}_k\|$ are deterministically related to the positions $\mathbf{p}$ of the agent and $\mathbf{p}_k$ of virtual anchor (VA) that describe the geometry of the propagation paths [WM12]. The signal $v(t)$ denotes diffuse multipath, i.e., everything that is not or can not be modeled by the deterministic part. It is modeled as a (Gaussian) random process with auto-covariance $\mathrm{E}\{v(t)v^*(\tau)\} = S_v(\tau)\delta(t - \tau)$, where $S_v(\tau)$ is a PDP accounting for the non-stationary variance of the diffuse multipath in the delay domain. Finally, $w(t)$ is Gaussian measurement noise at power spectral density (PSD) N_0.

Using this model, the Cramér Rao lower bound (CRLB) on the position error can be derived. The Fisher information matrix is the inverse of the CRLB, which quantifies the information about the position $\mathbf{p}$ contained in the received signal $r(t)$. Under the assumption of *separable* deterministic MPCs, it is written as

$$\mathbf{J}_{\mathrm{p}} = \frac{8\pi^2\beta^2}{c^2} \sum_{k=1}^{K} \mathrm{SINR}_k \mathbf{J}_r(\phi_k), \tag{3.7}$$

where, β denotes the effective (RMS) bandwidth of $s(t)$ and c is the speed of light. Furthermore, $\mathbf{J}_{\mathrm{r}}(\phi_{\mathrm{k}})$ is a unitary rank-one matrix with an eigenvector pointing along the angle-of-arrival (AoA) of the k-th MPC, the line connecting $\mathbf{p}$ and $\mathbf{p}_k$. According to this equation, each MPC adds position-related information. The *amount* of this information is quantified by the signal to interference plus noise ratio (SINR) of the respective MPC, which is defined as [WM12]

$$\mathrm{SINR}_k = \frac{|\alpha_k|^2}{N_0 + T_\mathrm{p} S_\mathrm{v}(\tau_k)}. \tag{3.8}$$

It evaluates the ratio of the MPC's energy $|\alpha_k|^2$ to the noise density N_0 and the interference term $T_\mathrm{p} S_\mathrm{v}(\tau_\mathrm{k})$. The latter is quantified by the product of the PDP $S_\mathrm{v}(\tau)$ of the diffuse multipath and the pulse duration T_p, the bandwidth inverse of $s(t)$. Equations (3.7) and (3.8) quantify the contributions of individual deterministic MPCs to the position information available from the received signal $r(t)$. This result clearly highlights the beneficial influence of a large bandwidth through β and T_p [WML+14].

3.3.1.2 Channel analysis

In Meissner and Witrisal [MW12], an estimation technique has been proposed for the SINRs Equation (3.8) to evaluate the available position-related information from real measurement data. The method works with signals captured at known locations in the environment. Figure 3.16 shows a few such signals and the SINR estimates. MPCs with stable amplitude tracks have large SINRs, indicating high reliability of their position information. These reflections are less impaired by interference through diffuse multipath.

The figure also shows the energy capture (EC) of these MPCs, the relative amount of energy they carry. A method to extract this parameter is discussed in Meissner et al. [MAGW11]. These numbers show that about 70% of the energy can be attributed to geometrically modeled MPCs. The LoS component already carries almost 50%. Clearly, a large fraction of energy originates from the deterministic components. However, energy does not directly translate into position-related information, as seen from a comparison of the SINR and EC values in Figure 3.16.

3.3.1.3 Channel modelling

The evaluation of the *robustness* of a localisation system requires extensive experimental campaigns that span a sufficient range of target environments. These are, however, tedious to perform. Alternatively, geometrically consistent channel models may be used to generate *synthetic* signals in test environments; a possible way forward to facilitate a robustness analysis [JHK+11]. The degree of realism can be evaluated based on (i) the position-related signal

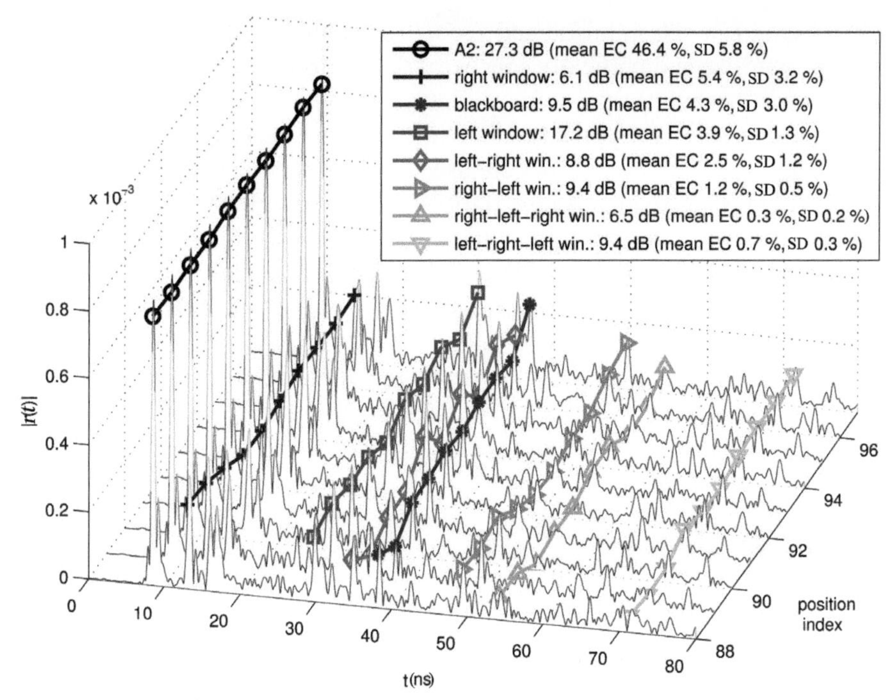

Figure 3.16 Signals obtained at closely spaced locations (5 cm spacing) in a seminar room [MLLW13]. Estimated amplitudes of geometrically modelled MPCs are indicated, together with their estimated signal to interference plus noise ratio (SINR) and energy capture (EC).

parameters (time-of-arrival (TOA), AOA, receive-signal-strength (RSS), etc.) and (ii) also in the light of position information, as defined by Equation (3.8). In the latter, also the level of diffuse multipath is accounted for. It is paramount to consider LoS, NLoS and OLoS situations, since the LoS component is generally considered to carry most of the position-related information [JHK+11]. All these effects are related to the geometry of the environment.

A useful channel model for the performance analysis of indoor localisation systems will thus be a geometry-based stochastic channel model (GSCM), in which the stochastic part needs to account for the geometrically non-resolvable components of the channel. The deterministic part needs to be spatially consistent and the overall model should be parametrised by measurements. Also, factors such as the mobility of the agents on a sufficiently large scale and cooperative multilink schemes must be supported to reflect the needs of future applications [AMAU14].

Ray-tracing is a deterministic and geometric technique that is by default suitable to syntesise geometrically consistent CIRs. However, a realistic consideration of diffuse multipath is often difficult and computationally intensive. To tackle this, surfaces are typically subdivided in tiles whose size depends on the wavelength and scattering lobes are used for each tile to create a number of weighted outgoing scattered rays [DEFV+07, GXH+15, MGM+13]. As Figure 3.17(a) shows, this can model the diffuse multipath for small excess delays reasonably well. But for larger delays, the power level drops significantly below the measured reference. A promising alternative is the combination with stochastic propagation graphs [SPG+14]. A graph structure is used to model recursive interactions between scatterers that may be placed according to stochastic or geometric considerations. A closed-form solution can be given for the channel transfer function, allowing for realistic levels of diffuse multipath as illustrated in Figure 3.17(b).

3.3.2 Estimation of Position-Related Parameters

The estimation of position-related parameters (position measurements) is usually the first step towards localisation and tracking. That is, parameters such as the TOAs, time-difference-of-arrival (TDOAs), AOAs, or RSSs are extracted from the received signals. These parameters are related to the geometry at hand through known or learned functional models.

This subsection summarises approaches towards the estimation of these parameters. We first discuss papers that deal with the estimation of the TOA and AOA of the received signal or its components. We then put our focus on the important problem of detection and mitigation of NLoS conditions.

3.3.2.1 Ranging in multipath channels and multipath parameter estimation

A conventional estimator of the TOA estimates the CIR from the received signal from which then the leading edge can be extracted. This approach is particularly problematic when separability conditions are violated, e.g., at non-UWB bandwidths. Furthermore, popular algorithms like estimation of signal parameters via rotational invariance techniques (ESPRIT) and SAGE require knowledge of the number of MPCs.

An approach that avoids these complications is discussed in Jing et al. [JPF14]. A stochastic point-process model of the channel is used to obtain a *direct* estimation of the TOA from the received signal. A ML estimator has been formlated and evaluated for LoS and NLoS conditions. Simulation results

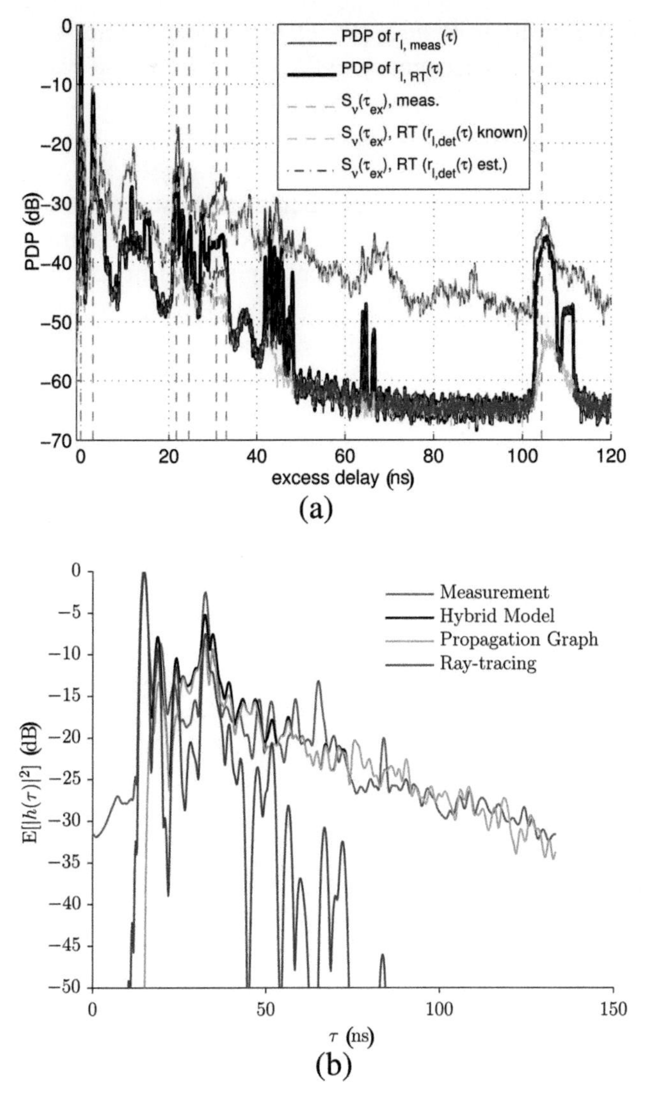

Figure 3.17 Channel modelling concepts developed within COST IC1004: (a) Ultra wideband (UWB) Ray-tracing [GXH+15] accounting for diffuse multipath compared to measurements [MGM+13], and (b) modelling of deterministic and diffuse multipath with a stochastic propagation graph combined with ray-tracing [SPG+14].

show the superior performance over conventional estimators. An OFDM signal with a bandwidth of about 7.5 MHz has been used. The impact of model mismatches is rather small if an NLoS scenario is assumed.

Section 3.3.1 has introduced an approach to quantify the position-related information contained in deterministic MPCs by their SINR values. This discussion highlighted the importance of a stochastic characterisation of diffuse multipath that interferes with useful components.

The same basic finding has been reported in Tanghe et al. [TGJ+12], which compares the performance of classical multipath estimation algorithms, namely SAGE and ESPRIT, to the more recent RiMAX algorithm. Only the latter accounts for the presence of dense multipath and hence outperforms the others, as shown in Figure 3.18. The results are based on synthetic data that are generated by a ray-tracing algorithm. A parameteric model is used in the estimator for the dense mulipath, consisting of an exponentially decaying PDP. Four-by-four-antenna arrays are used at TX and RX, the data have a bandwidth of 40 MHz, and the estimated parameters include AoA, angle of departure (AoD), and TOA. A related positioning approach has been discussed in Zhu et al. [ZVK+15].

For high-accuracy indoor localisation, MPCs can be extracted from UWB channel measurements [MLLW14, MLW14, LMLW15]. Assuming that the

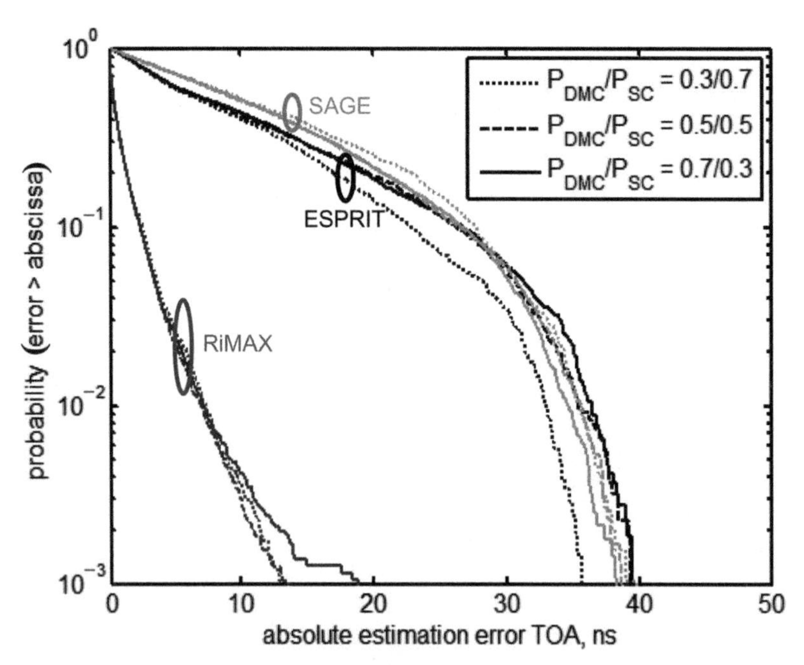

Figure 3.18 Complementary cumulative distribution functions (CCDFs) of absolute time-of-arrival (TOA) estimation error for different power ratios of deterministic to diffuse multi-path components (DMCs) [TGJ+12].

large bandwidth can provide separability of MPCs in the delay time, a relatively simple, iterative, correlation-based estimator can be used in a first step to estimate the TOAs. For example, the arrival time estimation for the k-th MPC is realised as an iterative least-squares approximation of the received signal $r(t)$

$$\hat{\tau}_k = \arg\min_{\tau} \int_0^T |r(t) - \hat{r}_{k-1}(t) - \hat{\alpha}(\tau)s(t-\tau)|^2 \, dt \tag{3.9}$$

using a template signal $\hat{r}_{k-1}(t) = \sum_{k'=1}^{k-1} \hat{\alpha}_{k'} s(t - \hat{\tau}_{k'})$ for all MPCs up to the $(k-1)$-st. The path amplitudes are nuisance parameters, estimated using a projection of $r(t)$ onto a unit energy TX waveform $s(t)$ as

$$\hat{\alpha}(\tau) = \int_0^T [r(t)]^* s(t-\tau) dt; \hat{\alpha}_k = \hat{\alpha}(\hat{\tau}_k). \tag{3.10}$$

The work in [MLLW14, MLW14, LMLW15] is based on this approach. It accounts for diffuse multipath only in the second step, the positioning/tracking algorithm. This is achieved by weighting the different MPCs based on their position-related information that is quantified by their SINRs (cf. Section 3.3.1).

In Froehle et al. [FMW12], a probability hypothesis density (PHD) filter is used for simultaneous tracking of several MPCs in UWB measurements. The PHD filter is a multi-target tracking algorithm that can cope with varying path visibility and an unknown number of paths. It uses input data extracted from UWB measurements (6–8 GHz) as described in Equations (3.9) and (3.10).

3.3.2.2 NLoS detection and mitigation

Non-line of sight conditions have particularly adverse impact on indoor localisation. In indoor environments, due to large number of walls, obstacles, and moving people, the first arrival path between a transmitter and a receiver is rarely LoS. Consequently, each localisation-aimed measurement, i.e., TOA, TDOA, AOA, or RSS will be significantly biased, leading to an erroneously estimated position.

As the NLoS effect is one of the main sources of localisation errors, a wide range of NLoS identification and mitigation techniques has been already proposed. A thorough survey of such techniques can be found in Kulakowski [Kul12]. The first common approach is to distinguish between LoS and NLoS anchors, i.e., to identify which anchors have a clear LoS to the localised object. If there is a redundancy in the number of anchors available, the object position can be estimated for each subset of anchors independently, and then the credibility of each estimation is calculated on the basis of residual errors resulting from the position estimation [Che99]. The second solution is to use advanced statistics of the received signal. Parameters like the kurtosis, maximum excess delay, and RMS DS can be efficiently exploited in order to decide if the path to an anchor is of LoS or NLoS type. In residential, urban and indoor office environments, the mean kurtosis is lower, while the maximum excess delay and RMS DS are higher in NLoS than in LoS channels. An example where the received signal statistics are used is the research work presented in Cipov et al. [CDP12]: the LoS and NLoS propagation conditions are distinguished calculating kurtosis. In Jing et al. [JPF14], which is also discussed in the previous subsection, a statistical TOA estimation method is proposed that is inherently robust against NLoS situations.

Another option is a cooperative approach: if the anchor signals are not sufficient for a node to determine its position, the network nodes can cooperate exchanging some data between each other. Wireless nodes perform ranging measurements and exchange the position estimates with their neighbours. Then, each node calculates its own position again, using the data from neighbouring nodes [SRL02].

When dealing with NLoS issues, fingerprinting techniques are also an interesting option [KK04]. They are very well suited to work without direct LoS visibility between the localised object and the anchors, because their basic idea is just to match data patterns (to look for their similarities) independently of the scenario these patterns describe in reality. However, gathering training data is troublesome and frequently not feasible. Moreover, it means the localisation system is not robust to environment or network topology changes.

A wide group of solutions is related to UWB systems. UWB systems are working in extremely large frequency bandwidth, which makes them potentially very accurate, but still susceptible to NLoS situations. The ability to distinguish MPCs at a UWB receiver opens the door for positioning techniques that can even take advantage of the multipath propagation phenomenon [GE11]. In combination with tracking filters, reflected MPCs can be utilised to retain robust localisation when the LoS signal gets obstructed.

3.3.3 Localisation and Tracking Systems

3.3.3.1 RSS-based systems

While in outdoor environments the satellite-based systems are a widely accepted standard, in indoor scenarios (where satellite signals do not penetrate walls) wireless localisation remains an open issue. Among other solutions, RSS techniques are gaining significant attention, mainly due to their simplicity and the ability of collecting data from the wide variety of heterogeneous wireless devices. These techniques, however, usually suffer from a lack of accuracy resulting from multipath propagation.

The issue of reliability of RSS-based systems is addressed in [KRG+14]. The results of a measurement campaign in a large single-floor laboratory show that the classical multilateration scheme does not perform well in such an environment: multipath propagation and shadowing effects are the reasons for very strong variations in the received signal levels and there is no clear dependence between the distance and the PL (Figure 3.19). On the other hand, a localisation algorithm based on calculating the weighted mean of the anchor positions (where the weighting coefficients are proportional to the signal power levels received from the anchors) results in much better accuracy, comparable with other indoor localisation schemes reported in literature. The weighting mean algorithm copes well with shadowing (NLoS) effects: if an anchor receives a reflected signal instead of a direct one, the weighted coefficient for this anchor is smaller. It weakens the impact of this anchor in final calculations, but it does not corrupt them, like in the case of multilateration. Using the frequency diversity also seems to be a promising solution: the authors report that it reduces the localisation error by about 30%.

These measurements are later continued with a mobile node to be localised (tracked) [KRO13]. With the weighted mean algorithm and Kalman filtering, a localisation accuracy of 2–2.5 m can be obtained.

An interesting solution is also proposed in Redzic et al. [RBO12]. A hybrid system based on camera images and RSS data from wireless local area network (WLAN) access points is described. The image data from the cameras and the RSS data from WLANs are used separately to calculate the position of the user. Later, the localisation results are analysed together taking into account their confidence levels, which significantly improves the system accuracy.

3.3.3.2 Passive localisation

A passive localisation system registers and analyses the impulse responses of the radio channels between a fixed transmitter and UWB receivers installed in an indoor area. When a new object or a person appears, some of the propagation paths are blocked and the modified set of impulse responses is again registered in the system. Evaluating the changes in the radio environment, the system can estimate the object position [PW10a]. Passive localisation has a few important

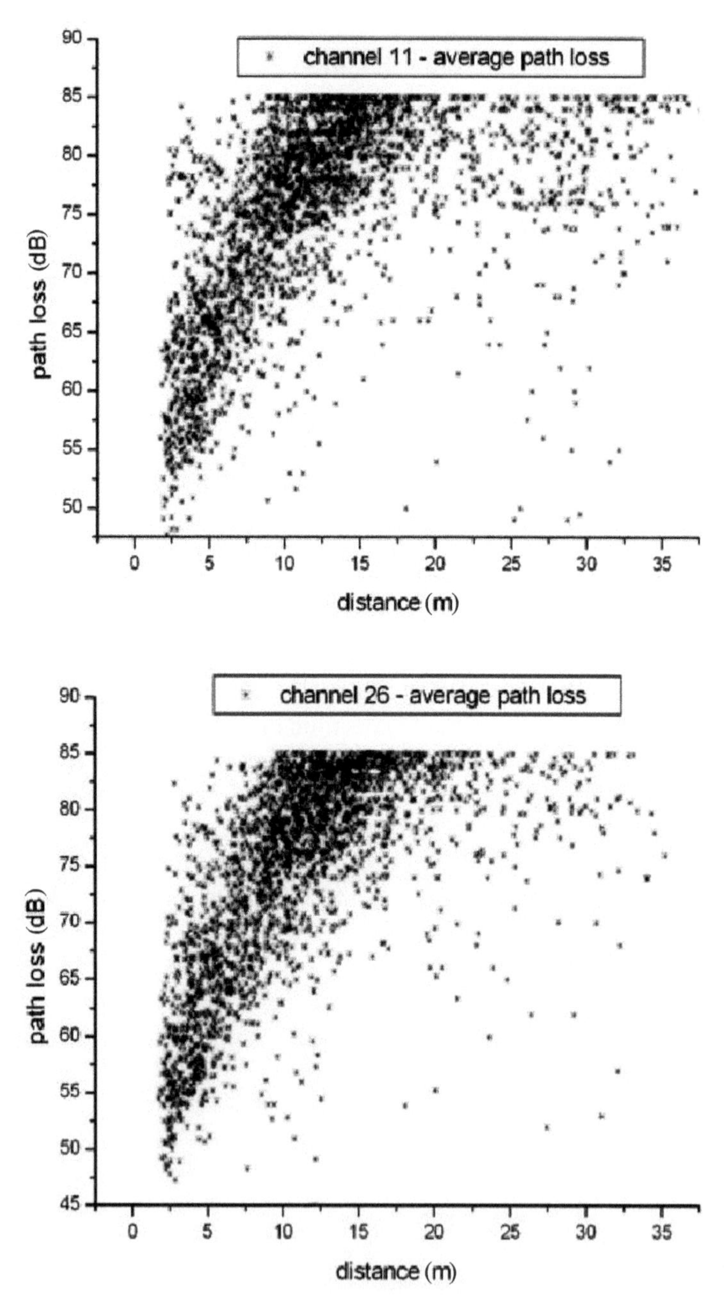

Figure 3.19 The PL as a function of the sensor–anchor distance for the channel 11 (2405 MHz) and the channel 26 (2480 MHz) [KRG+14].

merits that could decide about its future commercial success: (i) it can be used indoor, (ii) it does not need LoS visibility between a localised object and the system elements, and finally (iii) the localised object does not need to have any wireless device attached.

An example of a passive indoor UWB localisation system is analysed in Kosciow and Kulakowski [KK12] by means of computer ray tracing calculations. Two algorithms are used for calculating the object position. In the first one, an area where all the blocked propagation paths are crossing each other is considered as a possible position of the object to be localised. In the second approach, the sets of impulse responses resulting from all possible object positions (with a certain resolution) are analysed and juxtaposed with the set for the real object position. This research is continued in Kmiecik and Kulakowski [KK14], where not only localisation, but also simultaneous tracking of two objects is considered. This is later validated experimentally in Kmiecik et al. [KML$^+$15]. The aim is to detect the presence and to track the motion of a cardboard box with radio frequency absorbers on its surface, moving in a seminar room. With three transmitters and four receivers (12 fixed radio channels), localisation accuracy is quite satisfactory.

A related scenario is also investigated in Jovanoska et al. [JZG$^+$14], where measurements on a passive localisation system capable to work from beyond the wall are presented. The system is clearly motivated by radar solutions and is based on the signals reflected from the tracked people. The system consisting of a single transmitter and two receivers is able to track only one person; in order to track more people simultaneously the system should have more transceivers. The analysed system architecture has interesting security or military applications: a room under investigation can be inspected before entering it.

3.3.3.3 Radio-frequency identification (RFID) positioning on backscatter channels

Passive RFID is a technology used for the identification of tagged items. Significant effort has been spent on realising *positioning* methods for this technology, as this would bring added value to many applications. However, low-signal bandwidth and low-return power are two system properties rendering this capability extremely hard to realise. Multistatic (MIMO) readers and increased bandwidth are possible concepts towards the development of practical RFID positioning systems.

In Mhanna and Sibille [MS14], an experimental evaluation is shown of the power gain over the backscatter channel, when bistatic antenna setups and

UWB signals are employed. UWB signals would in principle allow for cm-level accuracy. A multistatic system with four reader transceivers (employing 16 channel branches) provides a mean gain of about 14 dB compared to a single monostatic reader. The gain over four monostatic readers is about 6.5 dB.

An appealing feature of an RFID system lies in the fact that the transponders sample the radio channel at defined locations. This fact may be used in a *cognitive* RFID positioning system [KLMA13] which learns the properties of the radio environment from past measurements in order to improve the efficiency of position measurements. The paper proposes a time-reversal processing for the transmitter side to focus transmit energy onto the tag. The analysis of the CRLB [LMFW14] demonstrates the impact of the transmitter-side processing on the information obtained from position measurements. This is an important property of a cognitive system that implements a perception–action cycle at the sensor level [HXS12].

3.3.4 Multipath-based Localisation and Tracking

In multipath-based localisation and tracking, MPCs can be associated to the local geometry using a known floor plan [MLFW13, MLLW14]. MPCs can simply be seen as signals from Vas/VSs. In particular in NLoS situations, a multipath-based method can increase the robustness of the localisation system tremendously. Competing approaches, discussed in Section 3.3.3, either detect and avoid NLoS measurements [MGWW10], mitigate errors induced by strong multipath conditions [WMGW12], or employ more realistic statistical models for the distribution of the range estimates [LMW13].

UWB signals are useful to separate MPCs because of their superior time resolution. Additional position-related information is thereby exploited that is contained in the multipath radio signals [MLW15] (cf. Section 3.3.1). Several contributions have been presented with in the COST action, which relate to this approach.

3.3.4.1 UWB techniques

The work in Meissner et al. [MLFW13] presents a comparison between multipath-assisted indoor localisation/tracking and conventional localisation/tracking methods that are based on TOA measurements. The former show superior performance in harsh indoor environments and NLoS situations. The presented algorithm uses the extracted multipath delays Equation (3.9) and the according position-related information introduced in Section 3.3.1 to properly weight the measurements. The estimated delays are associated to the VAs using an optimal sub-pattern assignment algorithm. These VAs are then used as additional anchors for the tracking filter. An extended kalman filter (EKF) employs the position-related information as a measurement noise model, quantified in terms of the SINRs of the MPCs, Equation (3.8).

Figure 3.20 shows an example of the multipath-based tracking of an agent position using the EKF with data association (EKF-DA). Here only one anchor was used. The SINRs were estimated at 60 measurement positions indicated by

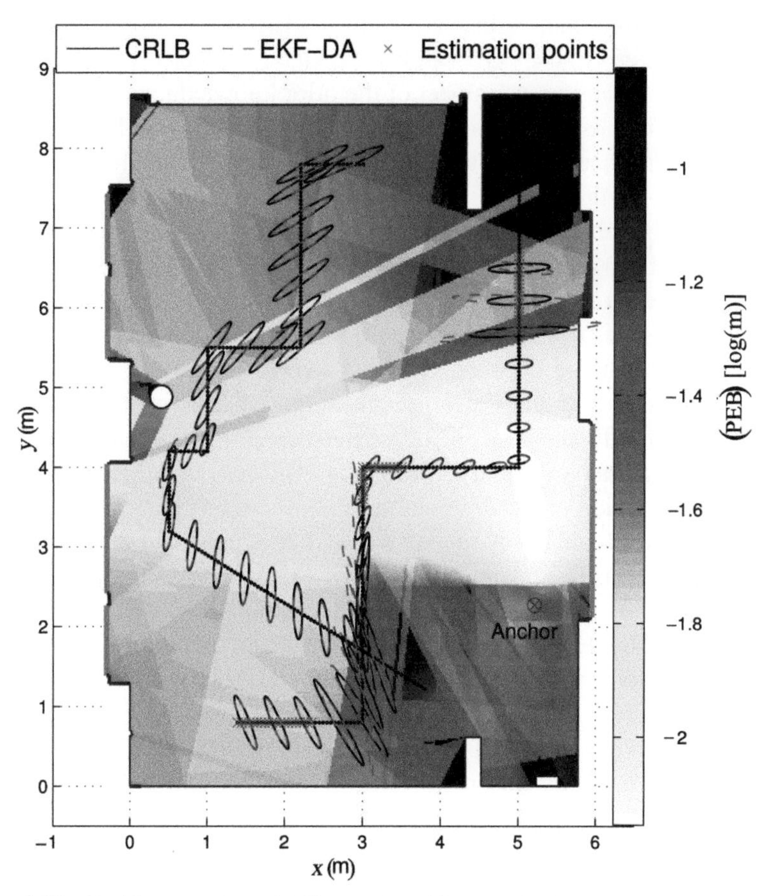

Figure 3.20 Position error bound (PEB) and tracking results for T_p = 0.5 ns, f_c = 7 GHz, and a single fixed anchor. The PEB Equation (3.7) has been computed from estimated SINRs Equation (3.8); *grey crosses* are the 60 estimation points for the SINRs. *Solid* and *dashed ellipses* denote the Cramér Rao lower bound (CRLB) and error covariance of the extended Kalman filter (EKF), respectively, at several points along the trajectories. All ellipses are enlarged by a factor of 20 for better visibility.

grey crosses. Also the position error bound (PEB) is shown over the entire floor plan. It has been computed using the estimated SINRs, providing a prediction of the spatial distribution of the achievable performance. We consider these maps as an indicator for the robustness of the localisation system.

A real-time implementation of this multipath-assisted tracking method has been described in [MLLW14] to demonstrate the practicality of this approach. This work presents an in-depth analysis of position-related information estimation and agent tracking for different environments. Therein, the estimated SINRs of the most dominant MPCs are analysed for two sample rooms and for centre frequencies of $f_c = 7$ GHz and $f_c = 8$ GHz. The performance of the EKF is analysed, evaluating the impact of the prior knowledge of the SINRs. A significant gain results from this additional model knowledge. In Meissner et al. [MLW15] it has been shown that the SINRs can be learned online during the tracking of the agent.

In contrast to this *tracking* technique, a multipath-assisted indoor *positioning* algorithm has been presented in Leitinger et al. [LFMW14], using the ML principle. The maximum of the highly multi-modal likelihood function has to be found, which derives from the channel model introduced in Equation (3.6). In order to cope with the multi-modality, a hybrid probabilistic–heuristic approach is used that combines a sequential importance re-sampling (SIR) particle filter (PF) with the concept of particle swarm global optimisation (PSO). The performance of the algorithm was evaluated using real channel measurements for which an accuracy better than 3 cm has been obtained in 90% of the estimates (using only one active anchor, a pulse duration of $T_p =$ 0.5 ns and $f_c = 7$ GHz). The results have shown that the knowledge of diffuse multipath significantly increases the performance of the positioning algorithm (see Figure 4 in Leitinger et al. [LFMW14]).

The above papers have assumed prior knowledge of a floor plan to compute the positions of the VAs, representing the geometry. In Leitinger et al. [LMLW15], a simultaneous localisation and mapping (SLAM) approach has been described that avoids this prior knowledge. The agent position and the VA positions are jointly estimated in a measurement-per-measurement manner. Only prior knowledge of the anchor positions is assumed. Here, the SINRs, the position-related information of the individual MPCs, are again used for estimating the agent position. Their estimation is also achieved during the position tracking. The position error performance virtually matches the performance of the method that assumes a known floor plan (see Figure 3 in Leitinger et al. [LMLW15]).

The paper by Kuang et al. [KAT13] present a structure-from-motion algorithm to simultaneously estimate the receiver motion, the transmitter position, and the positions of the corresponding VSs. The outcome of the structure-from-motion algorithm then serves as initial input for a non-linear least square algorithm to refine the estimated receiver, physical and virtual source (VS) positions. In comparison, a random sample consensus (RANSAC) approach for automatic matching of data has also been implemented and tested. Using real channel measurements over the whole-FCC bandwidth from 3.1 to 10.6 GHz both approaches were successfully tested resulting in an accuracy in the centimetrer region. Figure 3.21 shows the ground truth motion (red circles) and the estimated receiver positions (blue dots) and the directions to the source positions.

3.3.4.2 Non-UWB multipath-assited indoor tracking

All previous methods in this Section have in common that UWB signals are used for positioning. In Zhu et al. [ZVK$^+$15] the aim is to track user movements with accuracy down to centimeters using standard cellular bandwidths of 20–40 MHz by using the phase information of the MPCs and large antenna arrays to achieve separability. Using an EKF, the phases of

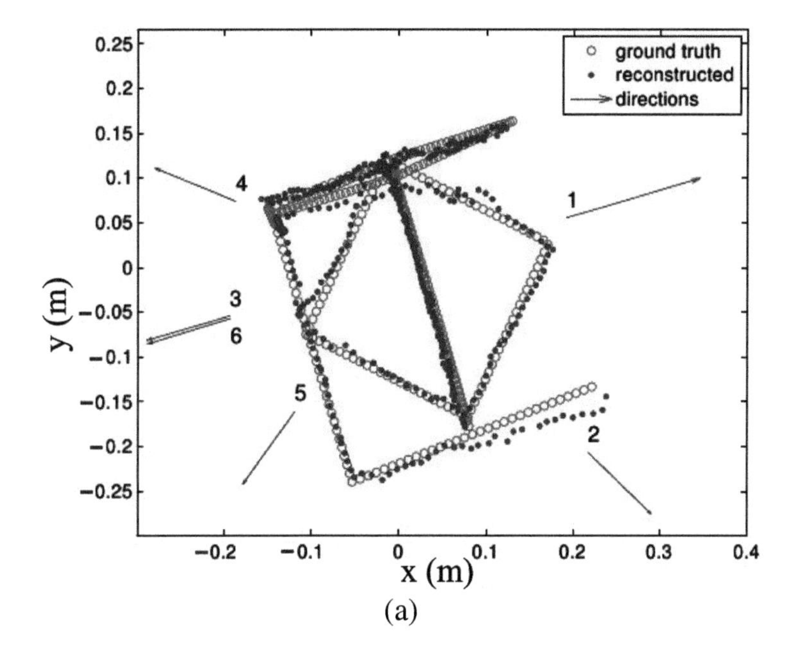

(a)

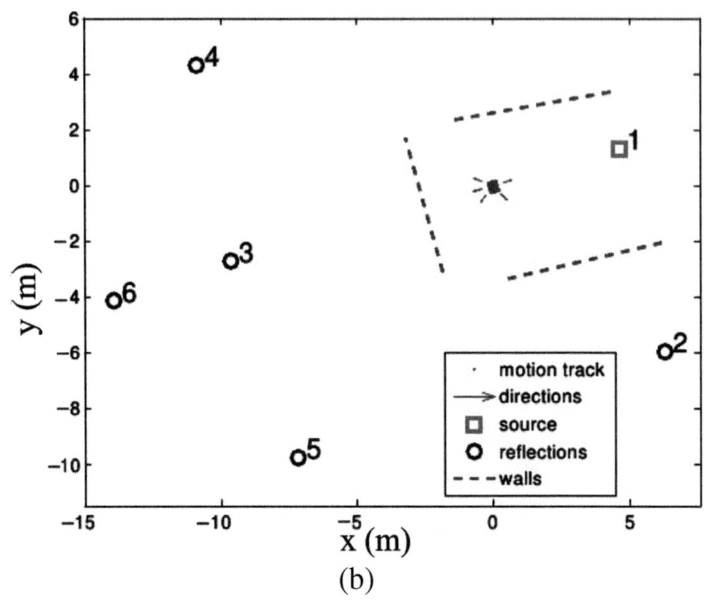

(b)

Figure 3.21 (a) Ground truth motion (*red circles*), the estimated receiver positions (*blue dots*) and the direction to the real and the virtual sources (VSs) positions. (b) Positions of the physical transmitter (red square), VSs (black circles), and hypothesised reflective wall positions based on the relative positions between physical transmitter and reflections.

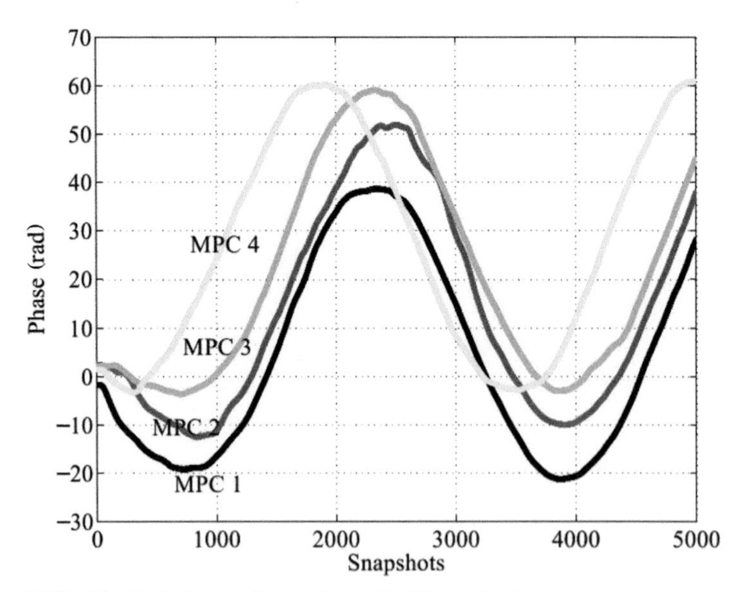

Figure 3.22 Tracked phases of a number of MPCs in circular movement measurements.

MPCs are estimated and tracked. The origins of the MPCs, the VA positions, and the according transmitter position and also the receiver position, are then estimated by translating the phase information into a propagation distance and using the structure-from-motion approach [KAT13]. By using the continuous behavior of the phases of the MPCs, a very high level of positioning accuracy is reached. In Figure 3.22 the tracked phases of a number of MPCs are shown for an agent (receiver) that moves on a circular trajectory (see Figure 1 in Zhu et al. [ZVK+15]).

4

Vehicular Communication Environments

Chapter Editors: Erik G. Ström and Levent Ekiz,
Section Editors: Taimoor Abbas, Ruisi He, Sławomir J. Ambroziak,
Veronika Shivaldova and Jörg Nuckelt

Communication to and between road vehicles (cars, truck, buses, trains, etc.) are of growing interest. This is partly due to the attractive services that cooperative intelligent transport systems (C-ITSs) provides, mainly in the areas of traffic safety and traffic efficiency. An enabler for C-ITS is wireless vehicle-to-vehicle (V2V) and vehicle-to-infrastructure (V2I) communication, collectively referred to as vehicle-to-X (V2X) communication. Another driver is the advent of *moving networks* in the context of 5th generation (5G) systems. Moving networks includes the use of vehicles as mobile or nomadic base station (BS), with the purpose of providing connectivity to both vehicle passengers and to users outside the vehicle.

In this chapter, we will discuss key issues in V2X communication: propagation, antennas, and physical (PHY) and medium access control (MAC) layer algorithms. Measurements, characterisation, and modelling of radio channel are reported for road, railway, and special environments in Sections 4.1.1, 4.1.2, and 4.1.3, respectively. Antennas are discussed in Section 4.2, while PHY and MAC layers are treated in Section 4.3.

4.1 Radio Channel Measurements, Characterisation, and Modelling

4.1.1 Road Environments

4.1.1.1 Measurement, characterisation and modelling

Vehicular connectivity will be rolled out via several radio access technologies; Third generation (3G)/Fourth generation (4G) long-term evolution (LTE), wireless local area network (WLAN), Bluetooth, 802.11p (ITS-G5/wireless access in vehicular environments (WAVE)), and possibly 5G in the future. Vehicular environments being highly dynamic pose new challenges for each

Cooperative Radio Communications for Green Smart Environments, 121–150.

of these technologies. Among all challenges faced, the radio channel is one of the most critical ones.

Understanding the properties of the radio channel is extremely important for robust and efficient system design. Typically, three main approaches are used to characterise the properties of the radio channel; deterministic, stochastic, and measurement-based approaches [Abb14]. Measurement-based characterisation and modelling for vehicular channels is of particular interest in this section.

A number of measurement campaigns have been performed and results have been reported to characterise the underlying propagation mechanisms, dynamics of the channel, and the antenna/environment interactions in scenarios such as line-of-sight (LoS), non-line-of-sight (NLoS) due to buildings at the street crossings and obstructed line-of-sight (OLoS) due to shadowing by other vehicles.

Double directional channel sounding is one of the techniques which are used to analyse the antenna/environment interaction. In Renaudin and Oestges and Martin Käske et al. [RO14, MKT14], double directional multiple-input multiple-output (MIMO) channel sounding campaigns were conducted on the campus of Aalto University, Finland and TU Ilmenau, Germany. The aim of the measurements was to separately investigate the impact of moving scatterers (e.g., other vehicles) from the inherent dynamic of the observed channel caused by the movement of transmitter (TX) and receiver (RX), which allows us to model both effects independently in the channel model. In Renaudin and Oestges [RO14] similar, but more extensive, measurements were performed in urban, sub-urban/campus, highway, and underground parking environments. Considered scenarios were, vehicles traveling in the same direction, vehicles traveling in the opposite direction, and vehicles approaching intersections from perpendicular directions with/without LoS obstruction by buildings.

Vehicles approaching intersections at an urban crossroad often experience NLoS conditions due to the buildings at the corners. Propagation loss in such a situation can be very high at 5.9 GHz, which can greatly impact the link reliability for V2V communications. In Schack et al. [SNG$^+$11], this issue is discussed and 3D ray-optical path loss predictions are compared with narrowband measurements at 5 GHz. It is found that the maximum communication range is quite low in the urban areas under NLoS conditions as compared to the LoS conditions due to worse signal-to-noise-ratio (SNR) levels at the RX.

Similarly, Paschalidis et al. [PMK$^+$11] address a measurement campaign focused on V2V communication at crossroads for two selected scenarios shown in Figure 4.1, e.g., vehicles traveling on perpendicular streets and non-traffic-light-regulated left turn.

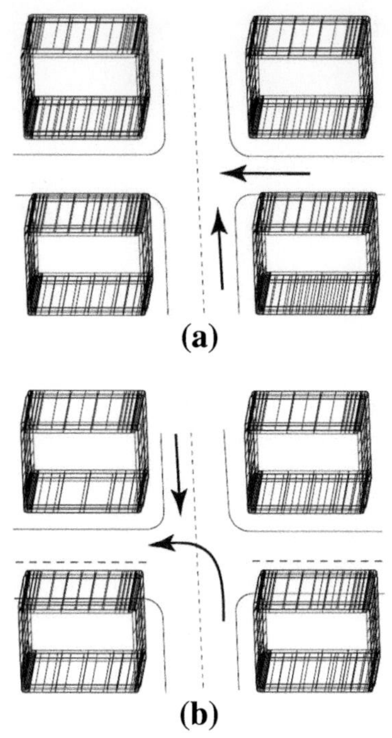

Figure 4.1 Illustration of measurement scenarios. (a) *Cross traffic* on a major-minor intersection, and (b) *Left turn* on a major-major intersection.

The measurements were performed in Berlin with the Heinrich Hertz Institute (HHI)-Channel-Sounder, a true 2×2 MIMO wideband radio channel sounder operating at a carrier frequency of 5.7 GHz. The antennas were positioned at the side doors, to achieve a high spatial separation. The focus of the investigation is set on the received power at each antenna.

It is found that the received power level fluctuates strongly and largely independently for each channel (Figure 4.2). This effect is more pronounced when the vehicles have larger distances from each other. The paper stresses that different effects such as shadowing or preferred angles of arrival/departure seem to be pronounced in different scenarios. As a consequence, the choice of an optimal MIMO technique for such a setup can not be universal.

In road environments, LoS between communicating vehicles can be blocked due to the presence of other vehicles. It is observed that the communication link in the crowded traffic scenarios experience severe fading, which is typically due to the obstruction by taller vehicles [BVF+11a]. In Abbas et al. [ATK12] a measurement-based analysis is performed to analyse the impact of LoS obstruction. The measurements were performed in varying traffic

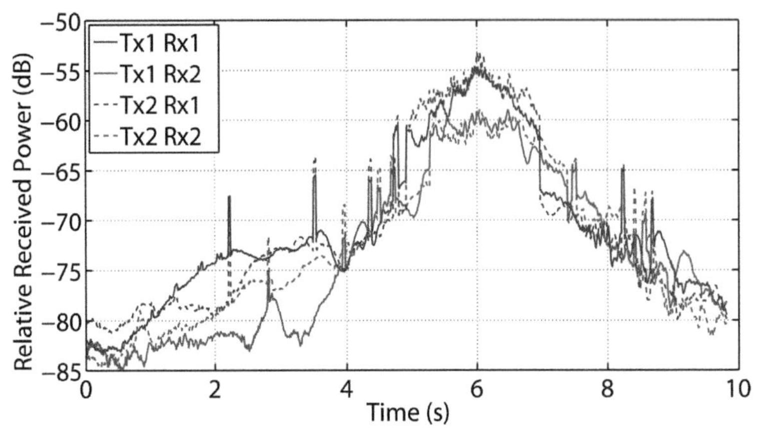

Figure 4.2 Relative received power for the four MIMO channels for the left turn scenario.

conditions from light to heavy traffic in the highway and urban scenarios. From the measurement results, it was observed that LoS obstruction by vehicles induce on average an additional loss of about 10 dB in the received power.

To validate the findings in Abbas et al. [ATK12], a measurement campaign was performed where a truck was used as an obstacle in a controlled way. Wideband 1×6 single-input multiple-output (SIMO) measurements were performed with the RUSK LUND channel sounder in rural, highway, and urban scenarios. In Vlastaras et al. [VAN$^+$14], the loss due to the shadowing of the truck (in dB) was observed to be Gaussian distributed with mean around 12 and 13 dB for the antenna with the best placement in rural and highway scenarios, respectively, confirming previous results (Figures 4.3 and 4.4). Parameters of path loss model, large-scale fading distribution, and auto-correlation were also provided. In Nilsson et al. [NVA$^+$15], multilink shadowing effects in measured V2V channels were investigated and it was found that the multihop techniques similar to the ones implemented in the

Figure 4.3 The two cars at TX/RX and the truck as obstacle used in the measurements. The total truck length was 980 cm with a container width of 260 cm and height of 400 cm from the ground.

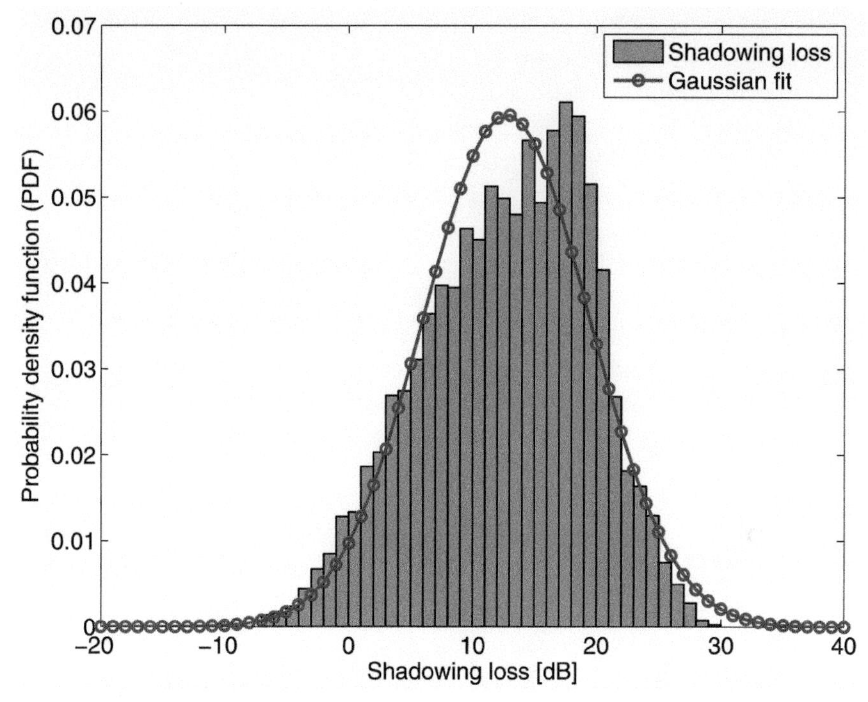

Figure 4.4 Shadowing loss due to the truck for antenna 1 in the highway scenario.

ETSI ITS-G5 standard can overcome the shadowing losses by relaying the information via another car.

Mobile terminals such as smartphones and tablet computers utilising direct high-data rate applications and services in cellular networks are often operated from inside vehicles such as cars and buses. The antenna systems in smartphones utilising multiple antennas for techniques such as spatial diversity, beamforming, and MIMO, also the directional properties of the channel, including the immediate environment, is of great importance for accurate performance evaluations. How this channel modelling could be done and what specific impact the car environment may have were investigated in Harrysson et al. [HMHT13] for frequencies around 2.6 GHz.

The investigation was based on channel measurements in two different static scenarios with an upper body phantom and a four-antenna handset mock-up located outside and inside a common type family car (station wagon), in the LTE band at 2.6 GHz for a synthetic 4×4 MIMO arrangement.

It was found from the measurements in the investigation that the penetration loss due of the test vehicle vary on average between 1.6 and 7.9 dB depending on the outer channel, i.e., the scenario and orientation of the car relative the BS, as shown in Figure 4.5.

From the measurements with a 128-port cylindrical array antenna both with the car absent and with the antenna placed inside the car, directional channel parameter estimation was performed to form the propagation channel model as a part of a composite channel model.

Combined with measured antenna patterns of the handset-plus-user it was found that this channel model produces channel properties such as path loss and channel statistics very similar to what was found by the direct measurements with the handset-plus-user in the same channel (Figure 4.6).

Similar analysis to find the vehicle penetration loss (VPL) was performed in Virk et al. [VHK^{+}14] based on empirical as well as numerical evaluations.

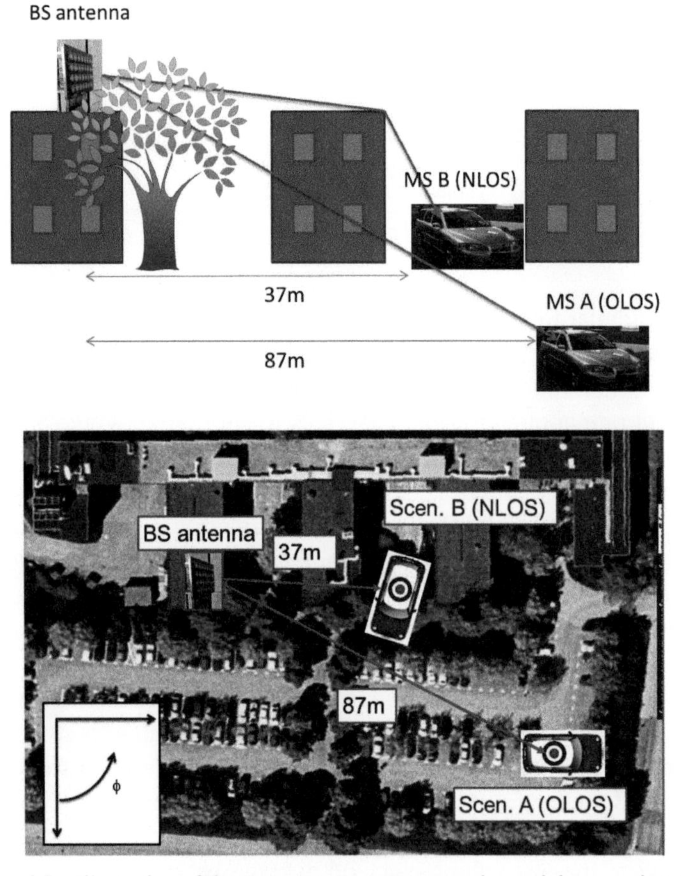

Figure 4.5 Illustration of the two measurements scenarios and the car orientations.

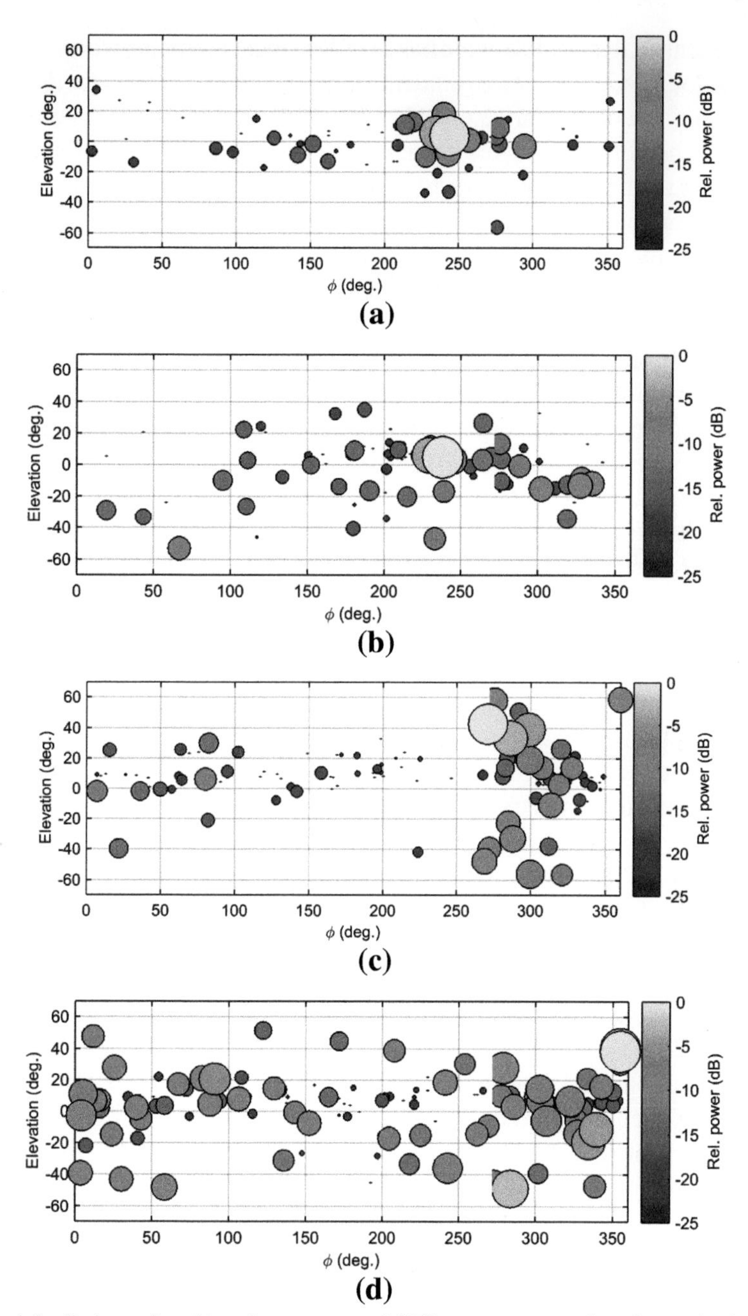

Figure 4.6 Estimated multi-path component (MPC) power versus elevation and azimuth at the MS (DoA) in the different scenarios: (a) A outside, (b) A in car, (c) B outside, and (d) B in car. The colour of the circles represent the relative total power of the MPCs.

For the empirical analysis, ultra-wide band (UWB) measurements were carried out in an indoor industrial environment at an isolated storage facility in Helsinki for the frequency range of 0.6–6.0 GHz. A regular-sized hatchback car was used, where windows were coated either by aluminium metal foil or by commercially available automotive window films. Typically, measurements are time consuming and require a lot of effort. Numerical analysis can thus be an alternative to measurements, given that adequate computational resources are available. Therefore, VPL was also investigated using an finite-difference time-domain (FDTD) method carried out at the discrete frequencies 900 MHz and 1.2 GHz. From the measurements, the mean VPL evaluated for no-coating, window film, and aluminium metal foil case was 3, 6.6, and 20.7 dB, respectively, for the complete observed frequency range. Similarly, simulated VPL values at 900 MHz and 1.2 GHz were 14 and 16.5 dB by considering the aluminium metal foil coating on the car windows. The deviation in the observed results could be due to the inaccurate description of the material properties used in car simulation model. It is one of the known limitations in the numerical methods.

In Shemshaki and Mecklenbräuker [SM12] path loss models are evaluated for future site planning of RSUs using LTE technology at 800 MHz and 2.4 GHz. Two types of vehicular environments were considered, V2V where scenario assumptions are car-to-car (C2C), truck-to-truck and car-to-truck, and V2I where scenario assumptions are car-to-roadside unit (RSU) and truck-to-RSU. It is found that for satisfying coverage and capacity requirements using LTE technology, the empirical models for median path loss, i.e., original Hata model and the extended Hata model for future V2I links, and the extended Hata-SRD model for V2V links, are applicable to RSU site planning.

Measurements are, of course, always subject to noise and interference. Methods for mitigate the effects of noise and interference when estimating path loss and fading parameters were proposed in Abbas et al. [AGT14]. Two new path loss estimation techniques for censored data based on expectation maximisation (EM) and maximum likelihood (ML) approaches were developed. The accuracy of the proposed methods is tested against censored synthetic data, which has shown promising results by providing accurate estimates.

4.1.1.2 Stationary issues

The propagation characteristics of V2V channels are quite different from those of traditional cellular channels. In traditional channel modelling, the wide-sense stationary-uncorrelated scattering (WSS-US) assumption has been widely used to describe random linear time-varying cellular channels. In V2V channels, due to the dynamically changing environment, the statistics of channel changes over moderate time scales (non-WSS); due to the several taps interacting with one-and-the same object, such as buildings on the road side, the taps at different delays show correlated fading property (non-US). Therefore, the WSS-US assumption is not accurate for V2V radio communication channels for large time durations or wide bandwidths.

In Li et al. [LAC+13] a non-WSS-US channel model is proposed for V2V communication systems. The proposed channel model is based on the tapped-delay line (TDL) structure and considers the correlation between taps both in amplitude and phase.

In Laura Bernadó [LB11], a detailed analysis of the time-varying channel parameters, RMS delay and Doppler spreads, and stationarity time are presented. The analysis is based on the data collected in a vehicular radio channel measurement campaign, named DRIVEWAY09 [Abb14]. It is shown that these time-varying channel parameters are statistically distributed and can be modelled as a bimodal Gaussian mixture distribution.

The non-stationarity of V2V radio channels is analysed in He et al. [HRK+15] using three metrics: the correlation matrix distance (CMD), the wideband spectral divergence (SD), and the shadow fading correlation. The analysis is based on measurements carried out at 5.3 GHz using a 30×4 MIMO system in suburban, urban, and underground parking areas. It is found that the stationarity distance ranges from 3 to 80 m in different V2V scenarios, and is strongly affected by several factors such as the existence of a LoS, the speed of cars, and the antenna array size and configuration. Based on the comparison of the equivalent stationarity distances estimated by the three metrics, it is found that: (i) a large electrical array aperture improves the angular resolution and thus results in a smaller estimated stationarity distance; (ii) strong LoS and a small difference of speed between a vehicle of interest and surrounding vehicles lead to large stationarity distances; and (iii) environments with a large number of scatterers exhibit smaller stationarity distances (Figure 4.7).

To statistically model the time-variant V2V channels, the dynamic MPCs are characterised in He et al. [HRK+14] based on suburban measurements conducted at 5.3 GHz. The CMD is used to determine the size of local WSS region. Within each WSS time window, MPCs are extracted using wideband spatial spectrum of Bartlett beamformer. A MPC distance-based tracking algorithm is used to identify the birth and death of MPCs over different WSS regions, and the lifetime of MPC is modelled with a truncated Gaussian distribution. The MPC characterisation considers both angular and delay domain properties as well as the dynamic evolution of MPCs over

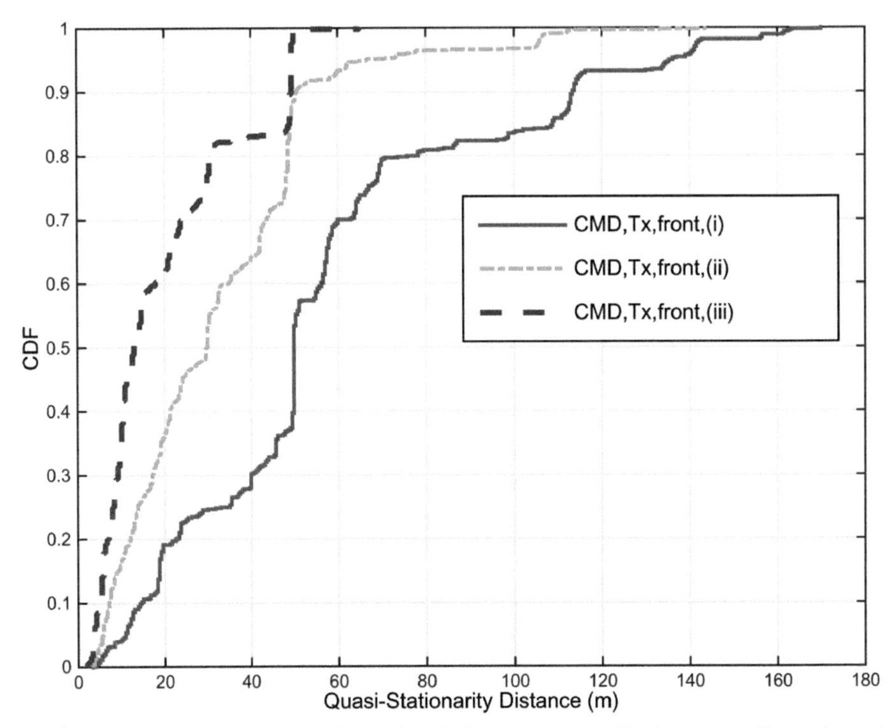

Figure 4.7 Results show that a large electrical array aperture leads to a smaller estimated stationarity distance.

different WSS regions. The results are useful for the scatterer modelling in geometry-based stochastic channel modelling.

Another approach to statistically evaluate the lifetime of multipath components for the V2V channel is presented in Paschalidis et al. [PMK+12]. A new identification and tracking algorithm, which follows a geometrical approach, is introduced that can identify the strong MPCs. The algorithm identifies the MPCs in the delay domain and tracks them in the time domain. Moreover, discrete and diffuse scatterers can also be differentiated during the tracking process.

4.1.1.3 Modelling, and simulations

Realistic models of the propagation channel are always beneficial especially if they are capable to reproduce most of the channel conditions while keeping the amount of computational efforts as low as possible. Typically, measurement-based channel models give a very realistic view of the propagation environment, but a major drawback with the measurement-based models is that they are scenario-specific and require a lot of time and efforts.

An alternative approach to that is the deterministic approach such as ray-tracing that aims to provide accurate or meaningful approximations of the channel for a specific propagation environment, given that an accurate description of the environment is available. Typically, the computational complexity of ray-tracing models is very high, but can be reduced at the cost of model precision.

For the ray-tracing models, there are a number of ways to develop a detailed and accurate topographic database including all objects on the surface depending upon the type of channel under investigation. One way is to use public-domain OpenStreetMap (OSM) database for the purpose of outdoors deterministic vehicular channel modelling. In Nuckelt et al. [NRJK13], a guideline to make use of OSM data is presented. It is also investigated that whether or not the available building data is capable to provide a satisfying accuracy required for an adequate channel modelling.

Furthermore, in Nuckelt et al. [NAT+13], a ray-tracing approach is used for V2V channel characterisation in an urban intersection scenario. For the simulations, relevant building data has been obtained from the OSM database as described in Nuckelt et al. [NRJK13] and converted into ray-tracing format. Additional objects such as roadsigns, street lamps, parked cars, and pedestrians have been identified and added manually. For accuracy analysis, the channel simulation results obtained from the ray-tracing have been compared with measurement results. In Nuckelt et al. [NAT+13] only two metrics, channel gain and power delay profile (PDP), were used for analysis and comparison. However, later on, in Abbas et al. [ANK+14] a more detailed analysis of the same scenario is presented in terms of PDP, path loss, delay and Doppler spreads, eigenvalues, antenna correlations, and diversity gains.

It is concluded from the analysis that ray-tracing simulations seem to have a reasonable accuracy in the sense that they are able to capture most of the specular reflections in the channel, i.e., LoS or first order only as shown in Figure 4.8(a,b). However, it is unable to capture most of higher order reflections and diffuse scattering, as these were not implemented in the ray-tracer. This analysis has revealed some of the limitations associated to the ray tracing model, i.e., to achieve good accuracy, it is required that the higher order reflections and diffuse scattering is implemented in the ray-tracing models.

For the ray-tracing simulation in Abbas et al. [ANK+14] 3D antenna patterns of the true antennas were used, measured in anechoic chambers after mounting on the car roof. Such realistic antennas patterns are very important

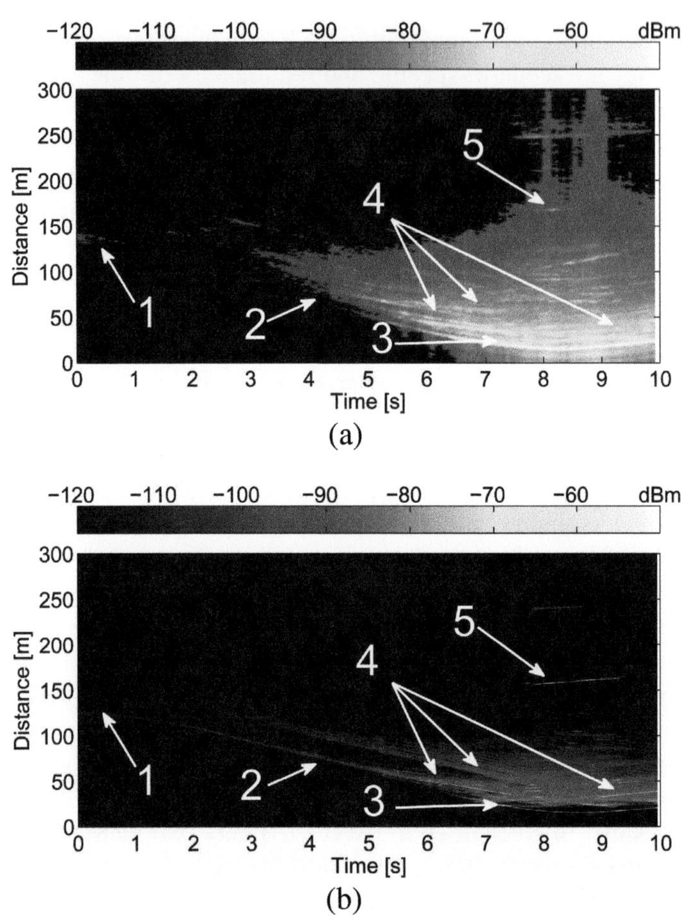

Figure 4.8 Power delay profile obtained from (a) the channel measurement data, and (b) the ray-tracing channel model. The physical interpretation of the multipath components (A)–(B) is detailed in [ANK+14].

for realistic simulations. This is also stated in Eibert et al. [MSLM13], in which a new method is described to increase the accuracy of ray-tracing algorithms by using an approximation on vehicle antennas. It is shown that antenna systems mounted on conducting surfaces, such as car roofs, must not be evaluated as isolated but together with their surroundings. The used approximations in this method are only true if the far-field condition is kept. When the far-field condition is not fulfilled, the accuracy of this simulation method decreases strongly. With a maximum dimension size of approximately 4 m, the far-field condition at 5.9 GHz is held in for distances larger than approximately 620 m.

To achieve a high accuracy also for smaller distances, the vehicle is subdivided into antennas with smaller apertures, called sub-TXs.

To achieve realistic radio channel models we have so far considered either complex ray-tracing approaches or time consuming measurement-based approaches. Another simpler, yet realistic, channel modelling approach is the stochastic approach. With this approach, instead of site-specific realisations, the statistics of the channel is modelled. More realism can be achieved by combining a stochastic channel model and simple geometrical aspects, so-called GSCMs. In Große [Gro13], a hybrid model applicable to V2V and V2I systems has been developed based on wireless world initiative new radio (WINNER) channel model. The model relies on a flexible and scalable layered structure in which the modelling task is separated into quasi-static and dynamic parts. The novel hybrid model ensures a flexible geometry-based stochastic channel model (GSCM) approach in which user-defined randomness can be included.

Another stochastic modelling approach is presented in Shivaldova et al. [SWM14a] where a V2I performance model at 5.9 GHz band is developed based on extensive measurements data. For analysis, the entire communication range is divided into regions of high, intermediate, and unreliable link quality, based on quantised model parameters of the Gilbert model. SNR performance is then analysed separately for each of these regions. Based on the analysis, a V2I performance model is developed, which is used to generate V2I SNR traces and associated error patterns. The accuracy of performance model is evaluated by means of comparison of model generated and measurement-based parameters. See Section 4.3.1 for more details on this model.

4.1.2 Railway Environment

4.1.2.1 Propagation scenarios

Radio propagation depends on topographical and electromagnetic features of the environment. A propagation scene partitioning for railways was first proposed in Ai et al. [AHZ+12]. We further derive seven typical railway-specific scenarios: urban, suburban, rural, viaduct, cutting, station, and river. Detailed descriptions of above railway scenarios can be found in He et al. [HZAG15].

4.1.2.2 Measurement campaign

Both narrowband and wideband measurements were conducted

1. Narrowband measurements were carried out along the "Zhengzhou-Xian" high-speed railway (HSR) lines of China, using a operational GSM-R system at 930 MHz. Details of the measurement system can

be found in He et al. [HZA+13b, HZAD11a, HZA+13a, HZAD11b]. To record sufficient power data, more than 6000 HSR cells in the above seven environments where measured. The train had a speed of approximately 300 km/h.

2. The wideband measurements were carried along the railway line "LGV Atlantique" in a rural environment. The campaign combines 4×2 MIMO and carrier aggregation between the 2.6 GHz band (two carriers with 10 and 20 MHz bandwidth) and the 800 MHz band (one carrier with 5 MHz bandwidth). The train passed the area with a speed of approximately 300 km/h.

4.1.2.3 Characterisation and modelling

Path loss

A statistical path loss model at 930 MHz is proposed using the extensive measurement data. The model is based on Hata's formula, expressed as [HZAG15]:

$$\begin{aligned} \mathrm{PL(dB)} = {} & \Delta_1 + 74.52 + 26.16 \log_{10}(f) \\ & - 13.82 \log_{10}(h_\mathrm{b}) - 3.2[\log_{10}(11.75 h_\mathrm{m})]^2 \\ & + [44.9 - 6.55 \log_{10}(h_\mathrm{b}) + \Delta_2] \log_{10}(d) \end{aligned} \tag{4.1}$$

where f is the carrier frequency in MHz (i.e., $f = 930$), h_b and h_m are the BS effective antenna height and the vehicular station antenna height in metres, d is the TX–RX separation distance in kilometres, and Δ_1 and Δ_2 are correction factors. It is found that Δ_i can be modelled as a function of h_b (Table 4.1). The estimated correlation factors for each environment based on a regression fit are found in Table 4.1. The proposed model is validated by measurements [HZA+14].

A heuristic semi-deterministic path loss model is also proposed for the 930 MHz narrowband channel. The following main propagation mechanisms affecting path loss are considered: (i) Free-space propagation and reflection from the track. (ii) Diffraction: This mechanism mainly occurs in the case of cutting. Finally, the Deygout model is chosen. The proposed model is validated by the measurements. Details of the semi-deterministic path loss model can be found in Guan et al. [GZAK13b].

Then, a heuristic deterministic model in railway cutting scenario is proposed. The deterministic model approach consists of a 3D ray-optical channel model using a vector database and a multi-edge diffraction model based on a raster database. The ray-optical approach takes into account reflections

Table 4.1 Path loss model (in dB) and shadow fading models for HSR Environments, from He et al. [HZAG15, HZAO15]

Environment	Correction Factor	σ (dB)	d_{cor}(m)
Urban	$\Delta_1 = -20.47$	3.19	57.1
	$\Delta_2 = -1.82$		
Suburban	$\Delta_1 = 5.74\log_{10}(h_b) - 30.42$	3.33	112.5
	$\Delta_2 = -6.72$		
Rural	$\Delta_1 = 6.43\log_{10}(h_b) - 30.44$	2.85	114.8
	$\Delta_2 = -6.71$		
Viaduct	$\Delta_1 = -21.42$	2.73	115.4
	$\Delta_2 = -9.62$		
Cutting	$\Delta_1 = -18.78$	3.63	88.8
	$\Delta_2 = 51.34\log_{10}(h_b) - 78.99$		
Station	$\Delta_1 = 34.29\log_{10}(h_b) - 70.75$	2.77	101.2
	$\Delta_2 = -8.86$		
River	$\Delta_1 = 8.79\log_{10}(h_b) - 33.99$	3.09	114.6
	$\Delta_2 = -2.93$		

and diffuse scattering based on the image method. All the railway structures are modelled in the simulator. By comparison with the measured data, the presented model has high accuracy. Details of the deterministic path loss model can be found in Guan et al. [GZAK13a].

Finally, the semi-closed obstacles (SCOs), such as crossing bridges, train stations, etc., densely appear and compose challenging scenarios for propagation prediction are analysed and a hybrid model for propagation in such composite scenarios is presented. The validation shows that the proposed model accurately predicts the propagation loss. Details of the model can be found in Guan et al. [GZAK14].

Shadow fading

Shadow fading is analysed using the narrowband measurements at 930 MHz. The measurements suggest that a zero-mean Gaussian distribution fits the shadowing data (in dB). The mean value of the standard deviation σ of shadowing is presented in Table 4.1. The auto-correlation coefficient $\hat{p}_{auto}$ of shadow fading components is found to follow an exponential decay function [Gud91]. The mean value of shadowing decorrelation distance d_{cor}, which is defined to be the distance at which the correlation drops to $1/e$ [HZAZ14], is summarised in Table 4.1. Moreover, the cross-correlation of shadow fading between two neighbouring BSs is found to depend upon the BS height and the tilt angle of the antenna [HZAO14], and a heuristic model is proposed. Details can be found in He et al. [HZAO15].

Delay-doppler spectrum

The delay-Doppler spectrum analysis is based on the wideband measurements. In a post-processing step power delay profiles and Doppler spectra and their evolution over time were derived. It is found that the near scatterers to the left and the right of the railway line are the poles of the gantries that support the railway electrification system. They are about 30 m apart and act as reflectors. The difference in the lengths of the LoS path and the first reflected path is smaller than the temporal resolution of the measurement and thus both rays appear to have the same delay. There are, however, also some reflections coming from gantries further away and thus show a longer delay in the Doppler-delay power spectrum. More details can be found in Kaltenberger et al. [KBA+15].

4.1.3 Special Environments

A few papers within COST IC1004 investigated propagation conditions in two kinds of special environments, namely maritime container terminal and in-vehicle environment. In this subsection, the main results of these investigations will be presented.

An extensive measurement campaign in the first of these environments (for frequency range from 500 MHz up to 4 GHz) have been carried out in the deepwater container terminal in Gdansk, Poland. The measurements as well as the investigated environment have been described in Ambroziak [AK14]. Since the Walfisch–Ikegami model [CK99] describes the environment quite similar to the container terminal, it was often used to predict basic transmission loss in this environment. In this model has been evaluated and then modified by tuning its coefficients. Evaluation revealed that the non-modified WI model cannot be used for path loss estimation in such environment, since the obtained mean error (ME), standard error of estimate (SEE), and coefficient of determination, R^2, 5.3 dB, 10.6 dB, and 0.01 dB, respectively. But its tuned version have achieved much better accuracy: SEE = 5.1 dB and $R^2 = 0.77$.

In Ambroziak and Katulski [AK13a] and [AK13b], an empirical propagation model for mobile radio links in a container terminal environment was proposed. This model, named mobile container terminal (MCT), takes into account the diversity of conditions occurring in different places of the container terminal. The terminal was divided into three sub-areas (LoS area, container area, and off-terminal area), where different propagation mechanisms have a crucial influence on the path loss. A full description of the MCT model with formulas for path loss estimation is presented in Ambroziak and

Katulski [AK13a]. The obtained SEE for the MCT is 4.7 dB. What is more, the obtained value of R^2 is 0.82. These results proves the accuracy and usefulness of the MCT model for path loss modelling in this special environment.

The second considered special environment is the in-vehicle environment. In Blumenstein et al. [BMM+14] wideband radio channel measurements carried out in the intra-vehicle environment were presented and discussed. Channels in the millimetre-wave frequency band (mmW) have been measured in 55–65 GHz (for different antenna placements and occupancy patterns) using open-ended rectangular waveguides. Blumenstein et al. [BMM+14] presented a channel modelling approach based on a decomposition of spatially specific channel impulse response (CIR) into large-scale and small-scale fading. The decomposition is done by the Hodrick–Prescott filter. The small-scale fading was parametrised utilising ML estimates for the parameters of a generalised extreme value distribution (GEV) and the large-scale fading was described by a two-dimensional (2D) polynomial curve. The comparison of simulated results with measurements has shown that for over 92.3% of cases p-value of the two-sample Kolmogorov–Smirnov test is higher than 0.1.

In Vychodil et al. [VBM+14] authors—based on an extensive measurement campaign—compared suitability of two frequency bands, namely UWB (3–11 GHz) and mmW (55–65 GHz) for ranging purposes in the intra-vehicle environment. It was shown that mmW band is more suitable for precise distance measurement, probably due to favourable material properties, therefore enhancing the distinctiveness of the first arrival multipath component. The average error and the standard deviation of error for UWB band equal 7.7 and 9.0 cm, respectively. For mmW band these errors are significantly lower and equal 1.6 cm (average) and 4.4 cm (standard deviation). Additionally, it was found out that the mmW band provides better distance measurement results for LoS as well as NLoS conditions, making it suitable for local positioning system deployment.

4.2 Antennas

The concept of C-ITS relies on high connectivity of the vehicles. In this context the vehicular on-board antennas, used to connect the vehicle to cellular or vehicular *ad hoc* network (VANET) become of particular importance. The main challenge in the development of such antennas is to satisfy both the performance and the esthetic constraints.

To evaluate the antenna performance in realistic conditions several approached have been proposed. In particular, the authors of Neira et al. [NCC+14] suggest a statistical method based on the radiation pattern of an antenna and a multipath simulation tool. By generating a number of incident waves with certain angle-of-arrival (AoA) distribution, the multipath simulation tool reproduces the realistic multipath propagation for highway, urban, and rural environments. The incident waves generate a voltage at the vehicle antenna ports, as shown in Figure 4.9(a). By analyzing the cumulative

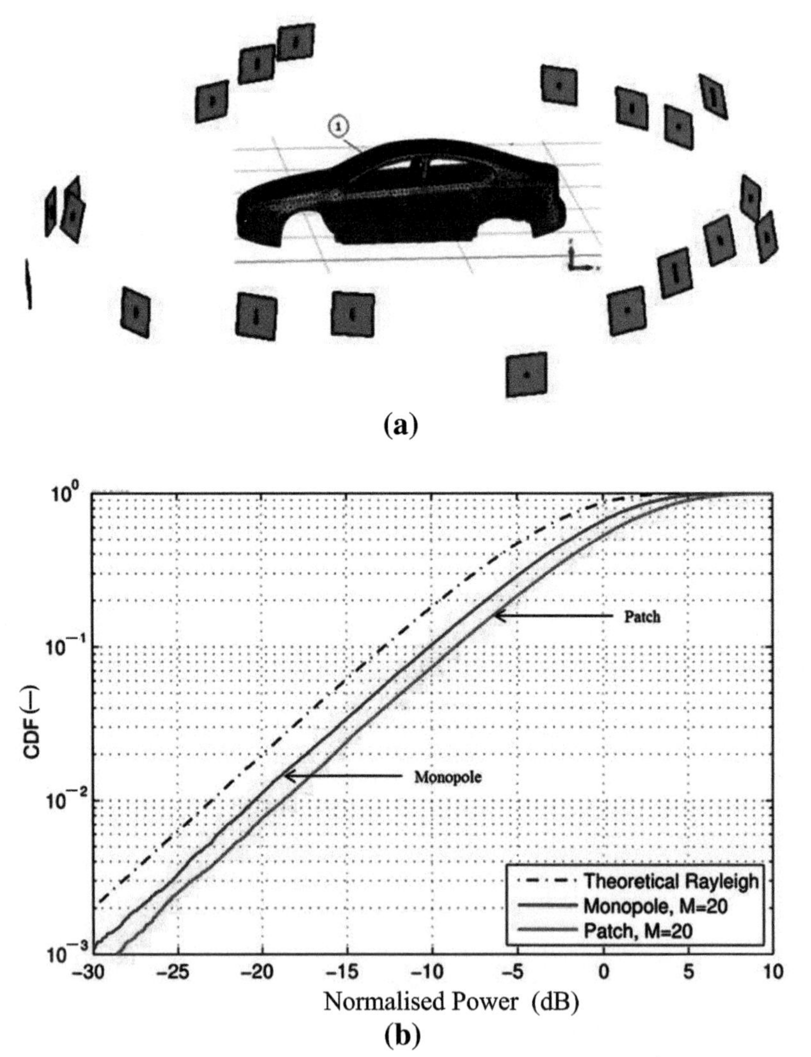

Figure 4.9 Comparison between a patch and a monopole antennas [NCC+14]. Both antennas are mounted near the vehicle windscreen. (a) Illustration of vehicle surrounded by sources of 20 incident waves, and (b) CDFs of the normalised power for 20 vertically polarised incident waves.

distribution function (CDF) of the received voltage the influence of the antenna types and positions on the system performance can be analysed. For example, Figure 4.9(b) shows the performance comparison between a patch antenna and a monopole.

Another efficient way to characterise the performance of vehicular onboard antennas is OTA multi-probe testing. The very first attempt of placing a vehicle (Volvo C30) in an OTA multi-probe test system was performed by the authors of Nilsson et al. [NHA+14] and is shown in Figure 4.10. To understand the uncertainty of this particular testing approach, the authors have performed three experiments including: measurement setup analysis, channel sounder measurements, and coupling measurements. The performed measurements have three significant outcomes: (i) over-the-air (OTA) multi-probe testing setup is capable of simulating a realistic radio environments; (ii) the measurement uncertainty due to a large test object, such as a car, in the OTA multi-probe ring is $\leq$1 dB; the uncertainty due to the coupling of transmit antennas in the multi-probering is negligible, as the disturbance level was below –56 dB. All these results suggest that the OTA multi-probe testing setup is a way forward for an efficient way of characterising the performance of vehicular on-board antenna systems.

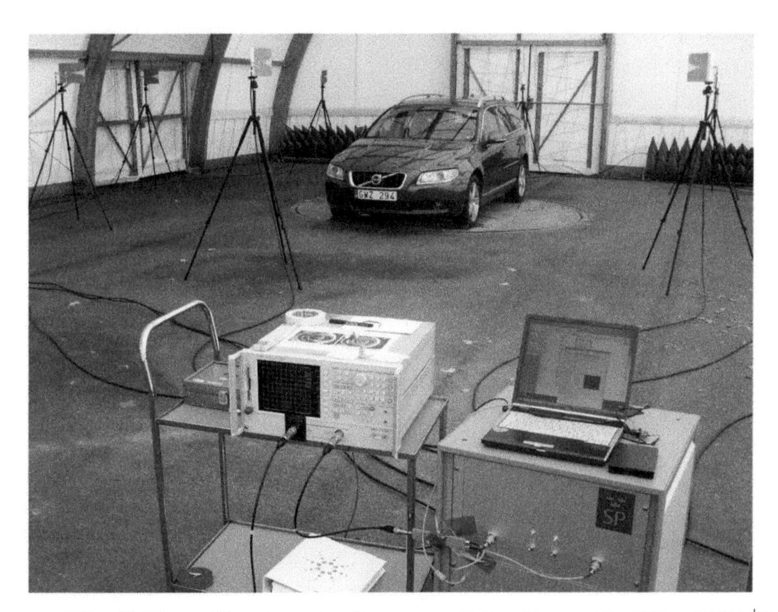

Figure 4.10 (OTA) multi-probe setup for a car at Volvo Cars test facilities [NHA+14].

For evaluation of the on-board antenna system performance, not only the testing methodology is of importance, but also the definition of the key performance indicators. To address the effects of antenna integration in a vehicle, the authors of Posselt et al. [PEK+14] have suggested to use the condition number and channel capacity. The condition number represents the channel correlation properties, while the channel capacity gives an insight on the quality of service. By employing these system level parameters, it is possible to capture the impact of any particular antenna system on the overall system performance.

The above-mentioned system level evaluation parameters were used to assess the performance of a vehicular MIMO antenna for LTE based on a volumetric 3D design, introduced in Posselt et al. [PFE+14]. The 3D antenna design was proposed to improve the utilisation of the available antenna housing space. In this context, the authors suggest to implement antenna functions directly on top of the vehicular antenna housing surface using MID technology. To assess the performance gain of a 3D antenna design relative a 2D antenna design, a reference antenna system has been developed, (Figure 4.11). Real-world LTE measurements have shown that a volumetric MIMO antenna design enables both efficient exploitation of the available integration space and improved system level performance.

For conventional 2D design antennas, integration of the 802.11p antennas along with the cellular antennas in the limited volume of the antenna radome is one of the main challenges. The cellular antennas are positioned in the rear and front regions of the antenna housing and the 802.11p antennas need to be placed in between of them such that the volume and interference constraints are met optimally. In [EPKM14] three possible 802.11p antenna positions are considered: (i) integrated in the cellular antennas, (ii) placed adjacent on one side of the housing, and (iii) placed on different sides of the housing. Evaluating the performance of different antenna prototypes on a ground plane and by performing measurement drives, the authors have concluded that 802.11p antennas placed on different sides of the housing, yield the best overall performance.

However, even under the optimal antenna placement, the antenna pattern will be influenced by the roof-top antenna housing material and its physical dimensions. Indeed, the authors of Ekiz et al. [EPKM13] came to the conclusion that the influence of the antenna housing on the pattern exhibits geometry and frequency dependent behaviour. Simulations and measurements in an anechoic chamber suggest that the performance loss due to dielectric antenna housing can be compensated by modifying the thickness of the radome.

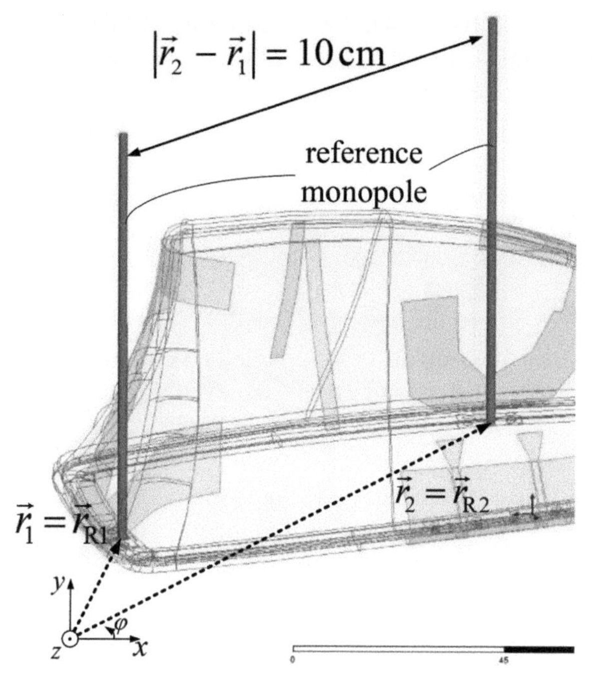

Figure 4.11 CAD model of the 2D reference monopole setup with 3D antenna shown for reference [PFE+14].

Along with the antenna radome, the ground plane materials has significant influence on the antenna pattern. In Artner et al. [ALM15] the performance of a simple monopole-antenna for 5.9 GHz on three different ground plane materials was compared. Following ground plane materials were compared: (i) an aluminium plate, (ii) a plastic plate with a metallised top layer featuring molded interconnect device (MID) design, and (iii) a carbon-fibre composite (CFC) plate consisting of unidirectional filaments with fiber snippets in random alignment on top. For both the MID and CFC plates, the relative efficiency and the maximum gain were reduced as compared to the aluminium ground plane.

Besides the antenna housing design and ground plane material, the antenna performance is significantly influenced by the environment. In this context, Reichardt et al. [RPJZ11] suggest an antenna design methodology for time-variant channels, that eventually leads to capacity maximisation. For antenna design, the authors have modelled five different scenarios in urban, rural and highway environments with a ray-tracer. Based on the radio channel parameters for these key environments, a 2×2 multimode MIMO antenna system including decoupling network has been designed and integrated into a car. The whole structure was measured in an anechoic chamber and the resulting performance increase has confirmed the advantages of the environment-aware antenna system optimisation.

4.3 Communication Systems

4.3.1 PHY Layer

The PHY performance of a wireless communication system mainly determines the robustness and reliability which can be achieved. For an efficient system design, a comprehensive understanding of the PHY data transmission is mandatory.

Basically, two different radio access technologies (RATs) are considered in the context of vehicular communications. On the one hand, *ad hoc* networks based on the IEEE 802.11 standards are indented to exchange data among vehicles or between vehicles and infrastructure (e.g., road-side units). For this purpose, the amendment p, also known as WAVE, of the 802.11 has been approved by the IEEE in the year 2009. On the other hand, 3G/4G cellular networks offer an attractive solution to enable wide-area connectivity of vehicles. In particular, LTE is a promising solution for relative low-latency and robust data transmission in vehicular environments. Both approaches have their specific advantages and disadvantages and a detailed evaluation of the underlying technologies is required.

The PHY of 802.11p suffers from the fact that is has not been designed from the scratch, rather it is a modified version of the 802.11a standard with only minor changes. According to the channel properties in typical vehicular environments the system bandwidth of 802.11p has been scaled to 10 MHz compared to the default 20 MHz option in 802.11a. In [Str13], the use of a larger channel spacing in 802.11p is investigated. The author motivates this option with a reduced channel congestion that affects or even simplifies the design of congestion control algorithms on the MAC layer. Due to the shortened orthogonal frequency division multiplexing (OFDM) symbol length, the main advantages are a reduced channel congestion and an increased robustness against the time variance of the channel.

To carry through robust and efficient design of V2V and V2I communication systems, a deep understanding of the underlying rapid channel propagation behaviour is required. In this context, real world link performance measurements, that are less expensive than channel-sounding experiments and more realistic than simulations, are in demand. A detailed overview of an extensive link performance measurement campaign carried out on the Austrian highways within the project ROADSAFE [ROA] is given in Shivaldova et al. [SPP+11]. During this measurement campaign, various combinations of the system parameters such as data rate and packet length were considered and the influence of system components, such as RSU antenna gain and placement, on the overall link performance were evaluated. For this purpose five RSU transmitters, each equipped with a set of two directional antennas were installed and V2I measurements with a sufficient number of repetitions were carried out.

Some outcomes of the ROADSAFE measurement campaign can be found in Shivaldova et al. [SWM13]. With respect to the system parameters, the authors were searching for a combination of data rate and packet length yielding the largest TP at a constant transmit power of 10 dBm. In this context, there are two possible approaches: (i) to increase the packet length or (ii) to increase the data rate. Increasing the packet length would decrease the total amount of non-payload overhead. However, the quality of the preamble-based channel estimates will be decreased due to the increased transmission time. While when using higher data rates, the time required to transmit a packet will be reduced, resulting in an improved quality of channel estimates. The measurements showed that, in contrast to theoretical considerations, using longer packets is more suitable approach for the TP increase in V2I communications, than increasing the data rate. Another set of measurements presented in Shivaldova et al. [SWM13] emphasises that not the change of the driving direction itself, but rather the mounting position of the RSU antennas with respect to the road geometry has significant impact on the performance. Based on the experimental results the authors suggest to mount RSU antennas in the middle of the highway, rather than on one side to avoid undesirable performance degradation.

Despite the high importance of the link performance measurements, they are rarely conducted due to their high costs and complexity. To make the results of such campaigns more accessible for the research community and to increase the reproducibility of the results, a computationally inexpensive link performance model was introduced in Shivaldova and Mecklenbräuker [SM14c]. This model is an extension of a simple two-state hidden Markov model introduced by Gilbert. As shown in Figure 4.12, Gilbert's model is fully described by only three parameters: the transition probability from the bad state to the good state, P_{BG}, the transition probability from the good state to the bad state, P_{GB}, and the probability of error P_{E} in the bad state.

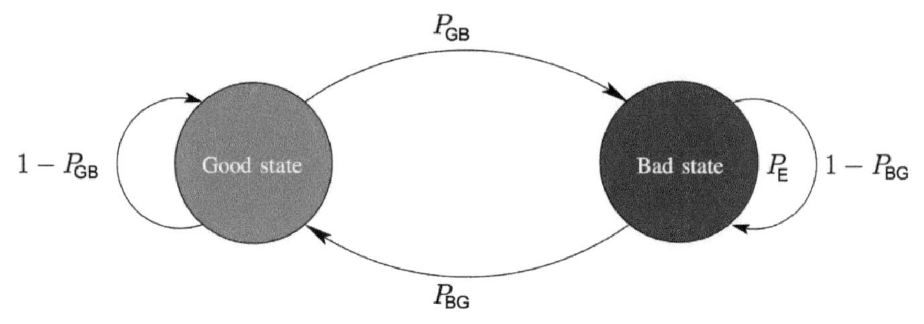

Figure 4.12 Schematic illustration of the Gilbert model [SM14c].

In this model, the good state is error-free. However, since the link performance is strongly distance dependent use of model with just two states would be insufficient and, therefore, the parameters of the proposed range-dependent modified Gilbert model should change depending on the TX–RX distance. To obtain the model parameters the whole coverage range is divided into equally large non-overlapping intervals and parameters are estimated for each of the intervals separately. Obviously, the size of these intervals constitutes a trade-off between the accuracy of the model and its complexity. The authors show that an acceptable level of accuracy can only be achieved by estimating the model parameters for intervals less than 10 m.

In Shivaldova et al. [SWM14b], a further extension of the range-dependent modified Gilbert model was proposed. With the proposed extension realistic SNR traces corresponding to the model generated error patterns can be produced. The SNR model is based on the superposition of three components: SNR trend, large-scale fading, and small-scale fading. The general SNR trend results from dissipation of the power radiated by the TX over the distance. It is constant in the bad state of the model and is exponentially decreasing in the good state. The amplitude and the power of the exponential function depend on the model quality levels. The large-scale fading caused by obstacles between the TX and RX that attenuate signal power through absorption, reflection, scattering, and diffraction is modelled as a correlated Gaussian random process. Finally, constructive and destructive superposition of multipath signal components resulting in a small-scale fading is modelled by a clipped Laplace distribution.

Channel estimation and equalisation significantly affect the PHY layer performance. The pilot pattern in an 802.11p frame is not optimised for channel conditions in high-dynamic scenarios (i.e., with small channel coherence times). Especially for long data packets, accurate channel estimation and equalisation is a challenging task. Nagalapur et al. [NBS14] propose a cross-layer approach, where special data bits are inserted in the higher layers before the packet is transmitted. A modified RX makes use of these a-priori known bits as training data for improved channel estimation. The RX is informed about the inserted training data using an unused bit in the SERVICE field. The algorithm has been evaluated against a channel estimator based on the standard pilot pattern. The novel cross-layer approach outperforms the conventional estimator and performs close to the ideal RX where perfect CSI is available.

A decision-directed channel estimator combined with a 1D Wiener filter operating in frequency direction is discussed in Nuckelt et al. [NSK11]. Different designs of the Wiener filter coefficients assuming either a rectangular-shaped or an exponentially decaying PDP of the channel are analysed with respect to the achieved performance gain. It is found that filter coefficients based on an exponentially decaying PDP provide robust and accurate channel estimates. The SNR loss of the proposed method compared to an ideal RX with perfect channel state information (CSI) is less than 1 dB.

A more general design of a novel estimator for V2V channels is presented in Beygi et al. [BMS15]. The authors propose algorithms to utilize the structure typically found in V2V channels, namely that specular and diffuse multipath components give rise to joint sparsity in the delay-Doppler domain. The fact that different regions in the delay-Doppler plane has different levels of specular and diffuse components is exploited by the channel estimator. In this work, the estimator performance has been evaluated by means of simulations and it is shown the mean squared error (MSE) of the proposed structured estimator is significantly below the MSE of unstructured estimators.

To improve the communication performance on the PHY, system configurations using multiple antennas are also considered and investigated in the context of vehicular communications based on IEEE 802.11p. In Shemshaki and Mecklenbräuker [SM14a, SM14b], an antenna selection algorithm for a 1×2 SIMO system is proposed. The decision which antenna link is chosen and used for the further decoding at the RX is drawn by means of a channel prediction method using either Lagrange interpolation or linear regression. The authors numerically evaluated the achieved benefit using computer-aided simulations and considered the bit error rate (BER) as figure of merit. A significant diversity gain of the SIMO system compared to the default single-input single-output (SISO) configuration was obtained. It is also found that the channel predictor based on linear regression slightly outperforms the Lagrange interpolation-based predictor.

Next generation cellular networks are attractive candidates to extend the possibly limited coverage of IEEE 802.11p and improve the connectivity of vehicles. Currently, several research activities are analyzing hybrid protocol stacks that allow a seamless connectivity using 802.11p and LTE, for instance. A key question is to find out in which kind of vehicular environments the PHY of 802.11p outperforms LTE and vice versa. The work presented in Möller [MNRK14] compares the PHY performance of the 802.11p and LTE in an urban intersection scenario. For this purpose, a virtual scenario with two vehicles and a BS on a roof-top has been created and a ray-optical propagation model has been applied to compute the channel properties of the radio links. Afterwards, an 802.11p and an LTE PHY simulator were utilised to evaluate the downlink packet error rate (PER) against the distance of the receiving vehicle to the centre of the intersection. It has been shown that the LTE PHY layer performance is predominantly limited by the impact of the local interference level. However, if the LoS path between two vehicles is obstructed, LTE outperforms 802.11p even under worst-case intercell interference conditions. Due to the elevated antennas of the LTE BS as well as large transmit antenna gains and a high transmit power, LTE networks typically achieve an almost seamless coverage over large areas while 802.11p links are usually very limited. The results further show the high potential of LTE networks even in time-variant scenarios that are typical for vehicular environments.

Due to the time-dispersive and time-variant nature of vehicular channels, also LTE RXs need to be equipped with channel estimators that efficiently and accurately allow the equalisation of the received signal. When designing an adequate channel estimator, the non-stationarity of V2V channels has to be taken into account. Hofer et al. [HZ14, HXZ15] have addressed this issues and proposed an iterative estimator of non-stationary channels for the LTE downlink. The algorithm utilises a modified subspace representation that was originally developed for contiguous training sequences. The subspace is adapted to delay spread and Doppler spread of the time-variant channel. For this purpose, a hypothesis test is performed that exploits the pilot pattern in the LTE OFDM time-frequency grid. The authors have found that the achievable performance increases if the observation region—in the time or the frequency domain—is increased.

The use of MIMO configuration has also been considered for cellular networks such as LTE or universal mobile telecommunications system (UMTS). In Ekiz et al. [EPL+13], the results of a measurement campaign at 800 MHz, which aims at a system-level evaluation multiple LTE antennas integrated into a vehicle, are presented. A monopole antenna was used as a reference antenna and compared against two integrated automotive qualified antennas. It was found that a similar performance could be achieved if the antenna spacing is about 7 cm. A further increased separation of the antennas, which would cause problems in terms of integrating the antennas into the vehicle, does not provide additional performance gains. In this paper, performance metrics such as capacity and condition numbers of the MIMO channel matrices are presented and discussed. The achieved results give important insights about the design and integration of multiple antennas into vehicles.

Further results of an investigation of the MIMO performance in LTE networks are shown in Ekiz et al. [EKM12]. The authors conducted a measurement campaign on a test track which was served by an LTE BS operating at 800 MHz. Two antennas—a front and back antenna—were integrated into a shark fin module, which was mounted onto the vehicle rooftop. The measurements were used to derive metrics such as received signal strength (RSS), SNR, and TP.

The concept of virtual MIMO is studied in Beach et al. [BGW11]. The basic idea is to use at least two conventional MIMO users and exploit the physically wide separation of their antennas. In this way, the complexity at the user terminals can be reduced whereas the overall system performance is increased. The authors have designed a virtual 4×4 MIMO scenario which is fed by channel data obtained from vehicular measurements conducted in urban environments. In this work, the Rician K factor and the achieved capacity are evaluated. It is found that the use of virtual MIMO improves the conditions for one of the cooperating users, but degrades the performance of the other at the same time. The benefit in terms of capacity was observed to be small, but the changes regarding the K factor were more significant. Further investigation of metrics such as spatial correlation and angular spread remain for future work.

Finally, the work presented in Reichardt et al. [RSSZ12] deals with the issue to what extend 802.11p can be used for both communication and radar sensing. The authors propose a system concept and present results of a measurement campaign to demonstrate the feasibility. Due to the small bandwidth of the signal the resolution is in the order of a couple of metres that is not sufficient for radar imaging. Furthermore, some OFDM subcarriers cause side lobes in the radar image. To improve the accuracy additional 60 MHz bandwidth from the adjacent industrial, scientific, and medical band (ISM) band is utilised. In this way, the resolution could be improved to 1.7 m.

4.3.2 Medium Access Control

To carry out practical and reliable design of communication systems, all system components need to be tested in realistic conditions. In Section 4.3.1, it has been shown that the performance of PHY layer is mainly influenced by the system parameters and components, by the environment, and the scenario. For MAC layer the performance is rather dependent on vehicular traffic density, network load, and particular type of message to be transmitted. In this context, careful definition of challenging vehicular scenarios becomes crucial. Alonso et al. [ASU+11] have identified following demanding scenarios for testing the performance of MAC layer protocols:

- start-up phase of a VANET;
- two merging VANET clusters;
- emergency vehicle approaching traffic jam.

The two merging VANET clusters scenario was chosen in Alonso and Mecklenbräuker. [AM12] for performance comparison of two MAC layer protocols. The first protocol is the default carrier sensing medium access with collision avoidance (CSMA/CA), as defined by the IEEE802.11p standard, extended by a decentralised congestion control (DCC) mechanism. In this context, the DCC mechanism reduces the channel load by adjusting the transmit power, reducing the repetition rate of cyclic messages or even by dropping packets according to their priority. As an alternative MAC protocol the authors suggest self-organising time division multiple access (SoTDMA), an approach that divides the channel into time slots and allocates these slots to the nodes of a VANET. Analysing effects of the co-located transmissions on the performance of both MAC protocols, it has been shown that SoTDMA performs collision-free regardless of the channel load and the node position. Whereas the performance of CSMA/CA with DCC mechanism is highly dependent on the node location.

More details on how DCC mechanism is realised can be found in Alonso et al. [ASPM13], where the authors compare the performance of IEEE802.11p CSMA/CA MAC protocol with and without DCC. It turns out that parametrisation of the DCC state machine is essential for the overall performance. In particular, a well-balanced and realistic choice of the transmit power for different states, ensures substantial performance improvements. Moreover, it has been shown that the MAC performance is highly dependent on the traffic priority. For high- and medium-priority traffic, systems implementing CSMA/CA with DCC perform significantly better than systems without DCC. Whereas, for low-priority traffic the use of DCC mechanism leads to longer MAC-to-MAC delays and higher amount of omitted packets.

Above-mentioned collision avoidance MAC methods are intended to provide harmonised and fair access to the channel, and are obviously essential for smooth operation of the whole communication system. However, the description of the MAC layer protocol would not be complete without routing algorithms that facilitate seamless information transition through multiple hops. In particular, routing algorithms search for a suitable relay among the neighbouring nodes to forward packets to the desired destination. One of the most challenging characteristics of vehicular networks is high mobility of the nodes, resulting in strongly time-variant links that are difficult to maintain. With this constraints in mind Bazzi et al. [BMPZ13] propose two novel routing protocols: a position-based algorithm and a hop-count based algorithm. The proposed protocols exhibit good performance in terms of average network coverage and number of hops, as the other more complex routing solutions.

Although the overwhelming majority of existing studies on MAC layer focus on the protocol design, such important issue as clock synchronisation should also be taken into consideration. The existing synchronisation algorithms under linear clock model assumptions perform well for short periods, however, they will become problematic for applications with long-term requirements. In this context, a novel clock synchronisation algorithm adopting a statistical signal processing framework has been proposed in Sun et al. [SSBS12]. The authors consider a simple network comprised by two nodes that exchange time stamps over a channel with random delays. The data collected by one of the nodes is then used to estimate the clock values of the other node. The proposed scheme outperforms the existing synchronisation algorithms in many scenarios and is robust against different distributions of the random channel delays, incurred during the message exchange procedure.

4.3.3 System Simulations

For deep understanding and evaluation of all functions featured by complex vehicular systems, the performance evaluation of PHY and MAC layer alone will not be sufficient. It is necessary to evaluate the performance of the complete communication system under exceptional situations, such as safety attacks or severe failures. The basis for investigation of the system performance are the discrete event simulators. However, most of the existing simulators of this type are either lacking microscopic details of the unique car mobility dynamics or significantly abstract the radio propagation aspects.

To carefully reproduce the movement of individual drivers in presence of other cars, traffic lights, road junctions, speed limits, etc., Uppoor and Fiore [UF12] suggest to use the travel and activity patterns simulation (TAPAS) methodology. This technique generates the trips of each driver by exploiting information on home locations, socio-demographic characteristics, points of interests in urban area and habits of the local residents in organising their daily schedule. The TAPAS methodology was developed based on the real-world data, collected from more than 7000 households within the TAPAS-Cologne project. It is capable to precisely reproduce the daily movements of inhabitants of the urban area of the city of Cologne for a period of 24 h, for a total of 1.2 million individual trips. Comparison of the synthetic traffic with its real world counterpart underlines high accuracy of the proposed urban traffic modelling methodology.

The radio propagation aspects can be realistically incorporated into system level simulators by following the V2X simulation runtime infrastructure (VSimRTI) approach with integrate car2X channel model simulation (CCMSim) component, as proposed in Protzmann et al. [PMOR12]. In this simulation architecture, the higher layers of the communication stack are simulated with OMNeT++ and the lower layers with CCMSim. CCMSim includes an IEEE 802.11p PHY layer implementation and a stochastic vehicular channel model. The empirical channel model is based on the data acquired with the HHI channel sounder [PWK$^+$08] during several extensive measurement campaigns.

For even more complete and detailed evaluation of the vehicular communication systems a virtual road simulation and test area (VISTA) has been built. The VISTA approach, details of which can be found in Hein et al. [HBK$^+$15], aims at the holistic investigation of vehicle, road, driver, and radio environment, thus enabling realistic and real-time end-to-end system evaluations. The facility includes a semi-anechoic chamber with a build-in chassis dynamometer for emulation of realistic driving scenarios, facilities for electromagnetic compatibility measurements, a custom-designed multi-probe nearfield antenna measurement system and channel emulators. Integrating all these features in a single testing area allows for an end-to-end evaluation of the system performance.

4.4 Summary and Outlook

Vehicular communication is and will continue to be a hot research topic for years to come. One reason for this is the increased interest for mobile communication at higher frequencies, approximately 6–100 GHz. There is still a lack of knowledge in radio propagation at these frequencies, especially in NLoS. To achieve a reasonable link budget, high-gain steerable antennas might be required. Finally, PHY and MAC algorithms will need to be adjusted for higher frequencies.

5

Wireless Body Area Communications

Chapter Editors: R. D'Errico and K. Yekeh Yazdandoost,
Section Editors: R. Rosini, K. Sayrafian, T. Kumpuniemi,
S. Cotton and M. Mackowiak

5.1 Introduction

With the increasing interest in implantable and wearable technologies for healthcare, diagnosis and health supports by means of wireless connectivity, we are entering in a new era of communications. The wireless body area network (WBAN) is a wireless network technology with purpose to transmit data by implantable or wearable devices to the point of needs [Rei12]. The fundamental idea is to transfer the vital signs of a patient, sportsman, fireman, and so on, collected by sensors to the respectful cotrol unit for further action [BDH+11, Dru07, Rei11, PDRR+12a, PW10b]. Wireless medical information and communication technology will revolutionise health diagnosis and monitoring, with its huge number of possible applications, in home, hospital, and any places where healthcares are provided.

Besides medical applications [PDRR+12b], WBANs are providing new functionalities in peoples' everyday life such as sport, leisure, gaming, and social networks. Watches, glasses, earphones, and bracelets are the most popular, but there are dozens of other wearable and implantable devices which a person could carry in the future. In some sense, grouping those devices into a small network coordinated by the central personal device, which today is the "phone", may change it to become the "router" of WBAN.

The radio channel is central to the paradigm of WBANs [RMM+], and its modelling demands for a different approach when compared to the other know radio propagations [VZ12, YYSP10]. In particular, ANTs play an important role in the context of WBANs (Section 5.2), being affected by the close proximity to the body.

Given the wide range of applications, WBAN communications could be categorised into:

Cooperative Radio Communications for Green Smart Environments, 151–194.

- In-Body (Section 5.3), where at least one node of the network is located inside the human body and it should communicate with one or more devices either in, on, or off the body;
- On-Body (Section 5.4), where all the nodes involved in the communication are placed on the human body, either directly stitched on the skin, integrated into textile and worn by the subject;
- Off-Body (Section 5.5), where at least one of the devices is placed outside the human body, located everywhere in a general area playing the role of a gateway or an access point (AP). A particular subset is body-to-body (B2B) communications where both transmitter (TX) and receiver (RX) are placed on two different subjects.

Based on the exact knowledge of the radio channel cooperative and energy-optimised PHY–MAC procotocol can be designed and evaluated (Section 5.6). This chapter provides an overview of the last advances in research on WBANs, carried out in the framework of IC1004 action, and proposes a radio CM whose parameters are reported in Section 5.7.

5.2 ANT Design and Modelling

Antennas have an important role in the wireless communication and in particular in WBANs, where small ANTs are required that allows integration into WBAN devices. Placing an ANT inside or close proximity to human body will have severe effect on the ANT characteristics. hence, designing an ANT with high radiation performance that is optimised for these application is very challenging [Wei13]. In order to obtain a reliable wireless communication link, the human body has to be characterised as amedium for wave propagation [YY11a, YY11b]. The influence of the body, which can be considered as an irregularly shaped dielectric medium, on the ANT depends on the body's electrical parameters [MOC12b]. Human tissues have their own unique conductivity, dielectric constant, and characteristic impedance, that determine different degrees and types, of radio frequency (RF) electromagnetic interaction [YM12]. There are large number of ANTs (electrical and magnetic [YY09]) for WBAN applications and it is rather difficult to specify an ideal ANT for every application. Since design of WBAN ANTs are implicated with human body tissues, numerical simulation tools can provide substantial insight into the ANT design mechanisms.

5.2.1 Implant ANTs

An efficient ANT is a primary requirement for reliable medical implanted communication. The implanted ANT should be well matched inside a highly dissipative medium [DEY09, YYM12a, YY12]. The bandwidth limitation at lower frequencies such as industrial, scientific and medical (ISM) and MICS bands force the designer to the higher band such as ultra-wide band (UWB) to provide enough bandwidth that's required to transfer data images for application such as wireless endoscopy. The requirements, complexities, and difficulties to design an inbody ANT, make it impossible to use conventional ANT design of free space for body implanted applications.

In Yazdandoost and Miura [YYM12a, YY13b, YY14a, YY14b], ANT performance analysis is based on simulations. A 3D full-wave electromagnetic field simulator based on finite element method (FEM) has been used. Fabricated prototype is presented in Yazdandoost and Miura [YY13b, YY14b]. Figure 5.1 shows an example of the implant UWB ANT prototype [YY13b]. The ANT has dimension of $10 \times 7.5 \times 1.91$ mm and made on Rogers 6010 substrate with dielectric constant of 10.2 mm and thickness of 0.508 mm. The ANT is covered by Duroid sheet on the top and bottom in order to prevent the effect of the human body tissues on ANT performance. Duroid sheet was sited with help of adhesive material with relative permittivity of 3.7 mm and thickness of 0.02 mm [YY13a, YY13b]. The 10 dB return loss is achieved for the entire UWB band.

5.2.2 On-Body ANTs

The presence of body leads to changes in the ANT radiation pattern, shift of the resonance frequency (due to changes in input impedance) and reduction of ANT efficiency. Hence, statistical analysis of ANT performance close proximity to the human body is required, as ANT performance depends strongly on the frequency of usage, on distance to the body, location, and type of body tissues [MOC12b]. Full-wave simulation with realistic models of the human body are a powerful tool to study WBAN scenarios. In Machkowiak et al. [MOC12a], the ANT radiation characteristics is studied with respect to

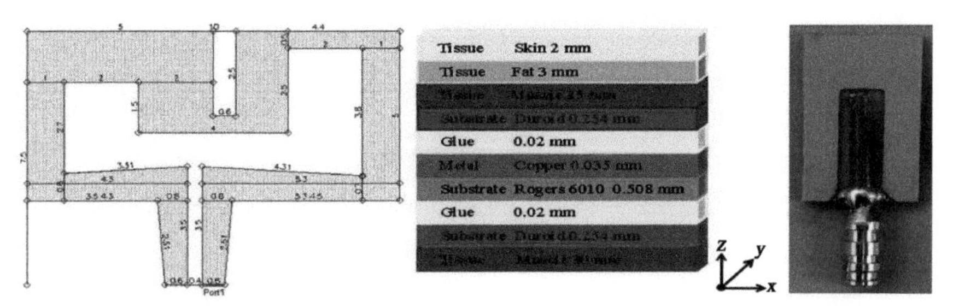

Figure 5.1 Implant ANT layout [YY13a] ©2013 and prototype [YY13b].

the ANT separation from body. Comparisons are done with simulation and measurements and a pattern average differences of 0.9 dB is found.

Different ANT designs are presented in [GHP+12, MOC12b, YM12, YM13] at 2.4 GHz with linear and circular polarisations. A wearable, electrically small loop ANT is presented in Giddens et al. [GHP+12]. The ANT is printed on Duroid substrate with relative permittivity of 10.2 and thickness of 1 mm. Substrate is separated from ground plane with foam which has a thickness of 4 mm. A large textile ground plane with dimensions of 160 × 230 mm^2 was attached to the ANT, in order to cancel out the effects of the human body. ANT geometry is shown in Figure 5.2. The measured results show that ANT is resonant at 2.42 GHz and has –10 dB bandwidth of 68 MHz.

The creeping wave is a surface wave along a curvature that reaches the receiving site by traveling along the surface of body. Creeping wave ANTs are receiving much attention in the recent days for on-body application in particular, when two devices are placed on the body surface, and the wireless communication is affected by the body itself [YM12]. To enhance creeping waves, it is necessary to have a vertically polarised ANT. A 2.4-GHz creeping wave loop ANT, on FR4-epoxy substrate with deictic constant of

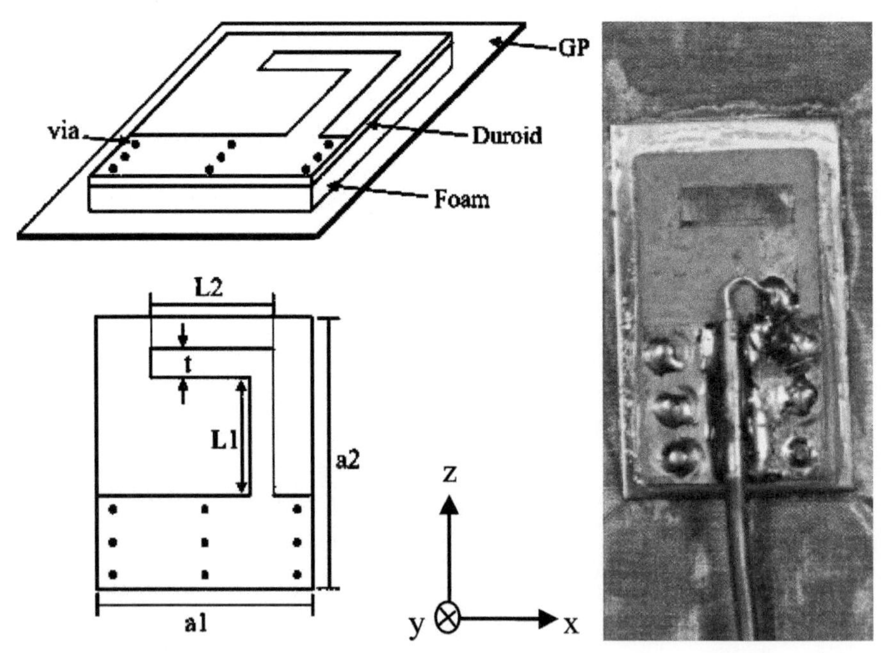

Figure 5.2 Wearable, electrically small, and magnetically coupled loop ANT: geometry and prototype [GHP+12].

4.4 and thickness of 1 mm, is shown in Figure 5.3. ANT has dimension of $34 \times 10\,\mathrm{mm}^2$ and no ground plane. In Figure 5.3, the simulated electric field is shown and proven to be at a 90-deg. angle with respect to the ANT surface [YY11a, YM12].

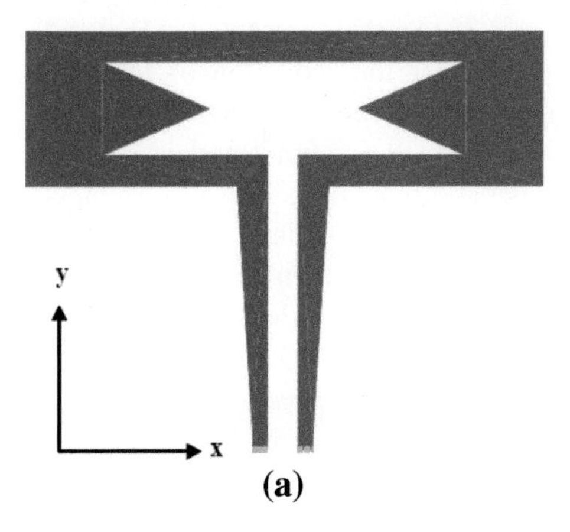

(a)

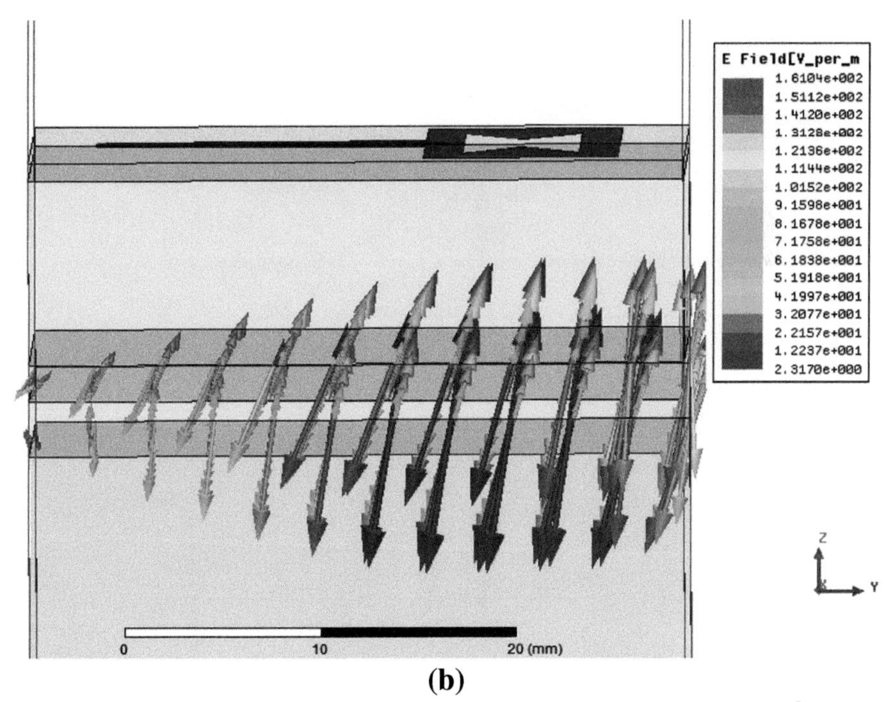

(b)

Figure 5.3 Creeping wave ANT and its vector electric field [GHP+12, YY11a] ©2011.

As narrowband ANTs, the performance of UWB are strongly dependent on the environment of operation. Hence, a challenge for the design of UWB WBAN ANTs remains their potential sensitivity to the proximity of human body, which is due to the strong coupling effect between body tissues and the reactive field of ANT [NC11, WR12, WR13, YY06]. A MSA-BP with operating frequency from 3.38 to 6.07 is presented in Wei and Roblin [WR12]. A back plane is introduced for the reduction of the backward radiation, producing a screening effect between the radiation and the body. ANT has overall sise of $68.1 \times 41.98 \times 4.445\,\text{mm}^3$ and is fed by a tapered CPW line, which provides wideband matching. The ANT geometry and its return loss are presented in Figure 5.4. Desensitisation of the MSA-BP was studied by measuring the refection coefficient on human body and measuring the radio links on a human body phantom. Measurement results confirmed that

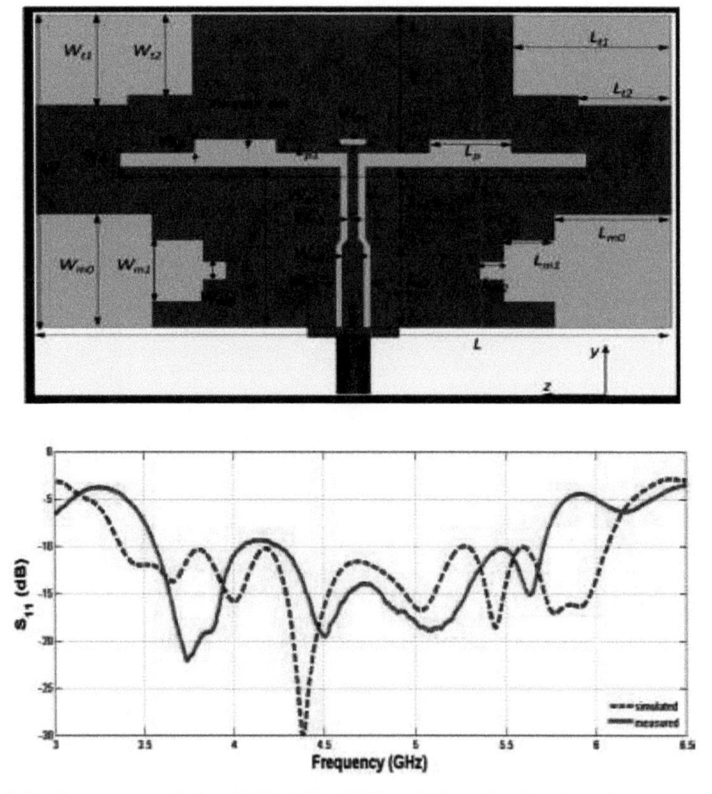

Figure 5.4 Geometry of the MSA-BP ANT and its simulated and measured return loss [WR12].

the MSA-BP is insensitive to the body proximity. A CFSA is presented in Figure 5.5 [Wei13]. The ANT sise is $21.06 \times 25.2 \times 0.8\,\text{mm}^3$ with bandwidth of 3–12.5 GHz. Further investigation on UWB ANTs for WBAN applications have been addressing the modelling of ANT–body interaction. A simple model based on spherical harmonics coefficient to represent an ANT pattern, allowing to predict a disrupted ANT behaviour by a human phantom with respect to the ANT-phantom distance is presented in [MB13].

WBAN devices are battery operated, energy recharging could be one of the option to avoid of battery replacement. In Barroca et al. [BSG$^+$13], an ANT with necessary circuits for ambient energy harvesting is presented. The ANT is designed for Global Systems Mobile, GSM900, and GSM1800, as they are most promising bands for RF energy harvesting. The ANT has efficiency of 77.6–84% and gain of 1.8–2.06 dBi.

The importance of polarisation in WBAN communication is due to fact that the user will change his positions in the different angles during his activities. A circular polarised ANT can be used when the opposite ANT (TX) polarisation is not defined. Since polarisation of a linearly polarised radio wave may be rotated as the poses of subject is changed. Circular polarisation will keep the signal constant regardless of these difference [YM13, YY13b]. Design and performance of linearly and circularly polarised ANTs are discussed in Yazdandoost and Miura [YM13, YY13b].

The WBAN CM includes ANT effects, hence, ANT will be part of channel [YY10]. De-embedding method for body area network (BAN) is aiming to remove this effect [Aoy13, ATK14b]. In this method, an ANT in proximity to human body is characterised by a small sphere, which surrounds human body (represented by dielectric material) and an ANT. Certain number of multi-poles which represents ANT radiation are set inside the sphere and expansion

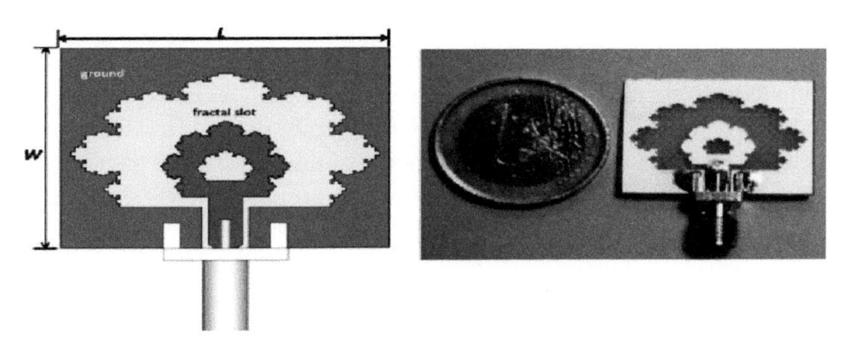

Figure 5.5 Configuration and prototype of CFSA [Wei13].

coefficients are determined to match boundary conditions on the surface of the sphere. To characterise an ANT in this method, radiated electric and magnetic fields are needed in certain number of matching points on the surface of the sphere. In Aoyagi et al. [AKT13], de-embedding is presented for WBAN ANT at 400 MHz, and in Aoyagi et al. [AKT14a] for an ANT at 430 MHz, and finally de-embedding at 2450 MHz is presented in Aoyagi et al. [ATK14b].

5.3 In-Body Radio Channel

Implantable devices offer a myriad of new and exciting medical applications such as implantable pills for precise targeted drug delivery, glucose monitors, bladder pressure monitors, smart capsule endoscopes, and micro robots operating inside the body for biopsy and therapeutic procedures [AMR07]. However, since the early stage of the WBAN standardisation process, it was recognised that modelling the radio channel for implant communication would be a great challenge. Because of a number of ethical and technical issues, measurement campaigns inside living human bodies are not possible. Hence, in order to characterise the propagation of radio signals through human tissues, researchers first carry out numerical simulations of electromagnetic waves propagation using a digital anatomical model or liquid phantoms. Then, the obtained data can be processed statistically to produce simulation or measurement-based models. This section outlines research studies related to implant radio channels that were presented during the COST IC1004 sessions.

Propagation models play a critical role in designing reliable and efficient wireless communication link. Authors Yazdandoost and Miura [YY11c] provide a theoretical investigation on the communication radio channel for body implantable devices based on specific absorption rate (SAR). Their study elaborates on the importance of factors such as near/far-field attenuations as well as ANT placement and orientation inside the body. By determining the average SAR over the entire mass of tissue between the TX and RX, they compute the total attenuation due to the human body. SAR is a standard measure of how much power is absorbed in the body tissue. The absorbed power is lost due to heat dissipation. The amount of this power depends on the strength of the E-field and H-field. In general, the propagation loss between the transmitting and the receiving ANTs (assuming that one is implanted) is dependent on thermal attenuation due to conductivity, reflection losses at tissue boundaries, and near-field and far-field losses. This loss is naturally a function of frequency and distance. Analysis in Yazdandoost and Miura [YY11c] indicates that these factors can not only affect the amount of radiation inside the human body, but also determine the optimum distance at which a reliable performance in the implant communication link can be achieved. UWB radio technology has inherently desirable characteristics (i.e., high data rate and low power) that make it highly suitable for the radio interface of implant sensors. However, there are currently few models that characterise the UWB radio propagation inside the human body. To address this problem, authors Chvez-Santiago and Balasingham [CS14a] present a statistical model for UWB propagation channels inside the human chest in the 16 GHz frequency range. The proposed statistical model is developed from numerical simulations using a heterogeneous anatomical model that includes the frequency-dependent dielectric properties of various human tissues. Their results show that the channel characteristics vary considerably at different depths inside the human body. Therefore, a depth-dependent statistical channel model (CM) for the frequency range 16 GHz has been proposed. This statistical model facilitates the realisation of a distorted pulse after propagating through human tissues. This is very beneficial for the design and evaluation of UWB transceivers, as well as generation of signal templates for correlator RXs and synchronisation schemes. As mentioned before, this study only considered the human chest. Due to the highly inhomogeneous structure of the human body, a customised model may be derived for other parts of the body. The work by Chvez-Santiago and Balasingham [CS14a] has established some guidelines for the development of such models. One of the most innovative applications of radio communication technology inside the human body is wireless capsule endoscope (WCE). This electronic device helps to examine hard-to-reach parts of the gastrointestinal (GI) tract with significantly less discomfort for the patient than traditional wired endoscopic methods. The use of an UWB radio interface, which can enable high data rate transmission, would significantly enhance the information (i.e., video) quality for a WCE device. However, the high attenuation of radio signals propagating through living tissues infrequencies above 1 GHz make the use of UWB radio links for this application quite challenging. Authors Chvez-Santiago and Balasingham [CS13] have presented several ideas to overcome this challenge. These ideas include: selecting the proper implantable ANT, increasing the transmit power, exploiting spatial diversity, and using a dielectric matching layer. Their simulation-based and experimental (i.e., *in vivo* animal test) studies suggest that the implementation of such a UWB communication link for the implantable device is feasible. Future medical applications may also require communication among two or more implanted devices, for example, to exchange information for better diagnostics or to relay data from sensors deep inside the human body. For such applications, authors in Chavez-Santiago et al. [CS14b] propose the use of UWB radios as a communication interface between in-body sensors.

An implant-to-implant statistical CM at the MICS band (i.e., 401–405 MHz) has been presented in Sayrafian et al. [Say10]; however, less information is available regarding the characteristics of implant-to-implant radio channel in the UWB spectrum. The work presented in [CGF14] provides insight into the behaviour of the implant-to-implant radio channel based on propagation measurements in the 3.1–8.5 GHz frequency range. As it is nearly impossible to conduct in-body measurements with human subjects, a liquid phantom that emulated the dielectric characteristics of the human muscle tissue has been used in these measurements. The initial qualitative analysis indicates the feasibility of UWB communication between two implanted sensors; however, the frequency-dependent attenuation of the channel imposes the need for a careful selection of the operating frequency band. Based on the measurements results, the authors in [CGF14] indicate that the most appropriate frequency band for general in-body to in-body UWB communications is 3.1–4.6 GHz. For specific applications in which the transmission range is very short (i.e., below 60 mm) higher frequencies can also be used. It should be noted that the liquid phantom used for the measurement is a homogeneous propagation medium with a single dielectric constant. This contrasts the human body which is a non-homogeneous structure of multiple biological tissue layers with different dielectric constants. Therefore, the validation of the propagation models derived from phantom measurements has to be done either through numerical simulations using a digital model of the humanbody or through *in vivo* measurements on chirurgical animal models.

5.4 On-Body Radio Channel

This section presents the results obainted in terms of channel characterisation and modelling for on-body communications. Several measurement campaings were carried out in different bands. Here for sake of clarity, we divide between UWB and narrowband channels. Final model parameters are reported in Section 5.7.

5.4.1 UWB On-Body Channel

5.4.1.1 Channel characterisation

In Llorca et al. [LBC12], two static measurement campaigns were conducted in an anechoic chamber and office environment by using VNAs in the 3–6 GHz and 1–12 GHz frequency bands. The TX was placed on top, right or left side or back of the head. Six arm positions with RX in hand, and cases when the RX lay in 12 spots along the front and back torso, right arm and foot were measured. The original 50 taps long CIRs were modelled with 24, 12, and 6 taps and the resulting error from the exact channel impulse response (CIR) was observed.

Yazdandoost and Miura [YYM12b] report simulation results for UWB on-body channels with an L-shaped loop ANT within the frequency range of 3.1–5.1 GHz. The RX positions were at the chest, and left and right ear, shoulder and waist. The TX was located either at the abdomen, right waist or left wrist. The path loss (PL) models were extracted both at discrete frequencies and the full frequency range with different TX locations.

Kumpuniemi et al. [KTH+13] present static measurements conducted in an anechoic chamber at a 2–8 GHz range using a four-port vector network analyzer (VNA) and two planar prototype ANT (dipole and double loop). The ANT spots were at the top of the head, left and right ear, right shoulder, upper left chest, left arm, wrist, waist, patella and ankle, right wrist and hip, and right hand first finger tip. From the CIR, PL models were derived CM, static. If the link distances are solved from the CIR, the propagation delay in the ANT must be compensated, or the results especially in short link distances may be affected.

In Kumpuniemi et al. [KHT+13], the data was utilised to produce CIR envelope models and small-scale CM. The statistical characteristics for the CIR amplitude and maximum CIR tap number were obtained.

Kumpuniemi et al. [KHT+14] develop pseudo-dynamic CM, a pseudo-dynamic with the measurement setup as in Kumpuniemi et al. [KTH+13]. The ANT lay at left and right wrist, and left ankle. Five static body postures describing a walking cycle were measured and the data was combined to model a dynamic movement. The attenuations of the first–arriving CIR paths were observed. Then the links were sorted into groups of line-of-sight (LOS), obstructed line-of-sight (OLOS) and non-line-of-sight (NLOS), and the distributions of the CIR taps were solved. Figure 5.6 shows the graphs of the average categorised CIR.

Kumpuniemi et al. [KHYI14] examines channel differences between humans with the equipment used in Kumpuniemi et al. [KTH+13]. Three males and one female were measured. The ANT were at the abdomen, left ankle, right wrist and lumbar region. The PL of the first arriving paths, excess delays, and cross-correlation coefficients between the links were solved.

Kumpuniemi et al. [KHYI15] presents dynamic CM, dynamic for discrete frequencies based on measurements with the setting as in Kumpuniemi et al. [KTH+13]. The ANT were located at the abdomen, left and right ear and wrist, chest, left shoulder and ankle. Both uniform and mixed ANT scenarios

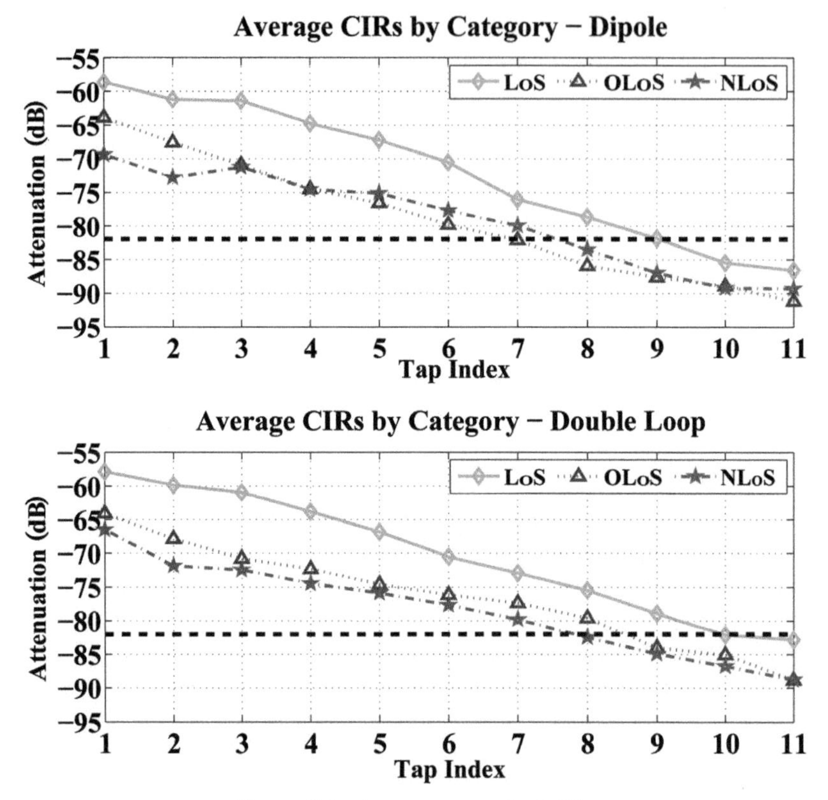

Figure 5.6 Average categorised CIR.

were examined. First, the links were examined at 10 frequencies. Then, the links were categorised into channel types with high dynamic (HD), medium dynamic (MD) and low dynamic (LD), and their probability distributions were found out.

Wang [Wan14] shows dynamic UWB MIMO measurements with a Medav channel sounder at 3.5–10.5 GHz. At first, two TX and four RX wearable fabric made ANT were attached on wrist, knee, foot, or chest. Then, two TX and two RX ANT lay at the eye, wrist or chest, and two omni-directional ANT acted as an off-body RX of an AP. Four body motions were recorded and the channel gain was considered.

Wang and Bocus [WB14a] extend the work by [Wan14]. Graphs of the channel gain, root-mean-square delay spread, ray number, level crossing rate (LCR), average fading duration, multi-link, and temporal auto-correlation function are visualised.

5.4.1.2 Channel model

In Yazdandoost and Miura [YYM12b], the PL models at a distance d follow the equation:

$$\mathrm{PL}(d) = \mathrm{PL}(d_0) + 10 \cdot n \cdot \log_{10}\left(\frac{d}{d_0}\right) + S, \tag{5.1}$$

where d_0 is the reference distance 50 mm, PL(d_0) is the path loss (PL) in dB at d_0, n is the PL exponent, and S is the normally distributed random scattering term with N (0, σ_S^2). The values lie between PL(d_0) = 18.90 ... 41.54 dB, n = 1.40 ... 2.85, and σ_S = 2.40 ... 9.72 dB depending on the TX location and whether full frequency band or discrete frequencies are considered.

Kumpuniemi et al. [KTH+13] explains PL models based on Equation (5.1) with the parameter values PL(d_0) = 31.6 (20.4) dB, $n = 3.3$ (3.1), and $\sigma =$ 12.8 (11.6) dB for the dipole, and PL(d_0) = 39.8 (25.1) dB, n = 2.7 (2.6), and $\sigma = 13.8$ (12.2) dB for the loop, depending on whether the first or all arriving paths (in brackets) are considered.

In Kumpuniemi et al. [KHT+13], the statistical CM are solved by fitting the amplitudes of the CIR against several continuous distributions and ranking them with the second order akaike information criterion (AICc). The CM for the dipole base on the Weibull distribution and the lognormal distribution. The double loop CIR amplitude follows the inverse Gaussian distribution. The probability density function (PDF) of the Weibull, lognormal, and inverse Gaussian distributions have the Equations (5.5), (5.4), and (5.3). Their parameter values are listed in Table 5.1.

Table 5.1 Parameters for the static UWB on-body CM

Tap	Dipole	Double Loop
1	$a = 0.0012$, $b = 0.8147$	$\alpha = 0.0011$, $\beta = 1.0413 \times 10^{-4}$
2	$a = 5.7829 \times 10^{-4}$, $b = 0.7795$	$\alpha = 6.3532 \times 10^{-4}$, $\beta = 3.5521 \times 10^{-5}$
3	$a = 4.8969 \times 10^{-4}$, $b = 0.7345$	$\alpha = 6.4157 \times 10^{-4}$, $\beta = 4.5918 \times 10^{-5}$
4	$\mu = -8.6100$, $\sigma = 1.4787$	$\alpha = 4.8655 \times 10^{-4}$, $\beta = 5.0342 \times 10^{-5}$
5	$\mu = -8.9040$, $\sigma = 1.4142$	$\alpha = 3.3322 \times 10^{-4}$, $\beta = 3.1961 \times 10^{-5}$
6	$\mu = -9.2348$, $\sigma = 1.4082$	$\alpha = 2.4969 \times 10^{-4}$, $\beta = 3.7238 \times 10^{-5}$
7	$\mu = -9.4845$, $\sigma = 1.2954$	$\alpha = 1.9503 \times 10^{-4}$, $\beta = 4.0707 \times 10^{-5}$
8	$\mu = -9.7374$, $\sigma = 1.3346$	$\alpha = 1.4940 \times 10^{-4}$, $\beta = 3.7844 \times 10^{-5}$
9	$\mu = -9.9844$, $\sigma = 1.2842$	$\alpha = 1.2693 \times 10^{-4}$, $\beta = 2.5836 \times 10^{-5}$
10	$\mu = -10.1519$, $\sigma = 1.3819$	$\alpha = 1.2102 \times 10^{-4}$, $\beta = 2.2066 \times 10^{-5}$
11	$\mu = -10.3447$, $\sigma = 1.4312$	$\alpha = 1.0046 \times 10^{-4}$, $\beta = 1.7603 \times 10^{-5}$
12	–	$\alpha = 7.7384 \times 10^{-5}$, $\beta = 1.5076 \times 10^{-5}$

The mean PL of the first arriving CIR path in Kumpuniemi et al. [KHT+14] range between 54.6 and 85.6 dB. The categorised CM describe the linearly scaled relative CIR tap amplitudes of the LoS, OLoS, and NLoS classes. The inverse Gaussian distribution distribution obtained the best AICc ranking and its parameter values are described in Kumpuniemi et al. [KHT+14].

In Kumpuniemi et al. [KHYI15], the mean PL lay between 32.8 and 80.6 dB and the standard deviation of the PL between 0.6 and 12.9 dB depending on the link and frequency. The dynamic HD, MD, and LD CM are based on the inverse Gaussian distribution, and they describe the linearly scaled relative amplitudes of the received discrete frequency signals. The distribution parameters for the models are gathered in Table 5.2.

5.4.2 Narrow Band On-Body Channel

5.4.2.1 Channel characterisation

Characterisation of the narrow band on-body channel within the remit of IC1004 TWGB has been performed using a range of methods including analytical modelling, simulation, and various channel measurement campaigns. Analytical modelling of on-body channels has been investigated in Chandra

Table 5.2 Parameters for the dynamic UWB on-body CM

			Link Type		
			HD	*MD*	*LD*
ANT Type	Freq. (MHz)	Param.	$\times 10^{-4}$	$\times 10^{-4}$	$\times 10^{-4}$
	2450.0	α	26.75	69.66	97.39
		β	3.55	17.69	33.25
Uniform	3995.0	α	14.45	85.16	65.96
Dipole		β	9.71	3.72	7.31
	7988.2	α	16.87	25.35	42.40
		β	2.98	1.50	3.63
	2450.0	α	20.51	21.62	68.59
		β	1.92	20.39	20.55
Uniform	3995.0	α	17.17	32.63	50.55
Loop		β	1.25	3.98	3.23
	7988.2	α	10.26	26.88	67.09
		β	2.22	1.30	10.23
	2450.0	α	46.48	144.46	none
		β	53.04	751.86	none
Mixed	3995.0	α	13.38	81.33	none
Dipole/Loop		β	14.08	44.64	none
	7988.2	α	4.04	60.33	none
		β	3.69	47.44	none

and Johansson [CJ13]. Here, the authors initially represented the human torso as an elliptical surface and studied the attenuation of the creeping wave. The ear-to-ear propagation channel was also investigated, and using FDTD-based simulations conducted at 2.45 GHz, good agreement was shown to exist between the proposed model and the simulations.

In the future, short-range high-bandwidth links with sensors and APs situated in the local environment will allow rapid data transfer to and from WBAN nodes. Representing the human body as a cylinder, an analytical model for the 60 GHz off-body channel has been presented in Mavridis et al. [MPDD+13]. Segmenting the analysis into 'lit' and 'shadowed' regions the proposed models have been shown to provide good agreement with physical (PHY) measurements, considering both transverse electric and transverse magnetic polarisations for illumination of the body by a 60-GHz wave. The authors of this study have also defined an extremely useful transition operator which enables the conversion of the IEEE 802.11ad indoor CM defined in Maltsev et al. [Mal+10] into an off-body CM.

Conducting body centric channel measurements at millimeter wave frequencies such as 60 GHz presents many challenges not least due to the high cable losses, but also sweep repetition rates if scanning wide bandwidths using VNAs. To overcome some of these limitations, a novel frequency-modulated continuous wave (FMCW) measurement system for 60 GHz channel measurements has been developed in Salous et al. [SNCC13]. Using FMCW channel sounder, on-body and short-range off-body channel measurements have been made at 60 GHz. To illustrate its effectiveness, the output of the new FMCW channel sounder was compared with the results of on-body channel measurements obtained from a VNA. The vastly superior sampling capability of the FMCW channel sounder meant that it was able to sweep 4.4 GHz of bandwidth in 819.2 μs compared to 411 s for the VNA. As a result, the authors noted more dispersion prior to the main peak in the power delay profile (PDP) for the VNA measurements compared to those acquired using the channel sounder for the same ANT configuration and body scenario. The authors of Salous et al. [SNCC13] reason that this was due to the exceptionally long acquisition time of the VNA, meaning that this dispersion could have been due to movement during the sweep time.

With the exception of Aoyagi and Takada [AT11] and Mhedhbi et al. [MAAU14], simulation of on-body channels has primarily been conducted using the finite-difference time-domain (FDTD) method [CJ11, CJ13, KJVM11, YYM12a]. For example in Yazdandoost and Miura [YYM12a], the FDTD method is used to investigate the creeping wave mode of propagation around a cylinder representing the human body at 2.4 GHz. To ensure that they were simulating the creeping wave component of the on-body channel, the authors Yazdandoost and Miura [YYM12a] placed their ANTs on opposite sides of the body to guarantee NLoS channel conditions. It should be noted that the ANTs used in this study were purposely designed to maximise coupling over the body surface by enhancing the creeping wave component.

Using the FDTD method to simulate wave propagation at 2.45 GHz between a pace maker acting as a central hub and multiple implant locations such as the liver, heart, spleen, and the kidneys, a PL model for a heterogeneous human model was developed in Kurup et al. [KJVM11]. For in-body communications, the authors state that increases in PL depend on two factors, namely the distance between the implants and the tissues and organs through which the propagation takes place with front-to-back propagation particularly susceptible to high PL. Adhering to SAR guidelines, a link budget equation for wireless communications between implanted devices has also proposed.

As an alternative to the FDTD method, the work presented in Aoyagi and Takada [AT11] proposes the use of the generalised multipole technique (GMT) technique [Haf90] to model WBAN propagation. More recently, graph based ray tracing for the purpose of on-body channel modelling has been investigated in Mhedhbi et al. [MAAU14]. The simulations used a motion capture sequence to model body movement and a perturbed model of a patch ANT operating at 2.4 GHz, taking into account the body presence. The output of the simulator was compared with open source channel data provided by NICTA [NIC12] covering a wide range of body positions such as the ankle, back, chest, hip, and wrist. While absolute agreement between the simulator and measurements was not achieved, the simulator was able to accurately model the variations observed in the channel due to the walking motion. The authors of Mhedhbi et al. [MAAU14] suggest that the discrepancy between the measured and simulated channels can be accounted for due to the trunk shadowing being underestimated by the perturbed ANT model.

In Rosini and D'Errico [RD12a] real-time On-Body channel measurements campaign was performed. At the transmitting side, the measurement test bed was mainly composed by a pulse step generator and a power amplifier, whereas at the receiving side four low noise amplifiers were connected to a digital oscilloscope. This configuration allowed the simultaneous collection of four CIRs, each one corresponding to a different On-body link composed by the i-th TX and the j-th RX. The four receiving ANTs (Rxs) were located on the right and left hand, on the right thigh and on the left ear. Instead, the device acting as the TX was placed alternatively on the right ear, on the heart and on the left hip position. All the acquisitions were repeated both in anechoic environment and in an indoor office equipped with some general office furniture. In order to account for the human body and movement variability, four human subjects (two males and two females) were involved in the measurements. Each subject performed different movements: still, walking, and standing up, and sitting down from a chair. In order to study the ANT effect on the channel characteristics, all the measurements were repeated with two sets of ANTs: wideband planar monopole (PM) and wideband top loaded monopole (TLM). PMs were placed parallel to the body surface in order to have a main polarisation tangential to the body and they were spaced from it by a 5-mm thick dielectric foam. TLMs, which present good monopolar behaviour, were actually designed to result into a normal polarisation with respect to the body surface.

In Van Roy et al. [VRQL+13], measurement campaign was performed by means of a multiple-input multiple-output (MIMO) 8×8 channel sounder. A WBAN consisting of 12 nodes, located over the whole body, as shown in Figure 5.7 was investigated. The positions of the transmit and receive nodes were changed between measurements to get a statistical characterisation of almost all of the 12×12. The measurements were taken successively on two male subjects in a quasi-empty room of approximately 5×10 m.

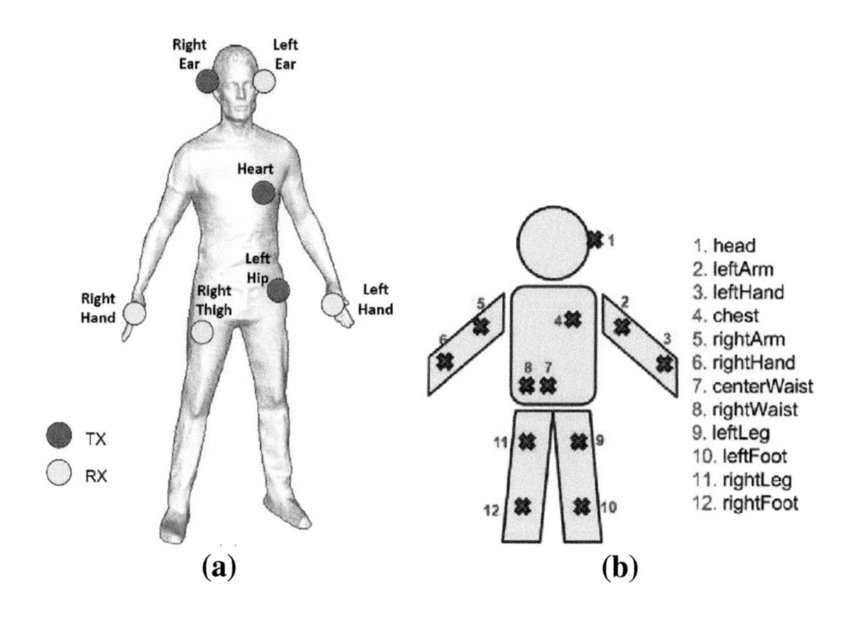

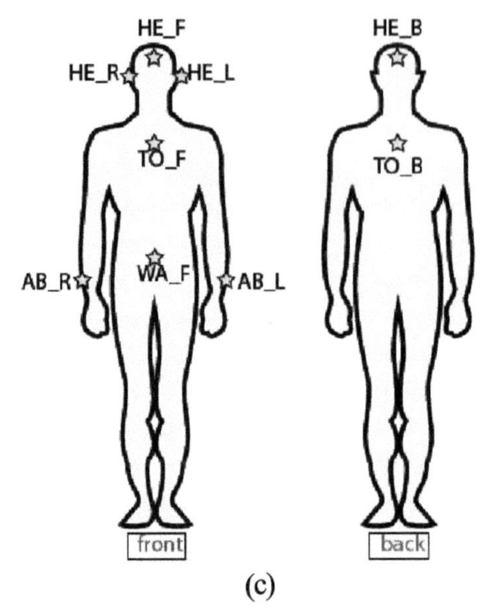

(c)

Figure 5.7 Investigated dynamic on-body channels: (a) Rosini and D'Errico [RD12a], (b) Van Roy et al. [VRQL+13], and (c) Oliveira and Correia [OC13c].

During the measurements, each subject was walking freely around the room. SMT-3TO10MA SkyCross ANTs were employed in two different orienations, respectively, vertically and normally to the body.

In Grosinger and Bosch [GB14], VNA measurements were performed considering different stationary and moving body posutre. Each posture was held 20 s, and the repetition rant of the measurement sas 5 s^{-1}, by employing monopoles at 900 MHz and 2.45 GHz.

Besides measurements a simulation approach has been employed in Oliveira et al. [OM12]. The authors exploited FDTD simulation results, using the transient solver of CST MWS, to characterise both static and dynamic on body channels at 2.4 GHz. The tissue electric parameters for the desired frequencies were obtained from the Cole–Cole Model described in [14]. A BAN topology composed of nine wireless nodes distributed over the voxel model in locations that can correspond to a realistic usage scenario was created. In a first study, the body was considered still and patch ANTs were place at different distances from the body (up to 25 cm). Than a realistic model of a female subject in different postures (replicating the body motion from the Poser animation software) was considered [OC13c, OC13a]. Each point of time in the human motion is key framed to create a smooth animation, where each frame is a 3D object (a posture) exported from Poser into CST. Two typical human day-to-day activities are reproduced, walk and run, each one being a combination of 30 time frames, with 1/30 s duration each.

5.4.2.2 Channel model

For each link under investigation, the time-dependent channel power transfer function in dB, $P_{ij}(t)$, could be computed integrating the channel transfer function, $H_{ij}(t, f)$ over the band of interest B:

$$P_{ij}(t) = \frac{1}{B} \int_B |H_{ij}(t, f)|^2 \, df = G_{0ij} \cdot S_{ij}(t) \cdot F_{ij}(t). \tag{5.2}$$

$G_{i,j}$ represents the *mean channel gain*, $S_{i,j}(t)$ is the *long-term fading* or *shadowing* contribution, mainly due to the masking effect of the human body and to its movement. $F_{ij}(t)$ is the *short-term fading* that accounts for the effect of multi-path component (MPC) arising from reflections and diffractions from the environment or from the body itself.

An example of the acquisition performed is shown in Figure 5.8, which presents the evolution over time of $P_{i,j}(t)$ in dB (dashed curves) while the subject is walking in indoor. Each colour refers to one of the on-body link, considering the *TX* placed on the user's left hip.

The comparison between Figure 5.8(a,b) (the latter is equivalent to the former but referring to a static case, where the subject is standing still in the anechoic chamber), highlights how the nodes position and the movement performed affect the dynamic evolution of $P_{i,j}(t)$. This effect particularly emerges when focusing on the slow-varying component of the power transfer function, $S_{i,j}(t)$ (continuous curves). It represents the *shadowing* impact of the body that dynamically masks the direct communication link according the subject movement. For example, following the trend of the continuous blue curve (hip/right hand link), it is possible to retrace the swinging movement of the arm during the walk. The dips of the curve refer to the moments where the arm is behind the subject's torso, i.e., the body completely shadows the communication, resulting in strong channel attenuations. On the opposite, *shadowing* phenomena are quite moderate in static conditions where no movement is performed (Figure 5.8a), and the slight variations of channel $P_{i,j}(t)$ are due to the involuntary breathing of the user.

The complete set of results, which are not reported here for the sake of briefness. In Rosini and D'Errico [RD12a], it was found that PMs generally present smaller values of $G_{i,j}$ as compared to TLMs. This can be explained considering that tangentially polarised ANTs, as PMs are, do not help creeping

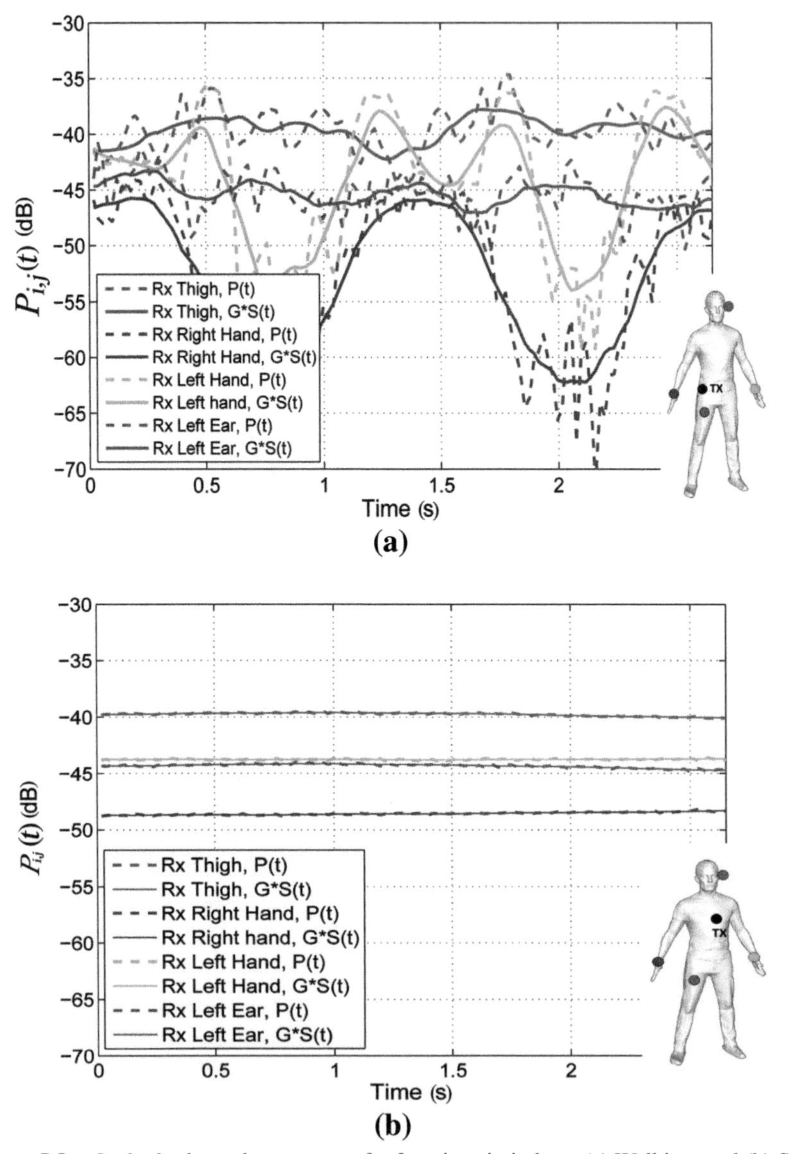

Figure 5.8 *On-body* channel power transfer functions in indoor: (a) Walking, and (b) Still.

waves propagation on the body, resulting in larger channel attenuations [LHOV09]. As a consequence, this effect is even more evident for the links with a strong on-body propagation component, such as chest/thigh, hip/thigh, or ear/ear.

In classical wireless systems, the *long-term fading* normally accounts for the shadowing effect of some obstacles that obstruct the direct link between nodes. For the specific case of BANs in *On-Body communications*, the definition of *shadowing* describes the condition where the human body itself acts as an obstacle to the main direct transmission path, dynamically evolving according to the movement, as already presented in Figure 5.8. This effect is particularly evident when one of the ANTs is placed on a limb (hand or leg). *Shadowing* values in dB results to be statistically described by a normal distribution, $S_{i,j}(t)|_{dB} \sim N(0, \sigma_S)$, where the standard deviation σ_S accounts for the slow variations of $P_{i,j}(t)$ due to the body shadowing effect. Generally, PMs present larger values of σ_S as compared to those found for TLMs, meaning that they are more affected by the shadowing effect of the body and by its movement. Indeed, normally polarised ANTs, helping creeping waves propagation around the body, present a dominant main path that overcomes the secondary paths generated by reflections on the body, resulting in a smaller *long-term fading* contribution.

As for the *short-term fading*, $F_{i,j}(t)$, it accounts for the effect of the MPC arising from reflections and reflections on the human body and/or on the surrounding environment. In narrowband, it was found that *short-term fading* is well modelled by a Rice distribution. Normally, polarised ANTs present larger *K* factors as compared to tangentially polarised, meaning that the channel experiments smaller amplitude fading episodes. Higher *K* factors are due to the presence of a main propagation path with stronger energy contribution, which confirms the hypothesis of an on-body creeping wave contribution enhanced by the use of normal polarisation. This difference is particularly relevant for the chest/thigh and chest/ear links, where the propagation happens mostly by on-body creeping waves, and the differences in ANT behaviour due to the ANT effect are even more stressed. Measements in which the user performed dynamic walking on the spot movements in both anechoic and reverberant environments have been exploited in order to address different scenarios and parameters have been extracted in Cotton [Cot15]. One of the main needs for cooperative protocol design is to take into account the space–time correlation properties. In the inter-link correlation properties of two different measurement campaings [PRDO14]. The analysis show that, depending on the walking scenario, the space–time correlation characteristics can vary from very stationary to not stationary at all. For instance, the stationarity cases correspond to the scenarios where the subjects walk uniformly and regularly along a straight line with periodic arm swings, whereas the non-stationary behaviour arises from random trajectories and irregular arm swings. In Oliveira and Correia [OC13d], the movement was generated by an animation software, using a realistic shaped homogeneous phantom, to investigate the inter-link correlation and power distribution, when the body is moving. Results shown that the movement decreases the correlation between the branches to values below 0.7. With respect to the targets for MIMO (correlation and imbalance below 0.7 and 10 dB, respectively), the running environment has 29% of its samples inside the target area, against 34% for the walk scenario. In particular, TXs on the ears are the ones with more links satisfying the minimum MIMO requirements [OC13b]. In order to reproduce the space–time correlation characteristics in walking scenario, an auto regressive model for long-term and short-term fading has been propsed along with a Matlab implementation [ARD14].

5.5 Off-Body Radio Channel

Off-body communications take place between at least one node *on* a human body and another one *outside* it, either worn by another user (*B2B* case) or acting as a static external gateway (*on-to-off body* scenario). Off-body communications have been just recently approached, with a limited number of contributions available in literature [ACS09, CS07, CS09, HMS+10, GBS08, KAY10, MPS+14a, RD12c, RVD14, SHJ+09, WBN+09, ZCSE04], mainly dealing with the *on-to-off body* case [Ros14]. In order to characterise this type of channels, specific investigations are needed, and a great effort has been done by the TWGB in this direction. Several TDs focused on the modelling of transmission channel's main components (i.e., distance-dependent mean PL, shadowing and fast fading), accounting for the dynamics of human movement and ANT/body joint effect. Both simulations and experimental approaches were adopted, covering not just the ISM band at 2.45 GHz, but also the UWB and the 60 GHz one. Moreover, some works studied link correlation properties, to be exploited in the perspective of realising MIMO configurations.

5.5.1 On-to-Off Body Channel Characterisation

The body presence on one side of the radio link significantly affects ANT's radiation properties [MC13]. Body movements result in time-varying propagation characteristics and variations of ANT's orientation (i.e., maximum radiation direction), which influence the overall channel. The authors Mackowiak and Correia [MC13] investigated the statistical behaviour of ANT's radiation pattern in dynamic conditions. Based on the description of the variation of the normal to the body surface during motion, a statistical characterisation of the changes in ANTs' radiation pattern (i.e., patch ANT at 2.45 GHz [MC06]) is provided. Considering that the influence of body dynamics on ANT's orientation depends on its on-body location, different node positions were taken. Simulation results show that when ANTs are placed in very dynamic body segments the standard deviation of the normal to the body surface is very large; the radiation pattern shape gets more uniform compared to the static case, and its standard deviation is also very high in all directions [MC13].

5.5.1.1 Channel modelling at 2.45 GHz

The work by Mackowiak and Correia [MC12] proposes the integration of the ANT statistical model in Mackowiak and Correia [MC13] with the description of the propagation environment, to develop a flexible simulation platform for modelling on-to-off body channels. Simulations were run for a human subject carrying four on-body nodes (patch ANT operating at 2.45 GHz) located on different positions, who moves along a street in front of an external isotropic device placed at 3 m height. The channel is characterised in terms of received power, delay spread, and direction-of-arrival (DoA). Propagation conditions are classified as *LoS, NLoS*, or *Mix*, according to the mutual position of on-body and outer devices. Results show that channel parameters depend on body movement, ANT position, and distance between devices. For instance, Figure 5.9 compares the evolution of the average received power while walking and running, considering four different situations: front side of the head (HE_F), left and right arms at wrists (AB_L and AB_R), and back side of the torso (TO_B). In the latter case, propagation conditions for the left arm node change rapidly from LoS to NLoS, resulting in deep fades. Moreover, arms present symmetric trends regarding the position of the outer ANT, while forehead and back nodes show similar values of the received power, but at opposite ends of the path [MC12].

Starting from Mackowiak and Correia [MC12], the same authors [Mc14] propose a PL model for dynamic on-to-off body channels. The model is given in terms of total PL as $L_{\mathrm{PT}}(d)_{[dB]} = \Delta L_{\mathrm{P}}(d)_{[dB]} + \Delta L_{\mathrm{B}[dB]} + \Delta L_{\mathrm{F}[dB]}$, where the mean PL, $L_{\mathrm{P}}(d)_{[dB]}$, is characterised by a log-distance linear fit of the channel values obtained for different body positions and orientations; $\Delta L_{\mathrm{B}[dB]}$ is a random variable modelling the body shadowing; while $\Delta L_{\mathrm{F}[dB]}$ accounts for the influence of the propagation environment describing the fast fading component. Simulations are run for a BAN at 2.45 GHz, with eight on-body patch ANTs carried by an user who changes position in an indoor

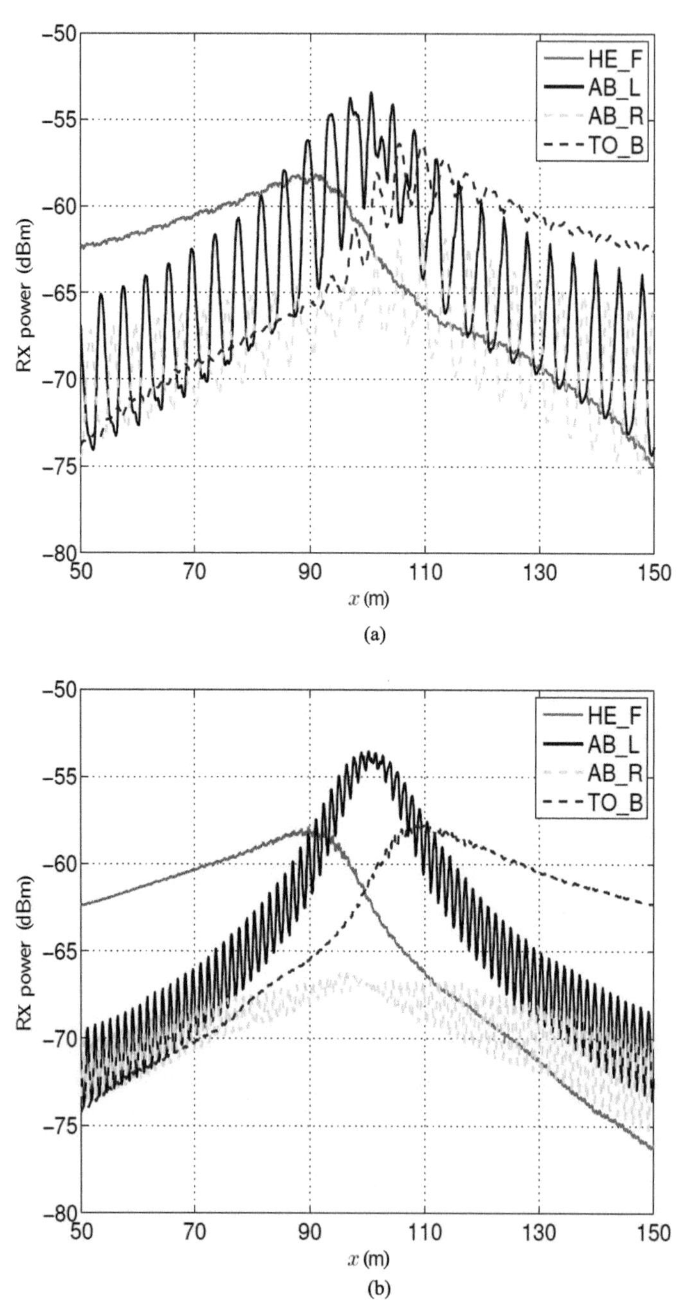

Figure 5.9 Average received power when a person is moving along a street [MC12]. (a) Walking, and (b) Running.

environment [MC14]. PL results dependent on the propagation conditions, with higher values in NLoS than in LoS. The fading amplitude is statistically modelled by a Nakagami distribution. As for the body shadowing, it is a dominant component in the CM, since the body can act as an obstacle to the direct communication link during movement, resulting in an additional attenuation to the mean PL.

Body shadowing characterisation is also the focus of Aoyagi et al. [AKT14, ATK14a, ATK15]. Simulations estimate the occurrences of shadowing (i.e., passage from LoS to NLoS), which depend on the movement type and the angle of the wave incidence on the body, according to nodes' mutual position [ATK14a]. Shadowing rate is modelled through a sigmoid or a normal distribution, considering the azimuth angle of the incident wave as the fitting parameter. Here, the shadowing rate means the probability of shadowing occurrence for a specific incident angle during motion. It is included in a closed formulation for the on-to-off body channel characterisation, to account for the body shadowing effect [AKT14, ATK15]. The effectiveness of the on-body node location is evaluated through the LoS ratio R^n, defined as the total number of LoS occurrences over the total number on LoS/NLoS ones. The head node gets the highest R^n, since it is less shadowed by the body and hence more often in direct visibility with the external device.

A channel modelling approach similar to the one in Mackowiak and Correia [MC14] is proposed in Rosini and D'Errico [RD12c], but based on experimental channel data. Measurements at 2.45 GHz were performed in anechoic and indoor environments, with one user carrying three on-body nodes. Both static and dynamic conditions were reproduced, with the user facing or giving his back to the outer device (i.e., LoS and NLoS case). Two set of ANTs with different polarisations relative to the body surface [RD12b] are used, to account for the ANT impact on channel characteristics. The model identifies two main components: a log-distance dependence and a fading contribution. The channel gain depends on body orientation, ANT type and on-body position, as shown in Figure 5.10 for measurements in standing LoS conditions in an anechoic chamber. Each colour refers to a specific node position. Generally, channel gain is smaller in NLoS with negligible distance dependence, due to the presence of the body shadowing the direct communication path [RD12c]. The body shadowing is also described as an additional loss contribution to the channel gain in LoS, given as a function of body orientation. The fading component accounts for the environment and motion effect, and its envelope is statistically characterised by a Nagakami distribution.

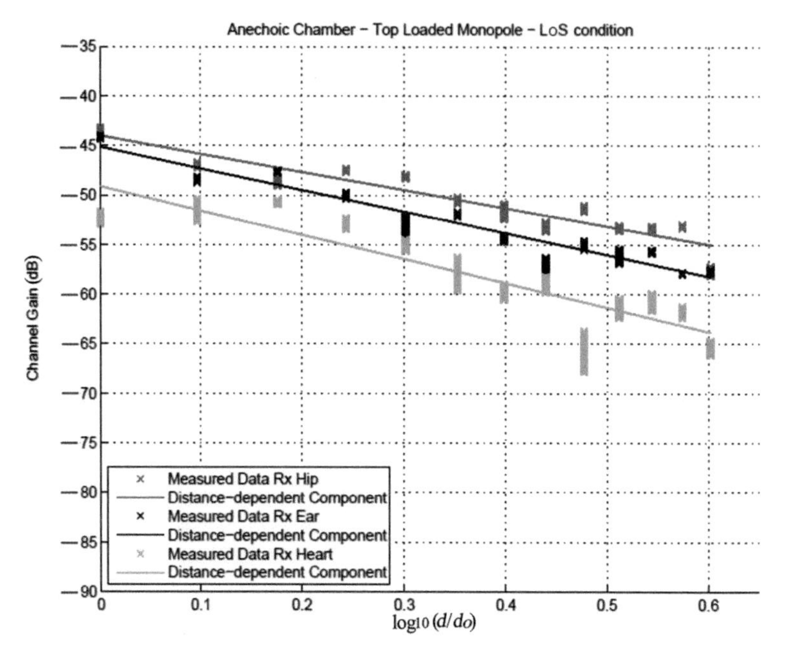

Figure 5.10 Channel gains for on-to-off body channels [RD12c].

Another experimental indoor off-body PL model at 2.45 GHz is derived in Ambroziak et al. [ARLM14]. Here, measurements are performed in both static and dynamic conditions with two users carrying three on-body patch ANTs. Acquisitions were repeated varying body orientation and for increasing distances from the external device (i.e., a dual polarised ANT). In accordance to the consideration drawn in Rosini and D'Errico [RD12c], results shows the strong influence of ANT on-body location, body orientation, and its PHY characteristics on PL and its distance dependence. The analysis of dynamic scenarios reveals also a great variability of PL values due to movement, when body shadowing episodes may take place, resulting in significant attenuation of the channel.

Paper of Mackowiak et al. [MRDC13] compares the results for the CM presented in Rosini and D'Errico [RD12c] with simulation ones obtained from Mackowiak and Correia [MC12, MC14]. The reference scenario includes indoor and anechoic environments, where an user walks towards/backwards an external device, carrying three on-body nodes [RD12b]. The comparison of the PL values shows a good agreement for both the anechoic and indoor case. An example is shown in Figure 5.11 for the head node, where PL values are given as a function of the distance from the external device. Curves follow the

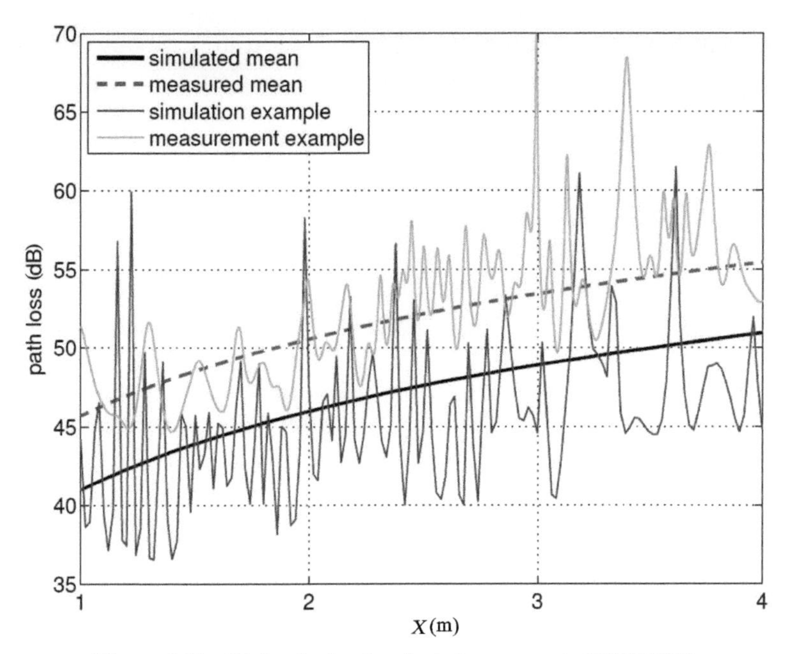

Figure 5.11 PL for the head node, indoor scenario [MRDC13].

same trend, and the gap can be due to differences in the simulated environment, exact node position and movement performed in the two cases. Moreover, the polarisation effect is not considered during simulations, while it is inherently included in measurements [MRDC13].

Another work comparing simulations and measurements is presented in Ambroziak et al. [AMS+14]. The validation of the PL model described in Mackowiak and Correia [MC14] is obtained using the channel data acquired as described in Mackowiak et al. [ARLM], considering a dual polarised ANT as the outer device. Results show a good agreement between measurements and simulations, even if some deviations are found especially in NLoS, due to ANT's depolarisation effect not modelled in simulations [AMS+14]. The same set of measurements is used in Ambroziak et al. [ARLM14] to study the depolarisation of radio wave propagation in indoor on-to-off body communications, which is mainly due to the presence of the MPCs generated by reflections and diffractions from the surroundings. The metric used in the analysis is a ratio between the power received by the vertically and the horizontally polarised outer ANT. A positive ratio means a stronger vertical component, which is desirable for on-to-off body communications [CLV98]. This parameter is extremely affected by node on-body position and user orientation, and the highest values are observed when the on-body ANT is co-directed with respect to the external one.

5.5.1.2 Channel modelling at other frequency bands

ISM is one of the possible frequency band for BAN operations, since it is worldwide available and licence free. The main drawback is that other existing communication standards are defined in the same bandwidth (e.g. IEEE802.15.4/ZigBee [IEEE06], and Bluetooth [blu10]), and coexistence issues may arise. To overcome these problems, other bands are proposed, among them UWB being a very promising alternative. The authors Pasquero and Errico. [PD14] present an experimental UWB on-to-off body indoor CM, in presence/absence of the human body. Measurements are performed by placing some UWB dipoles in different positions on a standardised human-like phantom, varying its orientation and distance relative to the outer device. Angles of arrivals and clusters are estimated via the SAGE [VRLO+09] and K-means algorithms [HTK06]. Results show that shadowing from the body has a strong impact not only on the PL, but also on angle of arrival and cluster distribution, due to the higher directivity of the ANT when on-body. Cluster centroids and dispersions are modelled as a function of the body orientation and the distance separating the on- and off-body devices. To complete the model, the dense MPCs component is also characterised [PD14].

The unlicensed 60 GHz band is also proposed for BAN applications for its high data rates. Paper by Mavridis et al. [MPS+14b] presents an indoor CM derived from the solution of the scattering of a plane wave on a cylinder at 60 GHz, the latter representing the human body. The scattering model is composed of two parts (i.e., the solution in the lit and the shadow region) and it is validated through some experiments with a real human body. To include the impact of the environment, authors consider the IEEE802.11ad 60 GHz indoor CM [IEEE12a] and find a mathematical operator that allows the conversion from that model to an on-to-off body one. It is shown that this operator depends on the radius of the cylinder, and on the polar coordinates of the on-body ANT relative to the outer device [MPS+14b].

5.5.1.3 Channel correlation properties and MIMO systems

Body area network performances are generally affected by the environment, body dynamics and posture, and ANT characteristics [MTT+13]. One solution to improve network reliability is the use of Virtual MIMO configuration as multi-ANT systems. Indeed, ANT diversity helps in overcoming the MPC propagation of the signal, increasing the robustness of the system [MTT+13].

A condition to exploit ANT diversity is the link independence, which can be measured by the correlation level among them and is related to radio propagation conditions [ML12]. Starting from the channel modelling approach by Mackowiak and Correia [MC12], Mackowiak and Correia [ML12] presents the results on the evaluation of the correlation between nine off-body links. The average correlation and standard deviation are calculated for all possible link combinations, results being presented in Figure 5.12. The highest correlation is found for co-directed ANTs (i.e., placed on the same side of the body), while the lowest (providing the best diversity) for opposite-directed ANTs (i.e., opposite sides). As for correlation's standard deviation, the largest values are for highly dynamic scenarios. These results helps in the definition of the optimal position for on-body ANTs in terms of diversity and decorrelation. To complete the model, paper of Mackowiak and Correia [ML13] analyses the capacity of a 2×2 MIMO system, starting from the same simulation scenario considered in [ML12], looking for the optimal node configurations maximising overall system performance. For the reference scenario, the pair of nodes located on the left side of the head and left arm realise the largest capacity gain, with 67% more data transmitted relative to a single-input single-output (SISO) configuration. The main reason is due to good propagation conditions during body movement, and low power imbalance [ML13].

		correlation								
		TO_F	WA_F	HE_F	HE_B	HE_L	HE_R	AB_L	AB_R	TO_B
standard deviation	TO_F		0.73	0.55	0.20	0.44	0.33	0.41	0.34	0.26
	WA_F	0.22		0.53	0.19	0.44	0.33	0.42	0.35	0.24
	HE_F	0.24	0.24		0.27	0.49	0.43	0.41	0.34	0.22
	HE_B	0.16	0.16	0.22		0.49	0.44	0.42	0.34	0.55
	HE_L	0.26	0.26	0.26	0.26		0.34	0.61	0.39	0.45
	HE_R	0.19	0.19	0.22	0.22	0.18		0.32	0.40	0.34
	AB_L	0.25	0.26	0.25	0.24	0.25	0.17		0.35	0.43
	AB_R	0.21	0.21	0.20	0.20	0.20	0.19	0.19		0.36
	TO_B	0.20	0.18	0.16	0.22	0.25	0.19	0.24	0.21	

CD co-directed OD oposit-directed XD cross-directed

Figure 5.12 Correlation average and standard deviation [ML12].

Correlation properties are also investigated in Rosini and D'Errico [RD13], starting from channel data acquired as described in Rosini and D'Errico [RD12c]. Links correlation is evaluated as the cross-correlation between the fading contributions of the channels, which is influenced by ANT's characteristics and measurement conditions. For instance, correlation levels in LoS strongly depend on on-body location, while this effect is reduced when in NLoS, due to the masking effect of the body [RD13]. Channels coherence over time is studied as its time correlation and is computed as the normalised autocorrelation function. ANT type and measurement conditions have a strong impact also on this parameter. In particular, moving from LoS to NLoS implies a faster decorrelation, and autocorrelation functions present a less smoother evolution, due to the stronger MPCs that makes the channel less stable over time [RD13].

The authors in Marinova et al. [MTT+13] propose a method for determining the minimum number of ANTs to be used in an UWB MIMO configuration in order to provide low correlation, hence higher diversity. Electromagnetic fields around a human body are generated for three sampled frequency representing the whole UWB (i.e., 3.1, 6.85, and 10.6 GHz). To solve this optimisation problem, the authors assume that if a combination of n ANT positions is optimal, then the same combination is used to find the next position for an optimal $n + 1$ configuration [MTT+13]. Once the positions are identified, simulations are run including ANT characteristics mounted on a human body phantom, and new correlation and diversity matrix are evaluated. Results are compared with those obtained for randomly positioned ANTs or a set of most commonly used on-body locations, as shown in Figure 5.13. The proposed methods generally achieves the best results, while the most common positions solution is the worst one, proving the need for an alternative method to chose the optimal ANT configuration.

5.5.2 B2B Channel

A limited number of contributions in literature focus on B2B communications, and just few of them propose a precise radio channel characterisation. Papers by Rosini et al. [RDV12, RVD14] present an experimental modelling of the B2B channel based on indoor measurements at 2.45 GHz. Two couples of subjects carrying some nodes placed on several locations and performing different type of movements are involved. To account for ANT's effect, measurements are repeated with two sets of ANTs. Similar to Rosini and D'Errico [RD12c], the channel is modelled through a mean channel gain

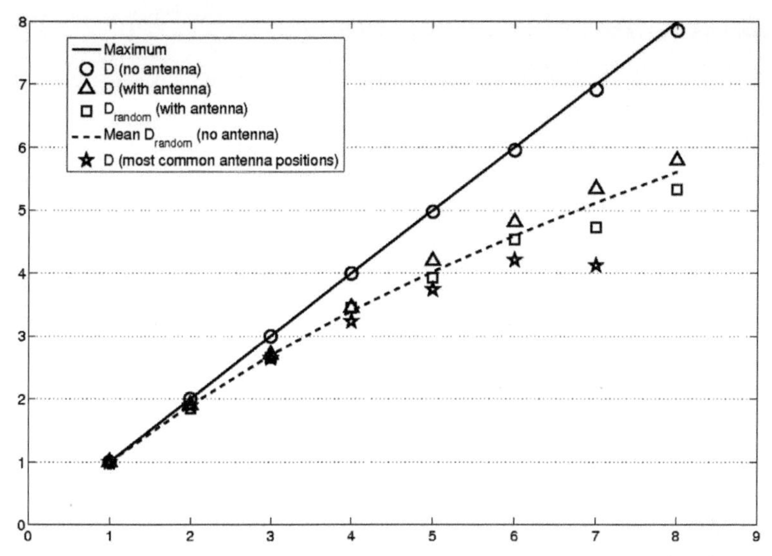

Figure 5.13 Diversity for different nodes configurations [MTT+13].

and a fading contributions. The mean gain is characterised as a log-distance dependent component, defined by the PL exponent and its value at the reference distance of 1 m. Results show that it depends on the node position and ANT type. The linear envelope of the fading component, accounting for the MPCs from the moving body and the environment, is described by a Nakagami distribution [RDV12].

The body shadowing effect is also characterised; it extracted from the channel power transfer function following the slow variations of the channel gain over its mean value, and it is accounted for as an additional loss contribution as a function of subject orientation [RVD14]. The distribution of shadowing values takes a two-lobe configuration, which is modelled using a mixture distribution composed of two gaussian distributions [RVD14].

Different type of movements result in specific effects on channel time-variant characteristics. For instance, Figure 5.14 shows the different evolution of the channel power transfer functions for the case of the two subjects walking side by side and when one of them walks towards the other. In the first case the oscillating trend of curves is due to the subjects' hand that cyclically masks the direct communication link (nodes are on both subjects' hip) while walking; while in the second scenario the curves smoothly decrease as the distance between users increases, presenting neither periodic behaviour nor strong supplementary attenuations, since nodes are always in LoS conditions.

5.6 MAC Protcols and Upper Layers

The peculiarities of BAN propagation channel, arising from the body presence, can impact the design of MAC and upper layer communication protocols. Network reliability and energy efficiency are among the most critical requirements [BSM+12] to be met, and common solutions used in generic wireless sensor networks (WSNs) may not be appropriate for *body-centric communications* [CGV+11]. Moreover, considering the proposed ISM operational frequency band at 2.45 GHz, BANs are expected to coexist with other co-located ISM-based wireless devices. As such, interference from coexisting wireless systems (e.g. Blutooth, ZigBee, and Wi-Fi) or even other nearby BANs could degrade the reliability of BAN communications [ABSV13]. A precise characterisation of interfering sources, and the evaluation of achievable performance for multiple BANs, is hence of outmost importance for the design of appropriate interference mitigation and management techniques [Ros14].

5.6.1 MAC Protocols for BANs

The wide variety of applications for BANs imposes a set of system requirements that none of the existing air interfaces are able to meet Rosini et al. [RMM+12]. As a result, the IEEE802.15.6 group was established to develop a standard (released in 2012) optimised for short-range communication in the vicinity of, or inside, the human body [IEEE12b]. A description of standard's main features, particularly those related to the medium access control (MAC) layer, can be found in Cavallari and Buratti and Subramani et al. [CB14, SCS13]. Performance of random and scheduled access in terms of packet drop, latency, packet delivery ratio (PDR) and reception occurrences for different data rates and transmit powers is evaluated and compared in Subramani et al. [SCS13]. Considering just the beacon access mode with superframe (SF) boundaries, authors Cavallari and Buratti [CB14] propose an analytical model to characterise the performance of IEEE802.15.6 carrier sense multiple access with collision avoidance (CSMA/CA), where nodes are assigned a different priority to access the medium according to the traffic type. Network performance is evaluated in terms of packet reception success probability, and no retransmission is allowed. The model is validated through experiments. This study can provide information on the expected performance of these networks, as well as criteria to optimise the assignment of node priorities [CB14].

Apart from IEEE802.15.6, other standards such as IEEE802.15.4 [IEEE06] and Bluetooth [blu10] are also considered for BANs communications. In particular, authors in Rosini et al. [RMM+12] present simulation

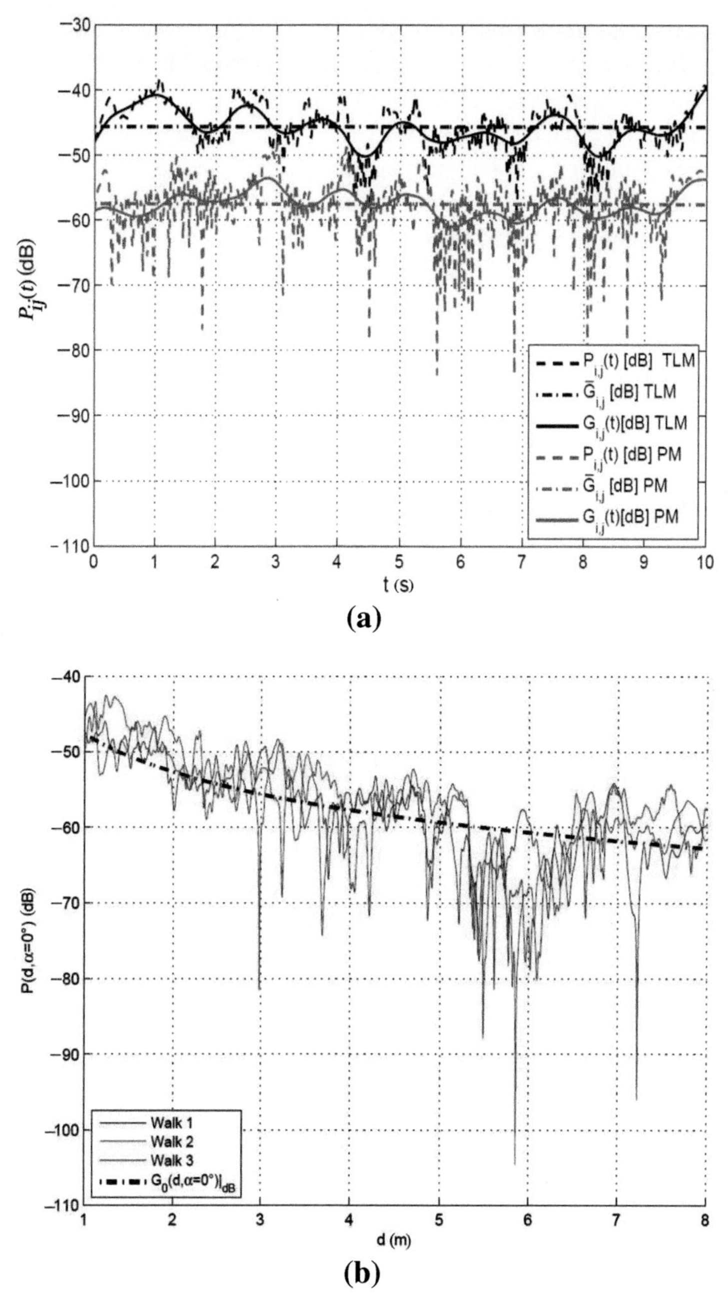

Figure 5.14 Channel gains for different movements [RVD14, RDV12] (a) Walking side by side, and (b) One subject walking towards the other.

results comparing the performance of two CSMA/CA protocols (as defined by the IEEE802.15.4 and 802.15.6 standards), and the IEEE802.15.6 Slotted Aloha algorithm, combined with three PHY layer solutions. The performance metrics considered are packet loss rate (PLR), average delay, and energy consumption. To account for realistic propagation conditions, real-time dynamic channel measurements obtained for two sets of ANT are included in the simulations [RD12c]. Figure 5.15 shows a sample of the results (i.e., PLR as a function of the packet payload); it is generally quite difficult to identify the best performing MAC, as the ANT type, the underlying PHY and the considered metric can greatly impact the outcome. For this reason, flexible protocols should be designed in order to adapt to different applications requirements. The results from Rosini et al. [RMM$^+$12] could serve as guidelines for the selection of the best protocol given a specific application.

Energy consumption, which is one of the critical issues in BANs, is the focus of Maman et al. [MDPRD12]. A well-designed communication scheme managed by an efficient MAC protocol can minimise the energy consumption of these networks. Authors in Maman et al. [MDPRD12] present

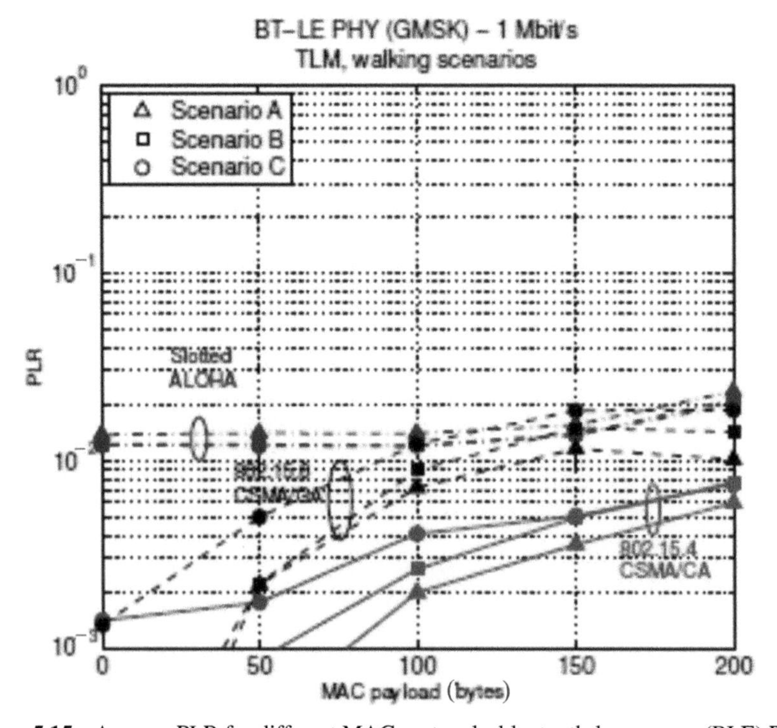

Figure 5.15 Average PLR for different MAC protocols, bluetooth-low energy (BLE) PHY.

the performance study of dynamic relaying and cooperative routing schemes to ensure energy efficiency. Simulations are run considering a coordinator node located on the hip and three potential relays on the wrist, shoulder and thigh. The MAC protocol considered (i.e., BATMAC [MO11]) uses the beacon packet error rate (PER) information in order to efficiently manage the cooperative relay mechanism. Packet reception decision is then based on PER value, which itself depends on the instantaneous signal-to-noise ratio (SNR) [DRM11]. Several relaying mechanisms based on beacon reception feedback are proposed and compared with the objective of ensuring energy efficiency, low latency and high data delivery ratio. Energy consumption levels are evaluated considering the characteristics of on-the-shelf Bluetooth-compliant devices. Both for an ideal case (i.e., stationary channel) and more realistic conditions (i.e., accounting for the impact of fast fading), the mix strategy proposed in Maman et al. [MDPRD12] provides the most efficient trade-off between energy consumption and reliability.

5.6.2 Interference Management and Coexistence Issues in BANs

One of the frequency bands envisaged for BAN operation is the 2.45 GHz ISM band. Considering that several existing communication standards such as Wi-Fi, Zigbee, and Bluetooth work in the same band, interference with such systems may take place and needs to be characterised. To that purpose, authors Martelli and Verdone [MV12] study some of these coexistence issues to evaluate the performance of a BAN operating in presence of nearby wireless devices working with IEEE802.11 (Wi-Fi) [IEEE11] and IEEE802.15.4 [IEEE06] protocols. Three modulation schemes at the PHY along with two CSMA/CA algorithms at the MAC layer are considered for performance evaluation [RMM+12]. Simulations are set up for a human subject wearing a BAN and walking in a room, while either a Wi-Fi AP and a notebook or a Zigbee network is active. BAN performance is assessed in terms of PLR, average delay and average TP as a function of the packet payload. Numerical results show that independent from the considered PHY/MAC solution, the PLR is the parameter mainly affected by the presence of interference, in particular when coexisting Wi-Fi systems are working. Therefore, to guarantee an acceptable level of performance, a dynamic selection of BAN operating channels could be extremely important [MV12].

As there are currently no coordinating mechanisms among multiple co-located BANs, co-channel interference from transmissions in adjacent BANs could impact the reliability and the QoS experienced by a node within an individual BAN. Authors in Alasti et al. [ABSV13] develop a flexible simulation platform that allows for statistical evaluation of interference in multi-BAN scenarios and performance of possible mitigation algorithms. The platform consists of a virtual room with variable number of BANs, moving around according to a defined motion scenario. Assuming a time division multiple access (TDMA) transmission protocol, several uncoordinated slot assignment strategies are proposed in Alasti et al. [ABSV13, BSA15] to mitigate inter-BAN interference. The best strategy, referred to as *"Minimum Required SINR Assignment"*, tries to re-allocate the slot assignment according to a protocol shown in Figure 5.16. Exploiting channel correlations and based on the experienced interference in the current frame, the protocol identifies the timeslots (i.e., those with SINR above the required minimum) that are least likely to collide with other BAN interferers in the next frame. The improvements in the experienced signal-to-interference ratio (SIR) and link

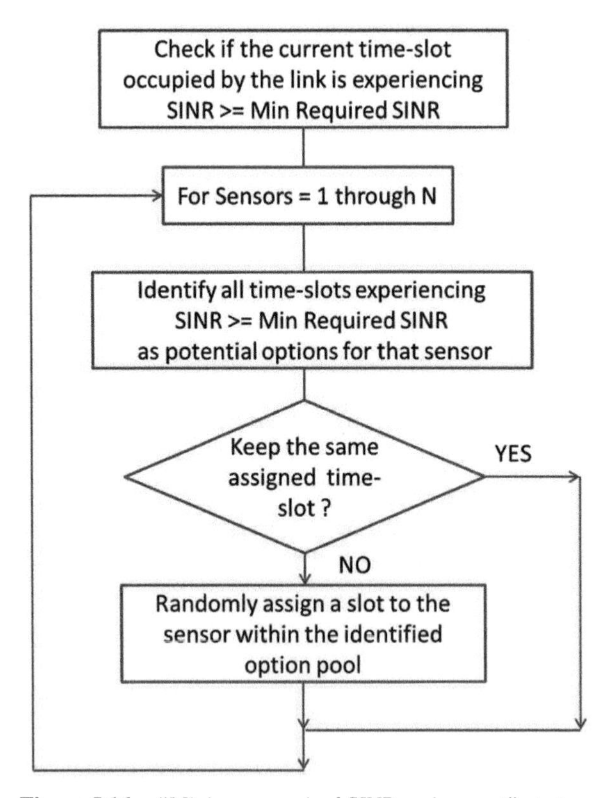

Figure 5.16 *"Minimum required SINR assignment"* strategy.

outage probability show the effectiveness of such uncoordinated mitigation strategies for several motion scenarios.

For off-body communication in dense BANs environment, authors in Mijovic et al. [MBZV14] propose cooperative transmissions to enhance link quality. It is assumed that the nodes of a BAN can cooperate and form a virtual ANT array (VAA) in order to transmit data to one of the sinks in an indoor environment. The latter are equipped with multiple ANTs, hence, a MIMO system can be established. Using beamforming, the ANT elements on both RX and TX sides set their weight coefficients as to enhance the received signal. Several scheduling techniques that sinks may adopt to optimise various performance metrics (e.g., block error rate (BLER), SNR, and Jain index) are considered. Simulation results show the advantages of cooperative beamforming in terms of SNR and BLER, over traditional SISO systems [MBZV14].

5.7 CM Parameters

In this section, we report the main parameters of the CM developed in the frame-work of the TWGB of IC1004 action. For the model explanation, the reader can refer to the previous sections and the references reported. CM parameters are derived from different measurement campaigns, involving different human subjects, ANTs and environments. For this reason some simplifications and merging of results were needed.

The on-body CM is derived for UWB and narrowband 2.4 GHz communications. The parameters here reported mainly refer to dynamic conditions, i.e., to walking scenarios. In UWB the results are reported for dipole and loop ANTs. At 2.4 GHz the results depend on the orientation of the ANT main polarisation orientation with respect to the body surface (i.e normal or tangential). Nodes naming are detailed in Figure 5.17. All the parameters avaialble, e.g. the space-time correlation properties, are not reported here for the sake of briefness, but Matlab implementations have been developed [ARD14].

For the off-body channel, node-specific parameters have been extracted (see Section 5.5). However here the basic idea is to extract a model able to account for the main aspects affecting channel characteristics and that are related to devices spatial diversity. For this reason the human body was partitioned in three areas, namely *Head, Upper Body,* and *Lower Body,* as shown in Figure 5.17. The *Head* part is composed of the whole skull and the neck. In practice the results here consider an ANT placed on the righ ear. It should be noted that for the *Head* part the proposed model applies to the case

Name	Position
LE	Left Ear
RE	Right Ear
LE	Left Elbow
RE	Right Elbow
LC	Left Chest
FC	Front Chest
RC	Right Chest
LHi	Left Hip
FHi	Front Hip
RHi	Right Hip
LTi	Left Tight
RTi	Right Tight
LK	Left Knee
RK	Right Knee
LF	Left Foot (Ankle)
RF	Right Foot (Ankle)

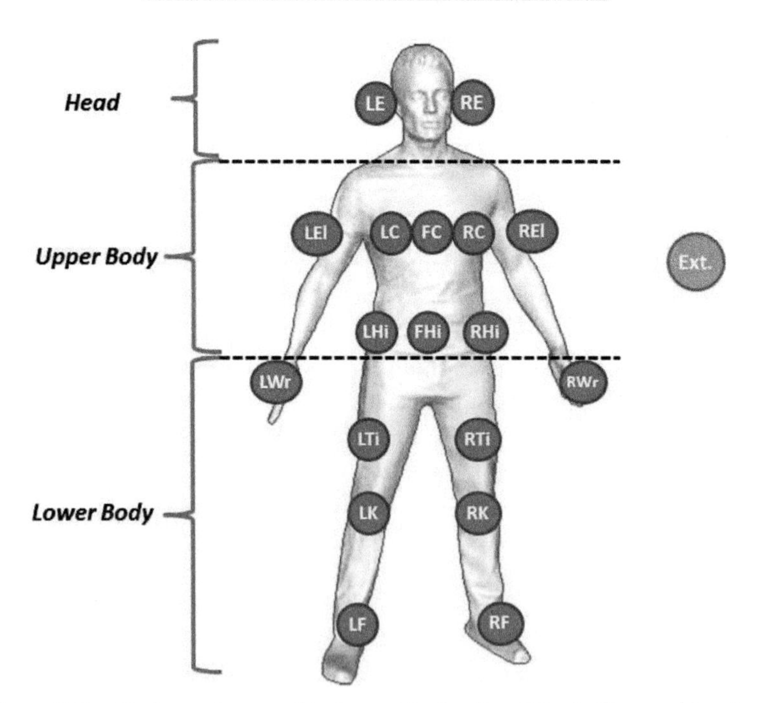

Figure 5.17 Nodes position, and body partitioning for *off-body* CM considerations.

where nodes are placed on user's ears, face, neck or back of the neck. Whean a node is placed on the top of the head, a classical indoor CM can apply, since there is no shadowing from the body. The *Upper Body* is the portion of the body included between the shoulders and the hip, plus the upper part of the harms, whereas *Lower Body* refers to the lower section of the body including feet and hands, hence the most mobile part of the body. The focus of this CM is on body dynamics and particulalry on walking conditions. When possible ANT topology-dependent parameters are given.

5.7.1 On-Body CM

5.7.1.1 Dynamic UWB on-body channel

In this subsection, we report the fading parameters for the UWB on-body channel, focusing on the static and dynamic, i.e., walking condition. For channel gain parameters the reader can refer to references in Section 5.4.1. The parameters refer to different distributions. The PDF of the Weibull, Lognormal and Gaussian distributions are:

$$f(x|a,b) = \frac{b}{a}\left(\frac{x}{a}\right)^{b-1} \exp\left[-\left(\frac{x}{a}\right)^{b}\right], \ x > 0 \tag{5.3}$$

$$f(x|\mu,\sigma) = \frac{1}{x\sigma\sqrt{2\pi}} \exp\left[\frac{-(\text{In}\ x-\mu)}{2\sigma^2}\right], \ x > 0, \tag{5.4}$$

$$f(x|\alpha,\beta) = \sqrt{\frac{\beta}{2\pi x^3}} \exp\left[-\frac{\beta}{2\alpha^2 x}(x-\alpha)^2\right], \ x > 0, \tag{5.5}$$

5.7.1.2 Dynamic on-body channel at 2.4 GHz

In this section, we report the On-Body CM parameters according to the description given in Section 5.4.2. The model is based on the separation of the *mean Channel Gain*, charaterised by its mean value μ_0 and the standard deviation σ_0, taking into account the human body variability. The *long-term fading* (or shadowing) is normmally distributed in dB (standard deviation σ_s), while the *short-term fading* follows a Rice distribution characterised by its *K* factor. The space–time correlation properties are not reported here, but can be found in Pasquero et al., Oliveira and Correia, and Arnesano et al., [PRDO14, OC13d, ARD14]. (Tables 5.3 and 5.4)

5.7.2 Off-Body CM

The proposed off-body radio CM describes the *off-body* channel power transfer function in *dB*, *P* (d, α), as a function of the distance *d* between devices

Table 5.3 CM parameters: normal polarisation

			Channel Gain		Shadowing	Fading
	Position 1	Position 2	μ_0 (dB)	σ_0 (dB)	σ_s (dB)	K
Anechoic	LHi	RTi	−38.78	2.06	1.43	41.32
	LHi	RWr	−52.09	4.58	4.05	15.43
	LHi	LWr	−40.32	3.17	3.73	15.43
	LHi	LE	−47.24	2.02	1.35	114.78
	LHi	RF	−63.26	5.13	2.03	2.73
	LHi	RC	−45.95	1.44	0.88	33.65
	LHi	REl	−58.28	5.79	4.11	9.27
	LHi	RE	−73.10	6.37	4.34	3.79
	LHi	RK	−57.44	5.17	3.24	3.45
	LHi	Rhi	−46.82	2.33	1.32	15.17
	LHi	RWr	−59.69	6.16	3.09	1.09
	FC	RTi	−38.45	1.06	1.07	130.07
	FC	RWr	−47.07	2.16	3.46	130.07
	FC	LWr	−43.86	2.46	3.67	122.07
	FC	LE	−48.62	3.62	1.06	171.47
	RE	RTi	−55.13	3.29	4.06	25.51
	RE	RWr	−53.82	3.92	2.25	105.02
	RE	LWr	−66.59	4.98	2.53	30.18
	RE	LE	−35.81	0.82	0.12	8433.13
Indoor	LHi	RTi	−38.4	1.17	1.01	29.59
	LHi	RWr	−53.09	2.89	3.98	5.83
	LHi	LWr	−40.74	4.32	3.45	6.72
	LHi	LE	−48.3	2.69	2.79	10.33
	FC	RTi	−38.65	1.15	1.4	69.75
	FC	RWr	−49.47	1.2	2.57	21.97
	FC	LWr	−42.97	3.19	1.97	54.67
	FC	LE	−46.82	4.57	1.5	17.34
	RE	RTi	−57.57	1.84	3.17	1.44
	RE	RWr	−53.31	3.19	2.89	7.40
	RE	LWr	−57.57	0.91	1.72	0.00
	RE	LE	−37.01	2	0.67	50.00
Indoor Dense	LHi	RF	−43.61	5.98	2.79	1.05
	LHi	RC	−40.42	5.40	2.81	2.33
	LHi	REl	−38.85	5.34	2.48	2.53
	LHi	RE	−40.75	5.74	2.58	2.07
	LHi	RK	−39.13	4.68	2.46	2.97
	LHi	Rhi	−40.01	5.22	2.78	3.31
	LHi	RWr	−41.42	6.06	2.36	0.93

Table 5.4 CM parameters: tangential polarisation

			Channel	Gain	Shadowing	Fading
	Position 1	Position 2	μ_0 (dB)	σ_0 (dB)	σ_s (dB)	K
Anechoic	LHi	RTi	−38.4	1.17	1.01	29.59
	LHi	RWr	−53.09	2.89	3.98	5.83
	LHi	LWr	−40.74	4.32	3.45	6.72
	LHi	LE	−48.3	2.69	2.79	10.33
	FC	RTi	−38.65	1.15	1.4	69.75
	FC	RWr	−49.47	1.2	2.57	21.97
	FC	LWr	−42.97	3.19	1.97	54.67
	FC	LE	−46.82	4.57	1.5	17.34
	RE	RTi	−57.57	1.84	3.17	1.44
	RE	RWr	−53.31	3.19	2.89	7.40
	RE	LWr	−57.57	0.91	1.72	0.00
	RE	LE	−37.01	2	0.67	50.00
Indoor	LHi	RTi	−60.77	4.89	3.05	4.79
	LHi	RWr	−61.69	6.9	3.76	5.56
	LHi	LWr	−49.62	2.95	7.22	8.68
	LHi	LE	−64.26	3.3	2.38	3.58
	FC	RTi	−62.72	2.41	1.82	3.00
	FC	RWr	−60.15	5.79	2.99	3.39
	FC	LWr	−58.35	1.5	4.21	5.56
	FC	LE	−60.48	3.5	1.96	5.56
	RE	RTi	−67.01	2.77	2.23	1.70
	RE	RWr	−59.18	2.27	2.58	3.39
	RE	LWr	−65.67	4.63	2.37	2.43
	RE	LE	−54.67	4.73	2.15	5.95

and the angle a representing the body orientation with respect to the outer device (i.e., the off-body one).

$P(d, \alpha)$, in dB, can be expressed according to:

$$P(d,\alpha) = G(d,\alpha) + F(d); \tag{5.6}$$

where $G(d, \alpha)$ is the *mean channel gain* and $F(d)$ represents the *fast fading* component, a random variable accounting for the multi-path contributions (MPCs) originating from the environment or from the body itself while moving. $G(d, \alpha)$ can be described as:

$$G(d,\alpha) = G_0\left(d_0,\alpha\right) - 10\cdot n(\alpha)\cdot \log_{10}\left(d/d_0\right); \tag{5.7}$$

where $G_0(d_0, \alpha)$ represents the mean channel gain evaluated at the reference distance $d_0 = 1\ m$ and $n(\alpha)$ is defined as the PL exponent.

Table 5.5 Model parameters for mean channel gain: off-body scenario

	LoS ($\alpha = 0°$)	
	$n(0)$	$G_0(d_0, 0)_{\|dBm}$
Ext./Head	[1/1.6]	[–45.7/–50.2]
Ext./Upper Body	[1.7/2]	[–45.4/–51.7]
Ext./Lower Body	[0.4/1.6]	[–45/–47.8]

Table 5.5 lists the numerical values derived for the mean channel gain, $G_0(d_0, \alpha)$, and the PL exponent, $n(\alpha)$. Results for the LoS case (i.e., a = 0°) are reported, referring to the scenarios where the subject was aligned to the external device. In order to characterise the *External Device/Head* communications, LoS values in Table 5.5 represents the case where nodes are placed on right ear while a generic subject is facing the outer device (Figure 5.18). Those values can be extended to other head parts, properly adapting α according to the subject head and trunk orientation.

When rotating, a channel gain correction factor can be applied according to the off-body ANT considered. Results in Table 5.6 are given normalised to the LoS case, representing in that way an additional attenuation to be added to the $G(d, 0°)$ values reported in Table 5.5. Given the disparity of results here the results are given by distinguishing between the ANT polarisation on the body and the kind of ANT off-body. Generally when considering omnidirectional off-body ANT this is limited to the azimuth plane, and the main polarisation is

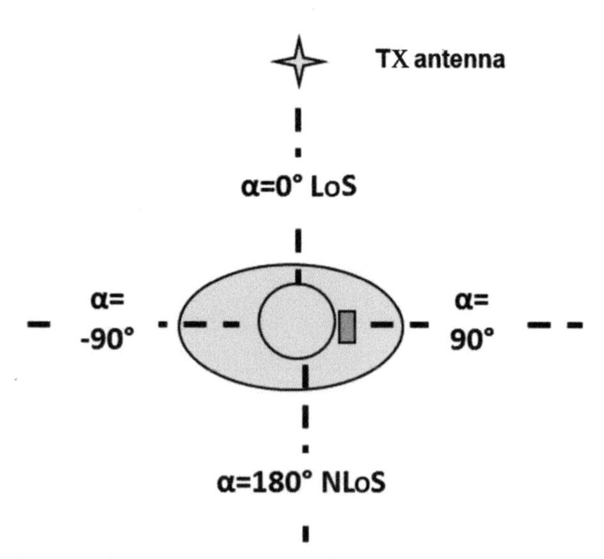

Figure 5.18 LoS and NLoS conditions and body orientation α.

Table 5.6 Channel gain correction factors for intermediate off-body orientation: Body Shadowing effect

Off-body ANT	Omni			Omni			Directive		
On-body ANT	Tangential Pol.			Normal Pol.			Tangential Pol.		
	90°	180° (NLoS)	−90°	90°	180° (NLoS)	−90°	90°	180° (NLoS)	−90°
Ext./Head	−5.4	−3	6.6	−5.2	−2	−8.4	−12.2	−5.7	9.9
Ext./Upper Body	−7.7	−20	−14	6.4	−12.8		−14.4	−22	−14.1
Ext./Lower Body	–	–	–	–	–	–	14.2	−0.4	−6.4

vertical with respect to the ground. On the other hand off-body directive ANT, also vertically polarised, is generally oriented towards the body, and typical gain is around 7 dBi.

As expressed by the general formula, besides the channel gain correction factor, the path LoSs expoinen is aLoS affected by the body rotation. The the PL exponent values in NLoS, i.e., $\boldsymbol{n(180°)}$, are generally vey low, between 0.25 and 0.6. Some specificic cases can be found with pathlosses closeto 1.5, for particular situations when the body orientation do not obstruct the LoS path, as for instance when the ANT is placed on a ear.

The statistic of the *fast fading* envelope is modelled by a Rice distribution, which is defined by the K factor, the noncentrality and scale parameters, ν and α, as reported in Table 5.7.

Table 5.7 Fading characterisation for off-body CM

	ν LoS	σ LoS	K LoS
Ext./Head	[0.8/0.9]	0.3	[3.6/4.5]
	ν NLoS	σ NLoS	K NLoS
	[1/1.1]	[0.6/0.7]	[1/1.7]
Ext./Upper	ν LoS	σ LoS	K LoS
	[0.9/1]	[0.5/0.6]	[1.1/2]
	ν NLoS	σ NLoS	K NLoS
	0.001	0.9	0
Ext./Lower	ν LoS	σ LoS	K LoS
	1	0.3	5.6
	ν NLoS	σ NLoS	K NLoS
	[0.9/1]	[0.5/0.7]	[0.8/2]

The results here reported on fading statistics refer only to those scenarios were omnidirectional ANTs are used as off-body node. When using the directive ANTs, the LoS path could be enhanced, giving a higher K factor (up to 35 in some scenarios). These results are not reported here, since they are strongly dependent on the angular selectivity properties of the off-body ANT, and cannot be generalised for a model, based on the measurements available.

6

Green and Efficient RAN Architectures

Chapter Editor: S. Ruiz, Section Editors: M. Garcia-Lozano, D. Gonzalez, M. Lema, J. Papaj, W. Joseph, M. Deruyck, N. Cardona, C. Garcia, F. J. Velez, L. Correia, L. Studer, P. Grazioso and S. Chatzinotas

6.1 Introduction

The explosive increase of capacity demand has resulted in more dense deployments involving a growing number of BSs. To make future mobile networks sustainable from an economic viewpoint, they should provide more while simultaneously cost less. Installing smaller cells in areas where infrastructure can provide the required backhaul connectivity is the current natural evolution of radio access network (RAN) infrastructure. But heterogeneous networks (HetNets) and small cells require new techniques for configuration, management, and optimisation oriented to self-organising networks (SONs), as well as new RRM algorithms, so that full capacity can be used in the most efficient way, considering not only services but also users' profiles. Another key concept to improve RAN efficiency is referred to as Cloud-RAN, a centralised processing, collaborative radio based in resource sharing and real time cloud computing to adapt to non-uniform traffic. The cloud radio access network (C-RAN) concept is expanded to include also the evolved packet core (EPC) functionalities, in what is known as network Virtualisation. Considering that wireless networks will serve not only a huge number of mobile phones and computers, but also a dense deployment of devices and sensors, effort has to be done to improve the overall energy and spectrum efficiency, combined with opportunistic access to certain frequency bands and cognitive radio (CR) devices.

Chapter 6 deals with all these topics and summarises advances done by European cooperation in science and technology (COST) IC1004 researchers. Sections 6.2 and 6.3 are devoted to explain advanced resource management ecosystem for both, cellular and wireless networks considering resource scheduling, interference, and power and mobility management. Section 6.4

Cooperative Radio Communications for Green Smart Environments, 195–270.

describes recent energy efficiency enhancements by analysing power consumption, cell switch off, and deployment strategies. Section 6.5 is focusing on spectrum management optimisation and cognitive radio networks with tools and algorithms specifically for TVWS. Section 6.6 defines and models the virtualised and cloud architecture by proposing new algorithms under realistic (RL) scenarios. Finally, Section 6.7 is addressed to re-configurable radio for heterogeneous networking by including recent progress in resource allocation, diversity, and adaptive antenna techniques to improve capacity and energy efficiency.

6.2 Resource Management Ecosystem for Cellular Networks

6.2.1 Radio Resource Scheduling

In the context of long-term evolution (LTE) networks, the packet scheduler can be defined as the selection of the user to be served in every 1 ms transmission time interval (TTI) and frequency resource block, that is to say, the resource allocation is made in both time domain and frequency domain. A good design of a scheduling algorithm should balance the trade-off between the maximisation of the system throughput and fairness among users.

The two basic types of scheduling that have been defined in LTE are dynamic and semi-persistent. In the first case, the scheduler reacts to user equipment (UE) requests opportunistically. Resources are dynamically adapted to users according to their buffer status and channel conditions, but at the cost of an important layer 1 (L1)/layer 2 (L2) load, carried by the physical downlink control channel (PDCCH). The semi-persistent option is more adequate for services making use of small but recurring packets, such as voice over IP (VoIP). In this case, only the first assignment needs to be signalled and the same allocation is kept in subsequent transmissions, with the corresponding savings in the L1/L2 signalling. The policy is *semi*-persistent in the sense that the allocation can be changed at some point, for example to provide certain link adaptation. On the other hand, the lack of flexibility implies a less efficient use of radio resources.

LTE networks are continuously being improved through the different 3GPP releases, which means new challenges and open issues to be considered when designing scheduling policies. Examples are the problem of L1/L2 signalling, self-optimisation of scheduling policies, specific uplink (UL) issues, and operation through different aggregated carriers. All these topics have been a matter of research in the context of COST action IC1004 and are covered in the following lines.

The impact of control channel limitations has been widely evaluated for VoIP but missed for non-real time services. González et al. [GGRL11] carefully modelled practical limitations of PDCCH and studied the trade-offs associated to control channel usage and the provision of quality of service (QoS) for this type of services. Thus, covering an aspect omitted in previous literature. The PDCCH is critical in LTE networks for a correct functioning of scheduling algorithms, since, among other information, it carries DL and UL scheduling assignments. The authors conclude that the selection of scheduling and QoS parameters is clearly affected by the limited amount of control channel resources in LTE. Results suggest that in order to let the system operate efficiently from the radio resource allocation perspective, it is important to carefully characterise the system performance. Not only considering traffic features, but also the scheduling policy and availability of resources for the PDCCH. This is particularly true for cases in which users mostly target low bit rates, in this case the system capacity would be seriously penalised.

Scheduling policies have been widely studied in the last years. However, the complexity of the LTE scheduler, typically divided in a time domain policy plus a frequency domain one, leaves a lot of margin to improve classic schemes. One of the novel ideas proposed by Comsa [CAZ+13] et al. is a new scheduling approach able to find the optimum scheduling rule at each TTI, rather than using one single discipline across the whole transmission as typically done. This idea is investigated along several research works that are interrelated and constitute a very detailed description of this novel strategy.

In the work of Comsa et al. [SMS+12b], the authors propose to assess the corresponding scheduling utility function in a TTI basis, rather than a global and unique evaluation. Thus, the problem to be solved is the selection of proper rules to achieve local optimisation that would yield a higher global utility. Since evaluating all possible rules in a 1-ms time scale would lead to an unaffordable complexity, a Q-learning approach is proposed and investigated. The algorithm uses reinforcement based on rewards from previous transmission sessions, so that prior experiences are turned into permanent. However, since convergence might require excessive number of evaluations, the authors extend the approach to work as a multi-agent (parallel) system.

The description of several scheduling rules is done in Comsa et al. [SMS+12c]. Also trying to capture the well known trade-off between system throughput and fairness, a candidate objective function being an aggregation of the Jain index and normalised system throughput is proposed. Initial performance results in synthetic scenarios (pedestrian and vehicular) indicate a successful operation. The evaluation is done in terms of new quality metrics named global utility, tradeoff utility, and supreme utility.

The scheduler should dynamically self-tune and respond to the network state. Hence, given a certain array of reported channel quality indicator (CQI) values and the distribution of normalised user throughput, the best scheduling rule is decided. However, one of the problems of Q-learning is the exploration–exploitation compromise. Since the number of possible states is extremely large, it is unfeasible that all states had been explored in the past, thus the authors introduce the use of a neural network working as function approximator [SMS+12a]. Results indicate a superior performance of this approach when compared to standard schedulers. The complete architecture of the scheduling policy is depicted in Figure 6.1.

The authors propose extensions in the form of a self-optimisation strategy that adjusts the level of system fairness [CAZ+13]. In this case the trade-off throughput versus fairness is studied by evaluating the cumulative distribution function (CDF) of the normalised user throughput, which provides a more complete picture of the system performance. The key point is that a level of accepted fairness should be provided and the scheduler should be kept on that range of tradeoff. A possible fairness criteria is guaranteeing that (100–x)% of users achieve at least x% of the normalised user throughput. This way, under or over fair situations can be detected and avoided. This is graphically depicted on Figure 6.2. The fairness level ranges from pure

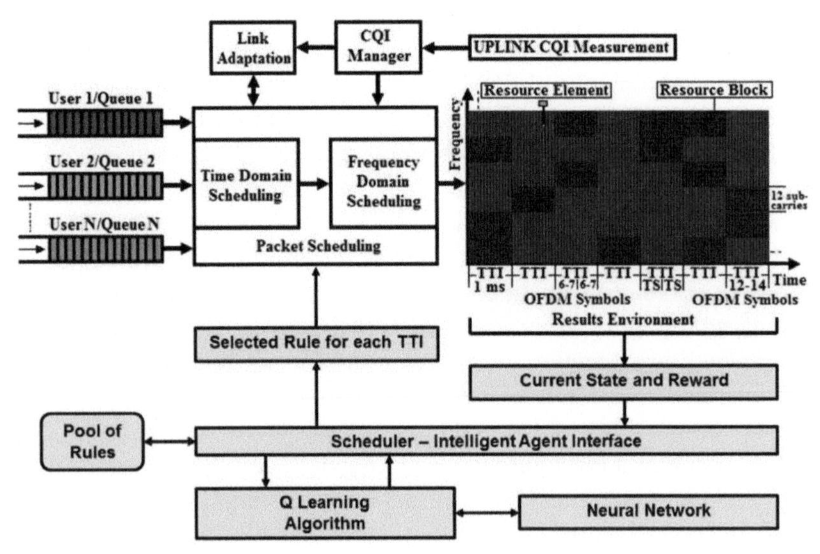

Figure 6.1 Long term evolution - advanced (LTE-A) packet scheduler framework from Comsa et al. [SMS+12a].

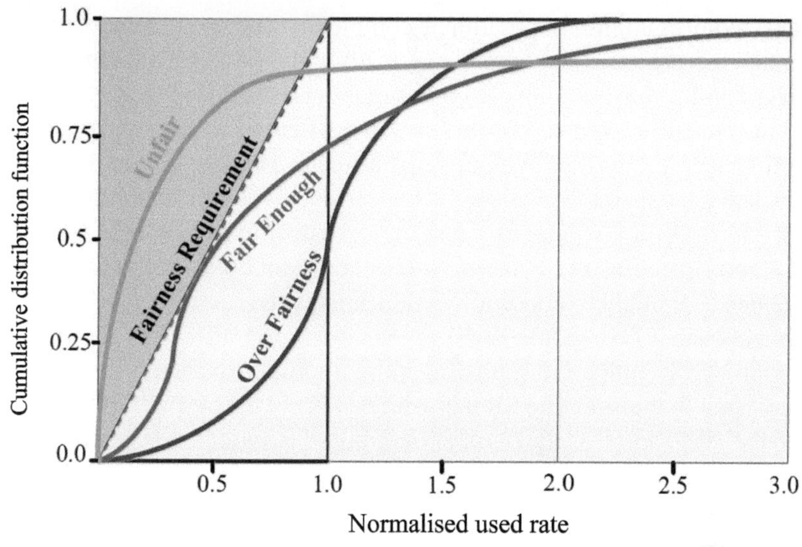

Figure 6.2 Fairness requirement from the CDF point of view [CAZ+13].

proportional fairness to a classic maximum carrier to interference ratio (CIR) selection, to be dynamically controlled by a parameter α. The evaluation of all possible α is infeasible at every TTI; however, this type of brute force search is not necessary since, as previously indicated, reinforcement learning is a component part of the approach. Four reinforcement learning techniques are proposed and described in this work. Results are further improved by Comsa et al. [CZA+14], where the previous idea is extended to a double parametrisation scheduling policy. This way, two parameters α and β are optimised to get the desired feasible state, situation with the required throughput versus fairness trade-off. Given the more restrictive domain of the parameters, a faster convergence time is obtained, which is an interesting feature for real time applicability. Improvements in terms of throughput and percentages of TTIs with a feasible state are achieved with respect to previous approaches. A second benefit is that lower fluctuations are obtained when the traffic load drastically changes, indeed the mechanism shows the ability of recovering a feasible state in less than 10 TTIs when severe changes in the traffic load and user channel conditions appear.

Allocation of resources in the UL poses new challenges. For example, in the first two releases of LTE the use of SC-FDMA imposes a contiguous allocation and equal power distribution. This implies that finding an optimal solution to the problem of ergodic sum-rate (SR) maximisation under proportional rate constraints requires high-computational complexity and existent proposals are unsuited for practical applications. Cicalò and Tralli [CT14a] address this problem and proposes a novel sub-optimal heuristic solution after using Lagrangian relaxation of rate constraints. In particular, in order to reduce the search space, the authors make use of a linear estimate of the average number of sub-carriers that are allocated to each user when the optimal rate is achieved. After posing the dual problem, it is possible to exploit an adaptive implementation to build the heuristic algorithm, which either dynamically tracks the per-user channel condition and estimates the number of allocated subcarriers at each TTI.

The approach is compared against an optimal solution and a quasi-optimal from the existent literature. The SR gap with respect to the optimal solution is limited to the 10%, whereas the number of required iterations is two orders of magnitude lower. The complexity of the algorithm just increases linearly with the number of users and number of resources, thus being a much more interesting option for practical operation. The algorithm also preserves long-term fairness for homogeneous full-buffer traffic, with a Jain Index greater than 0.99 in all the investigated cases. However, in order to improve the real applicability of the scheduler, short-term fairness evaluation is also required. The study by Cicalò and Tralli [CT14b] shows that the proposal provides excellent results when evaluated over time windows larger than 200 ms. Hence, the scheduler results in an attractive solution for applications where QoS requirements depend on the rate averaged over intervals of that order of magnitude. Finally, under heterogeneous traffic conditions, the scheduler is able to achieve a higher ratio of satisfied users with granted bit rate services when compared to proportional fair-based schedulers. All this, with a significant reduction of the packet delay and without starving non-granted bit rate users.

Abrignani et al. [AGLV15] present an approach of UL scheduling for LTE networks deployed over dense HetNets. In particular for the co-channel layout, in which macros and small cells operate at the same carrier, thus with strong inter-cell interference. The problem is tackled from two different viewpoints, first a three-step-based algorithm and, second, a method that solves the scheduling optimisation at a time. These algorithms are posed as mixed integer linear programing aiming at the maximisation of throughput, optimisation of radio resource usage and minimisation of inter-cell interference. Special care is put on the evaluation of their computational feasibility and so their implementation in the standard, considering that the LTE scheduler has to be executed every 1 ms. After solving the programs, the authors observe that the compact case performs better and executes faster, though still not being compliant with the execution time requirement. The algorithms are then compared against a heuristical greedy approach. Results indicate that just in the 90% of the cases, it reduces the performance with respect to the optimum by less than 10%. Besides, in this case the solution is met in less than 0.75 ms in more than 80% of the cases and in less than 1 ms in 100%.

The previous work was contextualised in the case of localised single carrier frequency division multiple access (SC-FDMA). With the introduction of LTE-A (release 10 and beyond) non-contiguous resource allocation is also possible in the UL, though with less flexibility than the downlink (DL). In particular, up to two separate clusters (sets of contiguous sub-carriers) can be allocated. The aim is to increase the spectral efficiency by exploiting users frequency diversity gain. On the other hand, this new transmission scheme brings an increase in the signals peak-to-average power ratio (PAPR) although still lower than pure orthogonal frequency division multiple access (OFDMA). Maximum power reduction (MPR) is introduced in 3GPP when multi-cluster transmission is used so that the general requirements on out of band emissions are met. On the other hand, this power de-rating implies and extra loss in the UL link budget that may result in throughput reductions.

Lema et al. [LGRO13] propose a novel scheduler that considers opportunistically the information on the MPR that is to be applied. The packet schedulers main task is to evaluate the gain or loss in throughput of the multi-cluster transmission over a conventional contiguous one. Based on the sounding reference signal (SRS) channel estimation, the eNB can predict the multi-cluster transmissions performance. In order to assess the performance of considering MPR wise scheduling decisions, it has been compared to other three benchmarks: Pure multi-cluster transmission, contiguous allocation, and a previous proposal of the literature in which multi-cluster transmission is pre-defined by a given path loss threshold. Also different cluster sises have been tested as the MPR to be applied strongly depends on this parameter. In all cases, the new approach has presented throughput gains. Enabling the MPR information in the scheduler adapts the transmission mode to each particular case, enhancing 30% cell edge throughput compared to a pure multi-cluster transmission. Average throughput is increased almost 18% when compared to classic contiguous allocation.

With the increase of supported bandwidths by modern cellular systems, spectrum aggregation (SA) has become mandatory. Indeed, it is a solution for the highly fragmented spectrum that operators typically have. However, this imposes new challenges over schedulers, that now must be able to handle inter carrier resource allocations. Acevedo Flores et al. [AFVC+14] analyses the cost/revenue performance of a mobile communication system in an international mobile telecommunications (IMTs) advanced scenario with SA over the 2 and 5 GHz bands. A system accounting for a general multi-band scheduler is compared against another in which users are allocated to one band without possibility of changing it after this first association, throughput gains of 28% are obtained. From the economic viewpoint, costs, and revenues are analysed on an annual basis for a 5-year project duration. Revenue and cost is evaluated, for different prices per MBytes (impacts revenue) and cell radius (impacts cost). For the worse study case, profits of 240% are obtained when the multi-band scheduler is introduced and 170% without it. Finally, an energy efficiency strategy is proposed and analysed. It opportunistically reallocates user to available bands with better propagation conditions. This yields a reduction in the transmission power, thus showing an additional benefit of the multi-band allocation scheme. It reduces down to 10% the required energy when compared to a solution with no possibility of link reallocation between bands.

Robalo and Velez [RV14] constitute an extension to the previous work, now in the context of LTE and for the 800 and 2.6 GHz bands. In this case, two multi-band schedulers are implemented and compared. The first one allocates users to a single carrier at a time, this is smartly done by using integer programming optimisation. The total number of users on each band is upper bounded by the maximum load handled, that is to say, the scheduler takes into account a per-band admission control constraint. The second scheduler is able to allocate users to one or both components and indeed this is the reason why the same integer programming approach is not feasible, due to prohibitive computation complexity. Thus, a per user scheduling metric is computed and the allocation is performed starting from the highest value. A complete evaluation is done from several viewpoints. The second case outperforms the first one and other benchmarks in terms of packet loss ratio and delay. A quality of experience (QoE) model for multimedia applications, computed as a function of packet loss ratio, delay and bit rate is also evaluated with maximum performance for the approach dealing with both bands jointly. Similar to the previous work, a cost/revenue analysis is performed, with similar profits for both approaches but far superior to the case without carrier aggregation.

One of the constraints in the previous work was considering admission control requisites in each band. More generally, the admission control in LTE has to consider the load situation in the cell when admitting a new UE in the system. A new request is only granted if the QoS for the new user is going to be satisfied without jeopardising the already connected ones. Lema et al. [LRFGL+12] present an scheme that estimates the new user demands but in the context of UL. Thus, the approach is based on the SRS. The pre-allocation of the sounding bandwidth is crucial for assessing correctly the user demands.

The criteria followed to allocate the sounding bandwidth is to update those spectrum regions that have oldest channel information. With this, a complete channel information across the system bandwidth is obtained. The proposed process is compared to an admission algorithm that pre-sets a maximum data rate per eNB. Each UE is assumed to consume the guaranteed bit rate. Simulation results show that the resource consumption estimation through SRS increases the number of UEs admitted to the system. There is an overall improvement of the cell rate at the expense of increasing the outage probability.

In the previous works, the research methodology closely relies on system level simulations. Indeed, the definition of new simulation strategies able to correctly model carrier aggregation deployments is a key issue. [RVPP15] describes an open source and freeware extension to the LTE-Sim simulator, which implements carrier aggregation functionalities. The work explains the degree of parametrisation that LTE-Sim allows and describes the three available DL multi-band schedulers. First, an integer programing based one that aims at maximising the cell goodput, second, a basic multiband scheduling that allocates users sequentially to the available carriers, and third an approach introduced in the work itself. In this last case, unlike the previous proposals, the user can be allocated to several bands simultaneously thus providing an improved frequency diversity. The authors close their contribution by providing a comparison of scheduling policies simulated with LTE-Sim. Results quantify average cell packet loss ratio, delay, and application level throughput in a reuse 3 deployment and indicate a superior performance of multi-band scheduling.

The allocation of resources in the data and control plane has been traditionally done by the same eNB. However, a plane split might yield additional cost savings [Zha14]. This way, data would be scheduled by cells without signalling, which opens the door to dynamic capacity activation, with the corresponding energy saving. This is graphically depicted by Figure 6.3. Zhang [Zha14] provides an insight on this issue with a particular emphasis on modelling of control signalling traffic and mobility management.

6.2.2 Interference Management

Without a doubt, interference mitigation is one of the most important elements in cellular systems. This subsection presents interference mitigation strategies and proposals for cellular networks based on OFDMA, in the DL, and SC-FDMA, in the UL, such as LTE and LTE-A. Both access schemes provide intrinsic orthogonality to the users within a cell, which results into an almost null level of intracell interference. However, intercell interference (ICI) is

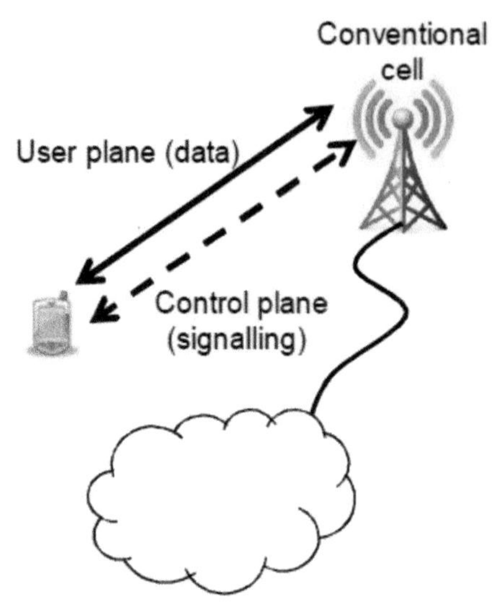

Conventional network:
Data and control signalling are served by a single cell. Coverage and capacity are always available.

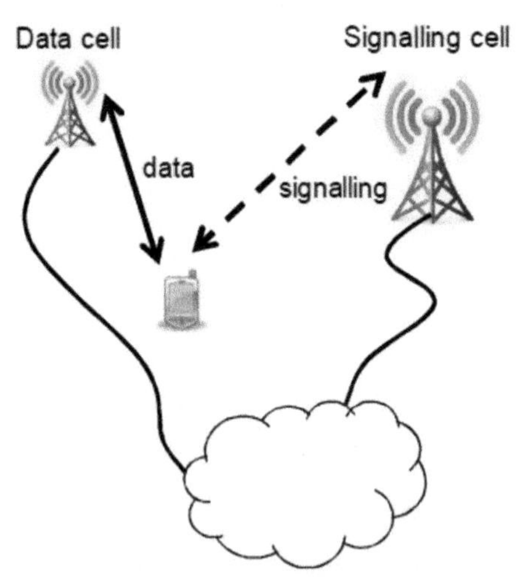

User/control plane separation:
Data and control signalling are served by different cells. Coverage is always available, but capacity is activated on-demand.

Figure 6.3 Brief concept for and control plane separation.

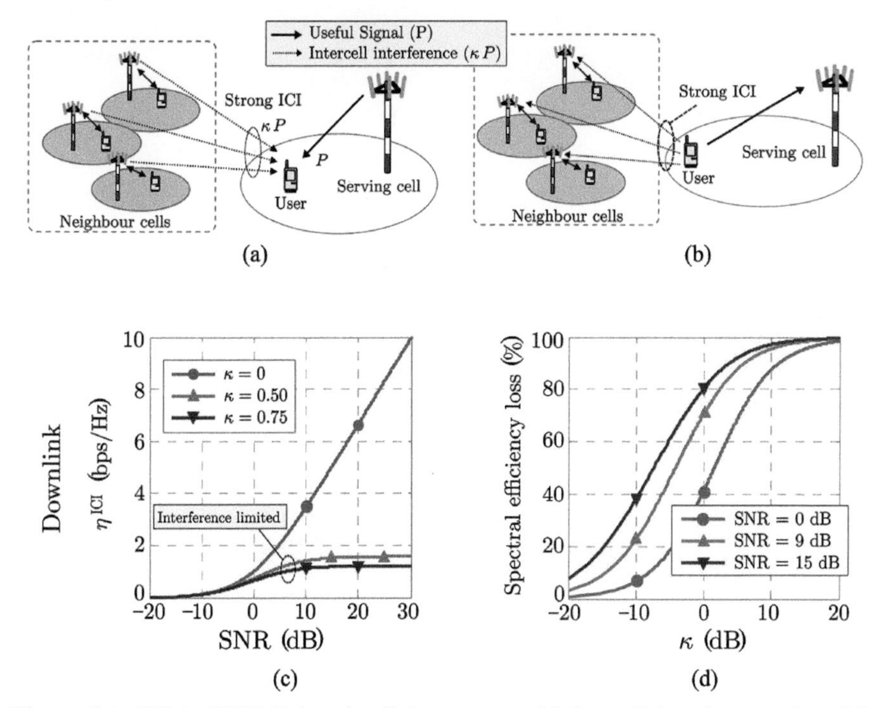

Figure 6.4 ICI in OFDMA-based cellular systems: (a) Intercell interference: downlink, (b) Intercell interference: uplink, (c) ICI: impact on capacity, and (d) ICI: losses.

created when the same channel is used in neighbour cells. Hence, the target of the strategies and techniques presented hereafter is to reduce or avoid ICI.

6.2.2.1 Essentials

In general, the rate at which the network is able to transmit depends on the signal-interference plus noise ratio (SNR). In order to introduce some fundamental notions related to ICI and its impact on the performance of cellular systems, the scenario shown in Figure 6.4 is considered. Figures 6.4(a) and (b) illustrate the case of DL and UL, respectively. As it can be seen, in the DL, ICI is generated by neighbour cells while in the UL, it is created by UE. The following analysis focuses on the DL, but it can easily be extended to the UL. According to the Shannon's formula, the spectral efficiency measured in bit/s/Hz of the *User* (Figure 6.4(a)) can be written as follows:

$$\eta^{\mathrm{ICI}} = \log_2\left(1 + \frac{\text{Useful signal received power}}{\sigma^2 + \text{ICI received power}}\right), \tag{6.1}$$

$$\eta^{\mathrm{ICI}} = \log_2\left(1 + \frac{P}{\sigma^2 + kP}\right). \tag{6.2}$$

σ^2 represents the noise power. In case that $kP \gg \sigma^2$, the scenario is interference-limited, and obviously, this is the case of interest from the interference mitigation standpoint. Note that the amount of ICI can always be expressed as kP, and hence, it can be understood that the goal of any interference mitigation strategy is to make k as small as possible. In order to illustrate this reasoning, the impact of ICI on the spectral efficiency is considered. Given that the signal-to-noise-ratio (SNR) and signal-to-interference ratio (SIR) correspond to $\frac{P}{\sigma^2}$ and k^{-1}, respectively, the spectral efficiency without ICI can be written as follows:

$$\eta^{\mathrm{NoICI}} = \log_2(1 + \mathrm{SNR}). \tag{6.3}$$

The spectral efficiency loss due to ICI will be given by:

$$\mathrm{Loss} = \frac{\eta^{\mathrm{NoICI}} - \eta^{\mathrm{ICI}}}{\eta^{\mathrm{NoICI}}} = \frac{\log_2\left(\frac{1+\mathrm{SNR}}{1+\left(\frac{1}{SNR}+k\right)^{-1}}\right)}{\log_2(1+\mathrm{SNR})}. \tag{6.4}$$

Figures 6.4(c) and (d) clarify the meaning of Equation (6.4) by showing the relationship among the involved variables. Figure 6.4(c) clearly shows that when ICI is high ($k \geq 0.50$), increasing P (to increase the useful signal received power) has a negligible effect on the spectral efficiency. Hence, ICI is the main capacity limiting factor. In addition, Figure 6.4(d) shows that in scenarios with relative low noise (SNR = 15 dB), the capacity is reduced between 40% and 80% when the SIR goes from –10 dB to 0 dB ($k^{-1} \in [-10, 0]$ dB), makes evident the severe impact of ICI. Indeed, the losses in the heavily interfered region (k $\geq$ 0 dB) are above 80%. In the light of the previous analysis, the need for effective interference mitigation is clearly justified. Additional practical aspects that further complicates the problem at hand, that of reducing or avoiding ICI include:

1. ICI is highly non-predictable (even in environments without mobility) due to the time-varying transmission patterns in neighbour cells. Moreover, frequency selective fading and practical limitations of channel state information (CSI) feedback schemes make harder to have accurate estimations of ICI. As it will be shown shortly, some of the proposals are aimed at overcoming this issue by employing predictive techniques, such as by Garcia et al. [GLG11], while other techniques are focused on

improving the *quality* of the CSI, see González et al. [GGRO12a] and Lema et al. [LGR14].

2. SINR levels are not uniformly distributed in the network coverage area. Indeed, users located near to cell edges receive not only weak signals from their serving base station (BS) but also high ICI. Thus, the QoS experienced by users strongly depends on the location in the network coverage area, and as a result, a *fairness* issue among users is created. An example of strategies focused on *protect* those *unlucky* users include [KMM11].
3. Finally, but no less important, the trend towards *densification* as a paradigm to increase the areal frequency reuse, and hence, as a mechanism to increase capacity, makes evident the pevalence of the techniques for effective frequency planning and interference mitigation. Indeed, as it is indicated in Gonzlez et al. [GGRO12b], the notion of frequency reuse is key in many of the proposals whose main target is to reduce the amount of ICI, mainly at cell edges, such as Peng et al. [PKHAM13] and Acedo-Hernández [AH13].

6.2.2.2 Classification of strategies for interference mitigation

Classifying the strategies for interference management is not an easy task because several criteria can be used. In addition, the boundaries in this context are blurred. However, a widely accepted classification is shown in Figure 6.5. In a nutshell, interference mitigation can be done by means of:

1. Coordination, where restrictions to the radio resources (mainly bandwidth and power) are applied at each cell (with more or less dynamism) to reduce ICI at cell edges,
2. Cancellation, where interfering signals can be either subtracted from the received signal,
3. Randomisation, where the interference is distributed uniformly across the system bandwidth through scrambling, interleaving, or frequency-hopping (spread spectrum).

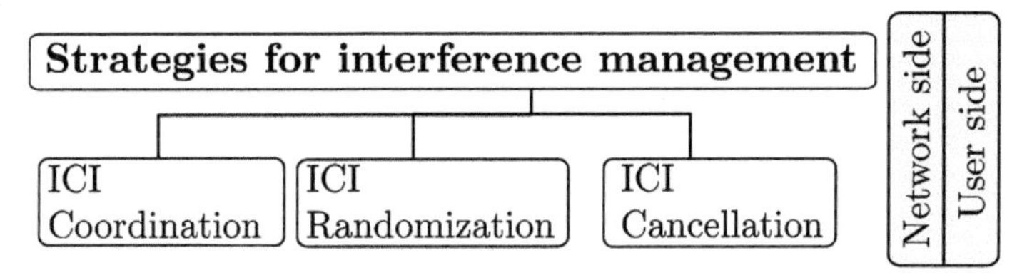

Figure 6.5 Strategies for interference management.

It is worth mentioning that, although most of the practical schemes are fully implemented in the newwork side, these strategies can also be implemented in the user side, and often in both. What technique and how is implemented mainly depends on the type of link (DL or UL). Other elements to take into account include: the type of environment, mobility conditions, and computational complexity, among others.

6.2.2.3 Static intercell interference coordination (ICIC): an approach to planning

As it was mentioned before, coordination implies apply a set *rules* in the network. This rules can be applied at different time scales, and hence, it is possible to talk about *static* and *dynamic* ICIC. Thus, in static ICIC the basic idea is to apply different frequency reuse factors to different groups of users based on a certain criterion (usually average SINR measurements are considered) over periods of times ranging from days to weeks. In static ICIC, there two fundamental schemes: soft frequency reuse (SFR) and FFR. The operational principle is depicted in Figure 6.6. In short, once users are classified (e.g., based on SINR_{avg} in the figure), the resource allocation is

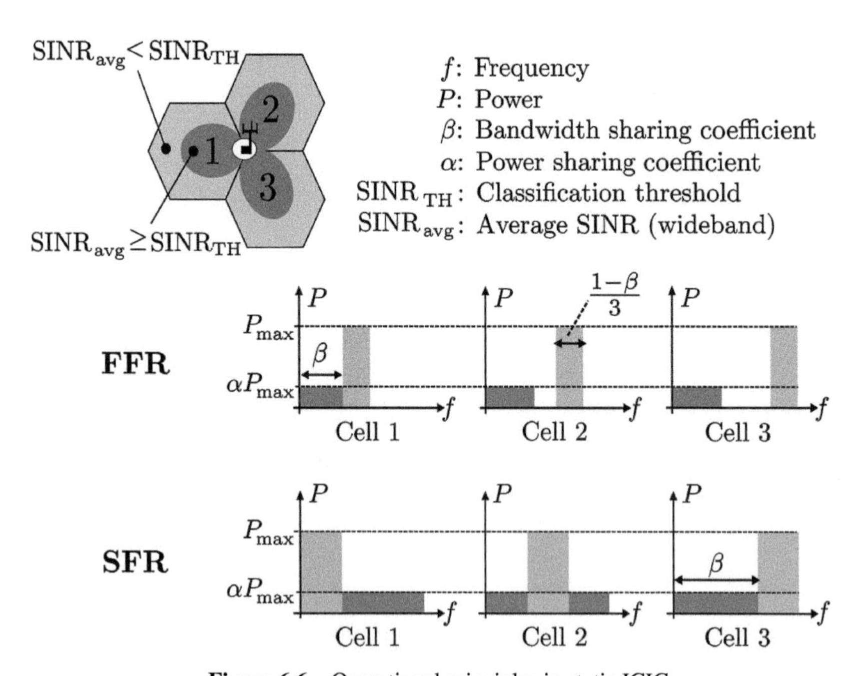

Figure 6.6 Operational principles in static ICIC.

done respecting the bandwidth and power allocation pattern that is defined by the paraneters α and β as it is shown. Further information can be found in Gonzalez et al. [GGRO12b].

One fundamental problem with static ICIC is the calibration of the aforementioned parameters. While such selection is quite straightforward is perfectly hexagonal cells [GGRO12b], the problem is much harder in RL cellular networks featuring irregular layouts [GGRL13] (Figure 6.7). In Peng and Kürner [PK13a], the authors proposed a method for SFR planning, in the DL, that takes into account an interference metric that is inversely proportional to the distance between interferers, i.e., BS. Thus, for two BS i and j, the interference metric is computed as follows:

$$m_{ij} = \begin{cases} \frac{1}{d_{ij}^2}, & i \neq j, \\ 0 & i = j. \end{cases} \tag{6.5}$$

Therefore, the objective of the proposed iterative scheme, based on the Gauss-Sidel algorithm, is to minimise the overall interference (*IS*) expressed as a function of the previous metric, thus

$$IS = \sum_{c=1}^{C} \sum_{i,j \in M(c)} m_{ij}, \tag{6.6}$$

where C is the number of available channels and $M(c)$ is the set of cells in which the cth channels is being used. In this manner, it can be seen that the

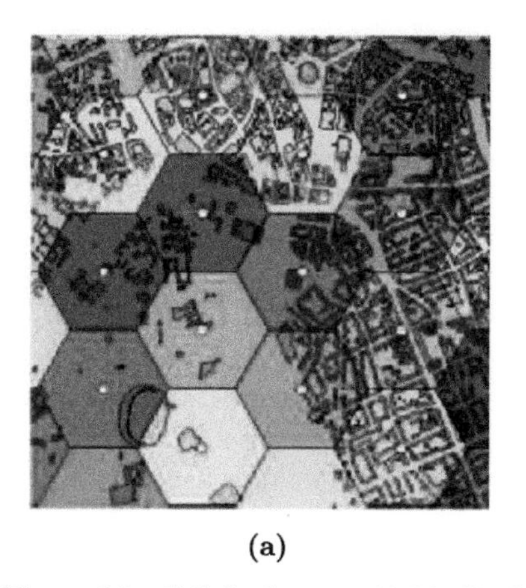

(a)

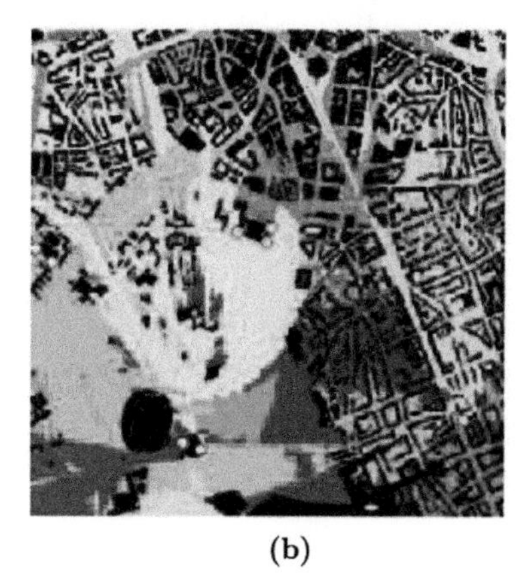

(b)

Figure 6.7 Cellular layouts: (a) Idealised hexagonal cells, and (b) Realistic irregular cells.

proposed method realies on the assumption that ICI is proportional to the path loss in order to update the channel allocation. This scheme is a clear example of how static ICIC can be applied in a planning-like fashion to determine frequency–power profiles. Indeed, taking ICIC as design criteria in more general planning tasks is quite common. For instance, in Acedo-Hernández [AH13], the planning of the physical cell identifier (PCI) is studied from an interference management point of view and potential gains are quatified. PCI planning is important because it defines the location of the cell-specific reference signals (RS). RS (often called *pilots*) are always transmitted in the same orthogonal frequency division multiplexing (OFDM) symbol, but in the frequency domain, RSs are shifted determined by the value of the PCI. Since RS are used for CSI feedback (among other thinks), an improper PCI planning may give inaccurate SINR estimates, which leads to inefficient data transmission in the DL. The analysis presented in Acedo-Hernández [AH13] considers several plan models for the primary synchronization signal (PSS) which in conjuntion with the PCI determine the exact location in time and frequency of subcarriers where RSs are transmitted. Key performance indicators (KPIs) include figures obtained from the distribution of the SINR observed in RSs. To be more precise, the median and fifth percentile, as measures of connection quality, and statistics from the CQI and DL throughput. Figure 6.8 shows one representative, yet important, result presented in Acedo-Hernández [AH13]. As it can be seen, the PCI planning has a significant impact of the resulting CDF of the cell throughput. Figure 6.8(a) shows that, for low network load, average cell throughput decreases by up to 30% with respect to

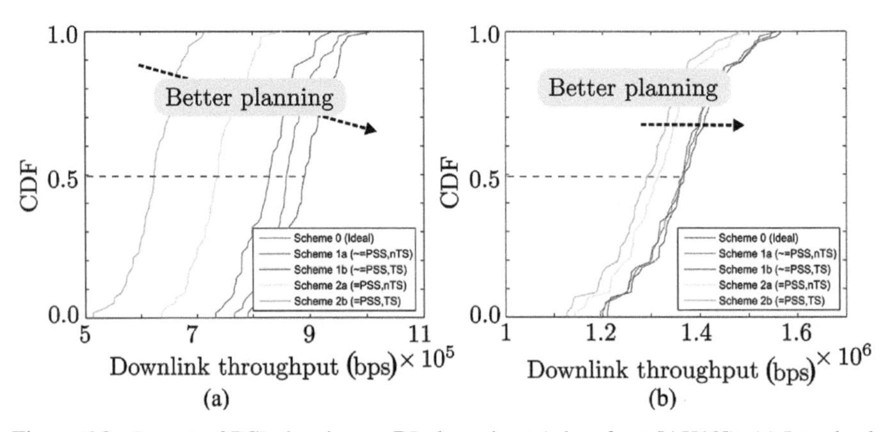

Figure 6.8 Impact of PCI planning on DL throughput (taken from [AH13]): (a) Low load scenario, and (b) High load scenario.

the best planning scheme. For high network load, shown in Figure 6.8(b), a similar behaviour is found, although gains are smaller. Further details about simulation setups and additional results can be found in Acedo-Hernández [AH13]. However, the study makes evident the impact of minimising the effect of ICI even in planning tasks. Obviously, as it was mentioned earlier, ICIC is also suitable to be applied in smaller time scales.

6.2.2.4 Dynamic ICIC

Dynamic coordination implies adapt the bandwidth/power allocation patterns in time scales ranging from milliseconds to minutes. This can be done by means of generic resource allocation strategies, but it can also be done starting from a predefined pattern, such as FFR. The latter approach was employed in Krasniqi et al. [KMM11]. In this scheme, a model based on two users is used to determine the optimal FFR-based resource allocation. In particular, three different cases are considered: (i) both users are in the *inner* region (high SINR), (ii) both are in the *outer* region (low SINR), and (iii) there is one user in each region. Closed form expression are obtained by means of Lagrange multipliers. Therefore, the proposed can easily be extended to scenarios with multiple users to achieve dynamic resource allocation. Details of the formulation and assumptions can be found in Krasniqi et al. [KMM11]. Figure 6.9 shows the simulation results for the average SR taken over uniform users positions in the inner and outer regions versus maximum BS power. The lower curve represents the average of all SRs when no re-allocation of the outer bandwidth to the inner users is carried out for the static method used. A better performance in terms of average SR is achieved when we

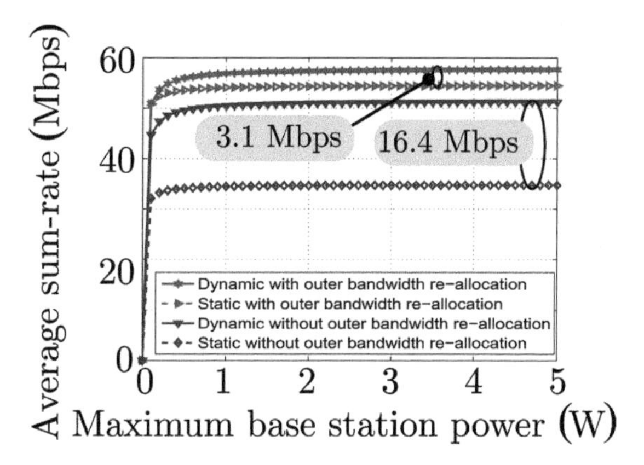

Figure 6.9 Maximum average SR (taken from Krasniqi et al. [KMM11]).

consider the dynamic method. Clearly, performance increase of approximately 16.4 Mbps is achieved when dynamic method is used. Considering also the outer bandwidth re-allocation to the inner users a performance increase of 3.1 Mbps is achieved when dynamic method is used. Hence, in the light of this results, the effectiveness of interference mitigation is demonstrated.

So far, strategies aiming at avoiding ICI have been reviewed. A completely different approach is just focus on cancelling its effect.

6.2.2.5 Topological analysis of ICI cancellation

In ICI cancellation, the basic concept is to regenerate the interfering signals and subsequently subtract them from the desired signal. In short, the approach is removing ICI rather than avoiding it. One good feature of interference cancellation is that the implementation at the receiver side can be considered independently of the interference mitigation scheme adopted at the transmitter, and hence, the coexistence with other techniques is not an issue. Some types of ICI cancellation include: interference rejection combining (IRC), successive interference cancellation (SIC), and interference alignment (IA).

One very interesing aspect of cancellation techniques is that they can modify significantly the reception area where transmitters can be *heard.* The work presented in Haddad [Had14] investigates this. Figure 6.10 shows the basic notions in this framwork. Transmitter s_x is supposed to be decoded by receiver r_x, and the region $H(s_x)$ is the region where transmitter s_x can be successfully decoded. Figure 6.10(a) shows zones $H(s_1)$ and $H(s_2)$ for s_1 and s_2, respectively, under the assumption of non-uniform power allocation, i.e., the transmit power of s_1 and s_2 are different. It can be proved that these two demands cannot be satisfied when both s_1 and s_2 transmit with the same power. The uniform power allocation is shown in Figure 6.10(b). Note that $r_1 \notin H(s_1)$, and hence, r_1 does not receive its information. In contrast, when

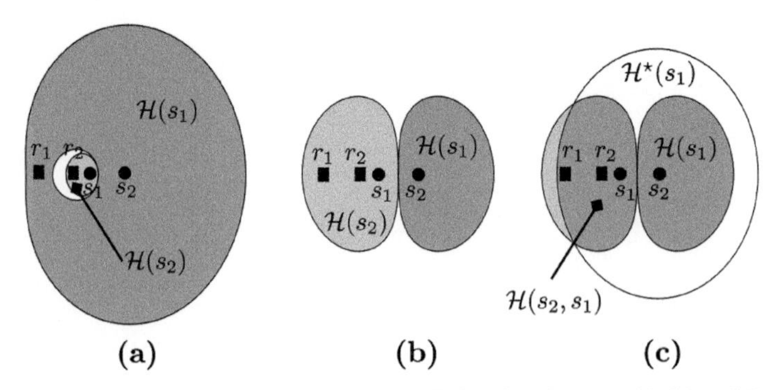

Figure 6.10 Topology with and without SIC (taken from Haddad [Hadl4]).

SIC is applied in r_1, it can first decode s_2, and then, it can cancel it to decode s_1. Thus, with SIC the two demands can be satisifed even with uniform powers as it is shown in Figure 6.10(c). Note that $H(s_2, s_1)$ is the intersection of two convex shapes, $H(s_2)$ and $H(s_1)$, where the latter (shown as an empty circle) is the reception zone of s_1 if it had transmitted alone in the network. It is clear that the total reception area of s_1 with SIC is much larger than without it. The most important contribution presented in [Had14] is that the authors present a tighter bound for the number of reception zones in terms of a novel metric called *Compacteness Parameter*, among other interesting results.

6.2.2.6 Enhancements through prediction and improved CSI feedback

As it was mentioned before, one fundamental problem with ICI is that it is highly unpredictable in real-life contexts. In this sense, several efford have been made to provide solutions to cope with this issue. The work presented in Garcia et al. [GLG11] introduce a model to predict the power variations in neighbour cells. Basically, the proposed framework try to avoid some drawbacks of distributed schemes, such as game theory that require static scenarios and sequential or per-round allocation to converge to equilibrium that may not be optimum solutions. Other practical radio resource management (RRM) schemes are difficult to implement due to their complexity or simply because they assume a global and perfect knowledge of the system. Thus, in order to avoid such difficulties, the proposed solution makes an assumption on systems evolution. The distributed power allocation, that does not require information sharing among transmitters, requires three phases:

1. *Estimation.* The channel is computed as a polynomial regression of degree 4, based on the five last measurements.
2. *Prediction.* The target is to determine the *a priori* values of the channels and interference for future steps. Finite-horizon-based prediction is considered.
3. *Decision.* Each BS produces a partially predictable interference, performing a trade-off between its inertia and its required power variations.

Another approach that can be taken to improve the performance of interference management schemes, and more precisely static ICIC, is to improve the CSI feedback. CSI is required both in DL and UL in order to allow for *opportunistic* scheduling. The schemes presented in González et al. [GGRO12a] and Lema et al. [LGR14] introduce mechanisms to improve the accuracy of CSI. To feedback the DL, UE transmit CQI-based reports through the UL to provide the network with SINR estimates at different subbands. In the UL, sounding signals are sent by the UE and configured by the network. The fundamental idea in González et al. [GGRO12a] and Lema et al. [LGR14] is to take advantage of the bandwidth allocation pattern that is used in static ICIC schemes, such as SFR and fractional frequency reuse (FFR), to focus the CSI feedback on the bandwidth portions that are relevant to each user. In this manner, CSI reports are updated more frequently, and hence, the modulation and coding scheme selection can be done in an effective manner. In case of Lema et al. [LGR14], the solution reduces the time delay between sounding measurements, while in González et al. [GGRO12a] also the accuracy of the wideband CQI is improved. In both cases, results show significant performance in terms of average user throughput and cell edge performance, thus achieving one of the main targets of interference management.

6.2.3 Power Management

Power control is a key degree of freedom that has always been of paramount interest in wireless communications. The intelligent selection or adjustment of the transmitter power may improve the general performance in terms of interference, connectivity, and energy consumption.

For network deployment planning and optimisation, it is necessary to determine the enhanced node-B (eNB)-transmitted power which depends on the average SINR, the desired cell radius and the frequency reuse being adopted [AFRJ14]. If the average SINR is maximised the BS transmitted power increases to very high levels, which compromises the eNB energy consumption. So, average SINR levels should be maintained lower than the maximum. Also, for a pre-set value of SINR as the cell radius increases in planning, the transmitted power demands also increase. In deployments with SA, with different frequency bands, if a constant average SINR and same cell radius across bands is desired, different power allocations must be considered.

In the UL, the 3GPP proposed an open loop power control (OPLC) which adjusts the power spectral density of the mobile terminal (MT) based on the estimation of the channel gain. The main advantage of this power control method is that the path loss can be partially compensated, which is known as *fractional* power control. Hence, UE with high path losses transmit at lower power (than classic full compensation) thus generating less interference.

The UE calculates the OLPC over the allocated bandwidth by estimating its own path loss. The algorithm mainly compensates for long term channel variations (i.e. path loss and shadowing), yet the performance degrades due to errors in the path loss estimation. For this reason, the algorithm also accounts of a closed loop component that aims to compensate these errors by sending feedback corrections to the UE periodically.

The power transmitted by a given UE in the Physical UL Shared Channel (PUSCH) is defined as:

$$P = \min(P_{\max}, P_0 + 10\log M + \alpha L + \delta_{\mathrm{TF}} + \delta_i), \tag{6.7}$$

where $P_{\max}$ is the maximum available power in the UE, P_0 is both cell and user specific parameter, M is the number of resources allocated, α is the path loss compensation factor, L is the estimated path loss, δ_{TF} is a parameter that depends on the transport format chosen, and δ_i is the UE-specific closed loop correction.

Concerning the OLPC, P_0 and α are the most important parameters to be adjusted. P_0 controls the SINR target at the receiver end. A rise in P_0 increases the transmitted power, hence, interference rises and degrading the system throughput. However, if P_0 is low the UE may not accomplish the bit error rate (BER) requirements. The path loss compensation α, brings the system into a trade-off. If a UE placed on the cell-edge corrects only a part of its path loss it generates less inter-cell interference. Also, as α approaches one, all UE compensate their path loss so all signals are received with the same strength, resulting in similar SINR values along the cell area. On the other hand, when α is reduced, MTs compensate only a fraction of its path loss. The resulting SINR distribution is less fair, and cell-edge users are received with less strength than cell centre ones.

Performance changes arise in terms of cell SINR distribution when modifying both variables as shown in Lema et al. [LGRO11] and Vallejo-Mora [VM13]. A rise in P_0 improves the connection quality if the number of power limited users is small. A low value of P_0 provides the system with a reduced value of inter-cell interference; however, transmission power can result insufficient to overcome path loss and shadowing.

The performance of the OLPC depends very much on the network deployment environment. One synthetic scenario and one urban deployment are tested and compared in Lema et al. [LGRO11] The first presents high sensitiveness to interferences while the RL one is clearly sensitive to the availability of the transmission power, given the high number of sources of attenuation. The possibility of obtaining the best performance despite the environment nature is due to the versatility presented by the algorithm.

On the one hand, the OLPC formula shows a fairly good performance with low computational effort; but, ont the other hand it has not been optimised for the network throughput as it does not consider the inter-cell interference. Work in Peng and Kürner [PK13a] presents an iterative algorithm that calculates each UE power that maximises data rate, which is estimated using the truncated Shannon formula. On each iteration, the transmission power is updated to optimise the network performance based on the last iteration resulting users transmission power. Results show a fast convergence rate, throughput is improved by 21.9% compared to the LTE formula with only two iterations.

An automatic tuning algorithm for the parameters of the UL power control in LTE is proposed in Fernández Segovia [FS13]. The study is based on an initial sensitivity analysis that characterises the impact of P_0 and sub-carriers utilisation limit, both with an impact on interference levels. Next, the proposed algorithm is composed of two parts. Initially, it iteratively evaluates decreasing values of P_0 and sub-carrier utilisation until cell edge throughput requirements are met. In the second stage, optimisation is oriented to maximise capacity, and so, the average cell throughput. An extension of the basic algorithm towards non-ideal scenarios requires certain simplifications, otherwise computational cost would be prohibitive. The authors propose building a simplified scenario from the irregular original one and two mechanisms are proposed and investigated. The underlying idea is to approximate a real scenario by many single and regular ones based on geographical relations with neighbour cells. This means that at the end of the optimisation several solutions are provided to each cell, as depicted by Figure 6.11. Four mechanisms are compared for the selection of a final solution. Results indicate a successful operation of the approach with similar network performance among the mechanisms and selection methods. Finally, even though the method is conceived for the planning stage, it shows a complexity cost which is one order of magnitude lower than an exhaustive simulation analysis.

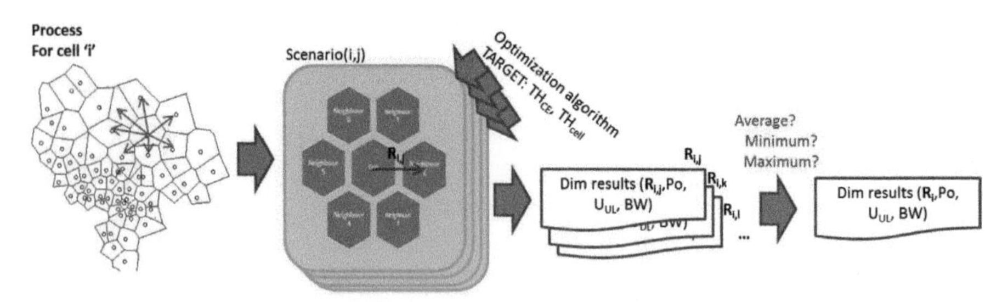

Figure 6.11 Statistical possible solution for cell *i*, from á [FS13].

6.2.4 Mobility Management

Handover is the mobility procedure that more directly affects the user experience, since it occurs during data transmission. More specifically, a sub-optimal setting of handover (HO) parameters could lead to a call drop or waste of network resources. Mobility robustness optimisation has gained attention in the research community, with several techniques for HO optimisation being proposed. Muñoz Luengo [MnL13] investigates the potential of adjusting handover margin (HOM) and time to trigger (TTT) to improve such robustness. An initial analysis performing a sweep of both parameters indicates that the network is more sensitive to variations of the HOM, with no additional benefits with the adjustment of the TTT. Hence, the authors propose the use of a fuzzy logic controller (FLC) that adaptively modifies just the HOM and being based on the values of HO ratio and call dropping ratio (CDR). Two fuzzy sets are suggested for the first input, on the other hand, three sets are defined in the second case. This allows having more granularity for the more critical performance indicator. The controller has been evaluated for different levels of traffic load and different user speeds. Results show an improvement in network performance, it is also shown that the network is more sensitive to margin variations for users with speed values above 10 km/h. Thus, the authors conclude suggesting the avoidance of changes in the margin greater than 1 dB.

6.3 Resource Management in Wireless Networks

6.3.1 Resource Management in Wireless Mesh Networks (WMNs)

A non-centralised, non-hierarchical, and distributed synchronisation process for WMN is introduced in [RPWD13, RPW14]. Each node has a time base that relative to an universal clock has an offset and a clock frequency difference. The purpose of the synchronisation is to correct the offset and the frequency so all the nodes will have almost the same time base. The synchronisation process is based on periodically transmission of messages with timing information by the all the nodes. The timing messages can be retransmitted to other nodes. Each node is correcting his time base according to the received messages. The synchronisation process has an acquisition phase when the clocks of the nodes have large variations and when the variations are reduced below a certain threshold the synchronisation process switches to the tracking phase. In the tracking phase, the variations among nodes are farther reduced and the changes in clock parameters are followed [RPW14].

Authors Ferreira and Correia [FC12] introduce a unified RRM strategy for multi-radio WMNs following self-organisation principles. This method built on an abstraction-layer, radio-agnostic in the operation of multiple radios and transparent to higher layers. This methods integrates *channel assignment, rate adaptation (RA), power control*, and *flow-control, mechanisms,* achieving a high-performing WMN. This strategy is proposed for self-organised WMNs. The stratedy integrate multiple mechanisms:

- a RA mechanism aware of WMN traffic load specificities,
- an energy-efficient power control mechanism addressing the non - homogeneity of nodes rates,
- a load- and interference-aware channel assignment strategy that guarantees connectivity with any neighbour.

For a WMN of 13 multi-radio mesh access points (MAPs) using IEEE 802.11a, a fair aggregated throughput per MAP of 4.8 Mbps is guaranteed by the use of four channels and a total transmitted power of 34.5 dBm and RRM strategy exploits 100% of the network capacity, guaranteeing fairness and minimising the spectrum and power usage. Several system improvements multiple-input multiple-output (MIMO), higher modulations (available in 802.11n) and receiver sensitivities could be considered, resulting in higher system physical data-rates and larger ranges [FC12].

In Ferreira and Correia [FC13], a hierarchical-distributed strategy, combining RA, power control, and channel assignment mechanisms to efficiently guarantee max-min fair capacity to every node in WMN is introduced. In this model, a multi-radio WMN is composed of MAPs equipped with various radios, providing Internet connectivity to end-user terminals via a mesh point portal (MPP) gateway. Each MAPs aggregated traffic flows between an MPP gateway and the end-user(s) connected to that MAP, crossing one or multiple intermediary MAPs. The adopted fairness concept aims guarantee that all MAPs achieve a max-min fair aggregated throughput. If any MAP is favoured, increasing its load and associated throughput beyond maximal capacity that still guarantees fairness, there will be disfavoured MAP(s) that will have their throughput decreased. The RRM strategy is proposed too. This strategy include RA and transmission power control (TPC), a max-min fair flow-control mechanism and CA mechanism. Each multi-radio MAP has a radio agnostic abstraction-layer that implements the RRM strategy for optimisation of mesh radios resources, based on a mechanism for monitoring and sharing resources. It is hierarchical-distributed, triggered by each MPP, being sequentially run by each child. Paths are computed in advance by a routing algorithm. It follows max-min fairness principles in the allocation of capacity to MAPs. The combined RA and PC mechanism is proposed. It is sensitive to the WMN fat-tree distribution of traffic flows. It uses the highest physical layer (PHY) bit rate possible only at MPPs (WMN bottleneck), and uses the remaining MAPs for minimum bit rates that satisfy their capacity needs. It is sensitive to the WMN fat-tree structure, using the highest possible bit-rates only at MPPs, and recurring, for the ramified links, to minimum bit rates that satisfy their capacity needs. This enables to reduce the transmitted power, an energy-efficient solution, and associated interference ranges, making channel reutilisation possible. A channel assignment mechanism is also proposed, aware of the load and interference of each MAP. It is concluded that with the proposed strategy max-min fairness (in the share of capacity) is guaranteed to every mesh node, interference is in existent, and resources such as power and spectrum are efficiently used [FC13].

In Cicalò and Tralli [CT13], a novel cross-layer optimisation framework to maximising the sum of the rates while minimising the distortion difference among multiple video users is introduced. The optimisation problem is vertically decomposed into two sub-problems and an efficient iterative local approximation (ILA) algorithm is proposed, which is based on the local approximation of the contour of the ergodic rate region in the OFDMA DL system. The ILA algorithm requires a limited information exchange between the application (APP) and the medium access control (MAC) layers, which independently run algorithms that handle parameters and constraints characteristic of a single layer. There is first formulated and discussed the feasibility of the optimisation problem showing that the optimal solution is achieved on the convex rate region boundary assuming subcarrier sharing. The problem has been then vertically decomposed into two sub-problems, each one characterised by parameters and optimisation constraints of a single layer. Numerical results have shown the fast convergence properties of the ILA algorithm and the significant video quality improvement of the proposed strategy with respect to optimisation strategies that only consider rate fairness.

6.3.2 Resource Management in Wireless Sensor Network (WSN)

In Sergiou and Vassiliou [SV12], the influence of source-based trees when they serve as topology control schemes in resource control congestion control algorithms and how they apply in a randomly deployed WSN are introduced and the simulation study is used to compare the performance of the two topology control schemes as well as the naive source-based tree when a resource control algorithm applies. Simulation results show that source-based trees could be more efficient when the data load is heavy, since they provide more routing paths from each node. On the other hand, sink-based trees provide better results in terms of delay and energy consumption. The major conclusion that we extract from this study is that source-based trees, if carefully tuned, can provide an efficient topology control solution for specific applications [SV12].

Problems with topology design and control of the large-scale system such as WSN is solved in Porcius et al. [PFMJ12]. There is designed a novel procedure for topology design and control and implement it in the TopoSWiM simulation tool. The procedure combines a mathematical approach based on graph theory with a physical model of the operating environment and of the radio propagation channel. The mathematical approach is based on a new *clustering algorithm for gateway positioning (CAGP)*. The procedure combines a mathematical approach based on graph theory with a physical model of the operating environment and of radio propagation channel. The mathematical approach is based on a new CAGP, which determines the minimum number of gateways that are needed and their positions in order to provide coverage and external connections for the nodes with predefined positions while maximising network accessibility. The procedure takes into account also the physical model of the environment where the network is to be deployed and the appropriate radio propagation channel model. The newly proposed CAGP algorithm is first compared to the benchmark k-means and k-means++ algorithms, in terms of execution time and number of isolated nodes or accessibility.

New block mechanism (BACK) for the aggregation of several acknowledgment (ACK) responses into one special packet for WSNs with contention-based random medium access control MAC protocol with multiple nodes sharing the same medium is introduced in Barroca et al. [BVF+11b]. The sensor block acknowledgement (SBAK-MAC) protocol that improves channel efficiency by aggregating several acknowledgement (ACK) responses into one special frame is introduced in Barroca et al. [BVF+11b] and Barroca and Velez. [BV13]. SBACK-MAC protocol protocol enables reduce the end-to-end delay and energy consumption due to the protocol overhead in S-MAC.

The block acknowledgment (BACK) mechanism improves channel efficiency by aggregating several ACK control packet responses into one special packet, the Block ACK Response. Hence, an ACK control packet will not be received in response to every data packet sent. The Block ACK Response will be responsible to confirm the data packets successfully delivered to the destination. This packet has the same length as a data packet [BVF+11b].

The BACK mechanism starts with the exchange of two special packets: RTS ADDBA Request and clear-to-send (CTS) ADDBA Response, as shown in Figure 6.12(a). ADDBA stands for Add Block Acknowledgement. Then, the data packets are transmitted from the transmitter to the receiver (10 messages, each one fragmented into 10 small data packets). Afterwards, the transmitter will inquire the receiver about the total number of data packets that successfully reach the destination by using the Block ACK Request primitive. In response, the receiver will send a special data packet called Block ACK Response identifying the packets that require retransmission. The structure of the packet is presented in Figures 6.12(b) and (c). The BACK mechanism finishes with the exchange of two special control packets: the RTS DELBA Request and CTS DELBA Response. DELBA stands for Delete Block Acknowledgment. The structure of the DELBA packets is the same presented in Figure 6.12(a). Figure 6.13 presents the message sequence chart for the BACK mechanism.

Barroca et al. [BV13, BBV14b, BBVC14] describe the novel mechanisms to reduce overhead in IEEE 802.15.4 based on the presence of BACK Request (concatenation mechanism), while the second one considers the absence of BACK Request (piggyback mechanism). The aggregation of ACKs aims at reducing the overhead by transmitting less ACK control packets and by decreasing the time periods the transceivers should switch between different states. In algorithm, there is introduced the mechanism

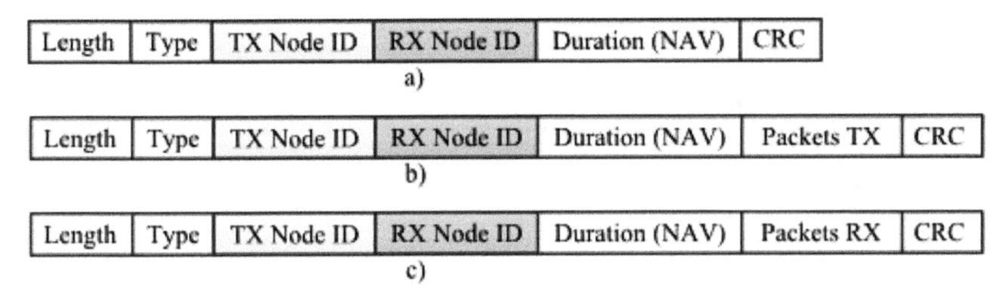

Figure 6.12 (a) RTS ADDBA Request and CTS ADDBA Response, (b) Block ACK Request, and (c) Block ACK Response packets format.

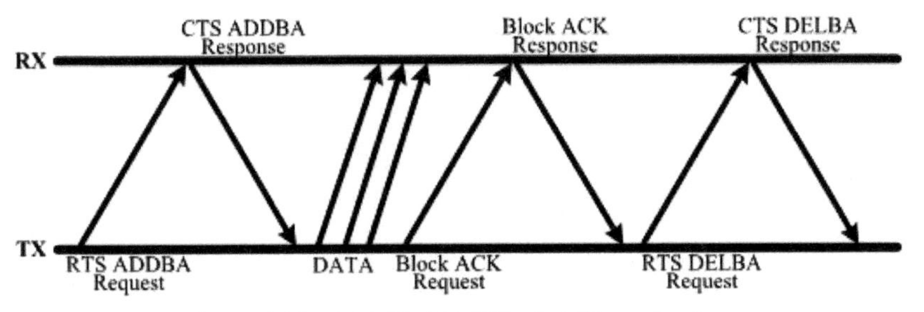

Figure 6.13 SBACK-MAC block ACK mechanism.

to increasing the throughput as well as decreasing the end-to-end delay, while providing a feedback mechanism for the receiver to inform the sender about how many transmitted (TX) packets were successfully received (RX). The mechanism also considers the use of the *request-to-send/clear-to-send (RTS/CTS)* mechanism, in order to avoid the hidden terminal problem.

A new mathematical model to characterise the interference in WSN based on the IEEE 802.15.4 standard is described in Martelli et al. [MBV12]. In this model, the consideration of the detailed operation of the carrier sense multiple access with collision avoidance (CSMA/CA) algorithm as described in IEEE 802.15.4 standard is integrated and included. There are assumptions that all nodes are distributed on a two-dimensional plane according to a homogeneous poisson point process (PPP), with a spatial density. They have to communicate with a central coordinator (CC), which is supposed to be located at the origin of the two-dimensional plane. In model a query-based traffic, the CC periodically sends a query to all nodes and nodes that receive it try to send their data to the CC, is considered. Authors mathematically derive the success probability for a generic node located at a distance r_o from the CC (reference transmitter TX_0) to correctly transmit its data and two factors can affect the success:

- connectivity, since only the nodes that receive the query from the CC can attempt their data transmission;
- interference, because all the nodes start the procedure to try to access the channel simultaneously.

Results collected during simulation show the impact of nodes density and sensing range on network performance evaluated in terms of success probability [MBV12].

Authors in Burati and Verdone [BV12] introduced a novel priority-based carrier-sense multiple access (P-CSMA) Protocol for Multi-Hop Linear Wireless Networks and it enables to increase throughput in comparison with traditional protocols like Slotted Aloha or 1- persistent Slotted CSMA. It partially enables solving the problem of interference among nodes simultaneously transmitting on the route. It's simple MAC protocol which exploits the linearity of the network topology. The protocol assigns to nodes different levels of priority, depending on their position in the route. Authors consider and describe the situations where all relays are (randomly) distributed over a straight line whose length (the source-destination distance) is fixed. The P-CSMA starts from the observation that, with traditional CSMA protocols, when a source nodes has many packets to send over an established route, they compete for accessing the radio channel with those that were previously transmitted and are still being forwarded by some relays in the route [BV12]. All nodes in route have assign different level of priority in the access to the channel. It means that nodes closer to the destination have higher priority with respect to those closer to the source. Authors impose nodes to sense the channel for different intervals of time: the smaller is the sensing duration, the higher will be the priority in the access to the channel. Algorithm gives priority to those packets in the route which are closer to destination stands in the fact that they will not compete with those transmitted in the rear. In this way, P-CSMA mechanism speeds up the transmission of packets which are already in the route, making the transmission flow more efficient. The performance analysis in the sense of success probability and throughput is introduced too and from the results it is clear that the performance in terms of success probability and throughput are better than the ones achieved in the case of 1-persistent CSMA and Slotted ALOHA. Moreover, even under a distributed approach, the protocol proposed tend to behave similarly to a TDMA scheme were all transmissions are scheduled by a centralised controller. In Fabian and Kulakowski [FK13], the performance of geo-routing is validated in radio conditions that are RL for sensor networks is refereed. It is shown that the protocols that are known to guarantee the packet delivery for unit disk graph model and perfect location knowledge are far from that in real sensor networks. A new modification to the angular relaying (AR) was proposed and tested. It was shown that with two new types of routing messages, the packet delivery ratio can be significantly improved in unreliable networks. At the same time the algorithm modification did not increase its mean algorithm cost neither the message complexity, which directly influence the battery usage of nodes. Even though the modified algorithm does not guarantee delivery in the conditions mentioned, it is a first step in that direction, showing others a way to propose further improvements. A routing algorithm, composed of two algorithms: *greedy distance-based algorithm* and *AR,* was implemented in a new Java-based computer simulator. In the greedy mode the algorithm utilise CTS and RTS messages with a timer adopted from the AR algorithm. When a message arrives to a concave node AR is launched. The algorithm switches from the greedy mode into the AR mode when after sending an RTS message, no CTS message is received. In the AR mode, the data message contains information about the concave node and a flag indicating the mode is set to AR. Every node that receives the data message with the AR flag set, checks whether it is located closer to the destination than the concave node, in which case the algorithm switches back to the greedy mode.

6.3.3 Mobility-based Authentication in Wireless Ad hoc Network

A novel approach for authentication of network nodes based on their mobility patterns observed from the radio-wave propagation is introduced in Skoblikov [Sko13]. It relies on the fact that propagation of radio waves is immutable by an adversary. During the authentication procedure the of nodes location in space as well as the mobility pattern are verified in the context of current communication situation using a number of plausibility tests. For the new mobility-based authentication scheme, this information is the knowledge about the current location and mobility pattern of communicating party. For the new mobility-based authentication scheme this information is the knowledge about the current location and mobility pattern of communicating party. If in the given example scenario *Bob* utilises a two-step authentication protocol, he will first verify that *Alice* is approaching the junction from southern direction. Only after this is proven, the password would be requested from *Alice* in order to distinct it from the other cars driving on the same road. Significant advantage of the mobility-based authentication scheme is its flexibility. It can, but does not have to be combined with some higher-layer authentication protocols. Furthermore, the number of plausibility tests and their strictness can be adjusted depending on many parameters, such as desired level of security, accuracy of the wireless channel parameter estimation (WPCE), computation capabilities of Alice and Bob. The core of the mobility-based authentication scheme are plausibility tests. But since these heavily depend on the PHY-layer parameters of the communication system, the next section address the key features of the IEEE 802.11p and the classification of the wireless channel in car-to-car scenario.

6.3.4 Hybrid Mobile *ad hoc* Network (MANET)–Delay Tolerant Network (DTN) Networks and Security Issues

In MANET, the security mechanisms are based on the assumption that there is/are a connection between source and target nodes (end-to-end connections). opportunistic networks (OppNet) and DTN, disconnected MANET are more general than MANETs, because dissemination communication is the rule rather than conversational communication [PDC13]. Security is an important issue in disconnected MANET, DTN, and OppNet. OppNets are formed by individual nodes that can be disconnected for some time intervals, and that opportunistically exploit any contact with other nodes to forward messages. Each nodes computes the best paths based on its knowledge. The messages are routed and transmitted by *store-carry-forward* model. The main philosophy of OppNet is to provide ability to exchange messages between source and destination nodes.

In OppNet, the security solutions need to reflect security for all nodes, all services and application that participate on routing and transmitting process. There is sporadic connectivity of nodes and we need to provide secure delivery of the messages from source node to destination node. In order to provide the effective communication between nodes, there is necessary to consider different aspects (disconnections, mobility, partitions, and norms instead of the exceptions [PDC12]).

The modification of the the reactive routing protocol *dynamic source routing protocol (DSR)* designed for MANET, which enables transportation of the packets between disconnected terminals is introduced in Papaj et al. [PDC13]. The main idea of the routing algorithms is to enable the routing mechanism not only if we have connected paths to the mobile nodes but when unable to find routes, if there are disconnected routes. If there are paths between the source and destination nodes, the standard routing algorithm DSR is use for selection of the paths. In the case that the DSR routing protocol cannot find the paths to the destination or when the disconnection of the paths are occurring then the new protocol is activated. Algorithm also uses the statistical information about all connections between its neighbours nodes and also provides useful information about how many times was the mobile node in contact with other mobile nodes. Based on these data the routing protocol can select the optimal forwarding mobile node. The modified algorithm has been implemented and tested in OPNET modeler.

In Matis et al. [MD14], the hybrid routing protocol is designed and described. This new routing protocol expands functionality of DSR routing protocol which implemented features of the DTN forwarding mechanisms, which enables delivering of messages in MANET networks also in the case when the height speed mobility of nodes causes that MANET network is fragmented to networks islands with zero connectivity between them. Because of the mobility a lot of new opportunistic connections between nodes are created. The proposed algorithm also enables to increase the performance of the network in the sense of the message delivery in disconnected environment [MD14]. The performance analysis of the proposed routing protocols was tested in MAT-LAB.

Hybrid MANET–DTN networks are new type of mobile networks, that provides the new way how the different application could be provided for end users. The main idea of hybrid MANET–DTN networks enable use the network not only for personal usage but for emerging applications and services. The trust-based candidate node selection algorithm for hybrid MANET–DTN is introduced [PDP14]. This mechanism enables the selection of the secure mobile node used for transportation of the data across the disconnected environment. Selected candidate nodes provide the secure mobile node selected to transport of the data between disconnected island of the MTs [PDP14]. The trust values are computed from collected routing and data parameters. Each mobile node have its own values about neighbours mobile nodes and then the routing protocol can select optimal node for transportation of the data. The proposed algorithm is implemented into OPNET modeler simulator tool.

6.3.5 Resource Managent in LTE Network

Authors in [Lue13] investigate the potential of adjusting HOM and time-to-trigger (TTT) for intra-frequency HO optimisation. There is described HO mechanisms and a sensitivity analysis of HOM and TTT is carried out for different system load levels and user speeds. Next the performance analysis considering both HO parameters for different situations, highlighting the variation of the user speed is discussed. There is also introduced the new incremental structure of the Fuzzy Logic Controller (FLC), that adaptively modifies HOMs for HO optimisation. The design of the FLC includes the following tasks:

- define the fuzzy sets, and
- membership functions of the input signals and define the linguistic rule base which determines the behaviour of the FLC.

The first step involved in a FLC execution is the fuzzifier, which transforms the continuous inputs into fuzzy sets. Each fuzzy set has a linguistic term associated, such as high or low. In particular, two fuzzy sets have been defined for the input HO Ratio (HOR): high and low. In the case of the CDR, since it is a more critical performance indicator, three fuzzy sets have been defined to have more granularity: high, medium, and low. Note that the number of input membership functions has been selected large enough to classify performance indicators as precisely as an experienced operator would do, while keeping the number of input states small to reduce the set of control rules. The translation between numerical values and fuzzy values is performed by using the so-called membership functions. Its main advantage is to allow addressing numerical problems from the human reasoning perspective, making the translation of the network operator experience to the system control easier.

6.4 Energy Efficiency Enhancements

6.4.1 Energy Efficiency in Cellular Networks

In Europe, the telecommunication market accounts for 8% of the total energy consumption and for 4% of the CO_2 emission. The problem of increasing energy consumption and CO_2 emissions in different industrial sectors led the European Commission to identify the so-called 2020 objectives which foresee to reach the 20% of renewable energy production, to improve energy efficiency by 20% and to decrease the CO_2 emissions by by 20% by the end of the year 2020 [BCE+11]. In both fixed and mobile telecommunication sectors, the access network is responsible for a large part of the energy consumption. As an example, in a typical mobile radio network, up to about 80% of total power consumption occurs at the BS. It is worth noting that the power consumption in cellular networks is steadily increasing due to the growing demand of broadband wireless internet access through the usage of new MTs such as smartphones, tablets, and other high-end terminal devices, as well as laptops with cellular connectivity.

For future networks, an energy efficiency improvement is expected. Litjens et al. [LTZB13] assess the energy efficiency of mobile networks in 2020 and compares it to a 2010 baseline by taking into account the trends of mobile traffic increase, the corresponding network deployments (including the BS density and use of small cells), and technological improvements w.r.t. mobile network equipment (reflected in, e.g., power consumption models and sleep modes). An energy efficiency improvement factor of 793 has been observed in 2020 over 2010, thanks to the traffic increase (leading to more bits transmitted per cell), hardware evolutions (lower power consumption of BSs and backhauling), network sharing (leading to higher resource utilisation especially in coverage-limited cases), micro-sleep mode of macrocells in 2020 dense urban scenario (energy saving at low traffic), and MIMO.

6.4.1.1 Power consumption of different telecommunication technologies

To determine the power consumption of the cellular network, it is important to be aware of the power consumed by the BSs active in the network. A power consumption model for a macrocell base station (MBS) is proposed by Baumgarten et al. [BJRK12]. A BS consists of different hardware elements, as shown in Figure 6.14, which can be grouped into two distinct categories: sector-specific hardware, which is exclusively used to operate one sector (blue box in Figure 6.14(a)) and shared hardware, which is either baseband

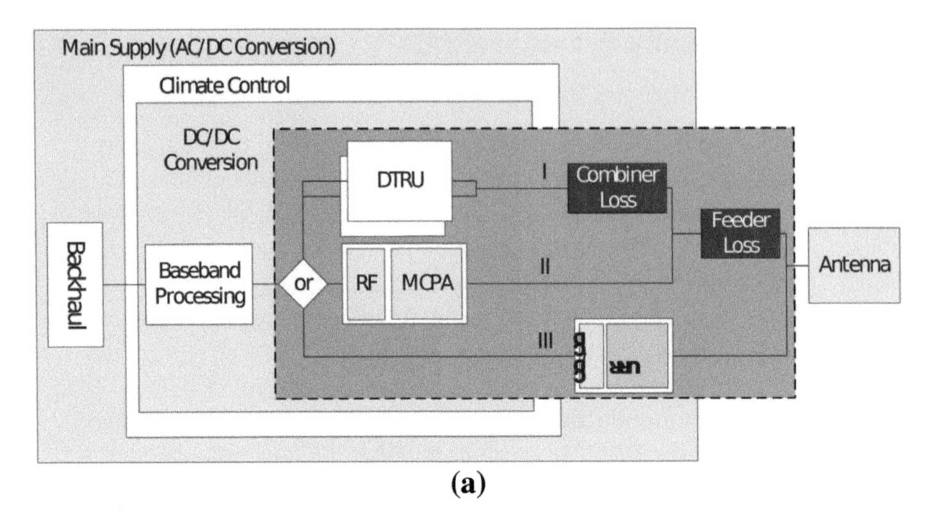

(a)

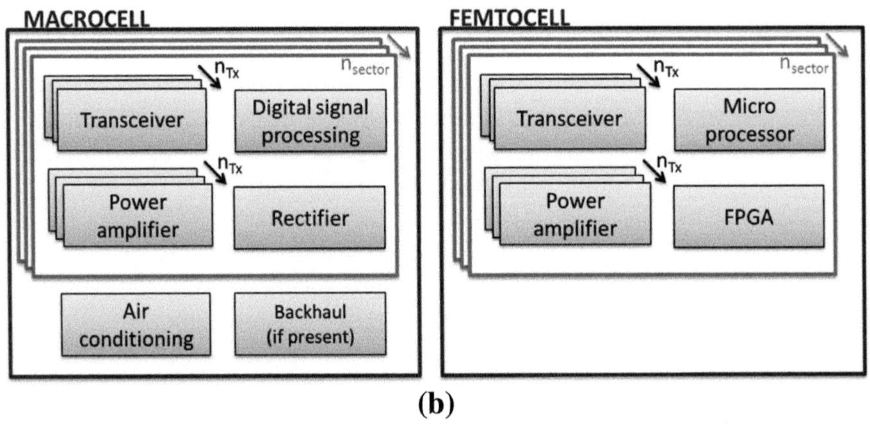

(b)

Figure 6.14 BS architecture as proposed by [BJRK12] (a) and [DJL+13] (b).

processing or site support. Based on this architecture, it is possible to set up a power consumption model for the BS. The power consumed by a BS consists of a part that is always used, even when the BS does not process any signals (idle state). The other part scales linearly with the load. Figure 6.15 shows the power consumed by a global system for mobile communications (GSMs) BS under different loads. A break-up between the power consumption of the different parts of the BS is also provided.

In Deruyck et al. [DJL+13], a BS similar power consumption model for a macrocell and femtocell BS is used for comparing the energy efficiency

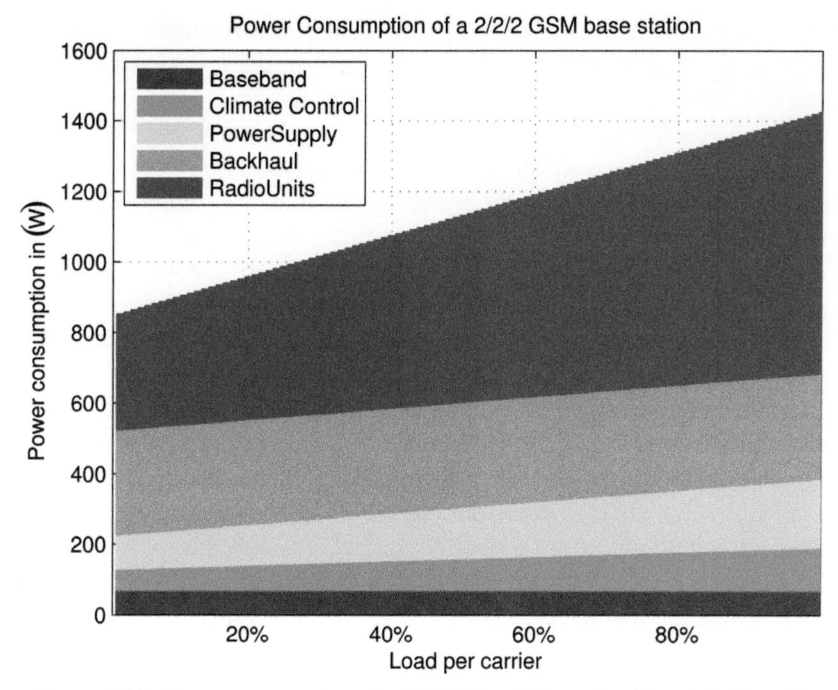

Figure 6.15 Power consumption of a GSM BS at different load levels [BJRK12].

between an LTE and LTE-A BS. The influence of three main functionalities added to LTE-A on the energy efficiency is investigated: carrier aggregation (to increase the bit rate), HetNets (whereby macro-cell and femtocell BS are mixed in one network), and extended support for MIMO (where multiple antennas are used for sending and receiving the signal). An appropriate energy efficiency metric is proposed, which takes into account the geometrical coverage, the number of served users, the capacity, and of course the power consumed by the network. In general, a higher bit rate results in a lower energy efficiency. However, due to carrier aggregation, LTE-A allows to obtain higher bit rates for even a higher energy efficiency. Heterogeneous LTE(-A) networks typically consist of macrocell and femtocell BS. For bit rates above 20 Mbps, the MBS is the most energy efficient. Below 20 Mbps, there is no unambiguous answer as it depends on the bit rate which technology is the most energy-efficient. For future networks, it is recommended to estimate accurately the required bit rate and coverage to decide which BS type or combination of these types should be used in the network. Finally, MIMO can also increase the energy efficiency, even up to 5 times by using spatial diversity and 8×8

MIMO. Future networks should thus support MIMO. It is recommended that future networks implement LTE-A as it will improve the energy efficiency compared to LTE.

Another way to improve the amount of radiated power (and thus the power consumed by the BS) is beamforming. Correia Gonçalves and Correia [GC11] developed two simulators, one for universal mobile telecommunications system (UMTS) and another for LTE, to evaluate in a statistical way the potential impact that adaptive antenna arrays have to reduce the radiated power, compared with actual BS static sector antennas. A logarithmic model was derived to represent the power improvement that is achieved by adaptive antennas in UMTS as a function of the number of users and the number of elements at the antenna array. When cells are near its top capacity, which was simulated with 70 users within the same cell, 90% of the power that is radiated by static sector antennas can be saved if adaptive antennas are used and, for antenna arrays with eight elements, more than 93% of radiated power saving. LTE power improvement, when compared with actual static sector antennas, does not change with the variation on the cell radius, for the same number of interferers in neighbour cells.

6.4.1.2 Cell switch off

A lot of power is consumed when it is actually not necessary. For example, during the night, all BS of the network are still active although there might be no activity taking place into their cells. Letting those BS sleep and turn them on when there are really necessary can save a significant amount of power. In literature, several cell switch off (CSO) schemes have been proposed [GYGR14]. Most of them address the problem of selectively switching off BS by means of heuristic algorithms. These schemes are preferred because CSO is a combinatorial NP-Complete problem, and hence, finding optimal solutions is not possible in polynomial time. However, González et al. [GYGR14] propose a multi-objective framework that takes the traffic behaviour into account in the optimal cell switch on/off decision making which is entangled with the corresponding resource allocation task. The exploitation of this statistical information in a number of ways, including through the introduction of a weighted network capacity metric that prioritises cells which are expected to have traffic concentration, results in on/off decisions that achieve substantial energy savings, especially in dense deployments where traffic is highly unbalanced, without compromising the QoS. The proposed framework distinguishes itself from the CSO papers in the literature in two ways: (i) The number of cell switch on/off transitions as well as handoffs are minimised. (ii) The computationally-heavy part of the algorithm is executed offline, which makes the real-time implementation feasible. The proposed multiobjective framework achieves significant gains with respect to the considered benchmarks in terms of power consumption (>50%), number of HOs (>85%), and the cell switch on/off transitions (>87%).

Litjens et al. [LTZB13] observe that dynamic switching off picocell BS at low traffic for energy saving has limited impact on the overall energy efficiency due to the low power consumption and low total number of picocell BS. However, a higher potential saving is expected for scenarios with more deployed picocell BS to serve hot zone traffic.

6.4.1.3 Cellular deployment strategy

Barbiroli et al. [BCE+11] show that the power consumption of a mobile radio network is strictly related to the cellular deployment strategy. For a Manhattan environment, whereby the city is composed of a regular grid of square building blocks separated by wide streets, a macrocellular and microcellular coverage architecture has been considered and results show that MBSs are power efficient for outdoor coverage and indoor, high floor locations, but become extremely inefficient when a coverage in indoor locations at the first floor is needed. Coversely microcellular layout is efficient for outdoor coverage and indoor location at first floor, but inefficient for coverage at high floors. Thus the maximum efficiency (i.e., the minimum transmitted power density per km^2) is achieved by combining the micro- and macrocellular architecture: using MBSs for indoor coverage at high floors and for outdoor coverage, acting as umbrella cells, and using microcell BS as gap fillers for outdoor coverage and for indoor coverage at lower floors.

The idea that the power consumption of a mobile radio network is related to the cellular deployment strategy is also considered by Deruyck et al. [DTJM12] in which the authors propose a coverage-based deployment tool. The goal of this tool is to design a network that covers a predefined geometrical area and bit rate with a minimal power consumption. Possible BS locations are provided and the algorithm selects those locations that result in the highest coverage with the lowest power consumption. 3D data about the environment is taken into account when predicting the coverage of the network. For a suburban area in Ghent, Belgium, an wireless fidelity (WiFi) 802.11n network and an LTE-A femtocell network are developed. The LTE-A femtocell network is about 2.5 times more energy-efficient than the WiFi network due to the fact that the LTE-A network uses about half of the BS than the WiFi network.

6.4.1.4 Joint optimisation of power consumption and human exposure

When we talk about green wireless networks, we should not limit ourselves to energy efficiency; also exposure for human beings is an important issue. People are becoming more concerned about possible health effects. In a Belgian (Flemish) study, about 5% of the respondents does not own a mobile phone. 6% of non-owners say that they are afraid of the health effects [MSM12]. Electromagnetic field exposure awareness has significantly increased in the last few years. International organisations such as international commission on non-ionizing radiation protection (ICNIRP) provide safety guidelines and national authorities define laws and norms to limit exposure of the electromagnetic fields caused by wireless networks. In Deruyck et al. [DJT+13], a capacity-based deployment tool is proposed which optimises the network towards power consumption or towards global exposure. The tool is capacity-based which means that it will respond to the instantaneous bit rate request of the users active in a considered area [DJTM14]. Appropriate distribution models for the number of active users, their locations, and their required bit rate had to be chosen. Preliminary results showed that when optimising towards power consumption, a network with a low number of BS with a high output power are obtained (leading to a low power consumption but high global exposure), while when optimising towards global exposure, the opposite situation is obtained: a high number of BS with a low output power (resulting in a high-power consumption but a low global exposure). As a compromise, one can optimise towards power consumption while satisfying a predefined exposure limit. This network shows a good compromise for the power consumption and the global exposure compared to the other two optimisations. However, further optimisation should be aspired. In Deruyck et al. [DTP+15], the capacity-based deployment described above is extended to optimise the network towards both power consumption and exposure. By choosing an appropriate fitness function, the required trade-off between power consumption and global exposure is obtained. Furthermore, the requirement to connect as much as possible to an already active BS is dropped as this might result in a good optimisation towards power consumption but is not necessary the right choice when optimising towards global exposure. The network obtained with the proposed algorithm has a 3.4% lower power consumption and a 37% lower global exposure compared to the network optimised towards power consumption while satisfying a certain exposure limit. A better trade-off towards both parameters is thus obtained.

6.4.2 Energy Efficiency in WSNs

6.4.2.1 Energy harvesting

Power consumption is also a major bottleneck in WSN due to the difficulty to provide a continuous or sporadic energy source in situ for the operation of the wireless nodes. A natural component of any comprehensive solution to this problem is to leverage energy-harvesting technologies, whereby the needed energy is collected from the environment by converting different forms of energy, such as solar, elastic or radio frequency, into electrical power. Castiglione et al. [CSEM11] address the problem of energy allocation over source digitisation and communication for a single energy-harvesting sensor. Optimal policies that minimise the average distortion under constraints on the stability of the data queue connecting source and channel encoders are derived. It is shown that such policies perform independent resource optimisations for the source and channel encoders based on the available knowledge of the observation and the channel states and statistics. The main drawback of these policies is that they require large energy storage system (e.g., battery) to counteract the variability of the harvesting process and a large data queue to mitigate temporal variations in source and channel qualities. Suboptimal policies that do not have such drawbacks are investigated as well, along with the optimal trade-off distortion versus backlog sise, which is addressed via dynamic programing tools. Using large deviation theory tools, Castiglione et al. [CSEM11] designed also policies that show a good trade-off between energy storage discharge probability, data buffer overflow probability and average distortion.

6.4.2.2 Power consumption in electronics circuitry

Another major issue to be addressed in WSNs is the power consumption of the electronic circuitry in transmitter and receiver and the related energy consumption. Dimic et al. [DZB11] address this problem by deriving an energy consumption model for low-power wireless transceivers as a function of transmitter power and packet length. Their results show that both variables can be set optimally, minimising energy per bit of data, for any channel attenuation and control data overhead. Up to 85% of energy can be saved compared to full power transmission of short packets. However, there are two limits to this approach: output power in existing transceivers is set with coarser resolution than necessary and typical packet length is shorter than optimal. Since optimal packet length increases with control data length, packet aggregation with fragment repetition (AFR) is applied to reduce optimal packet length. This approach shows further improvement of energy efficiency.

6.4.2.3 Routing protocol

The energy per bit received in WSNs can also be improved by using the mobility-based routing protocol proposed by Ferro and Velez [FV12]. Such a protocol forwards messages from the nodes to the sink in an effective manner. It calculates for each path a cost based on the probability of errors during transmission, and selects the path with the lowest cost. Constant information exchange about paths and their cost allow to keep the path information up-to-date, and avoid the use of dead links generated by node's mobility. The performance is compared with flooding, which is an easy-to-implement mechanism for forwarding packets in multi-hop networks. A node who receives a packet transmits it to all the neighbours. By 'flooding', the network with multiple copies of the same packet, eventually it will get to the destination. Simulation results of random networks showed that the developed protocol can deliver up to 77% more packets than Flooding, even with a lower energy cost as shown in Figure 6.16(a). With the protocol, one requires 34% less energy per bit received as shown in Figure 6.16(b). By introducing sleep cycles in the protocol, this energy saving rises to 47%.

6.5 Spectrum Management and Cognitive Networks

6.5.1 Cognitive Radio

According to Haykin [Hay05], spectrum utilisation can be improved significantly by making it possible for a secondary user (SU; who is not being serviced) to access a spectrum hole unoccupied by the primary user (PU) at the right location and the time in question. CR is an intelligent wireless communication system that is aware of its surrounding environment, and uses the methodology of understanding-by-building to learn from the environment and adapt its internal states to statistical variations in the incoming RF stimuli by making corresponding changes in certain operating parameters (e.g., transmit-power, carrier-frequency, and modulation strategy) in real-time, with two primary objectives in mind: (i) highly reliable communications whenever and wherever needed, and (ii) efficient utilisation of the radio spectrum. Given the importance of CR for future telecommunication systems such as fourth generation (4G) and beyond mobile technologies, the number of research projects and initiatives in this field is increasingly growing. Within COST IC

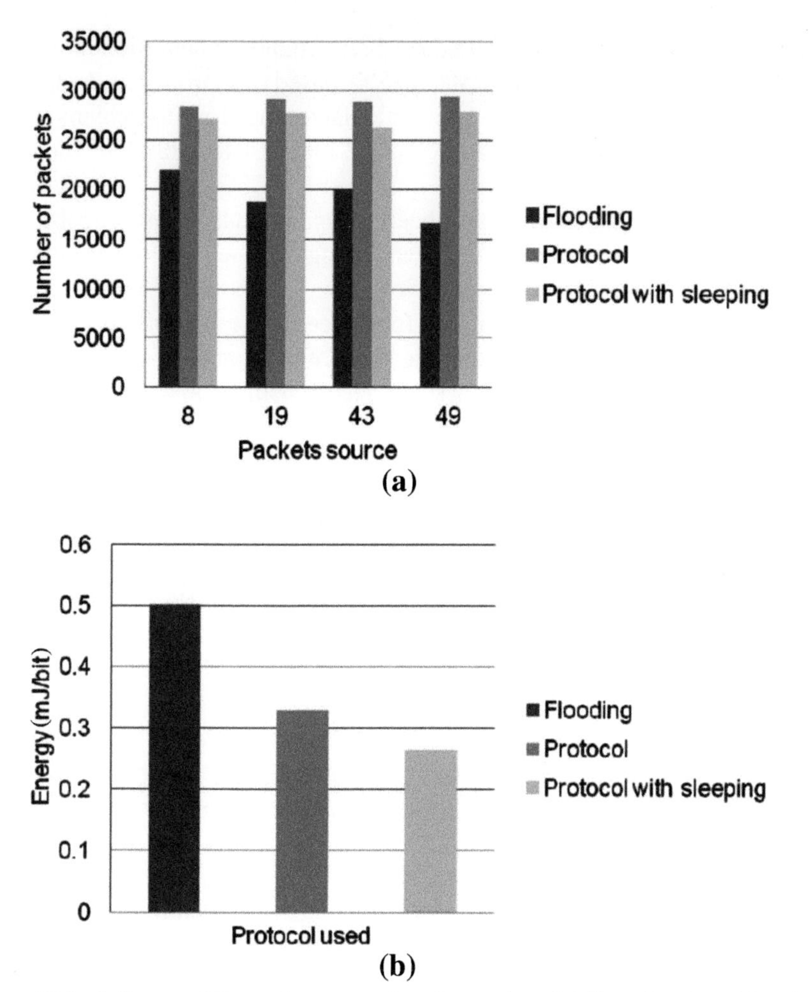

Figure 6.16 Influence of the routing protocol on the number of delivered packets (a) and the energy per received bit (b) [FV12].

1004, research has included aspects of how CR impacts on the performance of spectrum sharing and coexistence.

Cognitive radio terminals should be capable of acquiring spectrum usage information by means of some spectrum awareness techniques such as Spectrum Sensing (SS). Furthermore, in practical scenarios, the CR does not have *a priori* knowledge of the PU signal, channel and the noise variance. In this context, investigating blind SS techniques is an open research issue. To this end, Chatzinotas et al. [SKSO13] have studied several eigenvalue-based blind

SS techniques such as scaled largest value (SLE), standard condition number (SCN), John's detection and spherical test (ST)-based detection. The decision statistics of these techniques have been calculated based on the eigenvalue properties of the received signal's covariance matrix using random matrix theory (RMT). Several methods are considered in [CSO13a]. The sensing performances of these techniques have been compared in terms of probability of correct decision in Rayleigh and Rician fading channels for the presence of a single PU and multiple PUs scenarios. It has been noted that the SLE detector achieves the highest sensing performance for a range of scenarios.

The SS only approach ignores the interference tolerance capability of the PUs, whereas the possibility of having secondary transmission with full power is neglected in an underlay-based approach. To overcome these drawbacks, Chatzinotas et al. [SKSO14b] have proposed a hybrid cognitive transceiver which combines the SS approach with the power control-based underlay approach considering periodic sensing and simultaneous sensing/transmission schemes. In the proposed approach, the SU firstly estimates the PU SNR and then makes the sensing decision based on the estimated SNR. Subsequently, under the noise only hypothesis, the SU transmits with the full power and under the signal plus noise hypothesis, the SU transmits with the controlled power which is calculated based on the estimated PU SNR and the interference constraint of the PU [CSO13b]. Furthermore, sensing-throughput trade-off for the proposed hybrid approach has been investigated and the performance is compared with the conventional SS only approaches in terms of the achievable throughput.

In CR networks, the detection of active PUs with a single sensor is challenging due to several practical issues such as the hidden node problem, path loss, shadowing, multi-path fading, and receiver noise/interference uncertainty. In this context, cooperative SS has been considered a promising solution in order to enhance the overall spectral efficiency. Existing cooperative SS methods mostly focus on homogeneous cooperating nodes considering identical node capabilities, equal number of antennas, equal sampling rate and identical received SNR. However, in practice, nodes with different capabilities can be deployed at different stages and are very much likely to be heterogeneous in terms of the aforementioned features. In this context, Chatzinotas et al. [SKSO14a] study the performance of a decision statistics-based centralised cooperative SS using the joint probability distribution function (PDF) of the multiple decision statistics resulting from different processing capabilities at the sensor nodes. Further, a design guideline has been suggested for the network operators by investigating performance versus network sise trade-off while deploying a new set of upgraded sensors [SCO15].

Cooperative SS is a promising technique in CR networks by exploiting multi-user diversity to mitigate channel fading. Cooperative sensing is traditionally employed to improve the sensing accuracy, as shown in Figure 6.17 while the sensing efficiency has been largely ignored. However, both sensing accuracy and efficiency have very significant impacts on the overall system performance. In Zhang [Zha13a], the author first identifies the fundamental trade-off between sensing accuracy and efficiency in SS in CR networks. Then, several different cooperation mechanisms are presented, including sequential, full-parallel, semi-parallel, synchronous, and asynchronous cooperative sensing schemes. The proposed cooperation mechanisms and the sensing accuracy-efficiency trade-off in these schemes are elaborated and analysed with respect to a new performance metric achievable throughput, which simultaneously considers both transmission gain and sensing overhead. Illustrative results indicate that parallel and asynchronous cooperation strategies are able to achieve much higher performance, compared to existing and traditional cooperative SS in CR networks.

An opportunistic spectrum usage model facilitated with periodic SS and handoff in a two-hop selective relay network is considered in Wang [Wan13].

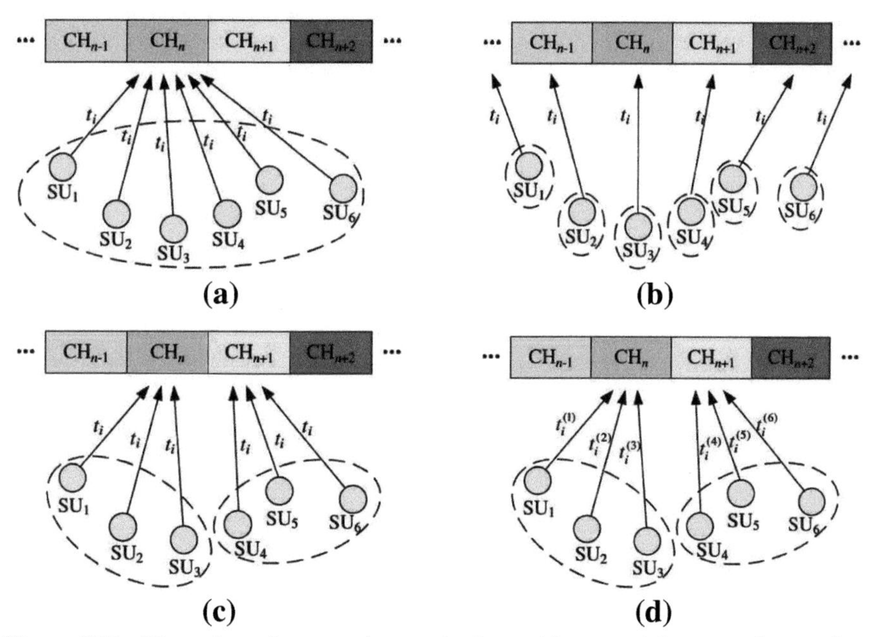

Figure 6.17 Illustration of cooperation mechanisms: (a) sequential cooperative sensing; (b) full-parallel cooperative sensing; (c) semi-parallel cooperative sensing; and (d) asynchronous cooperative sensing.

A novel sensing and fusion algorithm is proposed under a signalling bandwidth-constrained condition by introducing a quantised soft sensing and a test statistic restoration process. The reliability of secondary transmissions is studied, where expressions for the probability of collision, and the throughput of SU are derived.

Simulation results show the proposed sensing algorithm provides a better sensing performance with a lower signalling cost and a lower collision probability, compared to the conventional algorithm in selective relay networks in light of the missed detection probability and throughput. Finally, results show that the proposed algorithm offers a close performance to the throughput of a full soft algorithm in selective relay networks.

Wireless access networks based on dynamic spectrum access (DSA) in a densely deployed environment are promising solution to meet the future mobile traffic challenge with CRS capabilities. In such a network, not only the capacity improvement, but also energy efficiency (EE) are critical problems. It reported that in a multi-cell scenario the interference from neighbouring cells will degrade both EE and spectrum efficiency (SE). The challenges and solutions to achieve an energy efficient wireless communications are summarised in Li et al. [LXX$^+$11], and Hasan et al. [HBB11]. As illustrated in Figure 6.18, in Tao Chen et al. [TCZ12] the energy efficient resource allocation in a DSA-based wireless access network is explored. In such a network APS are densely deployed to provide open access to MTs. Spectrum of the network is opportunistically shared from PUs. In multiple channels divided from available spectrum, an access point (AP) selects one working channel and all MTs attached to the AP use the same channel to communicate with the AP. The work in Chen et al. [CZHK09] is extended by taking into account EE in the channel allocation and MT association. The main contributions from Tao Chen et al. [TCZ12] include: designing a local information exchange method to estimate inter-cell interference, proposing a local energy efficient metric used for distributed energy efficient algorithms, and developing the distributed energy efficient algorithms for AP channel selection and MT association, respectively. The original problem is divided into two sub-problems, i.e., the channel selection problem of AP and the AP association problem of MT. The distributed AP channel selection algorithm and MT association algorithm are developed, which allow an AP to explore spectrum with less interference, and an MT to connect to an AP with the most bit/energy gain. The approach is flexible and provides scalability to large networks.

Smart grid is widely considered to be the next generation of power grid, where power generation, management, transmission, distribution, and

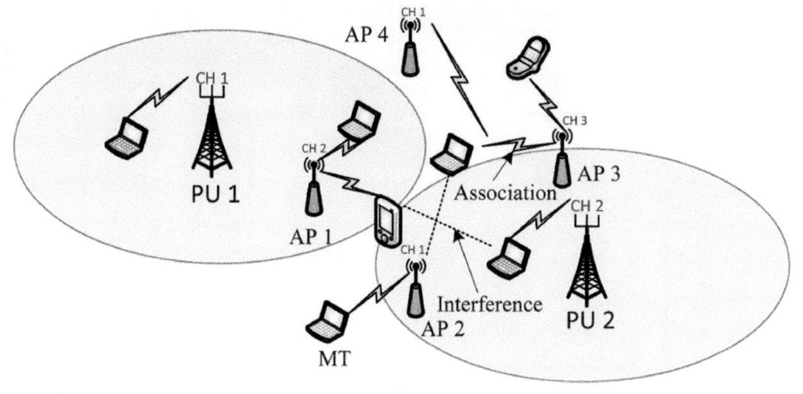

By re-association of MT and channel re-selection of AP, energy efficiency of SU network can be improved.

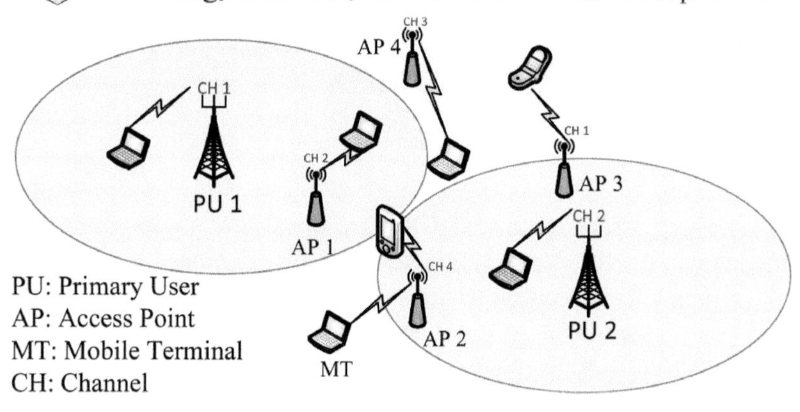

Figure 6.18 Reconfiguration of network by distributed MT association and AP channel selection algorithms to reduce interference and save network energy consumption.

utilisation are fully upgraded to improve agility, reliability, efficiency, security, economy and environmental friendliness. Demand Response Management (DRM) is recognised as a control unit of smart grid, with the attempt to balance the real-time load as well as to shift the peak-hour load. As shown in Figure 6.19, input of the system is provided by power plants, while feedback is demand of end users measured by smart metre. DRM acts as a control unit to balance and shape the realtime load. Output of the system is electricity delivered to each user through transmission and distribution. The forward path is *power flow*, and the bidirectional path is *information flow*, which provides two-way communications in smart grid.

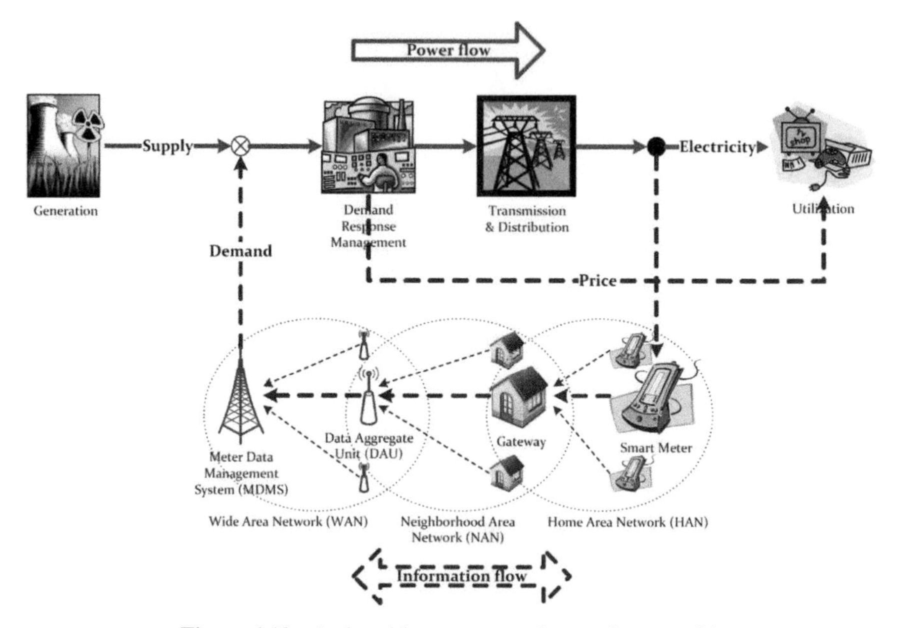

Figure 6.19 A closed-loop system scheme of smart grid.

Zhang [Zha13b] have introduced CR into smart grid to improve the communication quality. By means of SS and channel switching, smart metres can decide to transmit data on either an original unlicensed channel or an additional licensed channel, so as to reduce the communication outage. The impact of the communication outage on the control performance of DRM is also analysed, which reduces the profit of power provider and the social welfare of smart grid, although it may not always decrease the profit of power consumer. It is shown that the communication outage can be reduced. How the communication quality affects the control performance of DRM is also analysed. The work from Zhang [Zha13b] provides the guidelines of achieving better control performance with lower communication cost, paving the way towards green smart grid.

In the context of CR, spectrum occupancy prediction has been proposed as a means of reducing the sensing time and energy consumption by skipping the sensing duty for channels that are predicted to be occupied in future time instants [CAV13]. In a CR network, channel occupancy is not directly observable by the sensing nodes due to the wireless channel between the PU and SU. Therefore, SS can be modelled as a Hidden Markov Model (HMM) with a hidden process X_t and an observable process Y_t.

The hidden process Xt represents the PU activity and is modelled as a two-state Markov chain with state space $X = \{0, 1\}$, where '0' and '1' indicate an active and an idle PU, respectively. Similarly, Y_t is a random process with state space $Y = \{0, 1\}$, which represents the SS output, with '0' and '1' corresponding to an unoccupied and an occupied channel, respectively. Given the PU activity status x_t at time instant t, and y_t the corresponding sensing output, SS can be described by a two-state HMM with parametres $\lambda = (\pi, \mathbf{A}, \mathbf{B})$, where π is the initial state distribution: $\pi = [\pi_i]_{1\times 2}$, $\pi_i = P(x_1 = i)$, $i \in X$; $\mathbf{A}$ is the transition matrix that describes the probabilities of the PU activity status to change from active to idle: $\mathbf{A} = [a_{ij}]_{2\times 2}$, $a_{ij} = P(x_{t+1} = j|x_t = 1)$, $i, j \in Y$; and $\mathbf{B}$ is the emission matrix that describes the relationship between sensing output and the actual PU state: $\mathbf{B} = [a_{jk}]_{2\times 2}$, $b_{jk} = P(x_t = j|y_t = k), j \in X, j \in Y$.

Having an observation sequence of past SS outputs, O, of length T, the first step to HMM-based spectrum occupancy prediction is the model training process. During this process the HMM-based spectrum occupancy predictor estimates its parameters, using the Baum–Welch algorithm, so that the probability of observing O is maximised. After training the model's parameters, the Viterbi algorithm is used to determine the channel occupancy sequence X_t that is most likely to have generated the observation sequence, O. Given the estimated model parameters λ^* and the decoded channel occupancy states, the channel occupancy at a future time instant, $T + d$, is estimated by comparing the conditional probabilities of an observation sequence, O_T, to be followed by an unoccupied, $P(O_{T+d}, 0|\lambda^*)$, and an occupied channel, $P(O_{T+d}, 1|\lambda^*)$, respectively.

A performance comparison between the HMM-based spectrum occupancy predictor and a 1st and 2nd order Markov predictors is shown in Figure 6.20. The prediction performance is evaluated in terms of the probabilities of true positive predictions (upper end of the figure) and false negative prediction (lower end of the figure) and the channel occupancy status transition rate. With reference to Figure 6.20 it is shown that the HMM-based predictor has a higher adaptability to the channel status transitions which in turn results in an up to 50% higher prediction performance.

In the underlay CR scenario, Sharma et al. [SCO13b] has studied Interference Alignment (IA) as an interference mitigation tool in order to allow the spectral coexistence of two satellite systems. Furthermore, frequency packing (FP) can be considered as an important technique for enhancing the spectrum efficiency in spectrum-limited satellite applications. In this context, this contribution focuses on examining the effect of FP on the performance of multi-carrier-based IA technique considering the spectral coexistence scenario

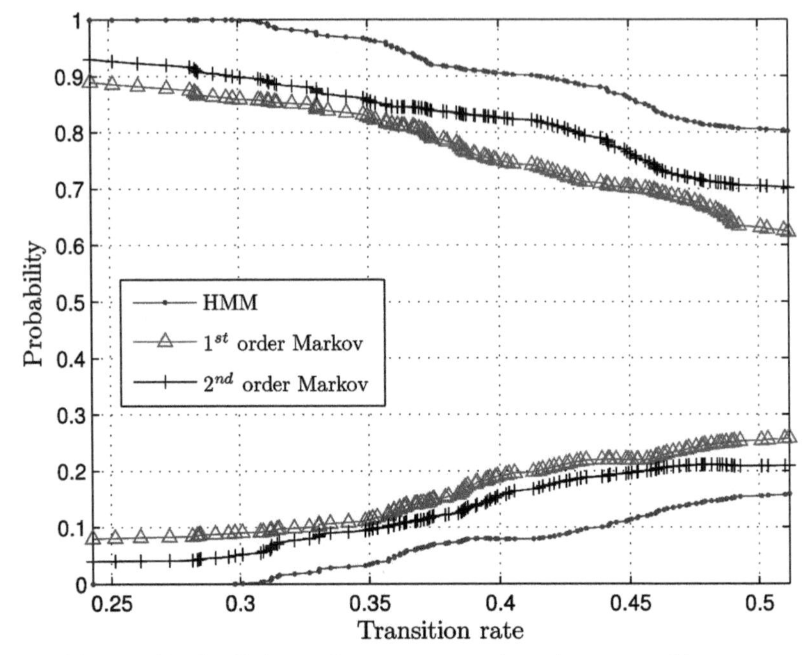

Figure 6.20 Prediction performance versus channel status transition rate.

of a multibeam satellite and a monobeam satellite with the monobeam satellite as primary and the multi-beam satellite as secondary. The effect of FP on the performance of three different IA techniques in the considered scenario has been evaluated in terms of system sum rate and primary rate protection ratio (PR). It has been shown that the system sum rate increases with the FP factor for all the techniques and the primary rate is perfectly protected with the coordinated IA technique even with dense FP [SCO14].

Author Diez [Die13] has studied different detection and hypothesis testing techniques that considerable influence on the performance of a cooperative CR network. Indoor trials with DVB T signals have been carried out at 690 MHz, using USRP-based devices. Measurement results proved that the wave-based form detector was more accurate than the cyclostationary feature detector. Besides, RF receivers with AGC and omni-directional antennas increased the signal detection probability. Taking into account the obtained results, a free channel detection algorithm has been defined based on installing some test points with a determined separation.

The cognitive device in X will be able to transmit a certain signal power with an omnidirectional characteristic if in the A, B, C, and D points the signal transmitted by the cognitive device is not detected. If the signal is detected in any of the points of a particular arm of the cross, the radiation pattern of the antenna situated in X could be modified. Installing all of these fixed points could lead to a dense cognitive device network. This disadvantage can be offset by creating real time data bases with the information given by all the mobile cognitive devices on the area.

With the profusion of low cost wireless platforms, such as the Ettus family of Software Defined Radios (SDRs), experimentation in wireless research became a reality. Such platforms have allowed many ideas to become a reality, providing researchers a never before possible way to demonstrate the feasibility of their concepts. Nevertheless, such platforms are usually limited to a few at a time, since they become very hard to handle in larger quantities without a proper underlying infrastructure. This fact, unfortunately, limits the kind and scale of techniques that can be implemented and demonstrated. In Cardoso et al. [LSCG12], authors have introduced the CorteXlab testbed, a large scale testbed comprised of heterogeneous platform nodes able to deal with distributed PHY design and CR.

CorteXlab will provide researchers all over the world with an up-to-date high quality PHY layer testing tool. Being part of the Future Internet of Things (FITs) project, CorteXlab will be accessible through an easy to use web portal, with possible future connection to the other testbeds integrating FIT. Advanced PHY layer reconfigurability also allow to address issues common to this kind of networks, such as synchronisation and power consumption. Note that the optimal way to address this kind of scenarios remains unknown in the academia, and is in the forefront of the research efforts. Ongoing work includes the development of a reference PHY design, based on PHY layer relay network techniques, the deployment of the middleware (heavily based on SensLab) that allows the management and remote activation of the nodes and the choice of the node platform itself.

6.5.2 Digital Terrestrial Television (DTT) Coexistence & TV White Spaces (TVWS)

With the arrival of digital television technologies and new compression systems, the concept of TVWS was introduced to define such regions of space–time–frequency in which a particular secondary use is possible [TMS09]. This is, such parts of the licensed spectrum (mainly for broadcasting service) that are not in use in a specific area and that can be used for secondary devices (known as TVWS devices) to provide other communications services in such region.

These new TVWS devices, also named white space devices (WSD) and considered as cognitive devices [MM99], should determine which frequencies are free and then only transmit according to an appropriate range of parameters such as power levels and out-of-band emissions. There are several strategies to determine when a frequency is unoccupied: **SS** (also called detection) where devices monitor frequencies for any radio transmissions and if they do not detect any, assume that the channel is free and can be used; or **geolocation database** where devices determine their location and query a geolocation database which returns the frequencies they can use at their current location [Ofc09]. Although WSD based on SS presents many advantages as its low-cost infrastructure, its major drawback is that sensing to very low signal levels is costly and possibly not achievable, giving rise to a not accurate detection and causing interferences to DTT transmissions [Ofc10].

In Barbiroli et al. [BCGP12], authors propose three different approaches to determine whether a frequency is free, based on geolocation databases: based on a threshold approach applied to a coverage map; based on the location probability; and a combination of geolocation approach and field strength measurements (sensing approach). Results showed that a combined approach could provide better protection to the incumbent services.

After the determination of a free TVWS, a WSD should ensure that its transmitting power is below the maximum equivalent isotropic radiated power (EIRP) allowable for this free frequency in order not to interfere to DTT transmissions. Authors Petrini and Karimi [PK13b] developed and improved method for the computation of the DTT location probability and the calculation for the maximum permitted EIRP with an improved accuracy and less computational complexity. The DTT location probability is introduced as a metric for quantifying the quality of the DTT coverage which will be degraded in presence of WSD. These algorithms are improved in Petrini et al. [PMB13], based on carrier-to-interference ratio (C/I) calculation.

A review about some trials performed in Cambridge about the coexistence of DTT with WSD is presented in Cataldi et al. [CL12].

The use of crossed polarisations between DTT and WSD has also been addressed as a strategy for increasing the protection of both the DTT and WSD [Bro14] when the broadcast transmitter is vertically polarised and the WSD horizontally polarised. Furthermore, regardless of the polarisation of the PU, an increase in path loss from the WSD to the PU should be accounted if the WSD has horizontal polarisation [BT14a].

Many standardisation groups have taken care about the use of TVWS for wireless personal area network (WPAN), wireless local area network

(WLAN) and wireless regional area network (WRAN) producing 802.15.4m, 802.11af and 802.22 standards, respectively. Moreover, the 802.19.1 standard published in 2014 regulates the coexistence among multiple TV white space networks. In Fadda et al. [FMV+13] and Angueira et al. [AFM+13], authors evaluated through measurements in laboratory the PR required at a DVB receiver in case of being interfered by an IEEE 802.22 WRAN signal (digital video broactasting (DVB) wanted signal). Regarding the influence of the parameters of the digital video broactasting-terrestrial 2nd generation (DVB-T2) OFDM signal on the PR, measurements showed that the PR decreases as the DVB-T2 code rate does [AFM+13]. The modulation type also affects the tolerance to interference, being 7 dB higher for 64QAM compared to 256QAM for a constant pilot pattern (PP) as shown in Figure 6.21. The use of rotated constellations did not have any influence on the PR and hence, on the interference levels.

Following the analogue switch off, the ITU in the 2007 World Radiocommunication Conference (WRC-07) decided to allocate the upper part of the former analog TV band, that is, from 790 to 862 MHz, to IMT technologies (with 1 MHz of band guard between broadcasting and mobile services in ITU Region 1). This allocation was also called the digital dividend (DD) [Sam12]. This decision was supported by studies carried out by the European conference of postal and telecommunications administrations (CEPT) reported in CEPT [CEP09a] and [CEP09b], where some technical conditions for IMT emission in the DD band were given.

However, the restrictions established by CEPT do not assure avoiding interferences in the DD. Thus, for a digital video broactasting-terrestrial (DVB-T) reception interfered by a LTE DL signal in adjacency (1 MHz of band guard) Barbioli et al. [BCF+13] obtained that for broadband receivers (mainly installed in blocks of flats in urban areas), an area around 1.000–1.500 from the BS will suffer from receiver saturation due to the high level of interfering power, impeding the successful reception of any DVB-T channel. In case of a typical set top box (STB) receiver the probability of interference is lower than in the previous case but not negligible.

In addition, the ITU 2012 world radiocommunication conference (WRC-12) concluded with a decision to allocate additional ultra high frequency (UHF) spectrum to mobile services and invited to perform further coexistence studies and-report the results to the next WRC-15. The new mobile allocation, also known as second digital dividend (DD2), is to be made in Region 1 in the 700 MHz band (the actual range is to be decided in 2015 world radiocommunication conference (WRC-15)). The main difference compared

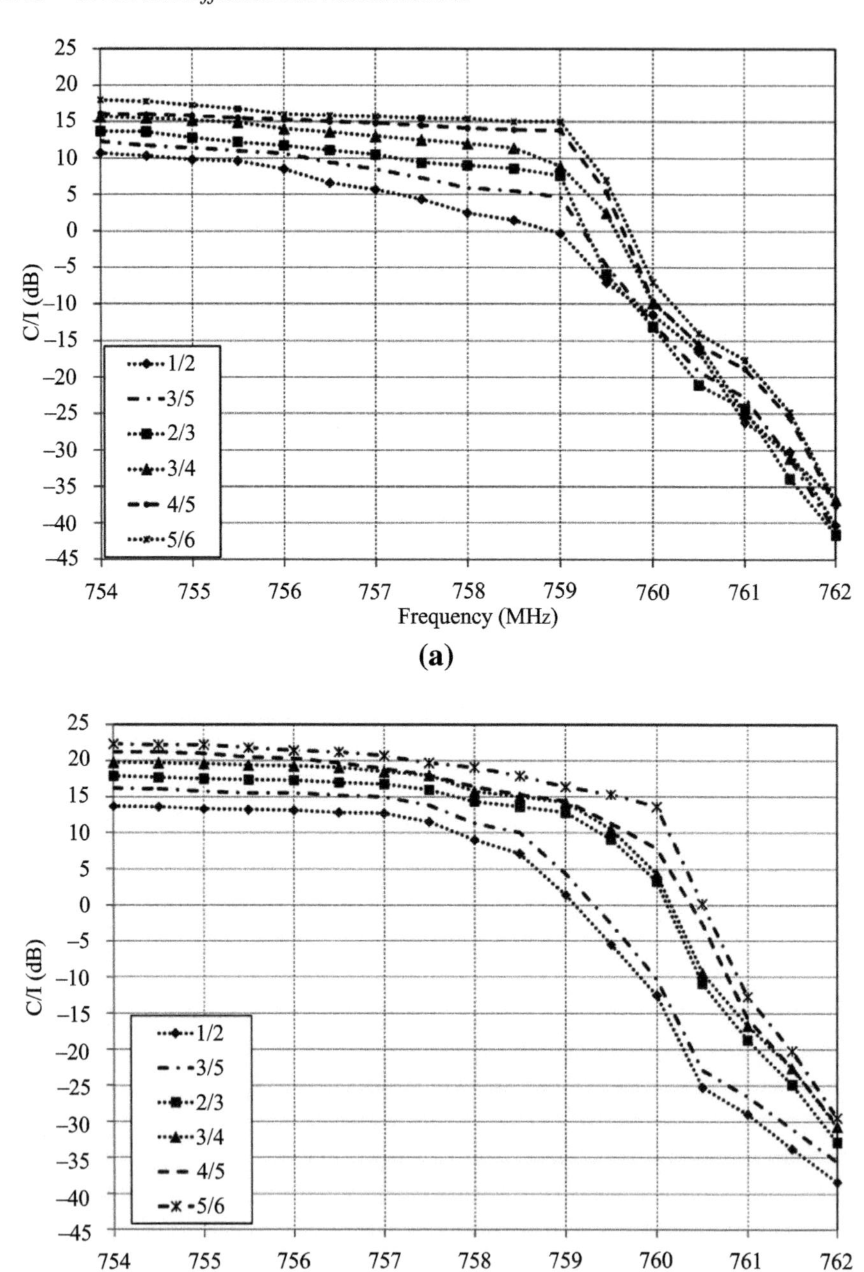

Figure 6.21 C/I requirements for all code rates for GI 1/16, PP4, and (a) 64QAM, and (b) 256QAM.

to the 800 MHz band lies in the fact that the UL is located in the lower part, instead of the DL.

In Fuentes et al. [FGGP+14b], authors present a thorough analysis of the the influence of the PHY parameters of DVB-T2 and LTE on the PR required for a-DVB-T2 wanted signal interfered by a LTE signal for both DL and UL cases and in the DD band. UL results to be the most interfering link as far as required PR are worse (higher) than those which are required for DL as observed in Figure 6.22. Furthermore, LTE signals with lower bandwidths are less interfering if LTE operates in adjacency (guard band higher than 9 MHz). In the same study, authors conclude that for LTE-DL operating in adjacency to DVB-T2, the higher the traffic loading, the higher the interference level. The contrary effect was observed for LTE-UL.

The PR obtained in Fuentes et al. [FGGP+14b], were used in Fuentes et al. [FGGP+14a] for planning studies in order to evaluate the coexistence of DVB-T2 and LTE in RL scenarios for DD and DD2. Thus, for the DD (DL in adjacency with 1 MHz guard band) case and fixed outdoor reception, with the receiver in the DVB-T2 coverage threshold (minimum DTT field strength), an average protection distance to LTE BS of 330 for urban and 580 m for rural environments must be leaved. In case of the DD2 (UL in adjacency with 9 MHz guard band) and DVB-T2 and fixed outdoor reception, a filter is not necessary if the LTE UE transmitted power is below 11 dBm. If portable indoor reception is considered, typical values of LTE UE transmitted power, assured a protection distance of 0.8 m in rural environments and 0.25 m in urban environments without the use of a filter in the DVB-T2 receiver.

The concept of micro TV white space (μTVWS) was introduced in Martinez-Pinzon et al. [MFGC15] to define the scenario in which the DTT signal is broadcasted to rooftop reception, and hence, obstructed for an indoor environment. Thus, these broadcast frequencies could be re-used in indoor for IMT technologies in the UHF band. In Martinez-Pinzon et al. [MFGC15], the authors perform some measurements in laboratory to evaluate the feasibility of the introduction of indoor LTE-A femtocells using DTT frequencies. The main conclusion of this study case is that, under the appropriate restrictions in EIRP, a LTE-A femtocell could operate in frequencies adjacent or even partially overlapping a DVB-T2 channel. The results fix the power limit for LTE in ranges from +20 dBm to –9 dBm depending on the proximity of the LTE-A central frequency to the DTT carrier. The co-channel case was disregarded.

Apart from the compatibility and coexistence between DTT and broadband and mobile communications, nowadays in Europe coexists the first and second

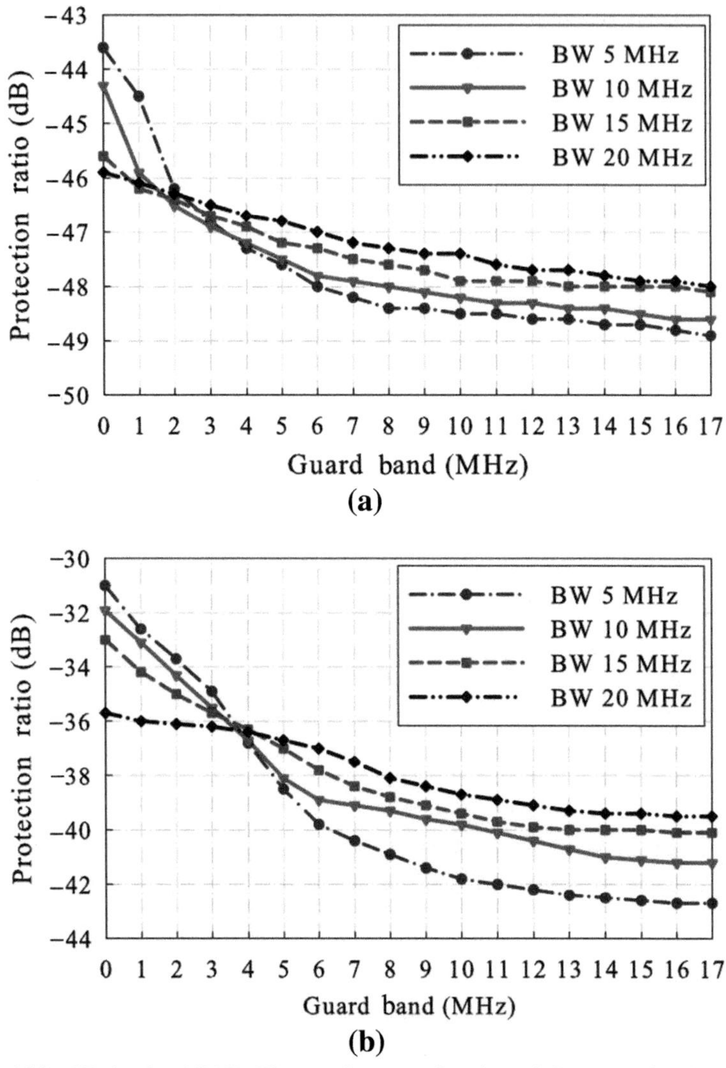

Figure 6.22 PR for fixed DVB-T2 reception, as a function of the separation between the carrier frequencies of DVB-T2 and LTE channels, for several values of the LTE channel bandwidth. (a) LTE-DL, and (b) LTE-UL.

generation of the European digital television standards, DVB-T and DVB-T2, with the mobile television standard, digital video broactasting-handheld (DVB-H). Thus, in Mozola [Moz12, Moz14], the author investigated the coexistence of these terrestrial and mobile broadcasting standards when a DVB-T2 signal interfered DVB-H reception.

6.5.3 Spectrum Management

Management of radio spectrum refers to the regulating process of RF which is necessary to efficiently use of such a limited spectrum resource. Because of increasing demand for more users and new services such as 4G mobile technology and beyond, spectrum management has become significant issue. During the last years, for efficiently use, sharing, coexistence, and aggregation of RF spectrum, several techniques have been introduced and different research initiatives have been conducted. In this section, recent researches results in the context of sharing, monitoring and aggregation of RF spectrum are presented.

When the primary is a rotating radar, opportunistic primary–secondary spectrum sharing can still be considered, as proposed in Saruthirathanaworakun et al. [SPC11] and as shown in Figure 6.23. A secondary device is allowed to transmit when its resulting interference will not exceed the radar's tolerable level, perhaps because the radar's directional antenna is currently pointing elsewhere, in contrast to current approaches that prohibit secondary transmissions if radar signals are detected at any time. The case where a secondary system provides point-to-multipoint communications utilising OFDMA technology in non-contiguous cells is considered. This scenario might occur with a broadband hotspot service, or a cellular system that uses spectrum shared with radar to supplement its dedicated spectrum, as described in Saruthirathanaworakun et al. [SPC11]. The secondary system considered provides point-to-multipoint communications in non-contiguous cells around the radar. Unlike existing models of sharing with radar, our model allows secondary devices to adjust to variations in radar antenna gain as the radar rotates, thereby making extensive secondary transmissions possible, although with some interruptions. Thus, sharing spectrum with rotating radar is a promising option to alleviate spectrum scarcity. Additional technical and governance mechanisms are needed to address interference from malfunctioning devices [Peh11].

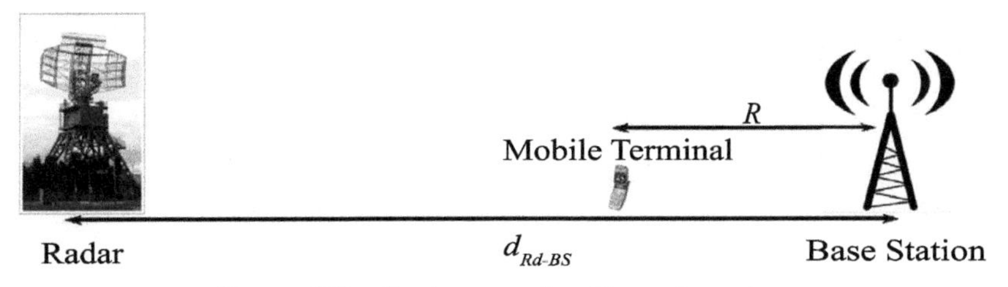

Figure 6.23 Sharing scenario with rotating radar.

It was also found that the secondary will utilise spectrum more efficiently in the downstream than in the upstream. Hence, spectrum sharing with radar would be more appropriate for applications that require more capacity in the downstream. The fluctuations in perceived data rate make sharing spectrum with radar attractive for applications that can tolerate interruptions in transmissions, such as video on demand, peer-to-peer file sharing, and automatic metre reading, or applications that transfer large enough files so the fluctuations are not noticeable, such as song transfers. In contrast, spectrum shared with radar would be unattractive for interactive exchanges of small pieces of data, e.g., packets or files, of which instantaneous data rate matters for the performance, such as VoIP.

We assess the system-level performance of non-orthogonal spectrum sharing (NOSS) achieved via maximum sum rate (SR), Nash bargaining (NB), and zero-forcing (ZF) transmit beamforming techniques, (Figure 6.24). A lookup table-based PHY abstraction and RRM mechanisms (including packet scheduling) are proposed and incorporated in system-level simulations, jointly with other important aspects of network operation. In the simulated scenarios, the results show similar system-level performance of SR (or NB) as ZF in the context of spectrum sharing, when combined with maximum SR (MSR; or proportional fair PF) packet scheduler. Further sensitivity analysis also shows similar behaviour of all three beamforming techniques with regard to the impact on system-level performance of neighbour-cell activity level and feedback error. A more important observation from our results is that, under ideal conditions, the performance enhancement of NOSS over orthogonal spectrum sharing (OSS) and fixed spectrum assignment (FSA) is significant.

Regulators face new challenges for radio spectrum monitoring in the near future, not only because the intensification of the radio spectrum usage associated to the growth of mobile technologies like LTR, LTE-A, or the new fifth generation (5G), but also because of the new cognitive systems that will be in use in the near future. These challenges (some of them mentioned in the ITU-R Recommendation SM.2039 about spectrum monitoring evolution) leads to the development of new spectrum monitoring systems, smaller and cheaper than the current system in use by regulators.

In Arteaga et al. [NAVA12] and Navarro [NAV^{+}14], a spectrum-monitoring system based on Open SDR systems is described. First, a basic spectrum analyser is implemented and after that, the procedures for accomplishing the ITU-R SM.1392 Recommendation is implemented. This system, named SIMONES implements a SS algorithm based on energy detection, to

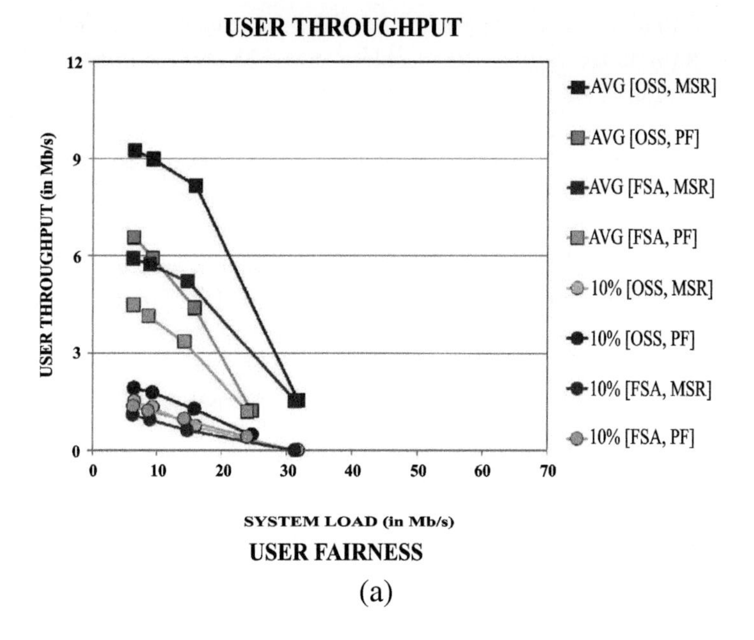

(a)

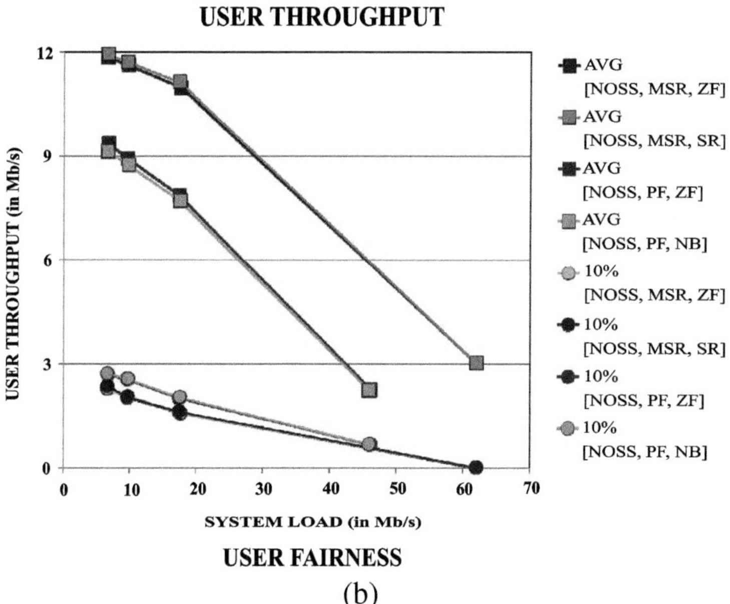

(b)

Figure 6.24 The performance comparison among NOSS, OSS, and FSA combined with different schedulers: (a) Average and 10%-tile throuhgput of OSS, and (b) Average and 10&-tile throughput of NOSS and FSA.

identify signals in a radio band, as well as bandwidth measurement, according to the $\beta/2$ method as recommended in ITU-R SM.443 and Chapter 4 of the Monitoring Handbook for fast fourier transform (FFT)-based systems. For frequency measurements, SIMONES has implemented the system in order to comply with ITU-R Recommendation SM.377, using the GPS locked reference oscillator for USRP. Because the system is based on FFT and Software Defined Radio, it is possible to measure variations in bandwidth and frequency for digital modulations. For occupancy measurement, the energy detection algorithm is adapted according to the ITU-R SM.1880 Recommendation, in order to obtain the occupancy parametre according to the recommendation. The spectrum monitoring system, SIMONES, has four functional components: the monitoring unit (SIMON); a set of drivers to interact with commercial monitoring software (TES Monitor suite); an independent web interface; and a user-based drive test unit.

In the Spectrum Monitoring Handbook 2011 version, a new section is introduced. In Section 6.8, the handbook describes the method for planning and optimisation of the monitoring stations. Since 2012, ITU Working Party 1C is developing a report on Spectrum monitoring network design for the future spectrum monitoring systems. In Navarro et al. [NA13], a modified method is proposed, which uses terrain-based propagation models as well as users density information to determine where the monitoring stations will attend bigger population (i.e., Spectrum Users). The population density criterion points to the prioritisation of monitoring in zones with more population density, which implies a bigger number of spectrum users covered by a monitoring station and therefore a bigger efficiency from the economical point of view, according to the recommendations of ITU-R Report ITU-R SM.2012.

The criterion of the use of semi-deterministic propagation models points to improve the coverage prediction using DTM and considering terrain information and diffraction effects for prediction and monitoring stations location. In mountainous regions, obstacles like big mountains could be quite important and are not considered by models like ITU-R P.1546 or Hata. Therefore, the risk of a miss location of a monitoring station using such models as proposed in the ITU Spectrum Monitoring Handbook 2011, is very high.

In [CMM+11], authors proposed an Integrated iCRRM. The iCRRM performs classic common radio resource management (CRRM) functionalities jointly with SA, being able to switch users between non-contiguous frequency bands. The SA scheduling is obtained with an optimised General Multi-Band Scheduling algorithm with the aim of cell throughput maximisation. In particular, we investigate the dependence of the throughput on the cell coverage distance for the allocation of users over the 2 and 5 GHz bands for a single operator scenario under a constant average SINR. For the performed evaluation, the same type of radio access technology (RAT) is considered for both frequency bands. The operator has the availability of a non-shared 2 GHz band and has access to part (or all) of a shared frequency band at 5 GHz, as shown in Figure 6.25. The performance gain, analysed in terms of data throughput, depends on the channel quality for each user in the considered bands which, in turn, is a function of the path loss, interference, noise, and the distance from the BS. An almost constant gain near 30% was obtained with the proposed optimal solution compared to a system where users are first allocated in one of the two bands and later not able to HO between the bands, as shown in Figure 6.26.

This work has been published in Cabral et al. [CMM+11].

6.6 Virtualised and Cloud-based Architectures

6.6.1 Architecture

Virtualisation and cloud computing are concepts widely used to enable the share of processing resources. This section presents several approaches on how these concepts have been adopted or extended in RANs, providing novel capabilities to these networks.

In Caeiro et al. [CCC12, 14a, 14b], the virtualisation of the wireless access is addressed. It is based on a network virtualisation environment that

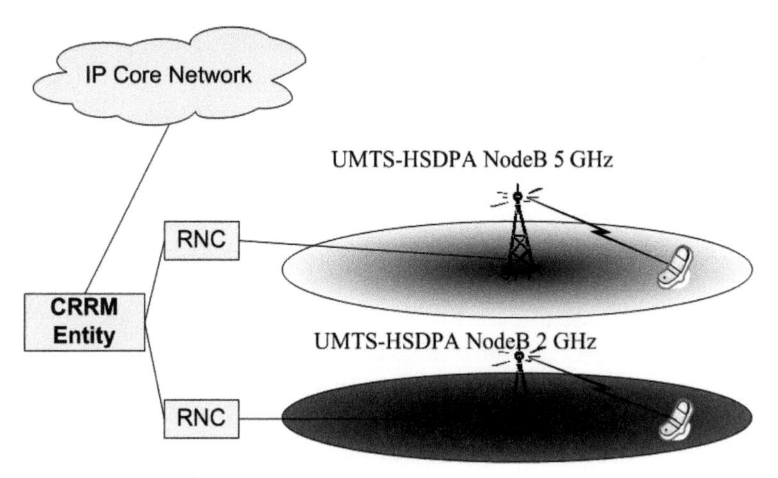

Figure 6.25 CRRM in the context of SA with two separated frequency bands.

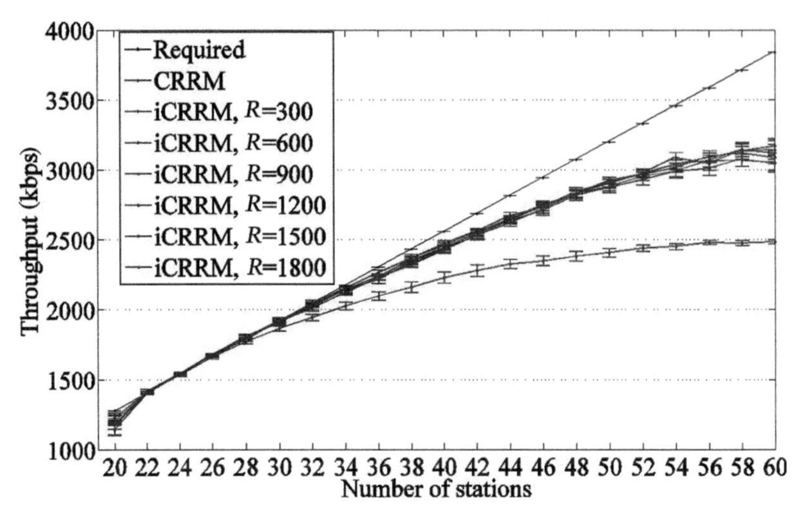

Figure 6.26 Average service throughput with the iCRRM with normalised power.

envisages the existence of multiple virtual networks (VNets) created by a VNet Enabler. A VNet is composed of one or multiple virtual base stations (VBSs), based on resources of heterogeneous RATs. In this environment, a virtual network operator (VNO) is a network operator that does not own any RAN infrastructure and needs wireless connectivity for its subscribers. For each VNO, a VNet is instantiated, settled on demand to satisfy its service requirements, in order to deliver services to their customers.

Similarly, the notion of virtual radio resources, as an aggregate of physical radio resources, and a model for their management is proposed in Khatibi and Correia. [KC14b]. It is a hierarchical management, as depicted in Figure 6.27, consisting of a virtual radio resource management (VRRM) on the top of RRM entities of heterogeneous RANs, designated as CRRM, while the classical local RRMs exist at the bottom. Given the VRRM, the RAN Provider, owner of the physical infrastructure, is capable of offering capacity to multiple VNOs according to certain service level agreements (SLAs).

On the other side, the concept of cloud has been adopted for RANs, where an architecture to offer cloud-based RAN radio access network as a service (RANaaS) is proposed in Ferreira et al. [FPH+14]. It follows the concept of C-RAN, which splits the BS into a remote radio head (RRH) (an antenna and a small radio unit) and a software-based base band unit (BBU) (baseband, control and management functions). Software-based BBUs are deployed and managed on general purpose platforms, by applying virtualisation and cloud

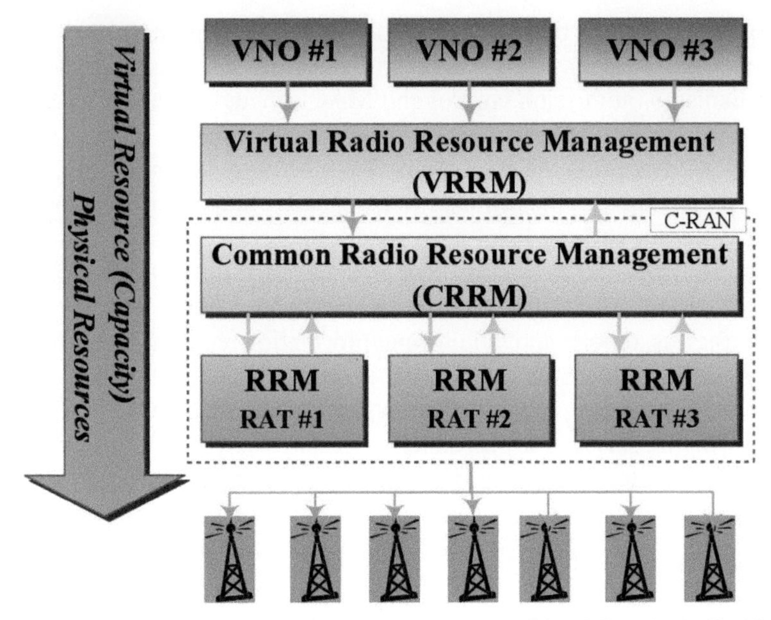

Figure 6.27 VRRM architecture (extracted from Khatibi and Correia. [KC14b]).

paradigms. RANaaS offers to operators an on-demand, scalable and shared RAN, efficiently adaptable to geographic and temporal load variations. It enables fast, dense, cost, and energy efficient deployments (1/3 roll out time, 15% CAPEX, 50% OPEX, 71% energy saving compared to traditional RAN). Also, new players may easily enter the market, allowing a VNO to provide connectivity to its users. One of the key research challenges is to match LTE processing requirements on general purpose processing platforms.

6.6.2 Models and Algorithms

Various models and algorithms are presented in this section to manage virtualised radio resources. An adaptive virtual network radio resource allocation (VRRA) algorithm is proposed in Caeiro et al. [CCC12], which optimises the utilisation of shared resources of heterogeneous RATs, in order to maintain the contracted capacity of VBSs. Initially, the algorithm pre-allocates radio resources units (RRUs) to each VBS. Then, sensitive to VBS changes in capacity, it dynamically uses compensation mechanisms to adequately allocate additional RRUs.

In Caeiro et al. [CCC14a,b], an on-demand VRRA algorithm is proposed, allocating resources elastically to end-users, depending on the current demand.

In order to compensate for the variability of the wireless medium and taking the diversity of available RATs into account, the algorithm continuously influences RRM mechanisms (admission control and MAC scheduling) to be aware of the VBSs state, to satisfy the SLAs.

A VRRM model is proposed in Khatibi and Correia. [KC14b] (Figure 6.27). It estimates the physical RAN capacity and allocates resources to support different service level agreement (SLA) contracts. It translates VNOs' requirements and SLAs into sets of polices for lower levels. VRRM optimises the usage of virtual radio resources, not dealing with physical ones. Nevertheless, reports and monitoring information received from CRRM enable it to improve the policies. VRRM is able to map the number of the available resources to the network capacity, optimising the network throughput and prioritising services with weights.

Obviously, offering the same data rate to a MT with low input SINR requires assigning comparatively more RRUs than to a MT of higher SINR. In Khatibi and Correia [KC15], the effect of channel quality in VRRM is studied. The capacity estimation technique is extended by considering three approaches according to pre-defined assumptions. In an optimistic (OP) approach, all RRUs are assigned to MTs of assumed high SINR. In a realistic (RL) approach, half of RRUs are assigned to MTs of assumed high SINR, while the other half of RRUs is assigned to MTs of low SINR. A pessimistic (PE) approach, all MTs are assumed to have low SINR. These assumptions will result in different boundaries for the possible data rates for each RRU.

An extension of the VRRM model considers traffic offloading support with WLANs [KC14a], where low data rate services are served by cellular RATs while high data rate ones by WLANs.

6.6.3 Scenarios and Results

A scenario with several physical BSs from multiple RATs (TDMA, CDMA, OFDMA, and OFDM) is considered. Three VNets are taken: A and C as Guaranteed Bitrate (GB), for services with stringent requirements; B as Best Effort (BE), for other services. Simulation results show that the adaptive VRRA algorithm [CCC12] satisfies minimum capacity requirements of VNOs. With the on-demand VRRA algorithm [CCC14b], the total capacity contracted for guaranteed VBSs should be limited, according to the average physical capacity. In Caeiro et al. [CCC14a], it is also concluded that by changing the quantity of created VBSs and the contracted data rate, the average data rate and efficiency may decrease if the number of GB VBSs (hence, the contracted capacity) is higher than the number of BE VBSs. Overbooking the capacity contracted by BE VBSs achieves 30% higher efficiency and data rates.

In Khatibi and Correia [KC14b], a scenario in which coverage is provided through a set of RRHs, is assumed. RRHs support multiple RATs. Similarly, three VNOs exist: VNO GB, where the contract guarantees the allocated data rate to be between 50 and 100% of the required service data rate; VNO BE with minimum Guaranteed (BG), where at least 25% of the required service data rate is guaranteed; and VNO BE. It is concluded that, with VRRM, VNO GB receives the biggest portion (59%) of resources, 7% for VNO BE, and the rest of resources (34%) for VNO BG, better served than the minimum guaranteed. When Wi-Fi coverage is used for traffic offloading [KC14a], results show an increase up to 2.8 times in network capacity, enabling VRRM to properly serve all three VNOs: not only guaranteed services are served adequately, but best effort ones are allocated with a relatively high data rate.

The effect of channel quality on VRRM is evaluated in Khatibi and Correia [KC15]. Figure 6.28 presents the allocated data rate to VNO GB in conjunction with minimum and maximum guaranteed data rate. As long as one has the data rates within the acceptable region (i.e., shown by the solid colour), there is no violation to the SLA and guaranteed data rate. It can be seen that in OP approach up to 600 subscribers, the maximum guaranteed data rate is

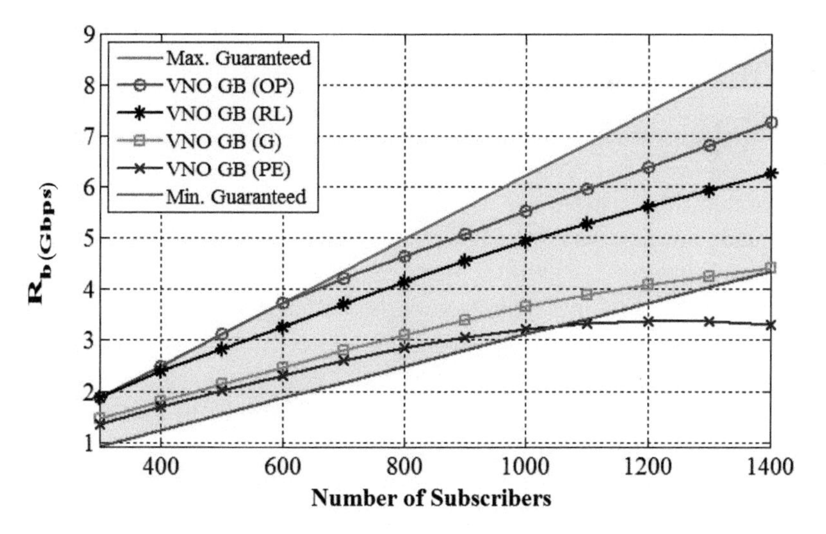

Figure 6.28 Allocated data rate to VNO GB for different approaches (extracted from Khatibi and Correia. [KC15]).

offered to this VNO. Considering the other approaches, as the number of the subscribers increases the allocated data rate moves toward the minimum level of the guaranteed data rate. Finally, in the PE approach, as the number of subscribers passes 1100, violations to minimum guaranteed data rate can be observed: this means that the network capacity is too low so that the model faced resource shortage and it has to violate the minimum guaranteed data rate of this VNO.

6.7 Reconfigurable Radio for Heterogeneous Networking

One of the main issues in performance evaluation of future mobile networks is the increasing mutual interaction among layers: thus, studying the performance of new systems requires an accurate cross-layer simulation, and possibly subsequent cross-layer optimisation, as the outcome of this interaction affects the QoE of users.

This issue attracted the interest of COST IC1004 researchers from the very beginning of the Action, when the authors of Werthmann et al. [WKBM11] developed a simulation tool to study these complex interactions.

They built an event-driven simulator that, as such, does not require a continuous time clock; the network simulator controls virtualised computers with unmodified operating system and unmodified applications. To model the interactions between the virtualised hosts and the network simulation, both need to share the same time source.

Since the network simulation performs all actions based on events, the simulated time follows from these events and is not related to the real-time. The virtualised computers need to adapt to this event-based time. Therefore the simulation has to control the time of these virtualised computers, and simulations may run slower or faster than real-time depending on the computational complexity of the PHY model. The tool is scalable, i.e., it can include a variable number of virtual machines.

To emulate the users' interactions with the virtual computers, the authors implemented an interface to the virtual computers keyboard. It is thus possible to schedule events in the simulation libraries calendar which send keystrokes to the virtual computers. Hereby it is possible, for example, to load different web pages in predefined intervals of time.

Thus, the authors built a framework for network performance evaluation, which can be used to evaluate applications performance in different situations as well as to evaluate the impact of PHY and MAC algorithms.

Cross-layer interaction is dealt with also in Litjens et al. [LGS+13], presented by representatives of the SEMAFOUR FP7 project. Their approach builds upon and extends the concept of SON, developing a unified management framework that integrates the existing and future advanced SON functions across several types of RAT.

The first phase of stand-alone SON use cases and possible solutions focused on self-organisation mechanisms (such as mobility robustness optimisation, interference management, coverage/capacity optimisation, energy savings) that were isolated from one another. Typically, these solutions target individual RATs and cellular layers, rather than addressing the network as a whole.

The configuration and the optimisation of individual SON functions becomes highly complicated and the likeliness of conflicts increases significantly with an increasing number of different SON functions operating in parallel and in a non-coordinated manner, covering multiple layers/RATs and coming from different manufacturers.

The paper presents the self-management system for the unified operation of multi-RAT/layer networks, defined within the SEMAFOUR project. As some key partners of the projects were also institutions participating in COST IC1004 Action, mutual exchange of experiences, and information (within the confidentiality limits required by the project consortium agreement) provided benefits to both parties.

Current and future mobile networks are different from their predecessors in that they support a variety of multimedia applications with different (and often contrasting) requirements in terms of quality parameters such as throughput, packet loss, delay, and jitter. Thus, QoS definitions and targets that originated from previous generations of mobile networks (or even from fixed telephony) are not sufficient for current and forthcoming networks. The research community is striving to derive suitable QoE metrics from measurable QoS parameters, and COST IC1004 is no exception to this.

The authors Robalo and Velez. [RV13] presented their work on this theme. They define QoS requirements for various classes of services, including context-based information, i.e., information belonging to location dependent services. The available multimedia applications are subdivided in four broad classes, namely gaming, video, audio, and web-browsing.

The authors then map QoS onto QoE for various service classes, utilising (when available in literature) third party experiments to evaluate mean opinion score (MOS) as a function of QoS; for web-browsing and audio applications, since MOS results are scarce, the model considers the ITU-T G.1030 and G.107 recommendations, respectively, which are themselves base upon experimental MOS measurements.

Finally, the authors propose a unified model, that computes QoE as a weighed average of QoE evaluated for different applications. This allows building an effective QoE control mechanism onto measurable QoS parameters for multimedia networks, i.e., for improving packet scheduling in mesh and/or CR networks. In particular, when considering the additional delay and possible packet loss induced by changes in spectrum utilisation in CR networks, the model allows to verify their impact on the actual overall QoE.

The variety of services and applications is not the only factor adding to the complexity of current and future mobile networks: also cells appear in a variety of sises and types, while different RATs - and hence different types of AP - are available, according to the HetNet paradigm. Furthermore, also the backhaul network plays a key role in determining the overall performance of a given system: transport capacity in the fixed segment of the network may constitute a bottleneck, if it is not adequately planned to match the radio part requirements in different areas and parts of the overall network.

This issue is tackled in Olmos et al. [OFGZ13], whose authors propose a new algorithm for cell selection. Their main assumption is that cell selection strategies must be extended to consider backhaul load conditions as an additional input.

The authors consider three different cell selection algorithms:

- Best Server Cell Selection (BS_CS): the call is assigned to the best server, if it has enough available capacity. It is used as a reference for the other two algorithms;
- Radio Prioritised Cell Selection (RP_CS): if a call cannot be served by the best server, the algorithm redirects it to another BS in the Candidate Set, if possible (often used in legacy networks);
- Transport Prioritised Cell Selection (TP_CS), proposed by the authors: if transport load in the best server is high, the call is served by another node to avoid congestion at transport level. Weights are used to balance load among nodes according to their transport occupancy.

The authors built an analytical model, based on multi-dimensional Markov chains, to assess the performance of these cell selection algorithms.

Of course, having more than one cell in the candidate set implies a reduction in the overall network capacity and performance, as one MT occupies resources in more than one cell.

The systems composing the analytical model are undetermined, since there are more unknowns than linearly independent equations; then, among all possible solutions, the one providing the minimum radio performance degradation should be chosen. In order to obtain such a solution, the authors propose an iterative procedure which is valid for arbitrary number of candidate cells.

A Monte Carlo simulator has been written in order to fully validate the results of the Markov model, resulting in excellent agreement with the analytical model and thus obtaining a mutual validation. One key result found is that the trunking gain achievable by TP_CS is greater in case of non-uniform capacity in the fixed network with respect to the uniform case. The proposed strategy is able to achieve the desired trunking gain while reducing the amount of radio degradation that we would have in case of using traditional cell selection schemes based only on radio metrics.

One powerful concept to cope with the multi-faceted complexity of current and future mobile networks is software defined network (SDN), whose opportunities are thoroughly discussed in Chaudet and Haddad [CH14]. SDN is a network paradigm that relies on the separation of the control and forwarding planes in IP networks. The interconnection devices take forwarding decisions solely based on a set of multi-criteria policy rules defined by external applications called *controllers*. It is possible to let multiple controllers manage each element of a given network, which allows creating independent networks on the same physical infrastructure.

The ability to separate the IP flows space in distinct subspaces is referred to as *slicing*. Implementing SDN requires at least being able to define slices and to limit interactions between these slices, and to let the network devices measure and report their status to the relevant controllers, which are non-trivial operations with the wireless medium.

The authors highlight the issues raised when trying to use the SDN paradigm in mobile networks: non-ideal separation between "slices" sharing the same band, coordination among different operators, interference, HO, changes in link conditions, etc. However, SDN can potentially bring tremendous benefits, listed below:

- *Improved end-user connectivity and QoS* by inter-operator cooperation (e.g., an AP of an operator routing packets of a customer of another operator when the latter is congested)-
- *Multi-network planning:*
  - SDN allows creating zone-specific controllers that transcend operators, suggesting channel selections and power control to participating ACs, based on sensed mutual interferences levels. Power control

helps reducing interferences by attaching users to the closest AC. The more ACs are available, the more efficient this process will be, and the capacity of wireless SDN to aggregate in a single VNet multiple BSs belonging to multiple ISPs eases the problem. Thus, the controller can create locally all the possible interferences graphs and select the one that maximises coverage while minimising interferences.

  – SDN can support soft HO between different technologies, by replicating packets in the core network and forwarding them through different radio interfaces.

- *Security:* the monitoring capacity of SDN allows detecting intrusions and malicious attacks; cooperation among APs enhances this capability.
- *Location*: if a terminal can connect to different networks, having a controller that supervises all networks will allow faster and more accurate determination of its position (even in indoor environments where satellite-based location is not available).

However, some issues shall be tackled prior to actual exploitation in commercial networks. There are confidentiality issues for the operators (e.g., they may not be willing to disclose technical details of their networks) and privacy issues for the users (e.g., a user may not want to let other operators know which websites he/she browses). For the time being, user data and network information can't be freely shared among different operators; however, a first step could consist in introducing the above discussed capabilities in different networks owned by the same operator.

One well-known technique to increase the capacity of mobile networks is the reduction of the cell sise. Having smaller cells, however, implies an increase in the HO rate, and hence of the probability of dropping a connection.

Previous experience shows that generally, in a small cell network (SCN), HO failure rate becomes excessive for speeds of about 30 km/h. The authors of Joud et al. [JGLR14] propose a solution allowing to use SCNs for speeds up to about 60 km/h without a significant degradation of HO performance.

Obviously, the most critical type of HO is the small cell to small cell handover (SS HO).

The principle of the proposed scheme is, therefore, to reduce the frequency of SS HOs and consequently handover failure (HOF) occurrence. This is achieved through dual connectivity with C-/U-plane split and DL reinforcement by means of cooperative multi-point (CoMP) in the SCN. The basic idea is transmitting the U-plane over more than one small cell for the users beyond a certain speed limit. Groups of small cells are dynamically created and perform coordinated scheduling just for these users, being likely to have high HOF rates.

The simulated scenario consists of macrocells operating at 2 GHz, that provide complete coverage of the service area, plus small cells at 3.5 GHz, deployed at random locations (Figure 6.29).

For the higher mobility case, the UE is initially connected to the best macrocell and establish normal communication in both C-and U-planes.When the cell receives a report indicating an A3 event towards a small cell, the U-plane is transferred, the UE keeps receiving C-plane from the macro-cell. Therefore, it does not perform a conventional HO since radio resource control (RRC) messages are still sent/received though the macrocell. The U-plane is transferred to a group of small cells that will schedule the UE in a coordinated manner. Note that cooperative groups are not created for pedestrian (or low mobility in general) users, which still perform classic HO among all cells.

The presence of many UEs requiring cooperative groups might well reduce the small cell layer capacity dramatically: for this reason, the authors limited the number of UEs in cooperative transmission manner served by a small cell to $k = 3$, and the number of small cells that can serve a single UE to $n = 3$ as well (one serving cell and two co-serving cells).

Simulations show that the proposed method allows an improvement of HO performance for fast moving users, at the cost of reduced throughput: more investigations are necessary to determine the optimum solution according to the actual traffic distribution and cellular layout.

6.7.1 Advances in Heterogeneous Cellular Systems

As services migrate from voice centric to data and multimedia centric, which requires increased link budget and coverage extension to provide uniform user experience, traditional networks optimised for homogeneous traffic face

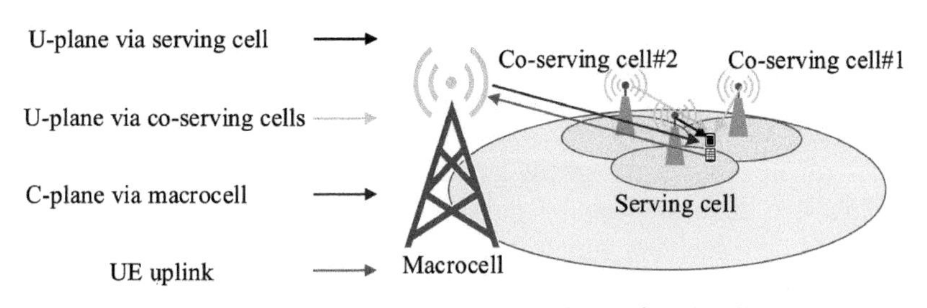

Figure 6.29 Coverage scheme proposed by Joud et al. [JGLR14]

unprecedented challenges to meet the various demands effectively. In this context, third generation (3G) 3rd generation mobile group (3GPP) LTE-A has started a new study to investigate HetNet [DMW+11] deployments as an efficient way to improve the system spectral efficiency as well as effectively enhance network coverage. Under this architecture, a number of wireless technologies such as small/micro cell deployment [SCO13a], Device-to-Device [DRW+09], and multicell cooperation [SWPR13] have been proposed and/or developed in order to enhance spectral and energy efficiency. Especially, in the LTE-A HetNet architecture, the deployment and configuration of small cells is considered a key technology to provide high capacity, good coverage, high spectral efficiency, and high energy efficiency. In the above context, some of the recent developments in resource allocation, energy efficiency, diversity, and adaptive antenna techniques for HetNet systems are discussed in the following subsections.

6.7.2 Optimal Resource Allocation

One of the main challenges in HetNet systems is how to optimally allocate the available resources such as frequency, power etc. in order to satisfy the users' service requirements as well as to guarantee the economic, environmental, and operational sustainability of the service provider. The resource allocation problem in wireless communication systems has evolved incrementally in the following three stages [NPGP13]: (i) performance management [SAR09], (ii) QoS management [ZY10], and (iii) interference management [KYRC07]. The first stage aims to optimise the meet a performance metric at the cost of denying the service to the worst users. In the second stage, the system tries to best-effort traffic users, subject to full satisfaction of the guaranteed traffic users. Subsequently, the third stage tries to manage the interference allocating by both frequency and power in an intelligent manner. Several authors have proposed different ways of handling resource allocation problem in HetNet systems [BD08, MH13, PNSR14].

The service providers point of view can be incorporated while designing a n acceptable resource allocation strategy in HetNet systems [NPGP13]. For a service provider, success or failure of the resource management task is directly related to the level of satisfaction of services demanded by users, on the basis of an efficient usage of available resources. To achieve this, the following two schemes can be considered while designing a model: (i) efficacy-oriented scheme which aims to find the best possible solution by improving the model performance in a variable time interval, and (ii) efficiency-oriented scheme which aims to find an acceptable solution in a defined time interval. Regarding the strategy for allocating frequency blocks to users, the following two schemes are considered: (i) a random allocation strategy with reuse factor 1, which allocates all available resource blocks, and (ii) a random allocation strategy with a flexible frequency reuse factor. Furthermore, the proposed resource allocation scheme in [NPGP13] considers the following aspects: (i) the resource allocation task solves the problem using a centralised scheme, (ii) it uses a super frame level scale of time, (iii) The solution is based in a scenario with heterogeneous services and architecture, (iv) it establishes a balance between efficacy and efficiency, (v) it tries to satisfy the SINR required to offer the service demanded by each user deployed in the scenario, and (vi) three different strategies to schedule users in picocells. Moreover, the SINR ratio can be used as a key parametre in order to evaluate the resource allocation task.

6.7.3 Energy Efficiency

In the coming years, the growth of wireless networks will significantly increase the energy consumption of information and communication technologies by approximately 15–20% [CZB+10]. This will cause an increase of environmental pollution, and impose more and more challenging operational cost for the operators. In this context, green communications, which aim to improve the energy efficiency of future wireless communications, has received important attention for researchers and wireless industries. Among many wireless equipments, the RAN consumes the most energy and it is necessary to control the DL power since BSs are the primary energy consumers of cellular networks. Deployment of low power BSs in traditional macrocells has been considered to extend coverage and simultaneously reduce the load of the macrocellular networks [3GPP10]. The integration of picocells in cellular networks is a low power, low-cost solution to offer high data rates to outdoor customers and simultaneously reduce the load of the macrocellular network. From the study of the effects of joint macrocell and residential picocell deployment on the network energy efficiency by Claussen et al. [CHP08], it can be noted that introducing picocells within macrocells can reduce the total energy consumption by up to 60% in urban areas with high data rate user demand. However, the massive and unplanned deployment of small cells and their uncoordinated operation may result in harmful crosstier interference and raise important questions about energy efficiency. In the following subsections, we provide various approaches considered for energy efficiency in HetNet systems.

6.7.3.1 Users' social pattern perspective for energy efficiency

In the last several years, social networks such as Facebook, Twitter, Microblog, etc. have attracted billions of active users and the number of users are increasing exponentially. In such networks, some users may exhibit a kind of social pattern/behaviour and tend to operate in similar characteristics. Several research works have focused on the users and traffic characteristics to optimise cellular networks. For example, the study of connections among users is considered as one approach [CSB10] in the context of traditional social networking and another approach can be to consider spatial and temporal characteristics of users by taking into account of their social characteristics [XZG14]. Due to the social nature and human habits, the probability that users close in vicinity (e.g., in the coverage of one or several BS) have similar habits, pattern/behaviour and mobility rules may be high. Furthermore, understanding and modelling a user pattern is crucial for the design of LTE-A HetNet systems and services, especially for the deployment and configuration of small cells. The user pattern can be used as the basis for the performance optimisation of the LTE-A system.

In the above context, two energy-efficient transmission control schemes in cellular networks for real-time and best-effort services have been proposed in [ASGL13, HWZJ12] exploiting the user pattern/behaviour. User Social Pattern (USP) can be considered as an important parametre which basically characterises the general user behaviour, pattern and rules of a group of users as a social manner. To mathematically describe the USP, Gini coefficient used in statistics and economics can be considered as a reference. The Gini coefficient (also known as the Gini index or Gini ratio) is a measure of statistical dispersion intended to represent the income distribution of a nation's residents. Similar to incomes, users and traffics in HetNets can also be described and modelled [XZG14].

6.7.3.2 Stackelberg learning framework for energy efficiency

In spectrum-sharing HetNet systems, the cross-tier interference between macro-cells and small-cells and the co-tier interference among small-cells may greatly degrade the network performance. Without effective interference management, the overall energy efficiency of HetNet system will become even worse than that of a network with no small-cell deployments. Various approaches have been proposed to mitigate the above interferences. Kang et al. [KZM12] formulated the resource allocation in two-tier femtocell networks as a stackelberg game (SG), where the MBS protects itself by pricing the interference from femtocell users. Similarly, Chandrasekhar et al. [CAM+09] proposed a distributed utility-based SINR adaptation algorithm to mitigate the cross-tier interference, under the assumption that macrocell BS knows the exact positions of overlaid users. However, in practice, the number and the locations of small-cells may be unknown, which results in unpredictable interference patterns. For such dynamic HetNet scenario, all cells tend to be autonomous. In this context, Lien et al. [LTCS10] applied the CR technique to identify available radio resource in a macrocell and authors in Chen [Che14a] formulate a SG to study the power allocation in HetNet systems with the aspect of energy efficiency. In the considered scenario, macrocells are considered as foresighted leaders while the small-cells as followers and the their objectives are to learn to optimise the power adaptation policies.

The SG is a strategic game in which there exists a leader that moves first, while a group of followers make their moves sequentially. In the macro-small cell coexistence scenario, the macrocell used can be referred to as the leader, while the small cell users as the followers. Accordingly, the formulated SG game can have two levels of hierarchy [Che14a]: (i) the leader behaves by knowing the reaction function of the followers, (ii) given the action of the leader, the followers play a non-cooperative sub-game. A Stackelberg equilibrium (SE) describes the optimal strategy for the macrocell user. In stochastic learning game, users are assumed to behave as intelligent agents in the formulated SG. The objective of each user is to maximise the expected utility, which reflects the users satisfaction of executing a specific power adaptation policy. Users with learning ability learn to adapt to the surrounding networking environment to maximise its individual expected utility. One of the possible ways of implementing adaptation mechanisms is Q-learning [WD92] where the users' power adaptation policies are parametreised through Q-functions that characterise the relative expected utility of a particular transmit power level. In Q-learning, users try to find the optimal Q-values in a recursive way.

6.7.3.3 Capacity and energy efficiency

There exists a trade-off between energy efficiency and system capacity when deploying small cells. The deployment of low power BSs may result in maximising the system capacity, however, a high number of lightly loaded small cells increases the network energy consumption. Therefore, to evaluate different cell topologies for reducing the energy consumption, it is important to use adequate energy efficiency metrics [CKY10]. The energy efficiency is usually measured by the traffic capacity divided by the power consumption of the BSs and it is considered as a key performance indicator. The impact of deploying picocells on the capacity and energy efficiency of macro-picocell two tier HetNets has been studied in Obaid and Czylwik [OC13]. Subsequently, an adaptive power allocation based on an iterative-water filling scheme is presented in order to control the DL transmit power for macro and pico BSs.

In a linear power model proposed by Richter et al. [RFMF10], the total power consumption of each BS changes linearly with respect to the average transmit power. Therein, the power consumption of different BSs is modelled as the sum of two parts. The first part describes the static power consumption, which is consumed by the regular operation of BS. The second part is the dynamic power consumption, which depends on the cell load. Let P^{M} and P^{P} denote the power consumption of macro and pico BSs, respectively. Then the relation between the transmit power and the total power consumption of each macro and pico BS is given by Richter et al. [RFMF10]

$$P^{\mathrm{M}} = N_{\mathrm{sec}} N_{\mathrm{ant}} (\alpha_{\mathrm{M}} P^{\mathrm{M}}_{\mathrm{Max}} + P_{\mathrm{CM}}) \tag{6.8}$$

$$P^{\mathrm{P}} = \alpha_{\mathrm{P}} P^{\mathrm{P}}_{\mathrm{Max}} + P_{\mathrm{CP}}, \tag{6.9}$$

where N_{sec} and N_{ant} denote the number of sectors in a macro site and the number of transmit antennas of a macro sector's BS, respectively, α_{M} and α_{P} represent the power consumption coefficients that scale with the transmit power due to amplifier, cooling of sites and feeder losses, P_{CM} and P_{CP} denote the fixed amount of powers consumed by the BS due to signal processing, battery backup and other auxiliary equipments.

Regarding energy efficiency metrics, although several metrics have been proposed in the literature [CKY10], the traffic capacity divided by the power consumption of the BSs can be considered a suitable energy efficiency (η_{E}) metric while taking into account of the power consumption of the BSs, which can be defined as by Obaid and Czylwik. [OC13]

$$\eta_{\mathrm{E}} = \frac{\text{System Capacity (bits/s)}}{\text{Power Consumption (W)}}. \tag{6.10}$$

6.7.4 Diversity and Adaptive Antenna Techniques

Distributed antenna system (DAS) is playing an important role to provide higher data rate and mass wireless access service for broad area coverage. The purpose of indoor DAS is to split the communication cell to different areas by several remote units [SRR87]. Therefore, the line of sight scenario is more frequently presented if the DAS is employed so as to improve the coverage

[NFSS03]; meanwhile the DAS also increases the received diversity. The LTE Femtocell home BS (eNodeB) is a low-power cellular BS that uses licensed spectrum and is typically deployed in residential, enterprise, metropolitan hotspot, or rural settings. It provides an excellent user experience through enhanced coverage, performance, throughput and services based on location [ACD+12]. Combined with the DAS system [OZW10], the LTE femtocell BS can be installed indoors with a flexible configuration to provide the indoor users who need multimedia services with better user experience. The LTE femtocell DAS eNodeB can be divided into two types [Tia]: (i) a combined eNodeB, which employs a combined adaptive process hub unit (HU) to the system, and (ii) un-combined eNodeB, which does not adopt the combined adaptive process HU in the system. Once the eNodeB adopts the combined HU, more remote units can be allocated to the distributed system. Since more signals from different transmitted channels can be combined to input to the eNodeB by using the combined HU, the combined eNodeB actually improves the diversity of eNodeB without combined HU.

In the above context, authors in Tian et al. [YTB13] study the 3GPP LTE femtocell BS evaluation test-bed considering the LTE system working in band 13 defined by 3GPP, with a centre frequency of 782 MHz for the UL. The baseband LTE signal is sampled at 15.36 MHz, and an oversampling factor of four is used, giving a sampling frequency of 61.44 MHz. The number of input channels to the HU is the same as the number of distributed remote units. The FPGA unit receives CSI values and computes a set of weights to apply to the four received signals in order to produce a combined signal for the BS. The signals are initially sampled at low intermediate frequency by the on-board analog to digital converters. Furthermore, the performance of two maximum ratio combining (MRC) combined techniques, space–frequency and space-only methods, are measured and analysed in Tian et al. [YTB13]. For the case of employing the BS with combined HU, the space–frequency MRC combined algorithm always provides a better performance than the space-only combined algorithm. However, the space-only combined algorithm is simpler to implement and cheaper than the space–frequency method; therefore, the space-only method is expected to replace the space–frequency method in industry if the performance loss is small.

7

Cooperative Strategies and Networks

J. Sykora, V. Bota, Y. Zhang and S. Ruiz

Cooperative strategies and networks play an important role in current state-of-art wireless communications. The cooperation can have many forms and can be utilised at various levels of the processing hierarchy. At the physical layer, the cooperative algorithms revolutionise the design of modulation, coding, and signal processing. It can be directly at the level of channel coding which is *aware* of and pro-actively *utilises* the *network structure* knowledge as is performed by physical layer network coding (PLNC). Apart of the coding itself, the cooperative and relay-aided processing can substantially increase the network performance, most notably by using virtual multiple-input multiple-output (MIMO), distributed processing and smart interference cancellation. The cooperative network layer algorithms allows efficient coordination of the network and its resources.

7.1 Physical Layer Network Coding

The PLNC is concept of channel coding aware of the network structure and fully respecting all the aspect of the physical layer constellation space and channel parametrisation. PLNC-based communication networks deliver the information from sources to destinations through the complex relay network. Each node demodulates, decodes, processes, and re-encodes the hierarchical information (many-to-one function of the component data) directly in the constellation space. Various aspects of the design ranging from the network-coded modulation (NCM) (in many flavors: compute and forward (CoF), Denoising, hierarchical decode and forward (HDF), etc.) over the relay node (RN) strategies (decode and compress) to the Herarchical Information and hierarchical side information (HSI) decoding strategies are discussed in number of works [FLZ$^+$10, KAPT09, LKGC11, NG11, SB11, SB13a].

This Section addresses selected topics from PLNC technique ranging from the basic information-theoretic fundamental limits and concepts through

Cooperative Radio Communications for Green Smart Environments, 271–304.

particular design technique of NCM (hierarchical constellations, hierarchical network code (HNC) maps) and finally including relay/destination (D) decoding techniques.

7.1.1 Basic Concepts and Fundamental Limits

In the study of Burr and Sykora [BS12], the application of PLNC in a uplink (UL) transmission of a hierarchical wireless network is analysed. The UL is modelled as two-terminal two-relay topology where both relays receive signals from both terminals and relay the information towards a common destination with a utilisation of PLNC. In particular, the outage probability (OP) versus throughput for network coded hierarchical network for various signal–noise ratio (SNR) is evaluated. The simulation is used to obtain the probability density function (Pdf) of the throughput over the random fading, and hence a plot of outage probability against throughput. It is shown that this scenario can increase throughput, giving a number of the benefits of network MIMO, but without increasing the backhaul load. As a baseline for comparison, the equivalent results for a conventional non-cooperative system, in which each relay serves one terminal only. It is observed that first order diversity is obtained, and that outage capacity is much improved in the network coded case: for example for SNR 15 dB the 1% outage capacity is increased by a factor of more than 3.

Uricar et al. [USQH13] address butterfly network topology—two sources SA, SB communicating with their two respective destinations DA, DB through a common relay without a directS-D link SA–DA, SB–DB however having a HSI link SA–DB and SB–DA. HSI link provides only a partial (imperfect) HSI. It is shown that the superposition coding (SC) provides a natural tool for implementation of PLNC in the butterfly network relaying under an *arbitrary* HSI assumption. By splitting the source information into two separate *basic* and *superposed* data streams (and optimisation of rate and power allocated to each particular stream) it is possible to adapt the the processing to actual channel conditions (and hence the available HSI at destinations). Under this optimisation, superposition coding represents a viable solution for the case where only partial HSI is available at both destinations. Classical zero and perfect HSI cases are shown to be a special instances of the proposed superposition coding PLNC. The basic message parts are processed using the minimum cardinality PLNC while the superposed part is carried by the full cardinality map. Destinations decode the desired message with a proper utilisation of the HSI. The information-theoretic bounds for the achievable rates are derived under various settings.

A fundamentally novel approach to solving the PLNC is presented in Sykora and Burr [SB12, SB13b]. It would normally be assumed in the design of a PLNC strategy that each node has full knowledge of all channels, and especially the connectivity between nodes, so that the HNC maps at each node can be selected to ensure that the composite map is invertible. A novel strategy is proposed that works *regardless* of connectivity knowledge which is modelled in terms of a *discrete random* parameter which defines the *channel class*. In the simplest case a binary channel class indicates connection or disconnection. The core concept is to treat the channel class as an additional degree of freedom (DoF) of the received signal, and, therefore, to treat the index of the channel class in the same way as an information symbol. Hence the overall HNC map must be invertible not only in terms of the data, but also of the class index of all channels. The constellation space NCM mapping function play an important role in the design.

The classical signalling solution (Figure 7.1) is based on sensing the channel and then *switching* the HNC map at the node depending on the channel class (connection/disconnection). This sensing and switching is relatively easily solved for links ending at the node where the map should be switched. It can be combined with channel estimation using a suitable preamble, which would be required anyway. Much more important and harder to avoid is that this map must be made available to all succeeding nodes and stages of the larger multi-hop network. This creates at least a *latency*. Since the signalling information must be forwarded reliably it must be coded by long codewords (even if it is itself very short). This cannot be done at the channel symbol level. Scenarios with multiple hops, connectivity loops, and short packet

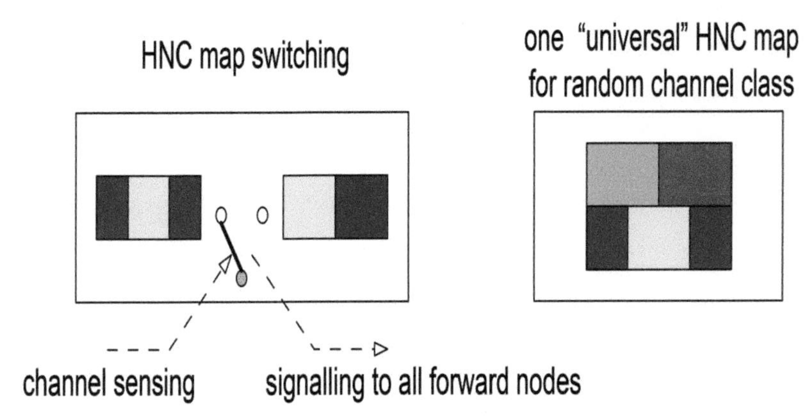

Figure 7.1 HNC map switching for classical signalling solution and "universal" HNC map for random channel class.

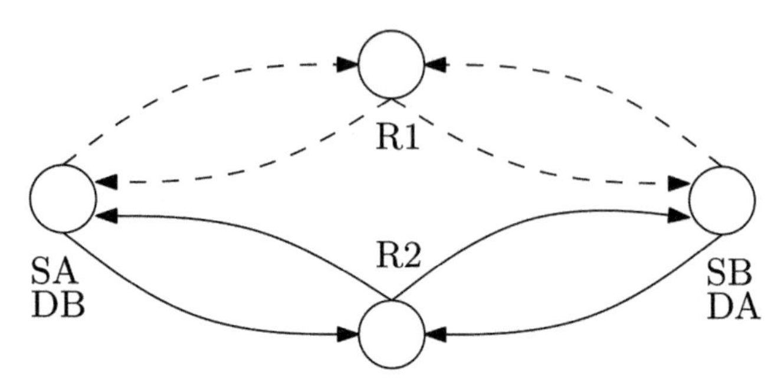

Figure 7.2 Two-way two-relay system example with random unreliable links (*dashed*) associated to relay R1.

services with rapidly changing channel classes are particularly vulnerable to this. In some cases, this reliable signalling is not possible at all. In Figure 7.2, all links at R1 have random channel classes and the destination has no reliable way to obtain the HNC map used at R1. The proposed approach with HNC map inherently containing the channel class does not need this explicit signalling. Of course it is at the expense of larger output cardinality. But since the bottleneck is usually the multiple access channel (MAC) stage, this may not restrict the overall capacity. Moreover, the channel class is contained in each payload channel symbol. This works like a long "repetition" code, so that a long channel class observation is spread over the whole payload.

The design needs to use *a non-linear HNC map* in order to avoid the mapping failures as shown in Figure 7.3. Detailed mutual information numerical results from Sykora and Burr [SB12, SB13b] show that all rates forming the cut-set bound for combined data-channel hierarchical symbols are better than the mean rate of the reference NCM scheme designed as if the channel class was always $a_{B1} = 1$. This result fully justifies this design approach.

7.1.2 NCM and Hierarchical Decoding

Non-coherent and semi-coherent schemes for PLNC in two-way relaying scenarios are investigated by Utkovski and Popovski [UP12]. We distinguish

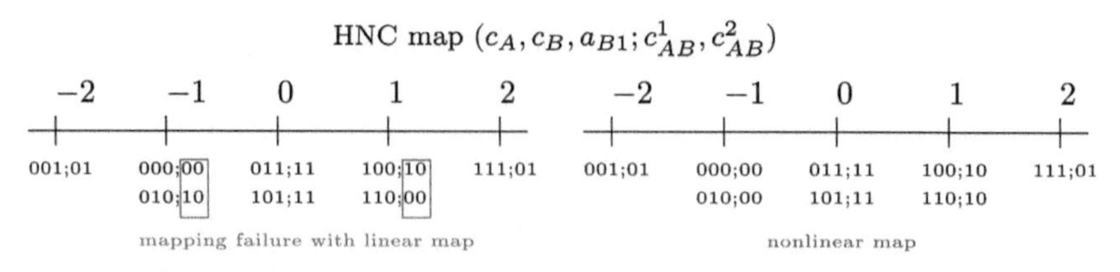

Figure 7.3 NCM hierarchical constellation at the relay for 2S–1R.

between scenarios without any channel knowledge requirements (non-coherent communication) and scenarios when either the relay or the users have receive channel knowledge (semi-coherent communication). We combine the paradigm of subspace-based communication originally developed for non-coherent point-to-point channels, with two-way relaying schemes based on physical-layer wireless network coding with denoise and forward (DNF). The aim is to demonstrate that denoising can be performed non-coherently and to investigate if these schemes offer an improvement over the schemes based on amplify and forward (AF).

The principle of non-coherent design is based on constructing the concatenated source SA and SB codebook fulfilling certain properties similar to the design criteria for non-coherent space-time codes, the chordal distance and the diversity product being the most important ones. The denoising map is based on the maximum likelihood (ML) decision which consists on the projection of the received subspace on all possible subspaces and the decision about the most probable one based on the Frobenius norm. The ML decision coincides with the search for the subspace which is at the smallest chordal distance from the received subspace, which justifies the choice of the chordal distance as one of the design criterion. The numerical error probability results show that the DNF scheme outperforms the AF scheme in most of the SNR region, and has a higher effective rate, due to the time slots saved in the broadcast stage.

PLNC systems where the final destination has only *imperfect/partial* HSI require relay hierarchical maps with extended cardinality in order to guarantee the overall global solvability. In Prochazka and Sykora [PS12], the solution based on the use of a specific sub-block-structured encoding matrices is proposed. Since the overall goal of NCM design is achieved by independent (layered) use of standard single-user channel outer codes and inner hierarchical maps, the scheme block-structured layered NCM. The problem is situated into 2-source 1-relay 2-paths scenario, with a straightforward way of generalisation. The principle of the proposed design consists in a predefined structure of linear channel encoders and hierarchical maps combining minimal and full cardinality maps. The layered design with *minimal* hierarchical cardinality is used on an equally long part of both source streams and the remaining part of the streams are encoded separately using a *full* cardinality map. Consistency conditions for this design are derived and the error rate performance is verified by the simulation.

The problem of designing NCM for the parametric channel is solved in Fang and Burr [FB12] by introducing rotationally invariant coded modulation for 2-way relay channel (2-WRC). Using a fully-adaptive adaptive soft demodulator and independent decoding levels, the proposed scheme can eliminate the effect of fading on the 2-WRC. To reduce system complexity at the relay, a low-complexity scheme with a series of fixed demodulators is provided. Based on maximising the mutual information between received signal and network coded symbol, the adaptive selection of the demodulators is optimised. The proposed simplified scheme exhibits advantages in terms of flexibility, complexity and performance.

The idea of using linear block codes for the construction of NCM is developed in Burr and Fang [BF14]. In the proposed design, each relay computes a linear combination of source symbols, namely, the HNC symbol and forwards it to the destination. The destination collects all HNC symbols and the original source symbols to form a valid codeword of the linear code. The resulting codeword can be decoded to reliably extract the original source symbols. The core principle is in using HNC linear maps having the same form as Reed–Solomon code which has t-symbol error correcting capability which introduces a diversity into the global HNC map. The numerical results show that our proposed design provides $(m + 1)$-th order diversity when there are m relays. Moreover, the proposed design provides a significant sumrate enhancement over the orthogonal multiple access scheme with network coding described in previous literature.

PLNC schemes based on lattice codes, particularly CoF, provide an excellent information-theoretic insight. However, these results assume idealistic perfect lattice codes. There only few attempts to bring the principles of CoF into a life in a practically realisable encoding schemes. One of them is the WPNC design using low-density lattice code (LDLC) [WB14b]. LDLC possesses high-coding gain and good algebraic structure which is inherently suitable for CoF. LDLC is a lattice code construction in which the lattice is generated by the inverse of a sparse matrix. LDLC was shown to be capable of achieving error free decoding within 0.6 dB of the Shannon bound. Inspired by its high coding gain and the algebraic property, a modified Gaussian mixture model (Figure 7.4) is used for the message representation in the low-complexity LDLC detection. A performance comparison for both non-fading and fading cases is demonstrated in the work. It is also shown that the ring-based constellation can be used to improve the average rate per dimension.

An alternative approach to solving the approximation of complicated density function in the factor graph- based iterative decoding is presented in Prochazka and Sykora [PS13]. The work proposes a message representation and corresponding update rules for both discretely and continuously valued

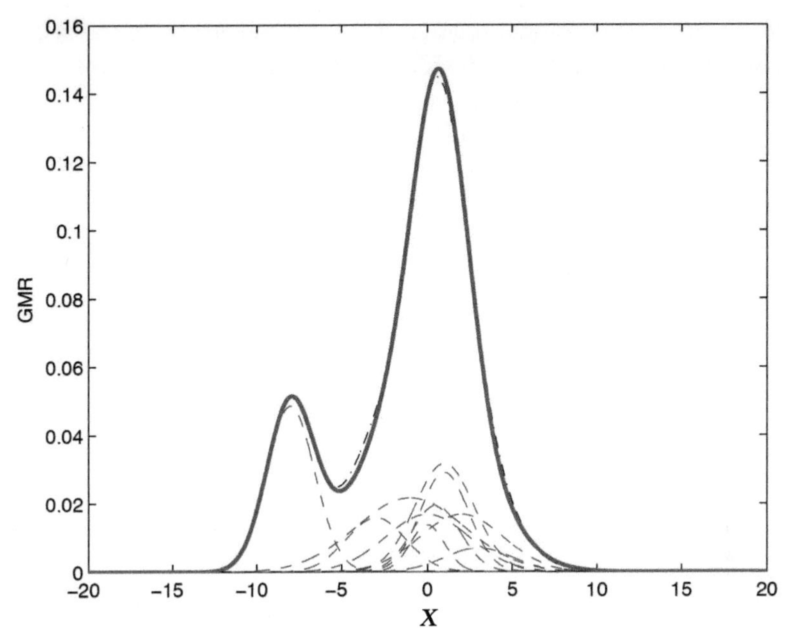

Figure 7.4 A Gaussian mixture (black dash-dot) comprised of $N = 10$ Gaussian components (*blue dashed lines*) is approximated by another Gaussian mixture (*red thick line*) with $N_{\text{max}} = 3$.

variables with an acceptable exactness—complexity-trade-off. It uses linear message representations with orthogonal canonical kernels and corresponding message update rules. The proposed update rule design is combined with an already known message parameterisation based on the Karhunen–Loeve transform (KLT) of the message. This combination of the proposed generic update rules with the KLT-message representation forms a generic implementation framework of the sum–product algorithm applicable whenever the KLT is defined. This framework preserves the properties of the KLT-message representation that is the best linear approximation in the mean square error (MSE) sense. A particular example on a joint detection and phase estimation is shown to verify the framework and to compare the proposed method with a conventional solution.

7.1.3 Hierarchical Constellation Design

A very specific problem in designing the NCM is the construction of the hierarchical constellation (H-constellation) as is visible at the relay in the perspective of its HNC map. Particularly, the H-constellation should have good performance under arbitrary relative channel parametrisation and also it should properly respect the desired HNC map of the relay. Mainly, it needs to correctly resolve the singular fading.

Work by Uricar and Sykora [US12b] shows how it is possible to improve the performance of the 2-WRC system (in a special case of Rician fading channels) by a design of novel 2-slot source alphabets. The proposed non-uniform 2-slot (NuT) alphabets are robust to channel parameterisation effects, while avoiding the requirement of phase pre-rotation (or adaptive processing) but still preserving the $\mathbb{C}^1$ (per symbol slot) dimensionality constraint (to avoid the TP reduction). Based on the analysis of the hierarchical (Euclidean) distance, a design algorithm for NuT alphabets is introduced and symbol error rate (SER) performance is compared to that of the traditional linear modulation constellations. The core idea of NuT is based on using a pair of properly individually and non-equally scaled symbols from a standard alphabet. The NuT constellation alphabet design could be generally characterised as an alphabet-diversity technique regarding the hierarchical min-distance. A suitable selection of the scaling factor is critical for alphabet performance, since it allows to trade-off the vulnerability to exclusive law failures with the alphabet distance properties.

Hierarchical distance and unresolved singular fading is influenced by many factors, one of them being the indexing of the component alphabets in the connection to a given HNC map. This situation is analysed in Hekrdla and Sykora [HS12a]. The goal is to find what kind of symmetry must be fulfilled by the constellation indexing in order to perform errorless modulo-sum decoding in a noiseless channel (because then the minimal distance of modulo-sum decoding equals to minimal distance of primary constellations). It is shown that the mapping from lattice-coordinates onto lattice constellation indices which is modulo-affine (i.e., constellation indices form modulo-arithmetic progression along each lattice (real) dimension) implies errorless WPNC modulo-sum decoding. The constellation indexing is denoted as an *affine indexing*. Some constellation shapes prevent the existence of affine indexing. These shapes comprise sphere-like shapes which possess maximal shaping gain. A modified greedy-sphere packing algorithm is used for the constellation shape design to maximise the minimal distance while keeping existence of affine indexing.

Work by Hekrdla and Sykora [HS14] targets a constellation design for adaptive PLNC strategy in a wireless 2-WRC. The relay requires an extended cardinality network coding adaptation to avoid all singular channel parameters at the MAC stage if the terminals are using standard 4QAM constellations. The cardinality extension is undesirable since it introduces redundancy decreasing the data rates at the broadcast stage. We focus on a design of constellations avoiding all the singularities without the cardinality extension. It is shown that such a constellation is 4-ary constellation taken from hexagonal lattice (4HEX) which keeps comparable error performance at the MAC stage as 4QAM, however without the cardinality extension (Figure 7.5). The similar properties has been found also by unconventional 3HEX and 7HEX constellations.

7.1.4 Receiver Processing

The NCM with HDF is known to be vulnerable to the mutual phase rotations of the signals from the sources. In Sykora and Jorswieck [SJ11], the additional degrees of freedom in SIMO channels are used to create a specific receiver beam-forming tailored for given NCM/HDF strategy. A closed form solution of the NCM/HDF specific beam-former is derived and applied on example

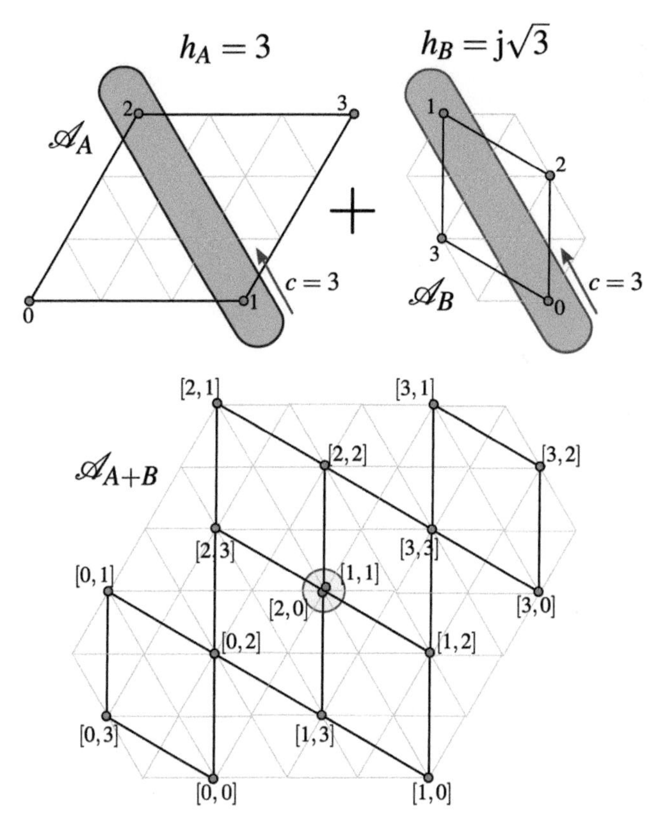

Figure 7.5 Singularity $\alpha = j\sqrt{3/3}$ is avoided by the modulo-sum network coding function if all indices in the critical lattice-dimension (emphasised) are indexed by affine indexing (here with coefficient $c = 3$).

NCMs. The resulting beam-forming is formed on a very different utility optimisation than it is in the traditional beam-forming which is optimised for Gaussian alphabet interference channels (ICs). The beam-forming strategy is equalising the signals at the relay to optimise processing of the *hierarchical* information to take the full advantage of HDF (confront that with classical interference nulling beam-forming). The performance is evaluated in terms of the mean and the outage hierarchical rates. It is shown that these rates significantly benefit from the beam-forming and that the resulting rate significantly outperforms classical (non-PLNC) channel sharing techniques.

A similar target and setup is investigated in Hekrdla and Sykora [HS12b]. In contrary with the previous approach, it does not use an explicit beam-forming equaliser, but the diversity is directly utilised in the receiver demodulator metric. The impact of relative fading on uncoded error performance and ergodic alphabet constrained capacity of representative QPSK alphabet in wireless Rayleigh/Rice channel is investigated. It is concluded that relative fading is sufficiently suppressed by systems with a reasonable level of diversity which is typically assumed if only absolute fading is present.

Successive decoding with interference cancellation (SD-IC) is well understood to be a capacity achieving decoding strategy in MAC. This decoding technique allows a consecutive elimination of interfering signals during the successive decoding of particular user data. The ability to perfectly remove the interfering signal from the subsequent decoder processing (after corresponding user's data have been successfully decoded) relies on a one-to-one mapping between the user data sequence and its signal space representation. However, this is not always guaranteed in PLNC systems, where only specific ("hierarchical") functions of user data are decoded. Work by Uricar et al. [UPS14] analyses this particular problem in a simple binary 3-user MAC channel, which is capable to demonstrate the issues associated with successive decoding in PLNC systems. It is shown that a hierarchical extension of conventional SDIC processing appears to be a viable decoding strategy, even though it does not allow a perfect removal of interfering signals from the receiver observation.

The principle of hierarchical interference cancellation (H-IFC) is analysed from the information-theoretic perspective by Sykora and Hejtmanek [SH14]. Classical CoF relay processing technique complemented with nested lattice type of NCM is believed to be "all-in-one" solution for designing PLNC-based networks. However, we see that it has several deficiencies. The major one is the fact that receiver lattice mismatch is aligned only through the scalar single tap equaliser. It does not provide enough degrees of freedom for multiple misaligned sources. The CoF strategy is generalised by introducing the concept of H-IFC using successive CoF decoding. This technique allows increasing the number of degrees of freedom in lattice misalignment equaliser while using all available hierarchical (many-to-one function) auxiliary codeword maps. The technique is not constrained to use linear map combinations and allows more freedom in choosing a given desired codeword map at the relay in a complicated multi-stage network. The standard CoF assumes essentially that the number of involved source nodes and relays is high and we have plenty of choices to optimise HNC map coefficients that maximise the computation rate. However, in practical situations with a small number of nodes, we are rather limited and frequently only few are allowed in order to guarantee final destination solvability. Respecting subsequent stages of the network (not just the MAC stage of the first stage) frequently dictates further constraints on the map that is required to be processed by the relay. A particular HNC map is typically desired to be processed by the relay. The H-IFC technique is developed in two, joint and recursive, forms. Numerical results show that the achievable computation rates can substantially outperform standard linear multi-map CoF technique.

7.1.5 Relay HNC Map Design

A majority of the research results on PLNC considers only a basic 2-WRC system scenario, where the perfect HSI is naturally available at both destinations. Although an extension of the PLNC principles from the 2-WRC system to the butterfly network could seem relatively straightforward, several new unconventional phenomena arise and need to be considered. First of all, the non-reliable transmission of the HSI can be overcome by an increased cardinality of the relay output. This opens a question how a suitable HNC mapper could be designed for a given quality of the side information link and source alphabet cardinality. Based on the work by Uricar and Sykora [US12a], we focuses on this problem. We introduce a systematic approach to the design of a set of HNC mappers, respecting the amount of available HSI. Observed capacity gains of this extended cardinality relaying pave the way for an adaptive butterfly network, where the achievable TP can be maximised by a suitable choice of the HNC map at the relay.

The core of the solution is based on the assumption that each destination can identify at least subset of bits from the constellation symbol received on the side link. The idea is to partition the source constellation alphabet into smaller subsets (according to the principles similar to Ungerboeck mapping rules) to increase the probability of successful partial side information retrieval at the destination. Hence, the intention is to maximise the distance between the particular subsets.

Linear HNC maps are attractive solution due their apparent simplicity. However the operations does not necessarily be constraint to algebraic field. A relaxation of this requirement allows more flexibility in the map choice. A sequence of works by Burr et al. [Bur13, Bur14b, BFW14] solves this problem first starting with a general formalism of the operation in algebraic rings and fields in the perspective of HNC map. Particularly, it is the global solvability (in some cases not requiring explicit inverses) and capability to give more freedom in resolving singular fades of PLNC in parametric channel. It is first investigated in the 2-WRC scenario [Bur13], then generalised into more complex scenarios in Burr et al. [BFW14]. Work by Burr [Bur14b] generalises linear PLNC HNC maps using constellations obtained from nested lattices of Gaussian integers. These constellations are isomorphic either to a ring or a field, either of integers or of first order polynomials. It shown how the coefficients of linear HNC maps which resolve singular fading states may readily be obtained in these structures, and also discuss unambiguous decodability in some simple networks.

The core idea of linear HNC function in these frameworks can be extended to various forms of ring, rather than being restricted to algebraic fields, provided care is taken with the choice of the coefficients from the ring. In particular, to ensure unambiguous decodability, the coefficients which appear in the matrix relating the symbols received at the destination to the transmitted symbols must not be zero-divisors within the ring; or equivalently, they must be uniquely invertible. Constructions with functions based on the ring of integers and also on the ring of square binary matrices were considered.

Also it was shown that care must be taken with the choice of network code function in order to minimise the effect of relative fading of wireless channels leading to a relay node. In particular, it should be as far as possible ensured that singular fade states, in which two or more different symbol combinations at the source nodes lead to the same (or nearly the same) signal at the relay, are resolved by the choice of network code function. In this context, resolution means that such combinations encode to the same network coded symbol, so that the singular fade state does not result in ambiguity of the network coded symbol.

The performance of linear PLNC is examined by Burr [Bur14a] using linear mapping functions based on rings over integers. In PLNC, the effect of the wireless channel is fundamental. The effect of fading coefficients of the channels between nodes has to be taken into account. In particular, some fade coefficients of incoming channels at a node will cause certain mapping functions to fail, even if none of the channels is subject to severe amplitude fading. It is the relative rather than the absolute values of the coefficients that give rise to this effect, and these relative fading coefficient values are referred as singular fade states. The primary objective of this work is to extend the investigation of the relative fading effects on the performance to the full range of possible relative fade states. The equivocation is determined as a function of the relative fade coefficient, and hence has determined the outage probability for the scheme. A significant improvement in outage performance, of 4–5 dB, through the use of non-minimal mappings. However, these have the disadvantage that they increase the cardinality required in the broadcast phase of the relay operation, potentially degrading the outage capacity at that point, and also that channel coding is not easily implemented.

An important aspect of the HNC design in large PLNC networks is initialisation and self-organisation. It should determine the local maps at nodes depending on the situation in their direct (Dir) radio visibility neighbourhood while optimising the performance utility of the network. The first step is the sensing and classification of the local signals received in the superposition in hierarchical MAC stage. The work by Hynek and Sykora [HS12c] addresses initialisation of the PLNC procedure in stochastic unknown connectivity network. The algorithm is based on a blind clustering of the received constellation by simple k-means algorithm. The algorithm provides the estimation of the number of the operated sources received by the given relay. The channel states can be also determined from the position of the centroids.

The self-organisation of the HNC maps is solved by Hynek and Sykora [HS13]. It provides an analysis of game theoretical approach to the distributed selection of the decision maps at each individual relay in complex multi-source multi-node network based on PLNC. The wireless network consist of arbitrary number of independent selfish source nodes, one layer of relay nodes (named a cloud) and a set of destinations. Network connectivity is globally unknown. The game uses a particular example of local relay utility function defined as cardinality of the output hierarchical symbol. It is shown that the distributed algorithm has promising properties in game theoretical sense and has atleast one Nash equilibrium. It can be also shown that the myopic better response dynamics converges to this equilibrium (or to some of them in case of the game with multiple equilibriums). By the straightforward extension it can be shown that those properties remain valid even in case of the games with more then two players. The proposed game is suitable for any decode and forward (DF) PLNC for wireless as well as wire-line networks.

7.2 Cooperative and Relay-Aided Processing

The research activities performed in this area were focused mainly in the three major directions, listed below, and are presented in the corresponding subsections:

- Virtual MIMO and RA distributed processing, which studied the reliability and/or spectral efficiency (SpE) improvements brought by the proposed algorithms based on these techniques in typical wireless environments.
- Relay-assisted (RA) algorithms that aimed at increasing the energy efficiency (EE) of the wireless transmissions under imposed reliability and SpE requirements.
- Interference analysis and cancellation algorithms in RA wireless transmissions.

7.2.1 Virtual MIMO and RA Distributed Processing

One of the approaches the improves bit error rate (BER) performance of RA transmissions is the space–time (frequency) block coding (ST(F)BC) technique. The technical report by Urosevic et al. [UVPD13] proposes a modified orthogonal STBC (OSTBC) code with a diversity order of 4 which requires two antennas only at the base station (BS). In order to reduce the processing at the RN, the proposed algorithm performs only permutation and forwarding of the signal sequences, omitting the coding procedure. This algorithm has a coding rate equal to the one of the quasi-orthogonal STBC (QOSTBC) schemes available and greater than the one of the OSTBC schemes and aims at ensuring a BER performance comparable with the ones of OSTBC. The BER performance provided by the proposed algorithm and by the "classical" OSTBC and QOSTBC schemes are compared by means of computer simulations that assume Ricean fading on the BS-RN channel and Rayleigh fading on the RN-(D) channel. The results show that the proposed scheme provides SNR gains compared to the QOSTBC scheme, at the same coding rate, and gets close to the BER performance of the OSTBC scheme, which has a smaller coding rate, only for high values of the Ricean constant, i.e., $K > 8$.

Paper by Bota et al. [BPV13] proposes and analyses the block error rate (BLER) performance provided by an alamouti adaptive distributed scheme (AADS) which implements a modified Alamouti SFBC encoding–decoding algorithm by using selective relaying (SR) at two fixed DF-RNs and only single-antenna devices over block Rayleigh-faded channels. The SR used at the RNs, replaces the wrongly decoded blocks by null blocks and signalises this fact to D. The receiver at D uses an adaptive alamouti decoding algorithm (AADA), which uses different decoding algorithms for the cases when the RNs have transmitted both or only one part of the Alamouti encoded message, according the RNs signalling. The paper derives the theoretical BLER expressions of the proposed scheme, of the distributed Alamouti scheme (ADS), which does not use SR at the RNs and does not adapt the SFBC decoding algorithm at D, and of the "classical" non-distributed Alamouti scheme. These theoretical expressions are derived by using approximate models of the Rayleigh-faded channel and of the channel code correction capability.

The BLER performance is evaluated and validated for both UL and DL directions versus the SNRs of the component links, showing that the proposed AADS ensures smaller BLER values than ADS or ANS, as shown in Figures 7.6 and 7.7. The paper also points out the different BLER versus SNRs behaviours of the DL and UL transmissions.

The employment of several successive VAA is considered in paper by Xie and Burr [XB12] to increase the capacity of the link between a source node S

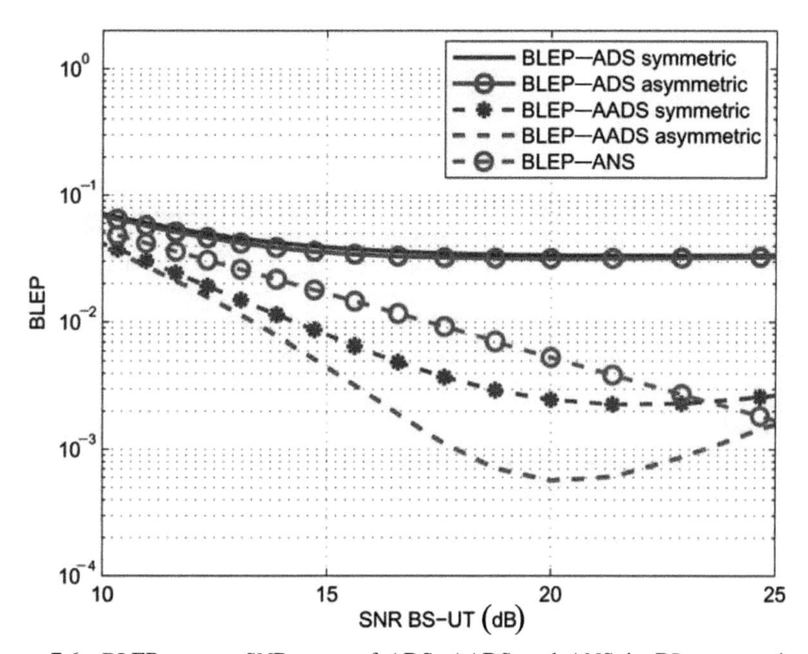

Figure 7.6 BLER versus $SNR_{BS\text{-}UE}$ of ADS, AADS and ANS in DL symmetric and asymmetric scenarios.

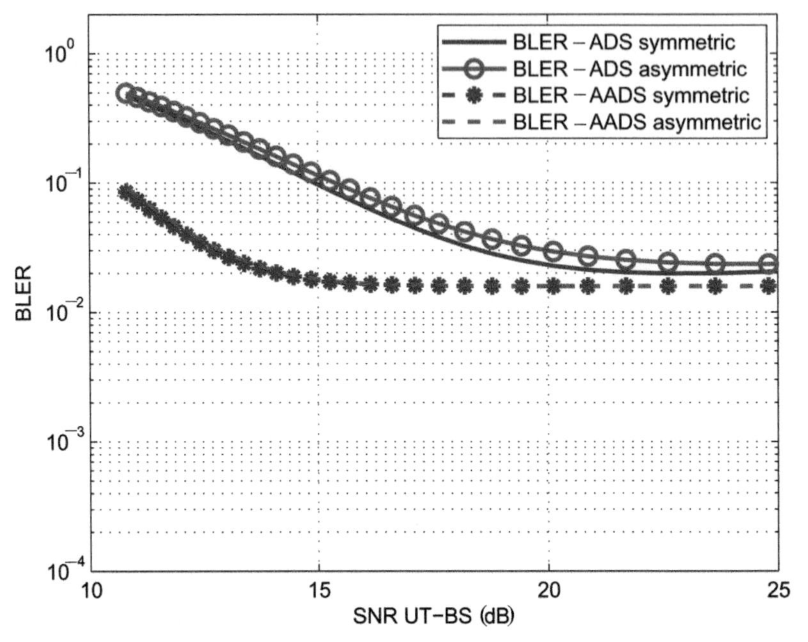

Figure 7.7 BLER versus $\text{SNR}_{\text{BS-UE}}$ of ADS and AADS in UL symmetric and asymmetric scenarios.

and a final destination node D. The paper focuses on the link between the virtual antenna array (VAA) at the receiving end and D and uses the Slepian–Wolf coding, implemented by the repeat accumulate repeat (RAR) code, to compress the information and thus to increase the SpE. In order to support the Slepian–Wolf coding, the paper analyses and evaluates several quantisation and de-quantisation methods.

The received signals at the destination-VAA are quantised, compressed by the RAR encoder and, finally, are transmitted simultaneously to D via a MAC. At D, signals are separated by the multiple user detector and decompressed by the Slepian–Wolf decoder, the output of which is then input into the de-quantiser. It either reconstructs the received signals or produces soft information directly for the outer decoder, according to the different analysed de-quantising algorithms.

The numerical results, obtained in a Raleigh-faded scenario, with two sets of two AF relays as VAAs, show that the quantiser–dequantiser implemented by a demodulator and log-likelihood ratio (LLR) combination ensures a significant SNR gain, compared to the one composed of a demapper–decoder and LLR combination.

Paper by Tao and Czylwik [TC13] analyses the optimal design of the source's S and RN's beamforming vectors and the RN selection if the channel state information (CSI) is available at all nodes, as a way to improve the reliability of a half-duplex MIMO network with multiple AF–RNs. To this end, the authors jointly optimise the beamforming vectors and RN selection matrix, and use the direct S–D link as well to increase the diversity gain.

To decrease the number of required feedback bits, the paper proposes a new scheme that extends to a multiple RNs scenario an available iterative algorithm of designing the beamforming vectors and selecting the best RN. This scheme uses Grassmanian codebooks which also maximises the total received SNR at the destination, by combining the signals from the best RN and from the source. Then the authors propose a sub optimal scheme with limited feedback in which D uses only the singular values of the S–RN channels to obtain their SNRs.

The proposed scheme's SER performance is compared to the ones of three previously proposed schemes, the optimal one, the Grassmanian codebook and the random quantisation schemes in a single RN transmission. The proposed scheme is shown to outperform the other schemes and require less feedback bits than the Grassmanian quantisation, with only a slight performance loss compared to optimal beamforming.

The two-user MIMO with two-way AF RNs scheme over block fading channels is studied in [UE13] in the non-coherent setting, where neither the user equipments (UEs) nor the RN have knowledge of CSI. It presents the derivation of a lower bound of the achievable sum-rate, assuming isotropically distributed input signals, and determines an achievable pre-log region, which is regarded as the main performance indicator of a particular relaying strategy in the high-SNR regime. The paper also makes a comparison of the analysed scheme to the time-division multiple access (TDMA) scheme and points out the coherence time values for which the two-way AF relaying provides greater sum-rates than non-coherent or coherent TDMA.

The BLER and SpE performance of two variants of the Two-Way Relay Channel is studied in paper by Bota et al. [BBV14a] over Rayleigh block-faded channels in a generic cellular scenario. The 2-WRC variant only uses the relay-path, while the Enhanced-2-WRC (2-WRCE) variant uses the direct link and the relay path between S and D and a hard network decoding (HND) method at Ds. To decrease BLER both methods use SR. The paper derives the theoretical expressions of the BLER and SpE provided by these schemes and validates against computer simulation results.

The BLER performance of the 2-WRC(E) is compared to the performance of the two-hop relaying (OwR) and of the Direct (Dir) transmission, for different modulation orders and for various positions of the UE versus the BS and RN. The theoretical and simulation results show that the proposed 2wRCE provides the lowest BLER for almost all positions of the UE within the cell. They also show that the use of higher modulations in the relaying phase of 2wRCE provides smaller BLER than the Dir, as shown in Figure 7.8, and increases its SpSE, under a BLER constraint.

In order to improve the reliability of RA links in which the S-RN link is not error-free, paper by Zhou et al. [ZCAM12] proposes and analyses a distributed joint source-channel coding (DJSCC) strategy to exploit spatial and temporal correlations simultaneously for transmitting binary Markov sources. This approach regards the BER on the S–RN link as measure of the spatial correlation between the S and RN nodes, estimates the spatial correlation and employs in the iterative processing at D.

Moreover, the temporal correlation of the Markov source is also utilised at D by a modified version of the BCJR algorithm, derived by the authors. This modified version performs successively horizontal andvertical iterations using the LLRs provided by two single-input single-output (SISO) decoders. The convergence of the proposed algorithm is studied by means

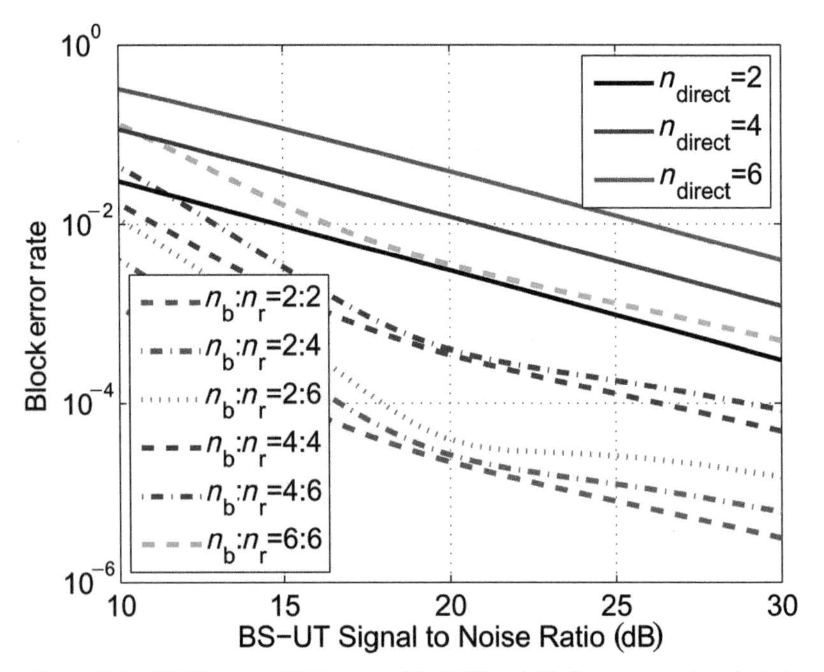

Figure 7.8 BLER versus $SNR_{BS\text{-}UE}$ of 2wRCE and Dir for n_b, n_r and n_d 2, 4, 6.

of extrinsic-information transfer (EXIT) charts. The paper also analyses the BER performance of the proposed scheme, pointing out separately the impacts of the spatial and temporal correlations upon BER at destination. The numerical results show that this approach brings very high SNR gains versus the performance of the same channel code decoded without using the two correlations; the SNR gains are reported to be of 1.9 dB up to 6.8 dB, depending on the error-rate of the S–RN link, for a memory-1, $\frac{1}{2}$ convolutional code.

Paper by Bota and Botos [BB13] analyses the BLER and SpE performance provided by two-level rateless-forward error correction (FEC) coding, which is used within two RA distributed schemes, namely the cooperative repetition coding (CoRCo) and separate source-relay (SSR) coding, within a two-phase cooperation scheme. Within CoRCo the RN retransmits the entire rateless-FEC-coded message received from the source. Within SSR, the RN transmits only additional symbols of the FEC-encoded rateless code symbols computed from the initial message received from the source, while the destination combines the two sets of rateless symbols to recover the S's message.

The paper derives expressions of the message non-recovery probability (p_{NR}) and of the SpE of the two schemes and studies the influence of the schemes' parameters, i.e., the coding rates modulation orders, upon the performance metrics.

The numerical results show that the CoRCo scheme provides smaller P_{NR} values, at the expense of lower SpE, while the SSR scheme provides better SpE, but requires channels of better qualities. The CoRCo is not affected by the S–RN channel's quality, while SSR exhibits an error-floor. The authors conclude that the two cooperative schemes should be used adaptively, adding a new dimension to the link adaptation (LA) process.

Paper by Anwar and Matsumoto [AT12] proposes an iterative spatial demapping (ISM) algorithm for simultaneous full data exchange in three-way AF relaying systems, without direct links between the UEs, which ensures simultaneous connections between three UEs using only a two-phase cooperation, i.e., the UEs transmit simultaneously in the first phase, while the RN transmits the received composite signal to all UEs in the second phase.

The proposed algorithm involves a convolutional encoder and a rate-1 doped accumulator (D-ACC), while the decoder includes an ISM block, providing the extrinsic LLRs of the desired bits from the received signal, and two turbo-loops, including a D-ACC decoder based on the Bahl, Cocke, Jelinek, Raviv (BCJR) algorithm. The BER performance exhibits a turbo-cliff behaviour at low SNRs, i.e., 2 dB for binary phase shift keying (BPSK) and memory-2, $\frac{1}{2}$ convolutional code, on additive white Gaussian noise (AWGN) channels, while in Rayleigh-faded ones the performance is affected by the phase offsets that cannot be completely controlled in three-way relay channel schemes.

The problem of determining the capacity bound of a multi-way relay network in the domain of high SNRs is analysed in Hasan and Anwar [HA14]. The authors propose a scheme that allows uncoordinated simultaneous transmissions between multiple users, who send their messages randomly according to a probability distribution, with the assistance of a single AF–RN which always forwards the received information to all users in the broadcast phase. To avoid interference and collisions, the authors adopt the concept of random access with a graph-based decoding algorithm. The proposed scheme employs at the receiving ends an iterative demapping technique to decode colliding packets, which uses alternatively the maximum *a posteriori* (MAP) decoding, successive interference cancellation (SIC) and the iterative demapping algorithm of Anwar and matsumoto [AT12] between the user and the slot nodes, to increase the TP, while working at relatively low SNRs.

The asymptotic analysis presented derives the nodes' degree distribution and the capacity bound of the proposed system. The numerical results show that it can ensure significantly greater traffic than the coded slotted ALOHA system.

Paper by Kocan and Pejanovic-Djurisic [KPD12] studies the BER performance of the orthogonal frequency division multiplex (OFDM) RA dual-hop scheme that uses a fixed gain (FG) RN which employs ordered subcarrier mapping (SCM), assuming that the RN has full knowledge of the CSI of both S–RN and RN–D channels.

The authors present an analytical derivation of the BER performance of OFDM AF FG relaying for the best-to-best (BTB) and best-to-worst (BTW) variants of SCM.

The theoretical BER expressions are validated against computer simulations and show a very good accuracy. Then, it is shown that the SCM–BTB scheme provides the best BER performance at high SNRs, while the SCM–BTW performs better at medium SNRs, when compared to FG AF relaying and the direct transmission.

The BER and ergodic capacity performance of an RA bi-directional scheme that uses analogue network coding (ANC), in the presence of phase noise (PN) is analysed in paper [LGAJ13] under the assumption of perfect CSI knowledge at all receivers.

The paper derives a PN model, which includes the inter carrier interference (ICI) at the RN and D, and derives the expression of the signal-to-interference plus noise ratio (SINR) degradation due to PN, the authors concluding that the ICI at D has the greatest impact on the SINR degradation. It also presents a method to compute the average BER and the ergodic capacity provided by this scheme. The paper concludes that the SINR degradation increases with the PN linewidth for medium $\frac{E_b}{N_0}$ values, and the BER and ergodic capacity performance are more sensitive to the PN at D.

The LA of DF RA transmissions in orthogonal frequency-division multiple access (OFDMA)-based systems is approached in paper by Varga et al. [VBB12] which presents an LA algorithm that selects the transmission type (RA or Dir), the modulation orders and amount and type of redundancy needed to ensure a target BLER at each phase of the two-phase cooperation, while requiring the smallest possible numbers of radio resource blocks. Within this algorithm, which uses convolutional turbo codes (CTC), the RN provides either repetition coding, or incremental redundancy or a hybrid redundancy, by using the rate-matching technique. The BLER is predicted by a method based on mean mutual information per coded bit (MMIB) and is used to select the Dir or RA transmission and the modulation orders and redundancies which ensure the highest SpE under the BLER constraint. The proposed LA algorithm has linear complexity with the number of available resource blocks and is applied to each type of supported service by only changing the specific target BLER.

The SpE performance, i.e., the number of information bits per quadrature amplitude modulation (QAM) symbol, is very good, thus emphasising the benefit of cooperative relaying, mainly at low and medium SNRs, as shown in Figure 7.9 for several sets of involved channels' SNRs.

The BLER performance abstraction in RA hybrid automat repeat request (H-ARQ) schemes is further analysed by Varga et al. [VPB14]. The paper proposes an analytical method to compute the MMIB for the bits which were repeated by the RN during the H-ARQ process. To this end the authors derive the expression of the pdf of LLRs of the repeated bits provided by the rate-matching algorithm, using an approach based on Hermite polynomials. The proposed theoretical method is validated against measurement results, showing that it provides good accuracies both for transmissions over Gaussian and frequency-selective channels.

Paper by He et al. [HZM14] analyses the BER performance of a cooperative practical coding/decoding scheme for data gathering in a WSN, which exploits the correlation knowledge among sensors' data at the fusion centre (FC),

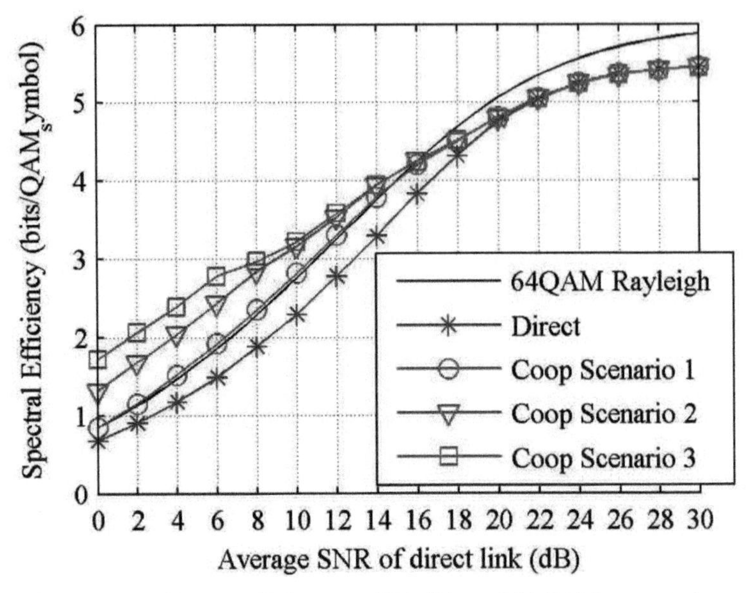

Figure 7.9 SpE versus $\text{SNR}_{\text{BS-UT}}$ of RA, Dir, and 64-QAM transmissions.

by modelling the wireless sensor network (WSN) by the binary chief executive officer (CEO) problem. Then, it analyses the theoretical limit of the sum rate based on the Berger-Tung inner bound and shows by simulations that the BER performance of the proposed scheme is very close to the theoretical limit. The error floor of the BER performance is further predicted by using the Poisson binomial process and validated by simulations. The authors conclude that the BER performance can be improved by increasing the number of sensors monitoring the same process, and that their coding decoding method reduces the energy consumption of the data gathering by exploiting the correlation knowledge.

7.2.2 Energy Efficient RA Algorithms

In paper by Tralli and Conti [TC12], the authors propose a multipurpose framework, composed of an useful cell and an interference generating cell, for the analysis of RA wireless networks on Rayleigh faded channels, which enables: (i) evaluation of the outage capacity for DF-RA links; (ii) computation of coverage and energy saving at given OP and SpE; (iii) analysis of the power allocation methods' impact on the OP and SpE.

By using the mutual information approach, the authors derive the expression of the OP and use it to compute the maximum SpE for which OP and transmitted power stay under imposed target values. Then, the paper analyses the OP performance for the uniform and balanced transmitted power's allocation between the S and RN, and concludes that in the presence of interference an optimal power allocation tends to the balanced power allocation. The numerical results presented point out the trade-off between EE, OP, and SpE of RA transmissions in the presence of interference.

The EE of RA and Dir transmissions is analysed by Dimic et al. [DZB12a]. The authors consider the SINR threshold and encapsulate the receiver parameters into the received power threshold, thus aiming at a generalised energy consumption model which includes non-identical transceivers, adaptive modulations and N-hop relaying. To this end, the authors propose a modified power consumption model which accounts for the received SINR. Using this modified model, the paper presents an EE comparison of the two-hop and Dir links, for non-identical transceivers at S, RN, and D, defining the regions of higher EE of the two transmissions. It then uses the ratio of their energy consumptions as an indicator of the border between their regions of optimality.

Moreover, the paper analyses the N-hop RA transmissions for identical transceivers, using adaptive transmission power control (ATPC), and concludes that the total transmitted power is dependent on channel loss, while the received power margin allows LA.

The EE of two-hop RA and D is further studied by Dimic et al. [DZB13], for heterogeneous transceivers, considering the SINR. The authors derive a power consumption model which expresses the transmitted power in terms of the link path loss, so that the SINR at the receiver antenna equals the minimum value needed to ensure the error-rate target. Then, they derive the energy consumption of the RA and Dir transmissions that use perfect ATPC, in the assumptions of the same modulations and transmission bandwidth. The energy expressions are then used to derive the conditions under which each of the two transmissions has a higher EE, leading to the determination of the respective regions of optimality. These expressions are used to evaluate the EE of two-hop RA and Dir transmissions in a generic cellular scenario.

In paper by Dimic et al. [DBB14a], the energy consumption model derived from the paper by Dimic et al. [DZB13] is applied to perform a comparison of the EE provided by a BS-RN-UE transmission via Type 1a RN and the BS-UE link in LTE-Advanced. The model is completed by the inclusion of the PHY and protocol parameters and is obtained by overlapping the set of channel losses with the set of transmit powers, which are linked by the link budget equation.

The analysis and numerical results are presented in two perspectives, i.e., in the space of channel losses, where the equipotential planes of the energy consumption ratio in the transceivers' operating regions are pointed out, and in the equivalent space of transmit powers, which indicates the required power levels to enable certain constellation and coding rate pairs, under the reliability constraint. Then, the paper analyses the trade-off between increasing the EE and SpE versus reducing the interference levels in the RA and Dir transmissions. They conclude that the trade-off EE versus generated interference can be tuned by adjusting the transmit powers, constellation size and coding rate.

Paper by Pejanovic-Djurisic and Ilic-Delibasic [PDID13] studies the increase of the EE in two-hop RA links provided by the use of a dual-polarised antenna at D. It analytically examines the BER performance of a standard DF RA scheme if the line of sight (LoS) S–RN is positioned to ensure LoS S–RN and RN–D links, i.e., Ricean-fading channels. Further on, it analyses the additional BER improvements obtained by using a dual-polarised antenna at D, in the assumption of two correlated and non-identical Ricean-fading channels for the RN–D link and performing maximum ratio combining (MRC) of the received signals at the D's receiver. The performance comparison of the proposed scheme to the standard DF–RN system with single antenna terminals shows that by using polarisation diversity only on Ricean channels, target BER values can be obtained at significantly lower SNRs, and the SNR gain increases significantly with the Ricean factor K, reaching 10 dB at BER = 10^{-4} for K = 10. The SNR gain is a measure of the energy saved by the proposed approach.

The paper by Moulu and Burr [MB14] tackles the problem of maximising the ergodic capacity of a MIMO RN network under low SNR on the RN–D hop. To this end the paper proposes and studies the use of the largest eigenmode relaying (LER) transmission in the RN. The study assumes a dual hop, half duplex non-coherent MIMO system, with one antenna at D, and with no CSI available at sources.

The authors analyse the transmission via the strongest eigenmode and derive a necessary and sufficient condition under which it maximises the capacity, concluding that whenever the necessary and sufficient condition is fulfilled, in order to achieve capacity, the entire power in the relay must be assigned only to the largest eigenvector. Then the authors discuss the optimality of the LER scheme at low SNR and high-SNR cases, and conclude that, regardless of the other system parameters, LER is the best relaying strategy, compared to the equal gain transmission and smallest eigenmode relaying.

Paper by Bagot et al. [BBN$^+$14] explores the increasing of the EE in a digital TV broadcast scenario by using adaptive beamforming techniques and antenna arrays at the transmission tower, according to the user's CSI. The proposed broadcast scheme involves TV sets with channel monitoring and internet connection capabilities, and a dedicated controller at the provider's end. The simulation results show that the percentage of served users in a given area could be increased significantly, compared to a non-adaptive broadcasting, at the same transmitted power. The paper also shows that CSI feedback provided only by 25% of users would bring significant coverage improvement.

7.2.3 Interference Analysis and Cancellation Algorithms

Paper by Clavier et al. [CWS$^+$12] analyses the BER performance of a generic RA-DF transmission in the presence of non-Gaussian multiple-access interference (MAI), which is modelled using α-stable distributions. The authors propose a non-linear receiver that uses normal inverse Gaussian (NIG) distributions to approximate the noise plus interference distribution, and compare its BER versus signal-to-interference ratio (SIR) performance to the one of a generic linear receiver, based on MRC, and to the ones of a non-linear receiver that uses the p-norm. Finally, the paper concludes that the non-linear receivers outperform the linear one in the presence of interferences, and that the NIG receiver is to be used when interferences can be modelled by α-stable distributions.

The generalised degree of freedom (GDoF) of the interference relay channel (IRC) is analysed in Gherekhloo et al. [GCS14] in the case when the S–RN link is stronger than the strong interference at D, within a transmission strategy which is a combination of DF, CoF and cooperative interference neutralisation (CIN), proposed by the authors, which includes block-Markov coding and lattice codes. The paper characterises the GDoF of the IRC in the strong interference regime, derives a new upper bound of the sum capacity and shows that the proposed transmission scheme achieves this new upper bound in the Gaussian channel. By comparing the GDoF of the IRC with that of the IC, the authors point out an increase in the GDoF even if the RN–D link is weak, and show that in the strong interference regime the GDoF can decrease as a function of the interference strength, a behaviour which is not observed in the IC.

Paper by Tian et al. [TNB14b] also analyses the DoF regions for the multi-user interference alignment, the analysis being made in a distributed RN system with greatly delayed and moderately delayed CSI feedback. It proposes some new transmission schemes for the RN interference alignment, based on the Maddah-Ali-Tse (MAT) scheme and space–time interference alignment (STIA), intended to operate either over delayed S–D channels and ideal CSI of the RN–D channels, or over non-delayed CSI of the RN–D channels. The proposed adaptive precoding scheme implementing the MAT interference alignment algorithm is based on a distributed cooperative relay system (DCRS), which includes a relay control station (RCS) and several distributed remote units acting as RNs.

The paper derives the sum DoF ensured by the two proposed schemes, for ideal or delayed RN–D's CSIs, expresses the sum DoF gain in terms of the number of delayed RN–UE CSIs, and shows that they ensure greater sum DoFs than the MAT algorithm, if the number of connected UEs is large. The proposed distributed precoding scheme has a lower complexity at the BS, some of the precoding matrices being computed at the RNs.

The impact of beamforming at the RN upon the destination's SINR is analysed in [SNB13]. The paper derives the analytical expression of the global SINR at the destination UE, provided by a fixed AF–RA transmission which uses the direct enhanced node B (eNB)-UE link, in a generic LTE-A cellular environment and evaluates the UE's SINR with variable 3 dB beamwidth antenna-pattern at the RN, aiming at the maximisation of the smallest UE SINR in the sector, to increase fairness and coverage. The presented results show that the optimum RN antenna 3 dB beamwidth is 96°, which improves by 2 dB the sector 10%-tile SINR at poor SINRs, and by 2.7 dB the 80% tile at good SINRs, relative to the direct eNB-UE link.

7.3 Cooperative Networks

In this section, we mainly focus on cooperative relay networks and cooperative cognitive radio networks (CRNs). Relay networks is a very promising cost-efficient solution to improve wireless network performance without adding much cost. The emerging M2M communications paradigm can be traced back to the concept relay. Further, cognitive radio adds intelligence into wireless communications and it is able to boost the performance in terms of high capacity, energy efficiency, and spectrum utilisation. It is envisioned that both M2M and relay as well as cognitive radio networks will be key enabling technologies for 5G and beyond.

7.3.1 Cooperative Relay Networks

In cooperative relay networks, we will concentrate on channel modelling, optimal relay selection, energy efficiency and cooperative schemes. M2M communication, as a promising technique, will be also studied.

The cross-correlation behaviour of the channels for designing and optimising performance. In particular, the correlation of the small-scale characteristics of channel has not been studied. To address this, we need to build model for the channel cross-correlation in the small-scale for realistic environments and application scenarios. The study Yin et al. [XPKC12] proposed a new stochastic geometrical model for the cross-correlation of small-scale-fading in the channels between the relay station and the mobile station, and between the BS and the mobile station. The underlying measurement data was collected using the relay-Band-Exploration-and-Channel-Sounder systems in urban cooperative relay scenarios. It has been demonstrated that the measurement system is capable of measuring two co-existing channels simultaneously. Six geometrical parameters are proposed as variables in the presented models. These key parameters describe the geographic features of a three-node relay system consisting of a BS, a relay station and a mobile station. The proposed model also reveals the variation of cross-correlation coefficients. The illustrative results indicate that the small-scale fading in the two channels are more correlated when the angle between the direct links from the BS to the mobile station and from the relay to the mobile station decreases.

Relay networks has a very promising benefit with respect to saving energy consumption. This is achieved mainly by reducing path loss due to short transmission range and using low transmission power. Reducing energy consumptions is very related to the problem on relay selection. Taking into account the coordination overhead, the energy efficiency of the cooperative communication may degrade with the increase of the number of cooperators. The relay can be selected according to different criterions, e.g., the maximum SNR, the best harmonic mean, the nearest neighbour selection, and contention based selection. The study by Ling et al. [LRYG14] focused on a network where a user is selected as a relay for primary user. Selection criteria are based on maximisation of the network energy efficiency. The associated three-phase cooperation procedure shown in Figure 7.10.

In this system model, secondary users act as relay to assist in the transmission of primary users with the reward to transmit its own information. On one hand, the secondary user is selected using the metric of the energy efficiency by considering the outage capacity and energy consumption in a decoding and forward relaying. On the other hand, optimal power allocation and secondary

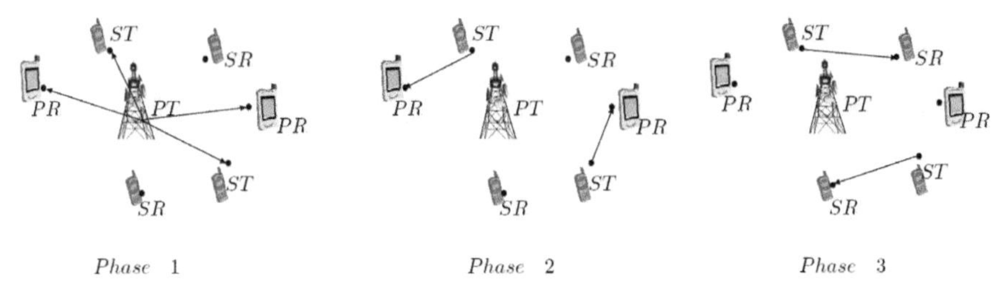

Figure 7.10 Cooperation in three phases.

user selection are performed in the scenario. The cooperative relay sharing procedure can be divided into three phases, i.e., primary users' direct transmission, secondary user's relay, and secondary user's own transmission. The optimal power allocation is obtained by solving an convex optimisation problem with the boundary conditions of QoS requirement. It is observed that cooperative communication performs better than non-cooperation transmission in saving network consumption. More candidates of the SUs may lead to lower power consumption of the network and better energy efficiency.

Relay is a flexible technology and can be applied in different networks. In the optimisation of cellular planning for fixed WiMAX, the use of relays reduces the necessary extent of wire-line backhaul, improving coverage significantly whilst achieving competitive values for system TP. Moreover, relays have a much lower hardware complexity, and using them can significantly reduce the deployment cost of the system as well as its energy consumption. Consequently, frequency reuse topologies have been explored for 2D broadband wireless access topologies in the absence and presence of relays, and the basic limits for system capacity and cost/revenue optimisation have been discussed. Relays are also amenable to opportunistic utilisation of power-saving modes.

The study by Robalo et al. [RHV+12] investigated the use of relays in WiMAX network deployments and concentrates on the cost/revenue performance and energy efficiency trade-off in such cases. Layered and cooperative elements such as femto-cells and relays can improve performance or energy efficiency in mobile networks. However, they consume energy *per se* and their durations in operational state must, therefore, be minimised. Specifically, the study investigates the performance achievable by networks that are deployed in various sectorisation configurations with and without relays, and matches this to varying traffic loads at different times of the day to maximise the use of sleep modes, where possible, by relays, also in consideration of coverage requirements. The study performs real measurement in the hilly area of

Covilhã, Portugal. Results show that through the maximal use of power saving by relays at low traffic times, considerable energy savings in the relays' power consumption are achievable, typically 47.6%. These savings are shown to map to a financial saving for the operator of 10% in the operation and maintenance cost.

Device-to-Device (D2D) has been known to be a key technology for 5G and beyond. Relay is the fundamental technique in D2D communications. The study by Komulainen and Tolli [KT14] proposed linear coordinated transmit-receive beamforming methods for spatial underlay direct device-to-device communication in cellular networks where the user terminals employ multiple antenna elements. A cellular system model includes one multi-antenna BS and multiple multi-antenna user terminals.

In the cellular mode, spatial multiplexing for both UL and DL user signals is employed so that the DL forms a MIMO broadcast channel and the UL forms a MIMO MAC. Furthermore, the data streams are transmitted via linear spatial precoding and that each receiver treats the signals intended to other receivers as coloured noise. The model is further extended by allowing spatial underlay device-to-device communication so that some terminal pairs are allowed to directly transmit to each other. As a result, the system model becomes a mixture of MIMO IC, broadcast channel, and MAC. For a D2D terminal pair, direct communication is a beneficial alternative compared to the cellular mode, where the devices communicate to each other via a BS that acts as a relay. For mode selection, spatial scheduling, and transmitter-receiver design, the study formulates a joint weighted sum-rate (WSR) maximisation problem, and adopts an optimisation framework where the WSR maximisation is carried out via weighted sum MSE minimisation.

Figure 7.11 shows the system sum rate utility in terms of average direct D2D channel gain. The results indicate that the joint design always outperforms the other designs. However, when the D2D channel becomes stronger, the rate provided by direct D2D transmissions increases rapidly, and the additional gain from the joint design disappears.

7.3.2 Cooperative Cognitive Radio Networks

Recently, cooperative networking technologies, which inherently exploit the broadcasting nature of wireless channels and the spatial diversity of cooperative users, are introduced as a powerful means to facilitate the design of CRNs. Two types of cooperation paradigms in CRNs have been studied up to date: cooperation between primary users (PUs) and secondary users (Sus; or called PU–SU cooperation), and cooperation only among SUs

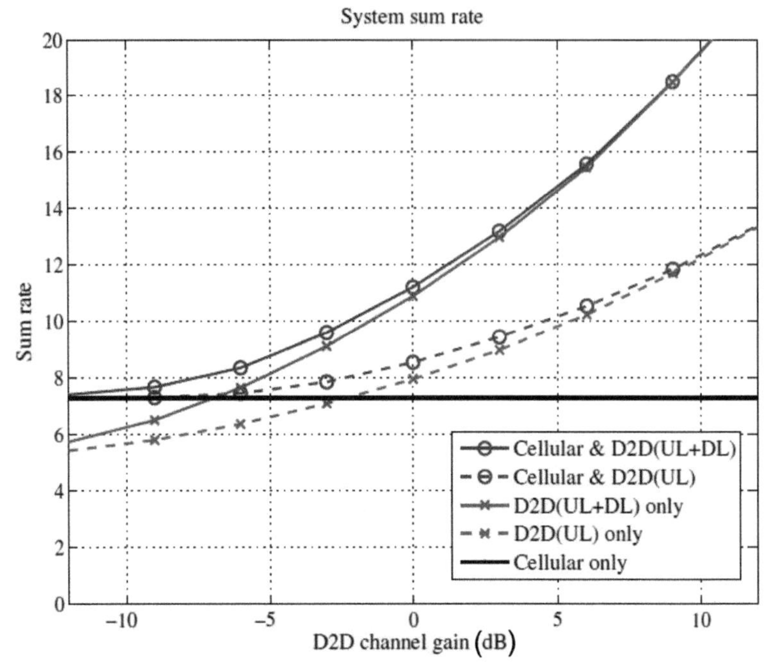

Figure 7.11 System sum rate in D2D.

(or called inter-SU cooperation). For the PU–SU cooperation, PUs are aware of the existence of SUs. The primary system regulates the spectrum leasing policy for the secondary system. The SUs utilise a small segment of the licensed spectrum, and in return, cooperate with PUs to improve the quality of primary transmissions. For the inter-SU cooperation, there is no interaction between primary and secondary systems. Cooperative spectrum sensing is a promising technique in CRNs by exploiting multi-user diversity to mitigate channel fading.

The study by Yu et al. [YZX^{+}13] identified that both sensing accuracy and efficiency have very significant impacts on the overall system performance. Then, several new cooperation mechanisms, including sequential, full-parallel, semi-parallel, synchronous, and asynchronous cooperative sensing schemes.

Figure 7.12(a) shows the sequential cooperative sensing, where the cooperative SUs (SU1–SU6) are scheduled to sense a single channel at the same time t_i. Figure 7.12(b) shows the full-parallel cooperative scheme scheme, where each cooperative SU senses a distinct channel in a centralised and

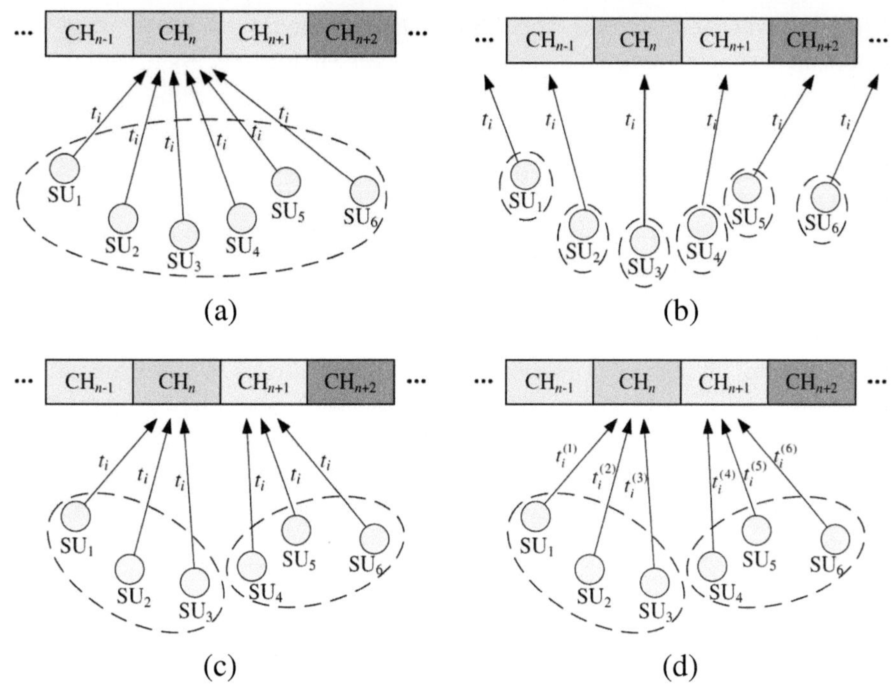

Figure 7.12 Illustration of cooperation mechanisms: (a) sequential cooperative sensing; (b) full-parallel cooperative sensing; (c) semi-parallel cooperative sensing; and (d) asynchronous cooperative sensing.

synchronised mode. To tradeoff the sensing accuracy and efficiency, the semi-parallel cooperative sensing is proposed and described in Figure 7.12(c). Figure 7.12(d) presents the asynchronous semi-parallel cooperative sensing. All the cooperative SUs are allowed to have different sensing moments. Although the cooperative SUs of a same group cooperate to sense the identical channel, their sensing moments are different.

The proposed cooperation mechanisms and the sensing accuracy–efficiency tradeoff in these schemes are elaborated and analysed with respect to a new performance metric achievable TP, which simultaneously considers both transmission gain and sensing overhead. Illustrative results indicate that parallel and asynchronous cooperation strategies are able to achieve much higher performance, compared to existing and traditional cooperative spectrum sensing in CRNs.

Spectrum management is an important mechanism in CRNs. Wang et al. [WCF13] noticed the transmission collisions between the SU and the PU and proposed periodic spectrum management for cognitive relay networks. In particular, unreliable sensing results lead to low successful switch rates, accompanied by intolerant SU's transmission collision with the PU. The study discussed a spectrum sensing and handoff model in a selective relay scenario. A partial soft sensing and fusion algorithm is proposed under the stringent signalling cost. The reliability of secondary transmission in light of the collision probability and TP is also derived analytically. The problem of spectrum leasing in CRNs is studied by Sharma and Ottersten [SO15]. This work proposed an overlay CRN where the primary users lease access time to the secondary users in exchange for their cooperation. The pairing problem among the primary and the secondary users is modelled as a one-to-one matching problem and a matching mechanism, based on the deferred acceptance algorithm, is applied in the CRN under consideration. The numerical results evaluate the performance of the proposed scheme, in terms of primary and secondary utility, compared to a random pairing mechanism.

Cooperative beamforming is the key concept for CRNs. Cooperative communication with spectrum sharing has been proved to be an efficient way to reduce energy consumption. The work by Wang et al. [WCF13] presented an optimal power allocation algorithm for mobile cooperative beamforming networks. The optimal power allocation scheme is obtained by solving a convex optimisation problem oriented to minimise network energy consumption while guaranteeing quality of service (QoS) for both PU and SUs. Illustrative results show that a cooperative beamforming network improved with our proposed power allocation algorithm performs better with more SUs in terms of higher energy efficiency and less power consumption per unit TP. Additionally, the proposed scheme outperforms the conventional relay selection approach with respect to high energy efficiency.

Cooperative cognitive radio can be applied in different environments. The work by Díez [Die13] discussed the performance of cooperative cognitive radio in indoor environments. The study shows the factors that have considerable influence on the behaviour of a cooperative CRN. Different detection techniques and the hypothesis testing techniques have been implemented in a basic and simple way, decreasing the detection times of the final testing model. The study also proved that the most accurate of the implemented detectors is the wave-based form detector and it has been studied the difference between both daughter boards with and without automatic gain control. Without automatic gain control, in some test points the level of the signal saturates the USRP and in this case, the occupied channels are mistakenly detected. This implies a reduction of the cognitive radio system quality.

A recent research by Sharma and Ottersten [SO14] focused on cooperative spectrum sensing for heterogeneous sensor networks. Existing cooperative spectrum sensing methods mostly focus on homogeneous cooperating nodes considering identical node capabilities, equal number of antennas, equal sampling rate and identical received SNR. However, in practice, nodes with different capabilities can be deployed at different stages and are very much likely to be heterogeneous in terms of the aforementioned features. In this context, the study evaluates the performance of the decision statistics-based centralised cooperative spectrum sensing using the joint PDF of the multiple decision statistics resulting from different processing capabilities at the sensor nodes. The performance is shown to outperform various existing cooperative schemes. The investigations in indoor and sensor networks are very important and can be straightforwardly extended to D2D scenarios.

8

Evolved Physical Layer

Chapter Editors: A. Burr, L. Clavier, G. Dimic, T. Javornik, W. Teich, M. Mostafa and J. Olmos

8.1 Introduction

In this chapter, we discuss progress on the physical layer (PHY) for next generation wireless systems, covering the range of "smart environments". However, we omit the PHY of cooperative systems for the majority of the chapter, since this has been covered in Chapter 7.

It is well-appreciated that interference is the major limiting factor in wireless systems of all kinds: the capacity of most networks is limited by interference, whether from outside the network or from within it. It is important to characterise this interference as well as to mitigate its effects, so in Section 8.1, we focus in particular on interference that does not necessarily follow the Gaussian distribution, but instead is impulsive in nature. We show that this may enable us to mitigate its effect on receivers. IA is another approach for mitigating interference within networks, also discussed in this section, especially in the context of realistic channel state information (CSI) at transmitters.

Also of increasing importance is the energy efficiency (EE) of wireless networks, and in keeping with the emphasis of the whole book, we consider this issue in Section 8.2. For completeness, the effect of relaying on EE is also included here, including modelling of the energy consumption of wireless terminals. The benefits of energy harvesting are also considered.

Multiple-input multiple-output by now has become a standard feature of wireless networks of all kinds, but weaknesses remain in multiple-input multiple-output (MIMO) techniques, which have motivated further research to improve performance in future network, as reported in Section 8.3. This includes pre-coding for adaptation on realistic channels, and also two novel approaches: beam-space MIMO and spatial modulation. The effect of certain channel imperfections, such as peak amplitude limitations, are also addressed.

Cooperative Radio Communications for Green Smart Environments, 305–340.

Significant effort has continued on the development of iterative methods in modulation and coding, given the power of the approach in wireless receivers, and this forms a major focus of the work reported in Section 8.4. Iterative interference cancellation (IIC) and iterative decoding and demodulation are considered, and non-coherent demodulation is also discussed.

In evaluating the performance of complete wireless networks, it is rarely feasible to fully simulate the operation of the PHY of all links involved. This leads to the requirement for link abstraction, that is, the provision of models of the PHY to incorporate into system-level simulations. Such models are discussed in the final section of the chapter.

8.2 Dealing with Interference

There are two ways to consider interference: first as a noise resulting from the activity of other devices in the networks or from other networks. It can be too complex for some simple devices to make advance signal processing to reduce the impact of this interference. It happens, however, that this interference in many situations does not exhibit a Gaussian behaviour but is rather impulsive. In this section, we first propose an overview of impulsive interference models as well as a discussion about the dependence structure. We also propose some receiver structures that outperform the classical linear receiver in such a context. The second approach is to minimise the impact of the noise. This can be done by properly shaping the transmitted signal in order that interference is in an orthogonal space to the useful signal at the receiver, which is called interference alignment (IA) and will be addressed in the second part of the section.

8.2.1 Impulsive Interference

Interference is, in many situations, not properly modelled with a Gaussian distribution. Pioneer works in communications can be found in Middleton [Mid77], who obtained general expressions based on series expansion. If is work is widely used, infinite sums are difficult to handle in practice and simplifications have to be made, for instance only considering the most significant terms [Vas84, GDK06, AB07, VTNH14]. More recently, many works have been done concerning time hopping ultra-wideband. Analytical solution have been proposed [FNKS02, SMM03] but to simplify the receiver design, many works have proposed empirical distributions to represent interference plus thermal noise (see Beaulieu and Young [BY09] for an overview). Even more recently, analysis of networks has attracted a lot of works relying on stochastic geometry. Although the first papers were published in the 90s [Sou92, TN95, IH98], an unbounded received power makes the interference fall in the attraction domain of a stable law. This can be seen as a consequence of the generalised central limit theorem [ST94, NS95]. The main advantage of the heavy-tailed stable distributions is their ability to represent rare events. In many communication situations, these events are in fact the events that will limit the system performance. The traditional Gaussian distribution ignores them leading to poor results. The proof of this result is generally done considering the log-characteristic function of the total interference (see for instance Sousa [Sou92], Win et al. [WPS09], and Ghannudi et al. [GCA+10]). This area of research is still active. Problems concerning the homogeneous position of users are studied, for instance based on cluster point process [GEAT10, GBA12] for general *ad hoc* networks or Poisson hole process for cognitive radio [LH12]. The dependence structure of interference is also attracting many works [GES12]: it is an important feature for the network analysis but difficult to handle.

8.2.2 Receiver Design

To illustrate the impact of impulsive interference, we consider a system where the symbol is repeated five times. We only represent here two significant examples: (i) a moderately impulsive case, represented by a mixture of α-stable ($\alpha = 1.5$) and Gaussian noises with, NIR = 0 dB (see Gu et al. [GPC+12] for more details); (ii) an ε-contaminated noise with $\varepsilon = 0.01$ and $k = 70$ reflecting rare but strong impulses (highly impulsive noise).

We show in Figure 8.1 the bit error rate (BER) performance of Gaussian, Cauchy, Myriad, symmetric normal inverse Gaussian (NIG), and p-norm receivers. When the noise involves an α stable impulsive interference, the BER is measured as a function of the inverse dispersion of the SαS distributions ($1/\gamma$) for network interference, since the increasing of inverse dispersion indicates the decreasing of the network interference strength, reflecting the conventional signal-to-noise-ratios (SNRs). For the ε-contaminated, the SNR at the receiver is used for the x-axis. The number of training samples for NIG, Myriad and p estimations is set to 1000 bits.

From the preceding figures and other approaches, we did not include here, we can make the following comments: linear receivers adapt poorly to impulsiveness even if optimal linear receivers [Joh96] are considered; a better approach is to find flexible families of distributions with parameters that can be easily and efficiently estimated. For instance, the Myriad approximation (based on Cauchy distributions) or the NIG distribution family [GPC+12] are

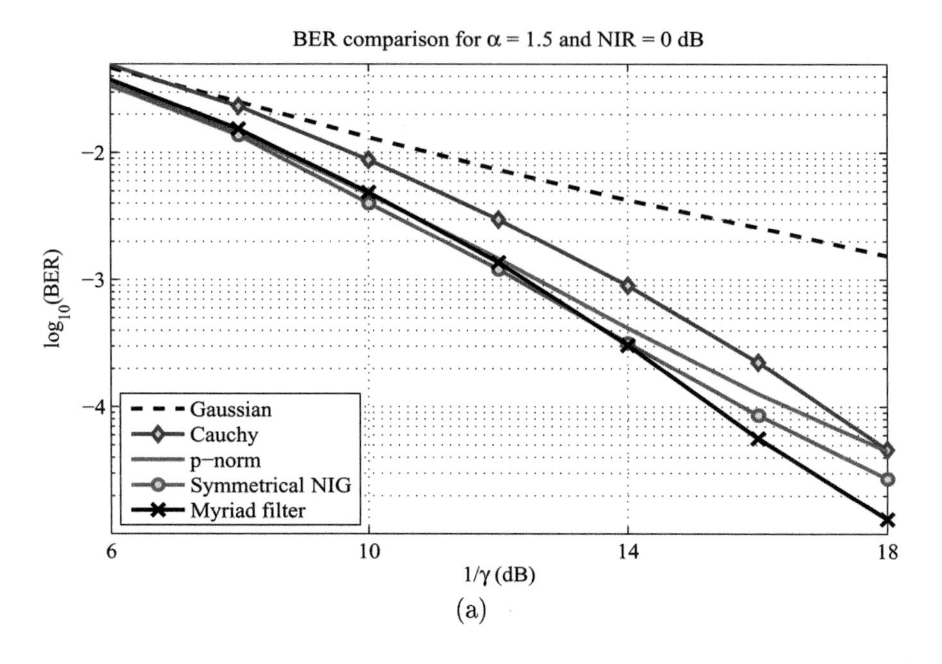

(a)

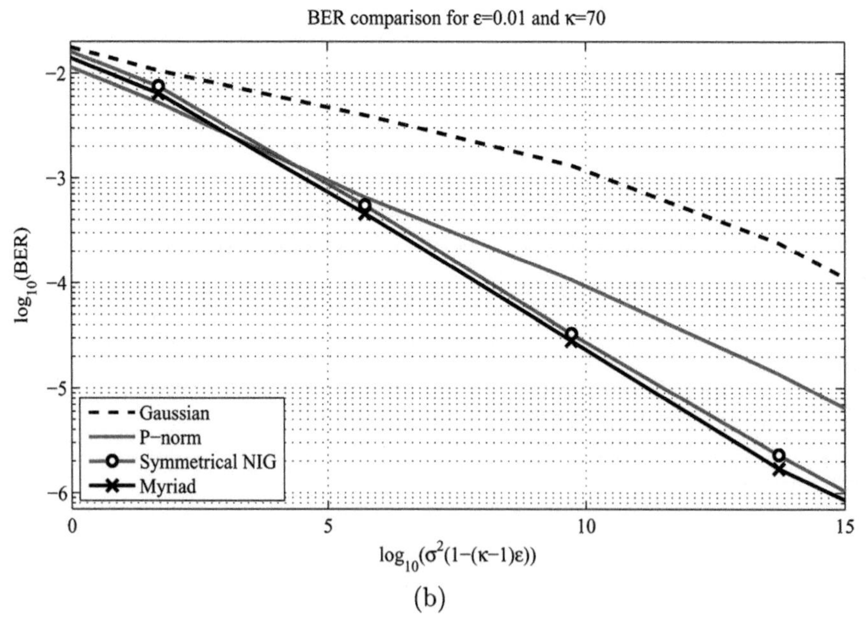

(b)

Figure 8.1 BER performance in (a) moderate impulsive environment and (b) α-contaminated noise.

good choices; log-likelihood ratio (LLR) approximation based approaches seem to have good potential. The intuitive approaches are the soft limiter and the hole puncher that limits the impact of the large values but remain less efficient than other non-linear solutions. On the other hand, the p-norm allows either a close to linear or linear behaviour when the thermal Gaussian noise is the main contribution to the noise and also approaches the *sharp* shape of the LLR when impulsiveness increases.

These works can be extended to more sophisticated receiver structures, for instance to channel coding [GC12b, MGCG13, DGCG14].

8.2.3 Dependence Structure

If many works consider independent and identically distributed (i.i.d.) interference random variables, space, time, or frequency diversity can result in vectors with dependent components. Mahmood et al. [MCA12] use the symmetric α-stable model, which, converted to its complex baseband form, is generally not isotropic, which means that the real and imaginary components are dependent. In Gulati et al. [GES12], the joint temporal statistics of interference in the network is derived along with an assumption of a bounded path-loss function. In Schilcher et al. [SBB12], closed-form expressions and calculation rules for the correlation coefficient of the overall interference are derived. Three sources of correlation are considered: node locations, channel, and traffic.

Whilst it may be possible to work with multivariate impulsive interference models, and especially multivariate α-stable models, it is very challenging in practice due to the intractable nature of the multivariate distribution function and because it requires the modelling of a high-dimensional spectral measure. We take a different approach [YCP$^+$15]: in order to maintain the suitable marginal behaviours offered by α-stable models and to consider a relevant dependence structure, we use a meta-distribution model (see discussion in Fang et al. [FFK02] and Abdous et al. AGR05]) and rely on a parametric dependence model based on a copula construction, which is justified by the Sklar's theorem:

Sklar's Theorem: Let X and Y be random variables with distribution functions F and G respectively and joint distribution function H. Then there exists a copula C such that for all $(x,y)\, \varepsilon \mathbb{R} \times \mathbb{R}$ one has

$$H(x,y) = C(F(x),G(y)). \tag{8.1}$$

If F and G are continuous, then C is unique; otherwise, C is uniquely determined on *Ran*(F) $\times$ *Ran*(G), the cartesian product of the ranges of the marginal cumulative distributions. Conversely, if C is a copula and F and G are distribution functions, then the function H defined by Equation (8.1) is a joint distribution function with margins F and G.

8.2.4 Interference Alignment

Interference alignment was introduced in Cadambe and Jafar [CJ08] as a means to allow spatial or temporal coordination among a group of users. The degrees-of-freedom (DoF) of the channel are exploited to confine all interfering signals to a common subspace, leaving the remaining orthogonal subspace for interference-free communication. Early IA studies showed promise of greatly increasing the number of co-channel users that could be supported with interference-free signal reception, but were based on unrealistic channel state knowledge assumptions: perfect instantaneous global CSI was assumed known at all nodes throughout the network, which is impossible in practice.

8.2.4.1 Reducing feedback requirements

In order to assess the feasibility of applying IA techniques to future communications systems, researchers have recently been looking at the impact of more realistic assumptions on IA techniques. One of the main hurdles to overcome in the implementation of IA systems is that of reducing the amount of feedback required to realistic levels. Several approaches in reducing feedback depend on quantisation using constellations of unit vectors uniformly distributed across the Grassmann manifold. In each of the methods described here, rather than directly quantising a vector or vectorised matrix, a vector that is representative of the desired information is selected from the constellation, and its index within the constellation is instead fed back, offering significant efficiency by reducing system overhead.

In Colman and Willink [CW11], a cooperative iterative algorithm was presented to reduce the overall signal-to-interference plus noise ratio (SINR) in a MIMO interference limited application using constellation-based pre-coding. This algorithm was designed not to require global channel information; the algorithm proceeds with information that each mobile station can measure locally. In every iteration of the algorithm, each receiver unit assesses the SINR impact of the transmitters using pre-coding vectors that are in the vicinity of the current pre-coding vector for each transmitter. Candidate pre-coding matrices that would improve the SINR are given a positive designation while candidate pre-coding matrices that would degrade the SINR are given a negative designation. This impact information is fed back to the transmitters where each transmitter selects the pre-coding vector resulting in the greatest positive system impact. This algorithm was shown in Colman and Willink [CW11] to provide improved SINRs when compared with an optimal IA algorithm that had its feedback quantised using the same number of feedback bits.

The same authors presented a second IA algorithm based on local information in Colman et al. [CMW14]. In this algorithm, the receivers use local CSI to determine desirable pre-coders for all interfering transmitters. These pre-coding matrices are then quantised, using a Grassmannian constellation of pre-coding matrices, and this information is fed back to the transmitters. The next pre-coding matrices used at the transmitters are then a linear combination of the quantised desirable pre-coding matrices. In studying these two algorithms, the authors demonstrated the significant degradation that quantisation and imperfect or delayed channel information make on the interference rejection capabilities of IA techniques. Furthermore, these degradation effects are exacerbated when the common assumption of equal power from all transmitters at all receivers is removed.

Rather than feeding back CSI information, it was proposed in El Ayach and Heath [EAH12] to return the differential in CSI to the transmitters, which requires less feedback in slowly varying channels. This work was extended in Xu and Zemen [XZ13] by tracking the basis expansion coefficients on the manifold instead of the channel impulse response. This difference allows the transmitters to predict future channel realisations, which can compensate for inherent feedback delay, improving performance. Furthermore, the differential information of these basis expansion coefficients was quantised using a Grassmannian constellation, showing that effective IA performance could be maintained even with as few as two bits per channel realisation.

Prediction techniques were also harnessed for improved robustness to time-varying channels in Xu and Zemen [XZ14]. In this work, effects on IA performance, due to mobility and feedback delays, were mitigated by enabling reduced-rank channel prediction. The time-varying channel was approximated in a reduced-rank representation of orthogonal bases in the delay domain. These delay domain coefficients were then quantised using a random vector code-book distributed throughout the Grassmann manifold. Simulation results showed that a significant improvement in sum-rate is realisable when compared with non-predictive feedback strategies.

By stripping away assumptions, a better idea can be made of the viability of emerging technologies. However, in order to obtain a realistic idea of the potential benefits of technologies such as IA, experimental implementations are of critical importance. Such an implementation was described in Zetterberg and Moghadam [ZM12] using a six-node setup to test MIMO, IA, and coordinated multipoint (CoMP) technologies, where each node had two antenna elements. System feedback was accomplished via cables and the nodes were fixed throughout the measurements. Experimental and post-processing SINRs show demonstrated a significant degradation between the ideal and achievable SINRs, due to hardware impairments.

8.2.4.2 IA applications

As IA becomes a more mature topic in the research community, the concept is being adapted and broadened from its initial concepts and considered as a method to mitigate interference in specific applications. In Kafedziski and Javornik [KJ13] the concept of IA, which is usually implemented in the space and/or time domains, was extended to the frequency domain with frequency–space IA using orthogonal frequency division multiplexing (OFDM). By using multiple antennas and multiple frequencies, the dimensionality of the signal space was increased, allowing exploitation to increase the number of signal streams or the number of users. By using IA techniques, both inter-cell and intra-cell interference could be removed effectively, demonstrating that IA may be a good technique to enable a frequency reuse factor of one in future wireless networks.

In Chatzinotas et al. [CSO13], IA concepts were extended by examining the effect of frequency packing on IA techniques. In a frequency-packed system, the carrier frequencies of adjacent signals are brought closer together in the frequency domain. This requires advanced interference mitigation techniques such as IA to suppress the additional interference. The work in Chatzinotas et al. [CSO13] was applied to a dual satellite coexistence scenario of monobeam satellites as primary and multibeam satellites as secondary users. It was shown that as the number of carriers in increased within a specified band, the overall sum-rate could be improved through effective interference mitigation. It was further shown that, with a coordinated IA technique, the primary rate could be perfectly maintained.

8.3 Energy Efficiency

8.3.1 On the Need for Improved EE

The EE in communications is of interest because of the increase in the number of communication devices. By increasing the EE, it may be possible to slow growth of the total energy consumption related to communications when the number of devices increases. The EE in wireless communications has been studied by a working group of the GreenTouch initiative, the EARTH-FP7 project, and other projects. The IC1004 Action has contributed to the EE analysis in cooperative radio communications. The results show that the EE can be increased by various advances at the PHY and medium access control (MAC) layer.

One studied method is the two-hop relay. The introduction of relay with no other improvements at the PHY and MAC layers comes at the price of an additional energy consumer—the relay transceiver, reduced spectral efficiency, and increased source-to-destination delay. However, several advances are proposed which improve the overall EE, in terms of [bit/J], while not compromising other performance metrics. A key is the improvement of the source-relay and relay–destination channels with respect to (w.r.t.) the source–destination channel (Section 8.3.2). To improve channel quality of the relay links, the focus was on appropriate relay placement [DBB14b], or the use of receive diversity [PDID13, IDPD13]. The better channels enable power gain/saving to improve the EE subject to performance constraints, or to trade-off the EE with performance.

Other methods, which do not assume relaying, propose the use of transmission policy optimisation when energy harvesting is used and partial information on energy availability is provided [ZBMP13], beamforming instead of sectoral antennas [GC12a], and a variation of automatic retransmission request (ARQ) [BDZ13, ZBD14].

8.3.2 EE Increase by Use of Two-Hop Relays

An EE comparison between the two-hop relay and direct link has been developed in Dimić et al. [ZDB14, DZB12a, DZB12b, DZB13, DBB14a, DBB14b]. The evolution of the model, dubbed Dimić-Zogović-Bajić, is described below.

The major power consumer in a wireless transmitter is the power amplifier. A wireless transceiver power consumption (WPTC) model for Class A, AB and B power amplifiers is proposed and justified [ZDB14]. It shows the total transmitter power consumption, p_{TX}, as a function of the transmit power, i.e., the power delivered to antenna, p_{t}: $p_{\mathrm{TX}} = f(p_{\mathrm{t}})$. It is a monotone, increasing function, which generalises the affine function between the minimum and maximum p_{t} points, $(p_{\mathrm{t,min}}, p_{\mathrm{t,max}})$. The receiver power consumption, P_{RX}, is modelled as a constant. The energy consumption is $E_{\mathrm{TX}} = P_{\mathrm{TX}}T$ and $E_{\mathrm{RX}} = P_{\mathrm{RX}}T$, where T is a unit of time.

The WTPC model is related to the aggregate channel loss along the link, L, by a simple link budget: $p_\mathrm{r} = l(p_\mathrm{t}, L)$, where p_r denotes the received signal power; and a simple, deterministic condition for successful reception: $p_\mathrm{r} \geq P_\mathrm{th}$, where P_th denotes a received power threshold [DZB12a, DZB12b]. The P_th encapsulates several receiver parameters into one scalar, in particular, interference power at the receiver and the SINR [DZB13]. A perfect adaptive transmission power control is introduced for link adaptation (LA), so that $p_\mathrm{r} = P_\mathrm{th}$, which yields $p_\mathrm{t} = g(L, P_\mathrm{th})$. Then, along the source-destination link

$$p_\mathrm{TX} = f(g(L, P_\mathrm{th})). \tag{8.2}$$

Equation (8.2) shows that the total power consumption has to adapt to a random variable L, where L is not under control of the link-or-network designer.

Equation (8.2) is applied to the direct link, and the first and second relay hops, denoted by subscript i = void, 1, 2, respectively. Then, P_TX, $P_\mathrm{TX,1}$, $P_\mathrm{TX,2}$ are evaluated in the space of channel losses (L, L_1, L_2). In (L, L_1, L_2), a feasible energy consumption ratio between the two-hop relay and the direct link is

$$\alpha = \frac{(E_{TX,1} + E_{RX,1}) + (E_{TX,2} + E_{RX,2})}{E_{TX} + E_{RX}}. \tag{8.3}$$

The ratio α is a scalar-valued function in (L, L_1, L_2), parameterised by the transceiver characteristics. This evaluation yields an equipotential plane in the space (L, L_1, L_2), which separates the regions of energy consumption corresponding to the given value of α. The DZB energy consumption model for non-identical, i.e., heterogeneous, transceivers at the source, relay, and destination is derived in [DZB13].

The operating region (OR) of the transceivers is the region in (L, L_1, L_2) where a triplet $(P_\mathrm{t}, P_\mathrm{t,1}, P_\mathrm{t,2})$ is feasible. The OR is bounded by $p_\mathrm{t,i,min}$ and $p_\mathrm{t,i,max}$, and respective $P_\mathrm{th,i}$, with i = *void*, 1, 2 [DZB13]. The DZB model analysis is applied to a comparison of EE of two-hop transmission via Type la relay and the direct link in LTE-advanced (LTE-A), between the same Base station (BS) and user equipment (UE) [DBB14a, DBB14b]. It is shown how the transceivers parameters and (L, L_1, L_2) influence α and that the supportive relay may provide energy savings.

In the OR of the space $(p_\mathrm{t}, p_\mathrm{t,1}, p_\mathrm{t,2})$, the required power levels to enable certain modulation and code scheme (MCS) (i.e., constellations and code rate pairs) are selected. These MCS levels define the data rate ratio between the two-hop relay data rate, r_R, and direct link data rate, r,

$$\beta = \frac{r_\mathrm{R}}{r} \tag{8.4}$$

Good channels L_1 and L_2 w.r.t. L facilitate increase in r_R w.r.t. r, and so β increases. Data rates of each relay hop must be 2 r_R

Overlapping the α-levels with the β-ratios quantifies the EE ratio

$$\eta = \frac{\beta}{\alpha}. \tag{8.5}$$

The increase in EE is achieved by maximising $p_{\mathrm{t},1}$ and $p_{\mathrm{t},2}$, and by selecting the MCS which maximises r_R, subject to the successful reception. In addition, a trade-off between increasing the energy and spectral efficiency versus reducing the generated interference when using a regenerative relay is proposed. Figure 8.2 shows α-levels in (L_1, L_2) space for a given L. Moving the operating point to the upper-right corner, by increasing $P_{\mathrm{t},1}$, $P_{\mathrm{t},2}$ improves EE.

In Tralli and Conti [TC11], it is shown that in the presence of other cells, the mutual interference can be a limiting factor and needs to be characterised. A framework for analysis and design of relay-assisted communications, which accounts for multichannel communications in each link, geometrical setting, interference, fading channel, and power allocation, while being general with

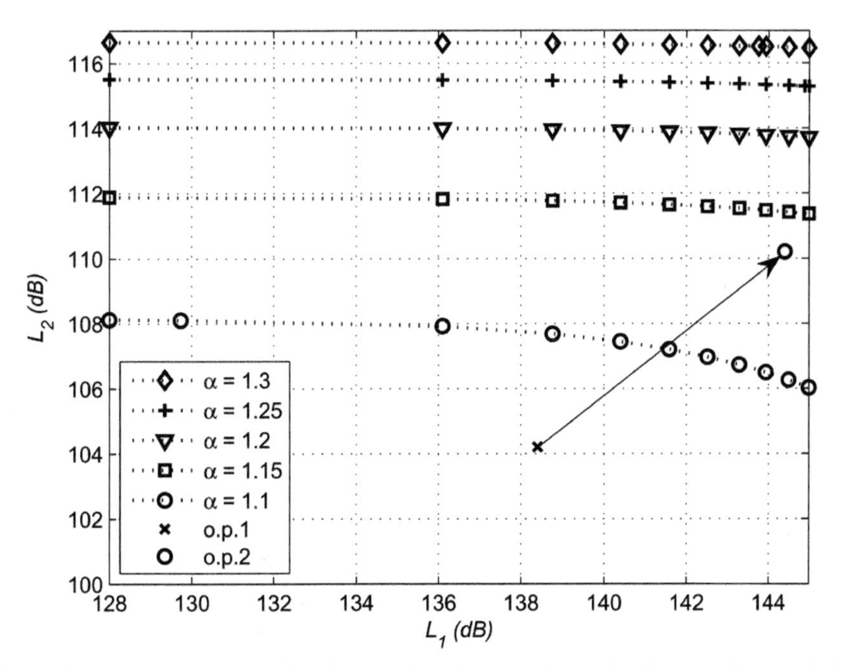

Figure 8.2 EE Example in (L_1, L_2): moving the operating point from o.p.l to o.p.2, by increasing $p_{\mathrm{t},1}$ and $p_{\mathrm{t},2}$ by 6 dB each, increases the constellation from 16QAM to 64QAM, which yields $\beta = 2$ and $\eta = 1.75$.

respect to MCS is proposed. The framework differs from the DZB model in the following:

1. There is a geometrical setting, with defined distances between the source, relay, and destination in the cell of interest and the interfering cell; DZB absorbs geometry in the channel losses (L, L_1, L_2);
2. The destination combines received signals from both the source and relay; DZB does not use receive combining;
3. Tralli and Conti [TC11] considers Rayleigh fading and outage probability (OP); DZB considers a simple condition for successful reception $p_{\mathrm{r}} \geq P_{\mathrm{th}}$;
4. Tralli and Conti [TC11] considers two power allocation techniques: (i) uniform (equal at the source and relay), and (ii) balanced (such that p_{r} is equal over source–destination and relay–destination links); DZB considers adaptive power control tightly linked to the total channel loss along a link;
5. Tralli and Conti [TC11] accounts for transmit power only, when evaluating EE; DZB considers total transceiver power consumption, p_{TX}.

The performance with and without relay-assistance is compared to quantify the benefits of relay and power allocation in terms of energy saving with respect to direct transmission [TC11]. The reference scenario assumes two source–relay–destination triplets. The first triplet is in the cell of interest, and the second is in an interfering cell. At each link, there are n_{T} transmitting antennas and 1 receive antenna.

The OP at the destination is evaluated for the source-destination data rate of *R* (*bit/s/Hz*). The OP optimal power allocation: in the absence of interference, with large enough link diversity, it tends to be balanced power allocation; in the presence of interference, it is not easy to find because of the mutual interference between the useful and interfering links. The average energy per bit required to support a given rate within the given coverage distance is evaluated. The use of relay with balanced power allocation provides power gains at medium and low rates for light frequency reuse, and only at low rates in heavy reuse. When power control is used, to guarantee the target OP, the average energy per bit is lower in both light and heavy reuse cases when relay is used Figure 8.3. The use of power control provides a 5-dB energy saving in all cases.

In Pejanović-Djurišić and Ilic-Delibašić [PDID13] and Ilić-Delibašić et al. [IDPD13], performance of two-hop relay system operating in Ricean-fading environment is studied. Similar to Tralli and Conti [TC11], both Pejanović-Djurišić and Ilic-Delibašić [PDID13] and Ilić-Delibašić et al. [IDPD13]

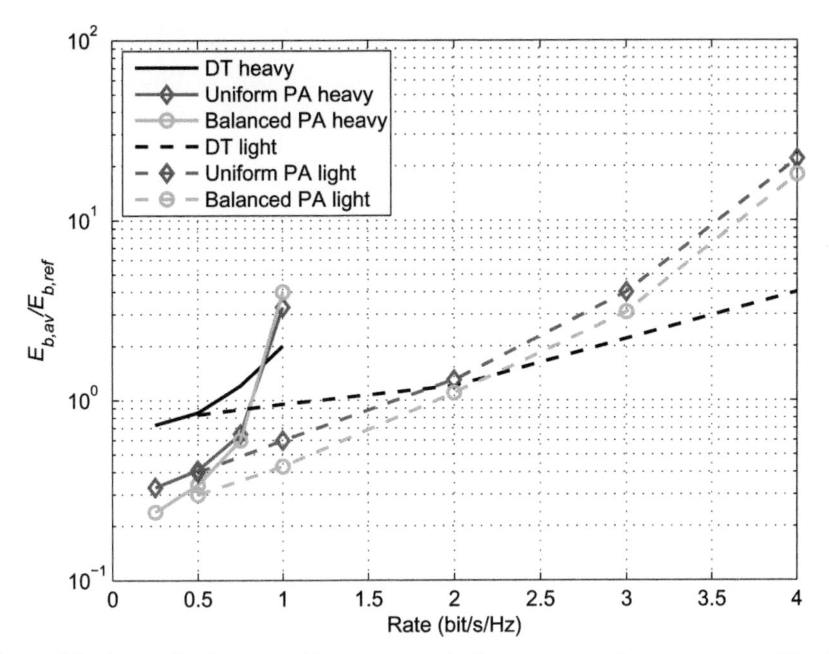

Figure 8.3 Normalised average bit energy required to guarantee the target rate at OP with power control.

account for transmit power only. Implementation of a dual-polarised antenna at the destination node is assumed.

The level of performance improvement for decode-and-forward (DF) relay systems is analysed, in terms of BER, accomplished by carefully planning the position of a relay node, to provide direct line-of-sight (LoS) signal component in both the source–relay link and in the relay–destination link. As the Ricean-fading K coeffcient increases, the SNR gain increases for the target BER. Therefore, it is desirable to provide LoS along both of the relay links. This result is aligned with the results of the DZB model.

If fading parameters along the source–relay and relay–destination links are different, the BER performance worsens. Typically, the relay–destination link has worse fading parameters, which motivates the implementation of polarisation diversity at the destination.

Two correlated and non-identical Ricean-fading channels are assumed for the relay-to-destination link, while maximal ratio combining (MRC) of the received signals is performed at the receiver. To determine effects of the polarisation diversity implementation, performance of the considered system is compared with the standard DF relay system with single antenna terminals. Using polarisation diversity, the same BER values can be obtained with significantly lower SNR, despite a certain level of correlation and power imbalance between the vertical and horizontal diversity branches. Compared to the system with no diversity, the total needed transmit power for achieving the same level of the relay system performance is reduced, Figure 8.4. Thus, increased EE is enabled.

8.3.3 EE Increase by Other Methods

When energy is harvested by wireless nodes from renewable sources, its availability becomes uncertain and its use for communications must be carefully designed. While the optimal power allocation has been derived in previous works when energy availability is fully known *a priori*, practical algorithms are needed when only causal and statistical information is available. In Zanella et al. [ZBMP13], the optimal transmission policy is studied

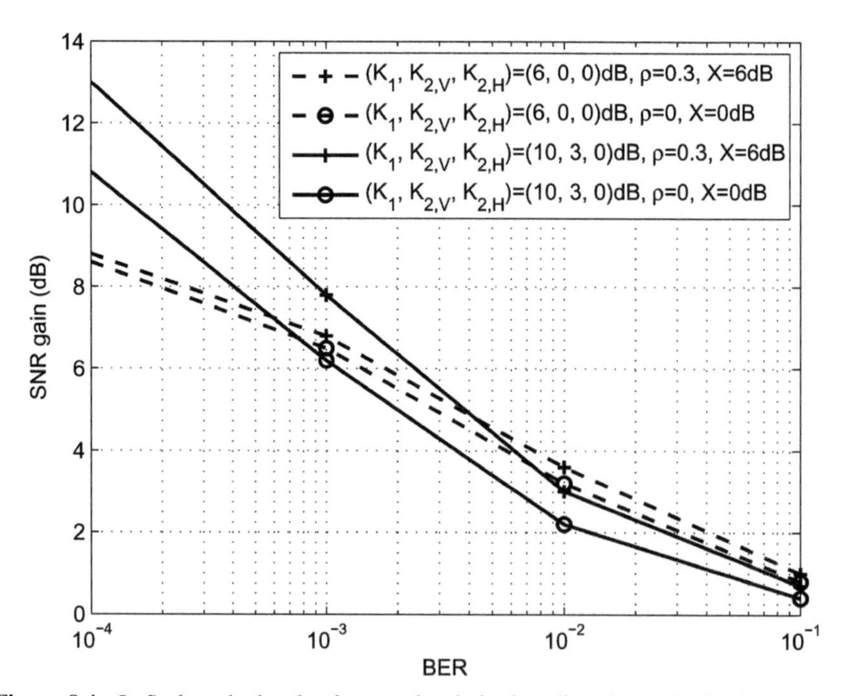

Figure 8.4 LoS along both relay hops and polarisation diversity at destination improve SNR gain and EE. K, Ricean coefficient; V, H, vertical, horizontal polarisation component; ρ, correlation coefficient between received signal envelopes; and X, cross-polar discrimination.

when only the statistical distribution of the energy arrival intervals is known and no information is available on the amount of energy that will be harnessed.

First, an exact solution for the case of a step-wise transmission power profile is obtained. This result is then extended to the time-continuous case. Within energy arrival intervals, the obtained power profile is shown to be non-increasing as a function of time, and non-decreasing as a function of the residual energy. For an exponentially distributed energy arrival process, the optimal policy can be expressed in a closed form. The numerical results for this case corroborate the main results. The optimal power allocation policy of Zanella et al. [ZBMP13] outperforms constant power allocation policy in every energy arrival interval, which increases the EE.

In Goncalves and Correia [GC12a], beamforming is studied to assess the efficiency in terms of radiated power that is possible to be saved when antenna arrays are placed instead of static sector antennas, for the universal mobile telecommunications system (UMTS) and long-term evolution (LTE) radio interface. Several multiple user scenarios were studied with different users' arrangements. Two simulators, one for UMTS and another for LTE, were developed to statistically evaluate the potential impact that adaptive antenna arrays have to reduce the radiated power, compared with actual BS static sector antennas. UMTS, besides signal improvement, has a lot of interference suppression potential due to its multiple access technique that separates users by codes in the same carrier frequency. LTE, due to the absence of co-channel inter-cell interference, is evaluated in terms of desired signal improvement. A model for UMTS is derived that describes the power improvement achieved as a function of the number of users and of radiator elements. For UMTS carriers near top capacity, a power reduction of the order of 90% is achievable. For LTE, significant power improvements are reached, especially for antenna arrays with eight elements, which can save near 65% of the radiated power.

The maximal number of transmission attempts in ARQ schemes is a positive integer, thus constraining the decision space [ZBD14]. A new retransmission scheme, r-ARQ, is proposed, where the expected value of the maximal number of allowed transmission attempts is a non-negative real number. Dependence of EE, throughput, delay, and jitter on the packet-loss and the expected value of maximal number of allowed transmission attempts have been derived. The packet-loss rate and the expected number of transmission attempts are compared for two examples of r-ARQ and traditional ARQ scheme. The examples show different behaviour, so that the r-ARQ should be designed specific to the optimisation problem.

8.4 MIMO Systems

This section addresses recent research in MIMO wireless communication systems. The research was focused on solving key weaknesses of MIMO systems, namely, (i) hardware and software complexity including needs for multiple RF chains and (ii) number of antennas and their spacing and size. Furthermore several documents looking at problems caused by distortions due non-ideal hardware and constraints due to limited feedback. This section is organised as follows. The first subsection looks at pre-coding techniques studying impact of realistic environment on its performance including limited feedback and realistic propagation environment. Two new approaches of applying pre-coding were also proposed: (i) considering CSI and data in pre-coding and (ii) usage of pre-coding for security in MIMO systems. The second and third subsection deals with spatial dimension of MIMO system namely spatial multiplexing and beamspace MIMO. Next section is devoted to the hardware impairments in MIMO system, while the topic of last section is adaptive MIMO systems.

8.4.1 Pre-Coding

The MIMO system consisting of very large number of antennas at the BS was evaluated in theoretical i.i.d Gaussian channel and measured residential-area channel in Gao et al. [GERT11] for three pre-coding scheme, namely dirty paper coding [APNG03], Zero-forcing pre-coding and minimum mean squared error (MMSE) pre-coding. It was found that the user channels, in the studied residential-area propagation environment, can be de-correlated by using reasonably large antenna arrays at the BS. With linear pre-coding, sum-rates as high as 98% of dirty paper coding capacity were achieved for two single-antenna users already at 20 BS antennas. This shows that even in realistic propagation environments and with a relatively limited number of antennas, we can see clear benefits with using an excessive number of BS antennas.

A new method was proposed to derive direction-of-arrival (DOA) information in a multiple-input single-output (MISO) or single-mode MIMO communications system that uses constellation-based limited feedback pre-coding in Colman [CWW12]. DOA estimation was accomplished by matching the distribution of selected codewords, $q(l)$, to a set of reference pre-coder index distributions generated using an idealised Ricean model under a range of K-factors. The DOA was estimated by minimising the Bhattacharyya distance between distributions. The method of estimating DOA was tested using time-varying channels generated with the WINNER2 rural LoS and non-line-of-sight (NLoS) models. The results presented in Figure 8.5 shows that good

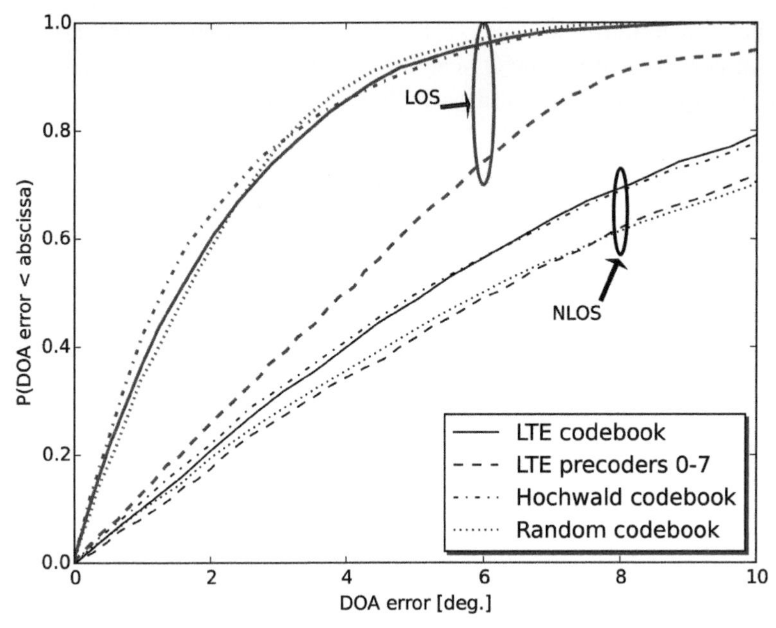

Figure 8.5 DOA error cdfs when generating $q(l)$ using 1000 samples over 4 m of mobile motion for the WINNER2 rural LoS, and NLoS scenarios.

direction estimates can be obtained with the algorithm presented, with median DOA estimation errors of less than 2°, regardless of the constellation used. Although the LTE codebook was designed with several directional pre-coders, the Hochwald constellation yields slightly better LoS error estimates.

The pre-coding approach was also applied to reduce the ability of a third-party eavesdropper to successfully receive pre-coded MIMO signals [Col13]. Instead of transmitting the pre-coding matrix chosen by the intended receiver, a secondary constellation of pre-coding matrices clustered around this selected matrix is used in a pseudo-random sequence, effectively transmitting with a spatially hopped signature. Polar-cap constellations proposed in Choi et al. [CCLK12] were suggested as a good candidate for the secondary constellation, which would be shared only among a desired group of users. The results presented in the document show significant degradation of eavesdropper reception while maintaining of the performance of the desired user nearly unchanged. This effective method to control eavesdropping comes at little additional system cost; both transmission and reception require few additional resources other than the knowledge of the sequence and scale of the spatial hopping behaviour.

The focus of Farah et al. [NEGJ11] lays on various spreading and pre-coding techniques for the MIMO scenario in LTE wireless communication systems. Two spreading/pre-coding schemes were compared, namely large-delay cyclic delay diversity (CDD) and multi-carrier cyclic antenna frequency spreading (MC-CAFS). In CDD, cyclically delayed replicas of one symbol are transmitted from the different antennas at the same time. In an OFDM system the CDD approach increases the space diversity. While for MC-CAFS, one symbol is sent over all the available subcarriers and over all available antenna ports which results in diversity in space and frequency. The simulation results show that the MC-CAFS spreading scheme shows a better performance compared to the CDD scheme by capturing more frequency diversity.

The concept of jointly utilising the data information and channel state for designing symbol level pre-coders for MISO downlink channel was proposed in Alodeh et al. [ACO15]. The interference causing by simultaneous data streams is transformed to constructive interference, which improves the signal to interference noise ratio at the receiver. The document proposes a new algorithm, which is based on maximum ratio transmissions algorithm. The simulation results plotting EE show the proposed algorithm superior performance to constructive rotation zero forcing pre-coding algorithm [Masll] on the expense of higher complexity due to the need for solving a non-linear set of equations.

8.4.2 Spatial Modulations

The spatial modulation (SM), introduced by Mesleh et al. [MHS+08], was proposed to overcome inter-channel interference and to increase the spectral efficiency. This is attained through the adoption of a new pragmatic transmission approach which employs (i) the activation of a single antenna that transmits a given data symbol (constellation symbol) at each time instance and (ii) the exploitation of the spatial position (index) of the active antenna as an additional dimension for data transmission (spatial symbol). Both the constellation symbol and the spatial symbol depend on the incoming data bits. The spectral efficiency is increased by the logarithm to base 2 of the number of transmit antennas. The number of transmit antennas must be power of two. This approach resulting in the simplified hardware requirements of MIMO communication systems and avoids inter-channel interference, while still benefiting from increased spectral efficiency.

The simple receiver hardware requirements of the spatial modulation is well suited to mobile terminals and initiated a research of its application as a fourth generation (4G) wireless communication techniques [WMJ+]. A sensitivity of spatial modulation to variations in sub-channel properties using both statistical channel models and measured urban channel data was investigated in Thompson et al. [WAM+]. The uncoded system performance was studied when transmitting over multi-antenna channels with variable correlations between spatial sub-channels and sub-channel power balance. The analysis aimed at finding out whether the variations in system performance over the measured channels can be recreated with the simple channel model used to create the synthesised channel model. The technique applied to create the synthesised channel is based on Kroneker model applying independent and identically distributed random variable, with complex normal distribution to create un-correlated MIMO channel and correlation matrix to introduce correlation between sub-channels, while the measured channel was obtained by extensive multi-antenna channel measurement in Bristol at 2 GHz with 20 MHz bandwidth using channel sounders configured as a 4×4 MIMO system. The simulation results when varying sub-channel correlation showed significantly more variation in BER results for the synthesised channels compared to the measured ones. The results for the spatial modulated system BER sensitivity to sub-channel branch power variations for the synthesised channels showed a strong correlation between the chosen measure of branch power imbalance and the SNR required for a specific BER performance. This strong relationship was not found for the measured channels (Figure 8.6).

8.4.3 BeamSpace MIMO

Beam-space MIMO system was proposed in Kalis et al. [KKP08] as a means to address the two key weaknesses of conventional MIMO systems namely the antenna size and the need for multiple RF chains. The research effort on Beam-space MIMO focuses on the development of functional MIMO transmission schemes with the use of a single RF chain while maintaining small antenna size using on electronically steerable passive array radiators (ESPARs). Beam-space MIMO applies a set of orthonormal basis antenna patterns to provide efficient multiplexing and beamforming, further using small-sized antenna arrays. Beam-space MIMO systems outperform conventional MIMO systems in terms of system capacity Kalis et al. [KKP08]. Research presented in Maliatsos et al. [MVK13a] and [MVK13b] makes first steps toward practical system design and focuses on basic channel estimation and a design and implementation of V-BLAST receiver for Beam-space MIMO. Exploiting the fact that there is equivalent algebraic representation of BS and conventional

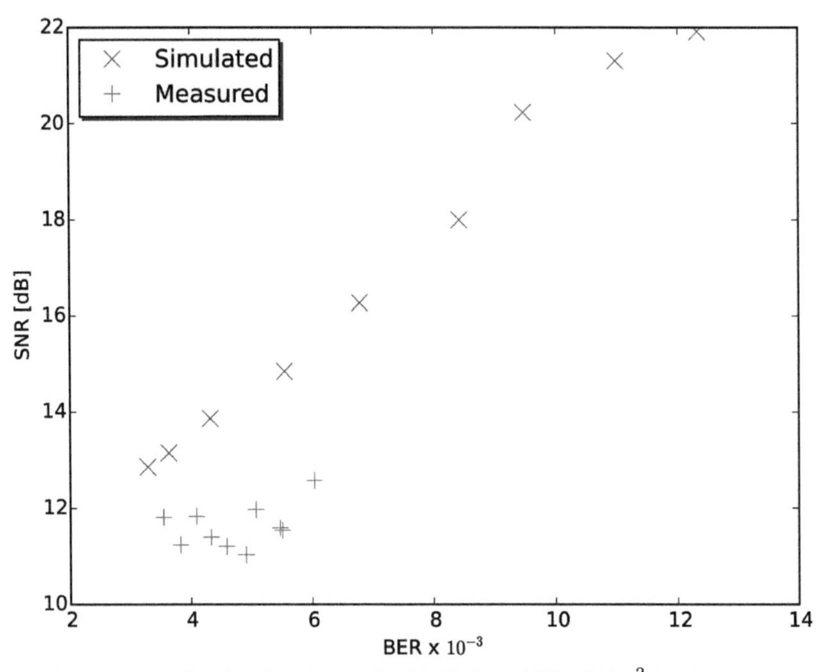

Figure 8.6 Scatter plot showing the required SNR for a BER of 10^{-3} for the measured and synthesisd channels.

MIMO systems, the algorithms and the training sequences applied in conventional MIMO systems are transferred in beamspace domain. Least Square and MMS estimators were applied to find channel estimates, which were used for basis antenna pattern reconfiguration. Beam-space MIMO system with channel estimation was evaluated applying system level simulations using rich scattering wireless world initiative new radio (WINNER) âĂŹB2âĂŹ channels. The performance comparison between three-element Beam-space MIMO and conversional MIMO system is shown in Figure 8.7. It is clear that Beam-space MIMO systems outperform conventional MIMO for small inter-element distances despite the fact that noise power is increased due to oversampling. However, it became clear [MVK13a] that this weakness limits Beam-space MIMO performance in low and medium SNRs.

In real-world systems the transmitter is not able to have full access to CSI. This fact initiated studies in a limited feedback technique in Beam-space MIMO based on conventional MIMO algorithm with limited feedback [MVK13b]. The codebooks from the beamspace channel partitioning study

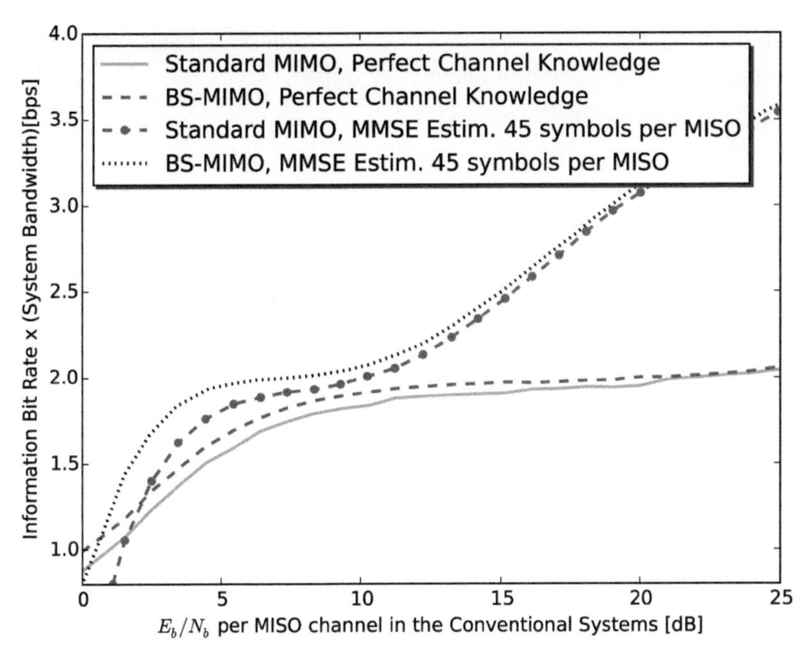

Figure 8.7 Performance comparison between three element Beam-space MIMO and conventional MIMO systems in terms of information bit rate in the PHY.

were analysed to map the computed patterns in reactance values providing a guide for the accuracy requirements in the antenna circuit design. Since CSI in the Transmitter (CSIT) is limited, the well-known V-BLAST reception (with MMSE detection) is adapted in the beamspace and adaptive pattern reception schemes were presented and analysed. The simulation results show a gap between system with complete CSIT (CSI at the transmitter) and the V-BLAST receiver without CIST, while the systems with limited feedback (4 and 5 bits) outperform conventional V-BLAST algorithm [MVK13b].

8.4.4 Hardware Realisation and Interference Impact on the Performance of MIMO Systems

Hardware implementation of MIMO system may significantly degrade its performance. Phase noise, non-linearities in transmission chain and I&Q imbalance have huge impact. The results in Zetterberg [Zetll] and Bjornson et al. [BZBO13] reveal that the MIMO systems distorted by hardware impairments have a finite capacity limit in the high-SNR regime—this is fundamentally different from the unbounded asymptotic capacity for ideal transceivers. Furthermore, the bounds in Equation (8.6)

$$M\log_2\left(1+\frac{1}{\kappa^2}\right) \leq C_{N_t N_r}(\infty) \leq M\log_2\left(1+\frac{N_t}{M\kappa^2}\right) \tag{8.6}$$

where N_r number of receiver antennas, N_t number of transmitter antennas, $M = \min(N_t, N_r)$, and κ is level of impairments. The bounds hold for any channel distribution and are only characterised by the number of antennas and the level of impairments.

The BER performance of MIMO-OFDM/TDM based on MMSE-FDE in a peak-limited and frequency selective fading channel was analysed in Ligata et al. [LGJ14]. The analysis was based on the Gaussian approximation of the residual intersymbol interference (ISI) after MMSE–FDE and the noise due to peak limitation. The theoretical MMSE equalisation weights for MIMO-OFDM/TDM in a peak-limited channel were derived to capture the negative effect of HPA saturation. The performance was evaluated by both numerical Monte-Carlo method and theoretical analysis. The theoretical results confirmed that the BER performance of MIMO-OFDM/TDM using MMSE-FDE in a peak-limited and frequency-selective channel is a function of the OFDM/TDM system design, i.e. K on number of block in OFDM/TDM system. This is also illustrated in Figure 8.8, where L denotes number of reflected ray and ϕ high-power saturation level.

An evaluation of the information theoretic capacity of MIMO communication system with the presence of interference is given in Webb et al. [WBN04]. Three different assumptions of CSI of both desire signal and interference signal was considered: namely (i) full information of channel and interference information are known at the transmitter, (ii) neither channel nor interference information are known at the transmitter, and (iii) only channel information without interference is known at the transmitter. Outage capacity with varies numbers of interferers, array size and SINR power was compared. Alignment of interferer subspace gives an upper and lower bound for the MIMO capacity with interference.

8.4.5 Adaptive MIMO Systems

Adaptive modulation is shown to be an effective approach to increase the effective data rate in multi-carrier communication systems [CCB95]. However; the transmitter and the receiver should have a perfect knowledge about the bit allocation per subcarriers. Transferring bit allocation per subcarriers from

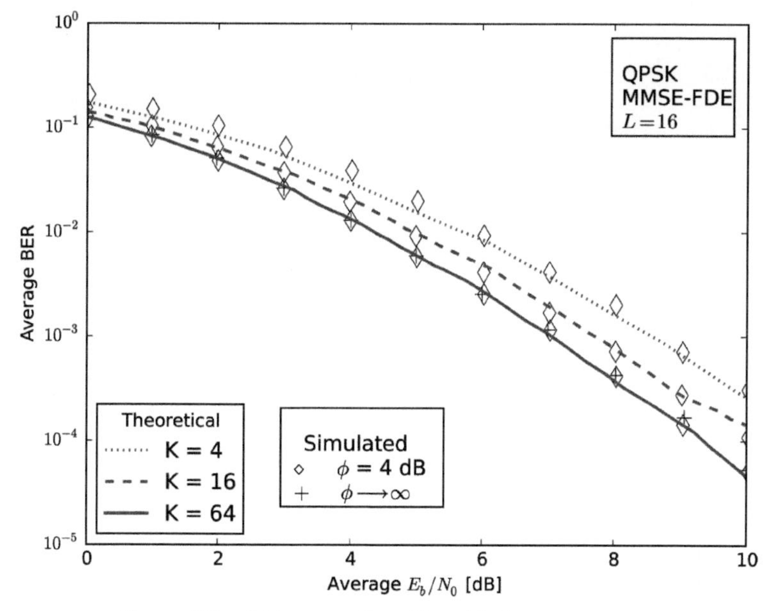

Figure 8.8 BER versus $E_b = N_0$ with K as a parameter.

transmitter to the receiver may significantly reduce the effective link throughput. A signal-assisted maximum *a posterior* (MAP) modulation classification algorithm was proposed in Haring and Kisters [HK14] in time-division-duplex wireless communications. The method utilises the received signal form, truncated signalling information and channel reciprocity in wireless TDD systems. Simulations show a significant system performance improvement of the proposed schemes over conventional schemes applied in typical indoor environment (Figure 8.9).

8.5 Modulation and Coding

This chapter summarises the results of the COST IC1004 project obtained in the fields of modulation and coding. An important aspect has been the application of iterative methods. In Section 8.5.1 iterative methods are utilised for equalisation, leading to a low complexity solution and a dramatic improvement of the power-to-speed ratio. Section 8.5.2 deals with iterative (turbo) detection for concatenated systems. In Section 8.5.3 advanced coding schemes and iterative decoding methods are analysed. Section 8.5.4 finally considers the non-coherent transmission methods differential phase shift keying (DPSK) and OFDM-MFSK.

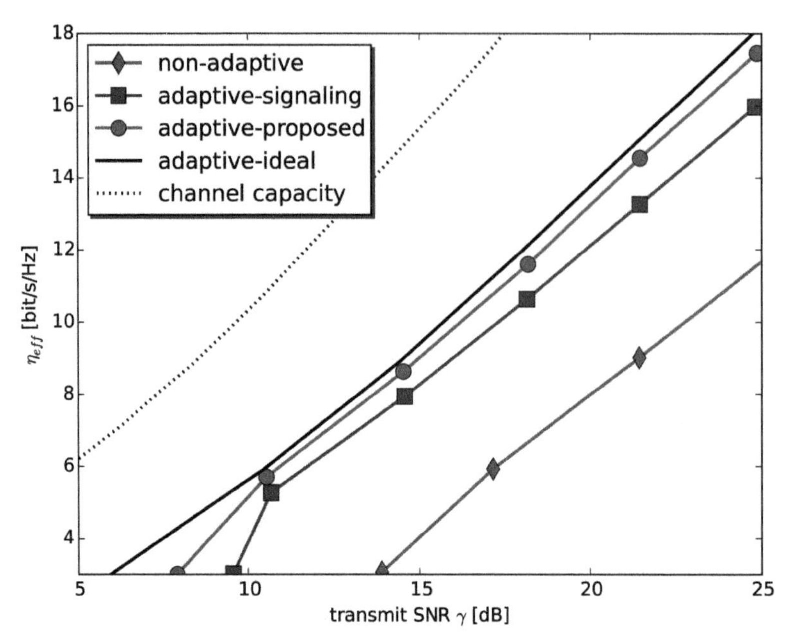

Figure 8.9 Effective bandwidth efficiency η_{eff} at target frame error rate (FER) $FER_{\text{target}} = 10^{(1)}$ versus transmit SNR for 4×4 SVD transmission.

8.5.1 Iterative Interference Cancellation

Multipath propagation on the physical channel between transmitter and receiver leads to ISI. In case of multiuser, multi-subchannel, multi-antenna transmission systems, and combinations thereof, additional interference emerge [Lin99]. To cope with this interference, equalisation has to be applied at the receive side. Because of the high computational complexity of the optimum equaliser, suboptimum schemes are usually utilised, mainly iterative ones because of their good complexity–performance trade-off. We introduce here recurrent neural network (RNN) as a suboptimum IIC technique.

Discrete-time RNN can be adapted to perform IIC without the need for a training phase. Stability and convergence proofs in this case are available. Properties without which an iterative scheme often is suspect. The same applies for continuous-time RNNs, which serve as promising computational models for an analogue hardware implementation. The later one combines reduced energy consumption with the fast signal processing required by the IIC for high-data rate transmission. An improvement of the power-to-speed ratio of up to four orders of magnitude compared to a standard digital solution has been obtained [OSM^{+}14].

8.5.1.1 Transmission model

The uncoded discrete-time (on symbol basis) vector-valued block transmission model for linear modulation schemes is shown in Figure 8.10.

In details:

- SRC (SNK) represents the vector-valued digital source (sink).
- $\mathbf{q}(\hat{\mathbf{q}}) \in \{0,1\}^{N_\mathrm{q}}$ is the vector of source (hard estimated) bits of length N_q.
- $\hat{\mathbf{s}} \in \psi^{N_\mathrm{t}}$ is the (hard estimated) transmit vector of length N_t. $\psi = \{\psi_1, \psi_2, ..., \psi_{2^m}\}$ represents the symbol alphabet where $m \in \mathbb{N}/\{0\}$. $M = 2^m$ represents the cardinality of the symbol alphabet. In this case, there exist M^{N_t} possible transmit vectors. The mapping from $\mathbf{q}$ to $\mathbf{s}$ is bijective and performed by $\mathcal{M}$. We notice that the number of possible transmit vectors increases exponentially with the block size N_t.
- $\mathbf{r}$ is the receive vector of length N_t.
- $\mathbf{R}_\mathrm{s}$ is the (discrete-time) channel matrix for the block transmission of size $N_t \times N_\mathrm{t}$. It includes the transmit and receive filter as well as the impulse response of the channel. Using a channel matched filter as receive filter, $\mathbf{R}_s$ is hermitian and positive semi-definite.
- $\mathbf{n}$ is a sample function of an additive Gaussian noise vector process of length N_t with zero mean and covariance matrix $\mathbf{\Phi} = \frac{\mathrm{N}_0}{2} \cdot \mathbf{R}_\mathrm{s}$. $\frac{\mathrm{N}_0}{2}$ is the double-sided noise power spectral density.
- IC: Interference cancellation (a suboptimum equalization technique).

The model in Figure 8.10 is a general model and fits to different transmission schemes like OFDM, code division multiple access (CDMA), Multicarrier code division multiple access (MC-CDMA), MIMO and MIMO-OFDM. The relation with the original physical continuous-time model can be found in Lindner [Lin99]. Mathematically it is described as follows

$$\begin{aligned} \mathbf{r} &= \mathbf{R}_\mathrm{s} \cdot \mathbf{s} + \mathbf{n} \\ \mathbf{r} &= \underbrace{\mathbf{D} \cdot \mathbf{s}}_{\textbf{Signal}} + \underbrace{\mathbf{C} \cdot \mathbf{s}}_{\textbf{Interference}} + \underbrace{\mathbf{n}}_{\textbf{Additive noise}} \end{aligned} \tag{8.7}$$

$\mathbf{D}$ contains only the diagonal elements of $\mathbf{R}_\mathrm{s}$. Off-diagonal elements of $\mathbf{D}$ are zeros. $\mathbf{C}$ contains only the off-diagonal elements of $\mathbf{R}_\mathrm{s}$. Diagonal elements of

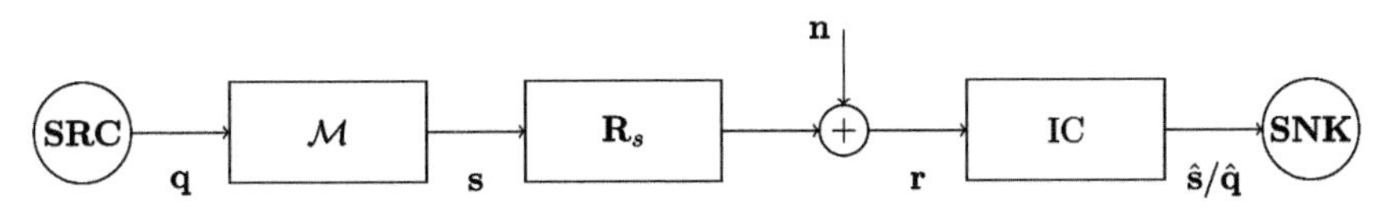

Figure 8.10 Uncoded discrete-time (on symbol basis) vector-valued block transmission model for linear modulation schemes.

$\mathbf{C}$ are zeros. We notice that the non-diagonal elements of $\mathbf{R}_\mathrm{s}$, namely $\mathbf{C}$ lead to interference between the elements of the transmit vector at the receive side (cf. Equation (8.7)). For interference-free transmission $\mathbf{C} = 0$.

The interference cancellation (suboptimum equalization) in Figure 8.10 has to deliver a vector $\hat{\mathbf{s}}$ which must be as similar as possible to the transmit vector $\mathbf{s}$ given that ψ and $\mathbf{R}_\mathrm{s}$, are known at the receive side. Mapping $\hat{\mathbf{s}}$ to $\hat{\mathbf{q}}$ for an uncoded transmission is straightforward. Equalisation can be seen as a classification process and represents a non-linear discrete optimisation problem.

8.5.1.2 IIC: A generic structure

The optimum equalizer, the maximum likelihood (ML) one, calculates for every receive vector $\mathbf{r}$ a distance (more precisely the Mahalanobis distance because of the coloured noise $\mathbf{n}$) to all possible transmit vectors $\mathbf{s}$ of cardinality M^{N_t} and decides in favour of that transmit vector $\hat{\mathbf{s}}$ with the minimum distance to the receive vector $\mathbf{r}$. Thus, the computational complexity of the optimum equalizer is too high for a realistic block length N_t. The computational complexity in general increases exponentially with the block length N_t. Therefore suboptimum equalizer schemes, mainly iterative ones, are applied. Figure 8.11 shows a generic structure of IIC.

The basic idea is to obtain an estimate (soft or hard) of the transmit vector. Based on this estimate, the interference is recreated (linear function of the estimate). The rebuild interference is then subtracted from the receive vector $\mathbf{r}$. This process is repeated on symbol or vector basis (serial or parallel update). Depending on the estimation function we distinguish:

- *Linear* IIC: The estimation function is linear, as an example the zero-forcing block linear equaliser (ZF-BLE) [HYKY03].
- *Non-linear* IIC: The estimation function is non-linear, as an example the multistage detector (MSD) [VA90].

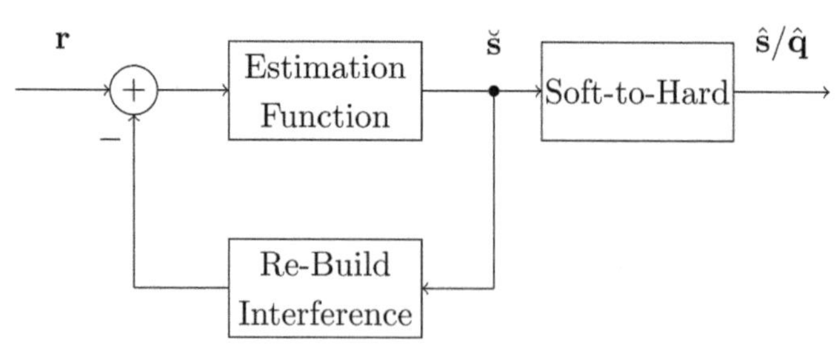

Figure 8.11 A generic structure of IIC, $\mathbf{r}$ is the receive vector, $\breve{\mathbf{s}}$ is the "soft" estimated transmit vector, $\hat{\mathbf{s}}$ is the "hard" estimated transmit vector.

- *Hard* IIC: The elements of $\breve{\mathbf{s}}$ belong to ψ. In this case, there is no need for the soft-to-hard block in Figure 8.11.
- *Soft* IIC: The elements of $\breve{\mathbf{s}}$ do not belong necessarily to ψ.

The superiority of non-linear soft IIC is widely accepted.

8.5.1.3 Recurrent neural networks

This class of neural networks has been attracting a lot of interest because of their widespread applications. They can be either trained to approximate a MIMO system ([GM04]; system identification), or they can be considered as dynamical systems. In the later case, one of the most important properties of these networks is their ability to solve optimisation problems without the need for a training phase, which is desirable in many engineering fields like signal processing, communications, automatic control, etc. In this case, the training process, always associated with computational complexity, time and free parameter optimisation can be avoided. This relies on the ability of these networks (under specific conditions) to be Lyapunov-stable. These conditions have been shown to be fulfilled for iterative interference cancellation-based recurrent neural network (IIC-RNN).

8.5.1.4 Discrete-time RNNs for IIC

The discrete-time IIC-RNN is well known in the literature [ETL+02]. The iterative process of the discrete-time IIC-RNN is described as

$$\begin{aligned} \mathbf{u}[\kappa] &= [-\mathbf{D}^{-1}\cdot\mathbf{C}]\cdot\breve{\mathbf{s}}[\kappa-1]+\mathbf{D}^{-1}\cdot\mathbf{r}, \\ \breve{\mathbf{s}}[\kappa] &= \boldsymbol{\theta}(\mathbf{u}[\kappa]). \end{aligned} \tag{8.8}$$

$\boldsymbol{\theta}(\cdot)$ is the non-linear optimum estimation function given for each component as [Eng03]:

$$\theta(u) = \frac{\sum_{j=1}^{M}\psi_j\cdot\exp\{-\frac{1}{2}\cdot(\beta_r\cdot\psi_{j,r}^2+\beta_i\cdot\psi_{j,i}^2)}{\sum_{j=1}^{M}\quad\exp\{-\frac{1}{2}\cdot(\beta_r\cdot\psi_{j,r}^2+\beta_r\cdot\psi_{j,r})}\;\frac{+\beta_r\cdot\psi_{j,r}\cdot u_r+\beta_i\cdot\psi_{j,i}\cdot u_i\}}{+\beta_r\cdot\psi_{j,r}\cdot u_r+\beta_i\cdot\psi_{j,i}\cdot u_i\}} \tag{8.9}$$

The subscripts *r*, *i* indicate the real and imaginary part of a complex-valued variable, respectively. β_r, β_i are positive free parameters, usually inversely proportional to the variance of the noise.

Successive over relaxation (SOR) is a well-known method to improve the convergence of linear iteration processes. SOR has been applied for the IIC-RNN Equation (8.8) in Mostafa et al. [MTL10, TW12]. Particularly in Teich and Wallner [TW12] SOR for discrete-time IIC-RNN has been applied for digital transmission schemes based on multiple sets of orthogonal spreading codes. It has been shown that the bandwidth efficiency can be increased by up to 50% without loss in power efficiency. Resulting interference between different non-orthogonal codes can be combated by IIC-RNN. By applying SOR to IIC-RNN the bandwidth efficiency can be increased even a bit further (without loss of power efficiency). Even more, the complexity of IIC-RNN (number of iterations) could be reduced substantially by the SOR.

The local asymptotical stability of the discrete-time IIC-RNN with/without SOR has been proven in the sense of Lyapunov in Mostafa et al. [MTL14c].

8.5.1.5 Continuous-time RNNs for IIC

For high data rates, the power consumption of iterative detection algorithms (equalization and channel decoding) is expected to become a limiting factor. The growing demand for jointly high data rate transmission and power efficient detection revives the analogue implementation option. Continuous-time RNNs serve as promising computational models for analogue hardware implementation [CU93]. The continuous-time IIC-RNN is described by the following differential equation

$$\frac{d\mathbf{u}(t)}{d_t} = -\mathbf{u}(t) + [-\mathbf{D}^{-1} \cdot \mathbf{C}] \cdot \check{\mathbf{s}}(t) + \mathrm{D}^{-1} \cdot \mathbf{r},$$
$$\check{\mathbf{s}}(t) = \boldsymbol{\theta}(\mathbf{u}(t)). \tag{8.10}$$

First work toward continuous-time IIC-RNN was limited to binary phase shift keying (BPSK) modulation [KM96]. This has been extended to square quadrature amplitude modulation (QAM) in Mostafa et al. [MTL12, Mosl4]. The local asymptotical stability proof in the sense of Lyapunov is given in Kuroe et al. [KHM02] and has been extended in Mostafa [Mosl4].

An initial analogue implementation of the continuous-time IIC-RNN for a BPSK symbol alphabet [OSM$^+$14] showed an improvement of the power-to-speed ratio of up to four orders of magnitude, compared to a standard digital solution, e.g., a low-power Intel XScale central processing unit (CPU): clock frequency 600 MHz, power consumption 0.5 W.

The analogue implementation of the continuous-time IIC-RNN requires an analogue implementation of the optimum estimation function Equation (8.9). It has been proven in Mostafa et al. [MTL14a] that the optimum estimation function for square QAM symbol alphabets can be approximated by a sum of a limited number of properly shifted and weighted hyperbolic tangent functions, which can be realised in analogue by differential amplifiers. A comparison between the dynamical behaviour of the discrete-time and continuous-time IIC-RNN can be found in Mostafa et al. [MTLllb, MTL12].

8.5.2 Concatenated Systems, Bit-Interleaved Coded Modulation (BICM) and Iterative Detection

The mobile evolution of the second generation digital terrestrial television broadcasting is the scenario considered in Vargas et al. [VWMG12]. digital video broadcasting-next generation handheld (DVB-NGF) is a MIMO system with BICM. Specifically an iterative receiver concept is employed, where the MIMO demapper and the channel decoder iteratively exchange extrinsic information (turbo detection). The focus is on implementation aspects such as suboptimal demodulators (soft MMSE demodulation with *a priori* information from the decoder) and quantisation effects (optimum LLR quantisation). FER simulations were performed for an outdoor mobile scenario with a user velocity of 60 km/h. For a low-code rate of 1/3, the soft MMSE demodulator shows no performance degradation compared to a much more complex max-log reference receiver. For a high-code rate of 11/15, the soft MMSE demodulator suffers from a performance loss of up to 2 dB compared to the max-log receiver. With turbo detection SNR gains between 1 and 2 dB can be achieved.

The system considered in Wu et al. [WAM14] is an interleave division multiple access (IDMA) uplink over frequency selective fading channels in the low-power regime. Performance close to capacity is achieved by using optimised block-interleaved coded modulation-iterative detection (BICM-ID) with a very low-rate single parity check code concatenated with an irregular repetition code together with a doped accumulator and extended mapping. The parameters of the system (code parameters, doping ratio, and modulation mixing ratio) have been optimised for an additive white Gaussian noise (AWGN) channel with an extrinsic information transfer chart (EXIT) constrained binary switching algorithm. As detector, a joint frequency domain turbo equalisation and IDMA signal detection algorithm is proposed. With this optimised system and for frequency selective fading channels, a performance close to the OP can be achieved. For a multiuser scenario, the detector complexity can be reduced substantially by implementing detection ordering according to the received power.

He et al. [HZAM13] considered a parallel wireless sensor network (WSN). A non-negative constrained iterative algorithm is presented to estimate the observation error probabilities for WSN with an arbitrary number of sensor nodes. The idea is to exploit the correlation among the observed data from different sensor nodes, representing the correlation of the links between the sensing object and the sensor nodes. To utilise these correlations, a multi-dimensional iterative receiver is proposed. The estimated observation error probabilities expressed as LLR are exchanged between the decoders (global iteration). As usual, each decoder performs an extrinsic LLR exchange (local iteration). For an AWGN channel Monte Carlo simulations show, that for large sensor networks the required SNR to achieve a given error rate can be improved by up to 9 dB by using a sufficient number of global iterations. For a Rayleigh-fading channel a SNR improvement of about 4 dB is obtained at a BER of 10^{-3}.

Az et al. [AAM13] also considered WSN. A detection technique based on factor graphs is proposed to estimate the position of illegal radio in a wireless system. To improve the geo-location estimation, received signal strength (RSS)- and DOA-based factor graphs are combined in a single factor graph with empirically found weighting factors. The performance of DOA is improved by introducing a modified variance approximation of the target location based on a Taylor series expansion of the tangent functions.

8.5.3 Advanced Coding Schemes and Iterative Decoding

An analogue realisation of iterative decoding algorithm based on high-order RRNs is the work presented in Mostafa et al. [MTLlla] and [MTL14b]. Besides a large decoding speed, the main advantages of an analogue very large-scale integration (VLSI) implementation are the improved EE (higher power-to-speed ratio) and, going hand in hand with that, the reduced area consumption compared to a digital VLSI implementation. This is similar as in the case of an analogue realisation of IIC (see Section 8.5.1). Starting from the iterative threshold decoding algorithm, Mostafa et al. [MTLlla] derives the specific structure and parameters so that a high-order RNN can serve as an analogue decoder:

- each code symbol is represented by one neuron;
- the *L*-value of each code symbol is represented by the inner state of the neuron;
- each neuron has an external input given by the intrinsic *L*-value of the corresponding code symbol;
- the non-linearity of the neuron is a tangent hyperbolic function with slope equal to 0.5;
- the code structure is represented in the non-linear feedback matrix and the non-linear feedback vector function of the network

The equation of motion of the resulting continuous-time dynamical system is given by a non-linear system of first-order differential equations. For a specific example (tail-biting convolutional code with memory one), simulation results show, that the BER performance of the analogue decoder (continuous-time dynamical system) is equivalent to the one of the digital counterpart (discrete-time dynamical system) and even matches ML performance.

In Mostafa et al. [MTL14b], the iterative decoding algorithm *belief propagation* and *iterative threshold decoding* are described as dynamical systems, where both, discrete-time and continuous-time systems are considered. Specifications for the stability of fix points or equilibrium points are given for the linear as well as the non-linear case. Furthermore, conditions are derived, so that the continuous-time dynamical system shares the same (stable) equilibrium points as the discrete-time dynamical system (twin dynamical system). This eventually allows to represent an iterative decoding algorithm (discrete-time dynamical system) by an analogue VLSI realisation. For repetition codes the dynamics of belief propagation and iterative threshold decoding is linear and a closed form solution for the stability of the equilibrium/fixed points is given.

Hu et al. [HKD11] considered the IEEE 802.11a WiFi system. Based on a modification of the BCJR algorithm the authors propose to improve the forward error correction (FEC) decoding in the PHY by exploiting known or predictable symbols from higher open system interconnection (OSI) layers. For the MAC header error rate an improved power efficiency of 1.5 dB was obtained. However, for the overall MAC FER the gain reduced to 0.35 dB. Introducing an additional interleaving to better distribute the known symbols, the gain could be increased to about 1 dB.

A theoretical performance analysis of a code based on the concatenation of a low-density parity check (LPDC) code and a rateless code is given in Botos and Bota [BB12]. The approximate analysis of various parameters is performed for a transmission over a Rayleigh-fading channel. The performance metrics considered are the probability of message non-recovery at a given SNR or the minimum SNR required to ensure an imposed probability of message non-recovery, the spectral efficiencies, and delay (time required to successfully transmit a message). Parameter of the evaluation are the message length and overhead and depth of the rateless code.

8.5.4 Non-Coherent Detection

In Zhu and Burr [ZB12], the authors consider a serially concatenated system composed of an (outer) recursive systematic convolutional (RSC) code concatenated with DPSK modulation. To compensate for the loss due to non-coherent detection, Zhu and Burr [ZB12] propose to employ a turbo receiver. To do so they view the differential encoder of the DPSK modulator as a rate one recursive non-systematic convolutional encoder (accumulator), which can be represented by a two-state trellis. In the iterative receiver, soft information is exchanged between the phase shift keying (PSK) soft demapper, the differential decoder of the inner code and the RSC decoder of the outer code. Simulation results are given for a correlated Rayleigh-fading channel. Depending on the number of iterations the power efficiency could be improved by up to 5 dB, compared to a coded PSK scheme with non-iterative detection. Compared to the case of having perfect CSI at the receiver, introducing a simple *a posteriori* probability (APP) channel estimator leads to a performance degradation of 0.5–1 dB.

Peiker-Feil et al. [PFWTL12] and [PFTL13] also consider a non-coherent transmission scheme. OFDM-MFSK has been proposed as a robust transmission scheme suitable for (fast) time-varying and frequency selective channel [Wet 11]. OFDM is concerned with the frequency-selective behaviour of the channel, whereas M-ary frequency shift keying (MFSK) allows an incoherent detection and is thus able to cope with a (fast) time-variance of the channel. Peiker et al. [PFWTL12] show that OFDM-MFSK can be described as a non-coherent transmission based on vector sub-spaces. This gives new insight in the design and analysis of advanced non-coherent transmission schemes based on OFDM-MFSK [PF14].

One of the major disadvantages of OFDM-MFSK is its poor bandwidth-efficiency. This problem is addressed in Peiker-Feil et al. [PFTL13]. Combining MFSK with multi-tone frequency shift keying (FSK) the bandwidth efficiency of OFDM-MFSK could be increased by 50%, compared to classical OFDM-MFSK. At the same time, the loss in power efficiency in an AWGN channel was limited to about 1.5 dB.

8.6 Link Abstraction

8.6.1 On the Need for Link Abstraction

System-level simulators of mobile communication systems often rely on simplified look-up-tables (LUTs) generated offline by a link-level simulator. Link abstraction techniques aim at obtaining LUTs to predict the block error rate (BLER) for multistate channels. A multistate channel arises when the received LLRs, within a given codeword, show very different reliabilities due to one or several reasons such as: (i) the frequency selective fading in OFDM, (ii) the LLR combination prior to hybrid ARQ (H-ARQ) decoding, and (iii) the unequal error protection with high order modulations in BICM. BLER prediction is needed at the UE and at the BS to decide about the most suitable modulation and code rate to apply for next transmission time interval (TTI) and also for large scale system evaluations.

8.6.2 Truncated Shannon Bound

The Shannon bound gives an upper limit on the throughput of a communication link. The state of the art in modulation and coding schemes is a few dB from this bound, since the Shannon bound can be reached only in the limit of infinite block length and decoding complexity. The throughput of practical modulation and coding schemes used in wireless standards, like LTE/LTE-A and worldwide interoperability for microwave access (WiMAX), can be approximated by a function similar to the Shannon bound, but shifted to the right and/or compressed on the vertical scale by a scaling factor less than 1.

Taking into account that in a real system the capacity is zero below a minimum SNR threshold (γ_0) and is upper bounded by the capacity of the highest MCS (C_{max}) for SNR higher than γ_{max}, the expression for the truncated Shannon bound (TSB) throughput is [Bur11]:

$$C_{\mathrm{TSB}}(\gamma) = \begin{cases} 0 & \gamma < \gamma_0 \\ W\alpha \log_2(1+\gamma/\gamma_{\mathrm{sh}}) & \gamma_0 \leq \gamma < \gamma_{\mathrm{max}} \\ C_{\mathrm{max}} & \gamma_{\mathrm{max}} < \gamma \end{cases} \tag{8.11}$$

where α and γ_{sh} must be adjusted to optimally approximate the actual throughput function.

Given that the SNR is random, the condition to apply to find α and γ_{sh} is that the average throughput given by the TSB should be as close as possible to the average throughput given by the true throughput function ($C_{\mathrm{thr}}(\gamma)$), which depends on the specific SNR thresholds of each MCS and its corresponding throughputs. The average throughputs are computed by averaging $C_{\mathrm{TSB}}(\gamma)$ (resp. $C_{\mathrm{thr}}(\gamma)$) over all possible values of the SNR.

In a MIMO link, we must also take into account the random nature of the channel matrix and the fact that the direction of the interference is relevant [Bur03]. To obtain the average throughput of a MIMO link the TSB can be applied on a stream basis (given the cumulative distribution function (CDF) of the SNR experienced on each stream). In general, the CDF of the stream SNR has to be estimated using link-level Monte Carlo simulation.

8.6.2.1 Link abstraction for multi-user MIMO (MU-MIMO)

Assuming an OFDM system with N subcarriers and a SINR vector $\{\gamma_1, \gamma_2, ..., \gamma_N\}$, the received bit information rate (RBIR) is defined as the average information capacity of a coded bit in the received codeword, i.e.:

$$\mathrm{RBIR} = \frac{1}{\beta} \frac{\sum_{\kappa=1}^{N} SI(\gamma_\kappa)}{\sum_{\kappa=1}^{N} M_\kappa} \quad \text{(bits/coded bit)}, \tag{8.12}$$

where $\mathrm{SI}(\gamma_\kappa)$ and M_κ are, respectively, the mutual information (MI) at modulation symbol level and the modulation order of subcarrier κ, and β is a parameter that accounts for the capacity loss of a practical code.

In a MU-MIMO link, the function SI(·) in Equation (8.12) can include the known interference statistics in the computation of the MI at modulation symbol level, thus modelling the performance of an "interference aware" receiver [LKK11].

From the RBIR the effective signal-to-noise ratio (ESNR), defined as the SNR that would give the same RBIR under AWGN channel [BAS+05] can be obtained. The ESNR allows BLER prediction using AWGN BLER LUTs.

8.6.2.2 Link abstraction for H-ARQ

In LTE, incremental redundancy (IR) H-ARQ each newly received redundancy version (RV) is soft-combined with the previous ones to increase the decoding probability. The key issue is to find an approximated method to compute the RBIR at the turbo decoder taking into account that any coded bit may have been received more than once and that in each transmission it may have experienced a different quality, due to the unequal error protection of 16 and 64 QAM and to the changing SNR conditions of the different subcarriers and H-ARQ RVs.

Let's assume that a sample coded bit is transmitted exactly twice using 16QAM in two different subcarriers with $\mathrm{SNR}_1 = \xi_1$ and $\mathrm{SNR}_2 = \xi_2$. In 16QAM the bit to symbol mapping creates two different channel qualities: A and B. Assuming, for example, that the first transmission uses bit channel A and the second transmission uses bit channel B, a possible procedure to obtain the RBIR contribution of the sample coded bit is [OSRL12]:

$$\mathrm{RBIR}(\xi_1, \xi_2) = I[I^{-1}(\mathrm{RBIR_A}(\xi_1)) + I^{-1}(\mathrm{RBIR_B}(\xi_2))], \tag{8.13}$$

where $I(\gamma) = 1 - exp(-\gamma/10^{-0.12})$ approximates the AWGN RBIR for BPSK and $\mathrm{RBIR_A}$ (ξ_1) and $\mathrm{RBIR_B}$ (ξ_2) are the AWGN RBIR of the 16QAM bit channels. Since both the subcarrier SNRs and the 16QAM channel quality are random, only the average value of Equation (8.13) can be computed.

The proposed complete methodology is to classify all the coded bits into disjoint subsets based on the number of times that the bit has been received inside each of the RVs. For each of these subsets we should know the set $\{\xi\}$ of SNRs and the cardinal of the subset relative to the codeword size (called "repetition factor"). The repetition factors may be computed off-line (only once) for each MCS and stored in a LUT. Finally, the explained methodology for the case of two transmissions is extrapolated and applied to compute the contribution of each subset to the global RBIR and the contributions are added together after being weighted by the repetition factors.

In order to simplify the computations for H-ARQ link abstraction it is also possible to compute a global ESNR for the whole set of received RVs [LKKO12], using:

$$\gamma_{\text{eff}} = \frac{r_m}{r_{\text{eff}}} \cdot \text{SI}^{-1}\left[\frac{1}{J \cdot M}\sum_{j=1}^{J}\sum_{m=1}^{M}\text{SI}(\gamma_{m,j})\right] \tag{8.14}$$

where J is the number of subcarriers, M is the number of received RVs, $\gamma_{m,j}$ - is the SNR of subcarrier j at RV m, r_m is the mother code rate (r_m = 1/3 for LTE) and r_{eff} is the effective code rate defined as the number of information bits divided by the total number of received coded bits (including repetitions) per codeword. The predicted BLER is then given by the reference AWGN BLER curve for the modulation in use and code rate r_m.

8.6.2.3 A system level simulator applying link abstraction

In Litjens et al. [LZN+13] the system level performances of non-orthogonal spectrum sharing (NOSS), based on transmit beamforming, are compared to those of orthogonal spectrum sharing (OSS) and fixed spectrum assignment (FSA). The ESNR concept is used to model the beamforming SINR performance and to aggregate a set of physical resource block (PRB)-specific SINRs to a single effective SINR which is then used to predict the user throughput. When transmit beamforming is in use two UEs are simultaneously served from two cooperating co-located BSs at the same PRB. In this case the link abstraction LUT provides an ESNR for each UE which depends on the beamforming technique being applied (SR, nash bargaining (NB), or zero forcing (ZF)).

9

Radio Channel Measurement and Modelling Techniques

Chapter Editor: C. Oestges,
Section Editors: C. Brennan, F. Fuschini, M. L. Jakobsen, S. Salous, C. Schneider and F. Tufvesson

This chapter is dedicated to radio channel measurement and modelling techniques for beyond 4G networks, in particular:

- new measurement techniques not only targeting radio channels, but also material properties in new frequency bands (Section 9.1),
- improved physical models, cowering full-wave as well as ray-based methods, with a specific sub-section dealing with diffuse scattering (DS) and complex surfaces (Section 9.2),
- progress in analytical models (Section 9.3),
- new channel estimation tools for model development based on channel sounding data (Section 9.4), and
- finally, updates on COST 2100 channel models, enabling to include new features such as massive and distributed multiple-input multiple-output (MIMO) aspects (Section 9.5).

9.1 Measurement Techniques and Material Characterisation

Radio noise and channel measurements are fundamental to channel characterisation and system design. While techniques using commercial off the shelf (COTS) such as VNAs offer a ready to use solution they tend to suffer from a slow repetition rate that does not permit capturing the dynamic behaviour of the channel. They also have limited range in particular in the higher frequency bands. VNA measurements can provide reference data and used to develop channel models for indoor environments in out of the office hours. Such measurements have been performed to study the possibility of extrapolating the channel parameters in space and in frequency [Nik12]. Using a three dimensional robot, both single-input multiple-output (SIMO)

Cooperative Radio Communications for Green Smart Environments, 341–382.

and MIMO measurements were performed with 50 MHz bandwidth between 2.45 and 2.5 GHz. In the SIMO measurements, the transmit antenna was fixed while the receiver (RX) antenna was moved in steps of 25 mm to cover a space of 1.5 × 0.5 × 0.5 m. The MIMO measurements emulated the downlink and measured the channel over 35 transmitter (TX) locations on a straight line while the RX was moved in the x, z domain over 8 × 8 locations. Assuming plane wave transmission, the amplitude, phase, and angle-of-arrival can be estimated at a particular location. Using this model, the channel characteristics were extrapolated from one location or frequency to another beyond the 0.5 correlation coefficient for the coherence bandwidth or coherence distance. Using a ray model, the extrapolation distance was extended by more than 0.5λ. The extrapolated channels in frequency were found to improve the beamforming results at separations larger than five times the channel coherence bandwidth.

To overcome the limitations of vector network analysers (VNA) custom designed radio channel sounders are developed. To address the move to the frequency bands above 6 GHz, collocated and distributed multiple antenna technology sounders with modular architecture using multiple frequency bands, and multiple transmit and multiple receive units (MIMO) have been reported in Konishi et al., Stuart and Laly et al. [KKCT12, SFRC15, LGT$^+$15]. The sounder in Konishi et al. [KKCT12] targets the multi-link scenario with a 24 × 24 MIMO configuration using three transmit units and three receive units with eight channels per unit as illustrated in Figure 9.1. The sounder operates in the 11 GHz band and uses multi-tone transmission to cover a 400-MHz bandwidth giving 2.5 ns time delay resolution. Each TX has

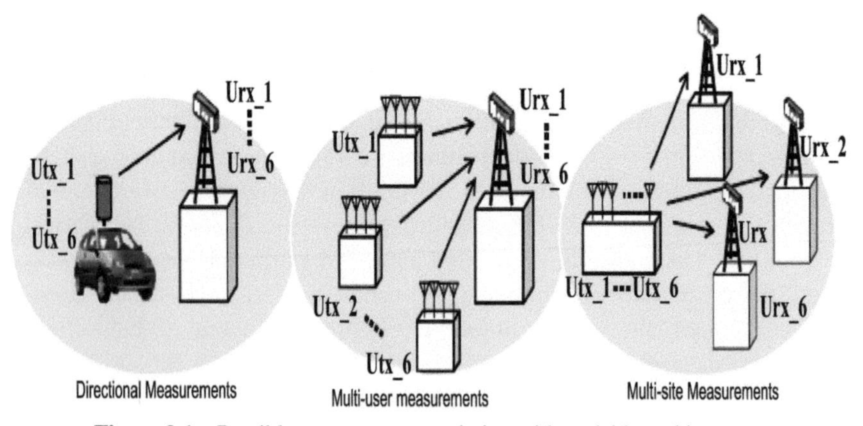

Figure 9.1 Possible measurement variation with scalable architecture.

10 mW per antenna and the RX has 4 GByte of RAM per channel. The overall time delay window is 5.12 μs and the maximum Doppler shift is 3.4 kHz corresponding to a frame length of 146.9 μs.

Using single 10-MHz Cesium clocks at the TX and at the RX for all three units at each site enables time synchronisation. However, the technique is challenging for distributed multi-link characterisation where each unit might require its own reference clock due to the physical separation of the units. Other issues associated with the architecture are the inphase and quadratue imbalance and the impact of phase noise on the estimated channel parameters. These two effects on the multitone signal used in the sounding have been studied in Kim et al. [KKT10, KWS$^+$15]. The impact of phase noise in the local oscillators leads to inter-carrier interference of multi-tone signals. To study its impact on the channel measurements simulations using two models of phase noise were performed. One model represents normal phase noise and has a noise level equal to –62 dBc/Hz, while the second model has high precision phase locked loops with phase noise of –103 dBc/Hz; at 1 kHz from the carrier. The results show that high precision oscillators would be needed for channel measurements and that the effect of phase noise increases as the number of units at the TX and RX increases for MIMO operation. The effect of IQ imbalance is also shown to be reduced by using quarter tone allocation.

To address the multi-tier, multi-band channel measurements, a frequency modulated continuous wave channel sounder with a modular structure (see Figure 9.2(a)) was developed at Durham University [SFRC15, SNCC13, SAG15]. The sounder has five frequency bands with increasing bandwidth as the frequency increases to enhance the resolution of multipath as the wavelength becomes smaller. The bandwidths achievable by the sounder are 750 MHz at 250 MHz–1 GHz and 2.2–2.95 GHz, 1.5 GHz at 4.4–5.9 GHz and 14.5–16 GHz, 3 GHz at 30 GHz and 6 GHz at 58–64 GHz. The sounder enables the configuration of modules as TXs or RXs and for the 30 GHz and 60 GHz bands the RF units are portable to enable on body measurements and avoid loss of cables as shown in Figure 9.2(b). In the mm wave bands, the sounder has an eight by eight MIMO configuration at 30 GHz and a 2×2 in the 60 GHz band which permits the estimation of channel capacity for multiple antenna measurements and for polarisation measurements.

Figure 9.3 shows the difference between the VNA measurements and that obtained with the channel sounder with waveform duration of 819.2 μs. The effect of the long duration of the VNA measurement is seen to distort the channel response in contrast to the clear response obtained with the channel sounder.

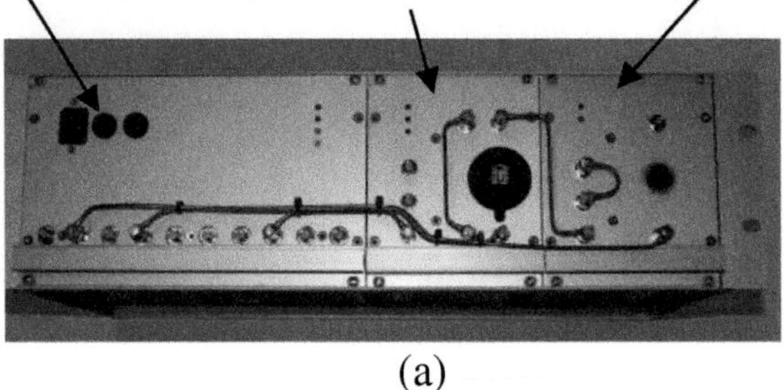

(a)

(b)

Figure 9.2 (a) TX/RX rack housing the three main units. The third unit housing the up converter up to 14.5–16 GHz can be replaced either by TX or RX units for the lower frequency bands and (b) portable 60 GHz units used in on body measurements.

Sounders using PRBS waveforms have been reported by Weile et al. and Hafner et al. [WPKW15, HDMS+15] with chip rates of 250 MHz [WPKW15] or 3.5 GHz [HDMS+15] with carriers at either 60 or 74 GHz, respectively. Using a two channel RX polarimetric measurements are enabled with manually switching the TX antenna [HDMS+15].

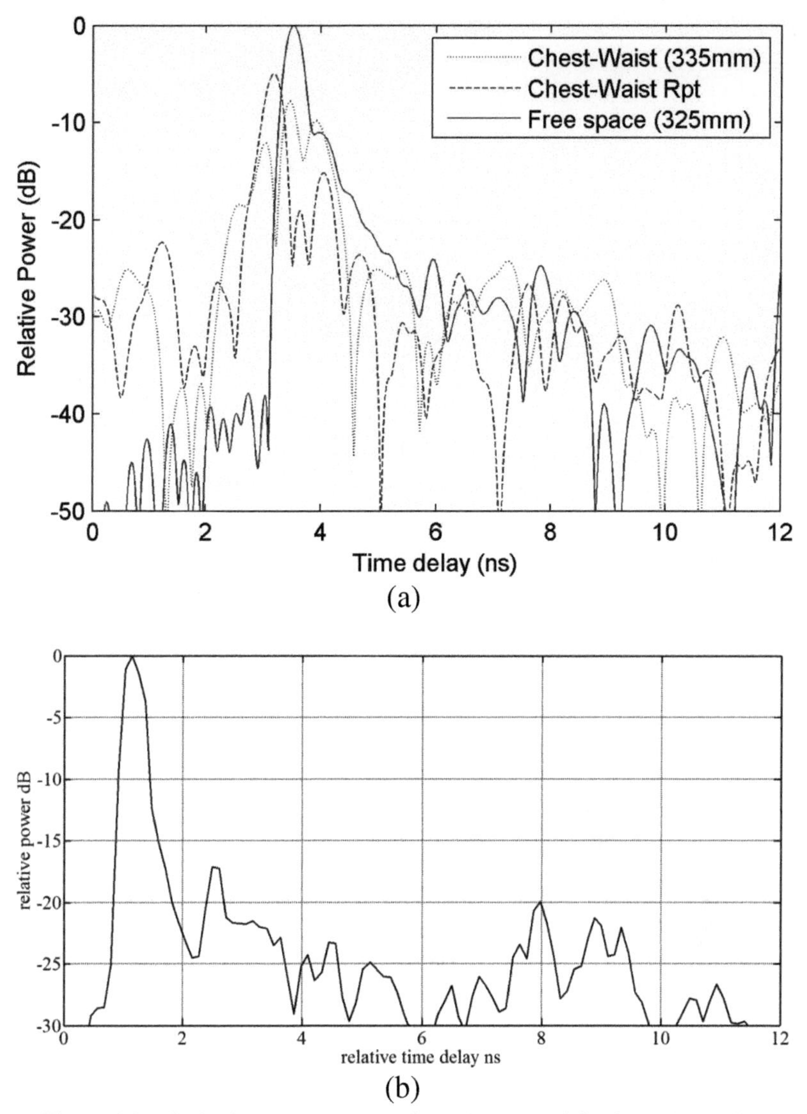

Figure 9.3 On body measurements using (a) VNA and (b) channel sounder.

Two other low cost channel sounders have been reported by Kim et al. and Laly et al. [KWS+15, LGT+15] which use digital techniques and configurable logic to implement either a mm wave sounder at 60 GHz as in Kim et al. [KWS+15] or an 80 MHz sounder at 1.3 GHz for 4×4 configuration [Pie15].

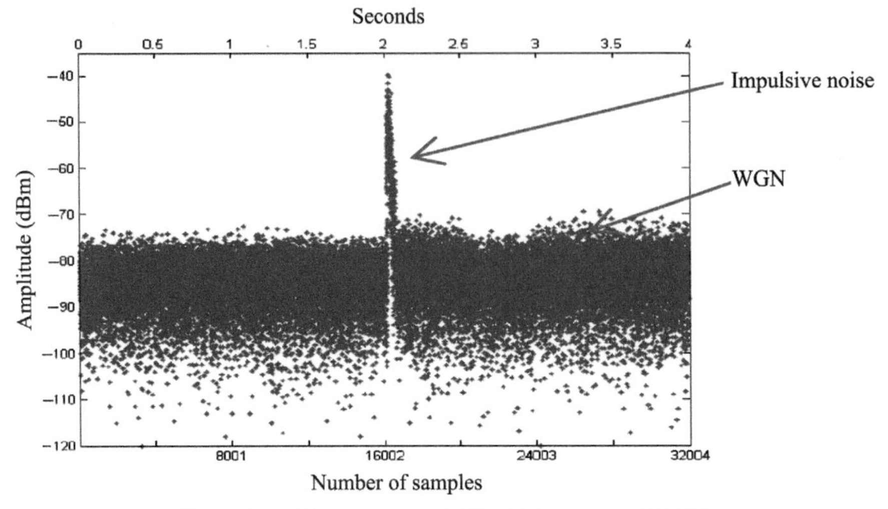

Figure 9.4 Measurement of IN with background WGN.

Another important aspect of radio measurements is that of noise. The presence of additive noise impacts the performance of radio RXs. Noise can be either white Gaussian noise (WGN) or impulsive noise (IN). Noise measurements are performed using an antenna connected to a high sensitivity RX and a programme to control the acquisition of data. To estimate IN, the Gaussian noise is first measured when the source of IN is switched off and subsequently when it is switched on [Mar14]. To capture IN which is short in duration, 8001 samples/s were captured in 4s to obtain the data in Figure 9.4. Parameters representing IN such as the number of bursts, burst amplitude level, burst duration, and burst separation can be evaluated from the data after setting up a threshold of 13 dB above the RMS value of the Gaussian noise as recommended by the international telecommunication union (ITU)-R report SM2155, 2009.

9.2 Physical Modelling Techniques

Radio channel physical modelling is based on the application of a *sound physical theory* to a proper *representation of the environment*. The model is *statistical* if the environment is simply described through few, general parameters; assuming the outdoor urban channel as a reference case, such parameters may be the mean buildings height, the mean street width, etc. On the contrary, if a somehow detailed and specific description of the environment is needed, the model is said *deterministic* or *site-specific*. Deterministic physical models are mainly addressed in the following.

A deterministic description of the environment refers to both its geometrical and electromagnetic properties. The geometrical description concerns the shape, the dimension and the position of each object inside the environment; these data are often stored into a proper digital *environment database* in vectorial or raster form which feed the model as input file. For a long time, environment databases had to be achieved by hand based on paper maps (e.g., cadastral maps); this procedure was cheap but also rather time consuming and practically limited to restricted environment like the indoor one. The final accuracy was often quite poor due to the lack of details in the starting map. Some automatic procedures have been then established (e.g., aerophotogrammetry), which allow a somehow detailed digitalisation of large areas but are also rather expensive. More recently, detailed digital description and representation of the environment are being more and more available on the web, and proper interface programs may be therefore able to extract the environment database from the on-line maps.

The electromagnetic description requires to associate each object with proper values of the electromagnetic parameters (e.g., relative electrical permittivity ε_R and conductivity σ) of the materials it is made of, so that its electromagnetic behaviour (e.g., reflection and transmission coefficients, absorbing properties, etc.) can be defined. Again, such information is often stored in a second input file somehow linked to the environment database. Unfortunately, permittivity and conductivity are little known in several cases, and compound materials are often accounted by means of rough effective electromagnetic parameters. Irrespective of the way they are achieved, both the environment and the electromagnetic databases are always affected by unavoidable imprecisions, which may impact on the model performance to an extent that must be evaluated case by case. In order to partly compensate for performance degradation due to inaccuracies in the environment description, empirical or statistical elements are often embedded into deterministic physical models. If, for example, surface roughness or building details such as windows, balconies, rain pipes, etc., are missing in the urban database, then an empirical/stochastic modelling of scattering generated by those details should be introduced in the deterministic model.

Deterministic physical models may potentially produce a wide spectrum of results, ranging from narrowband to wideband prediction, up to a full multidimensional characterisation of the channel, i.e., related to the polarimetric, space–time distribution of the received signal contributions in a multipath environment. This represents a valuable quality for the design and the deployment of next wideband, multi-gigabit radio systems. In fact, the effectiveness of multi-antenna technical solutions such as antenna diversity, spatial multiplexing, and beamforming, already included in the 802.11ad and long-term evolution (LTE) standards and seriously considered also for future 5G systems at millimeter-wave [RSZ+13, RSP+14], still depends on the received signal strength but also on others parameters strictly related to the multi-dispersive nature of the radio channel.

Physical channel modelling is commonly based on two main approaches: the full-wave electromagnetic approach and the ray approach. In the former case exact field representations such as Maxwell's equations or other similar analytical formulations are solved through numerical methods (e.g., finite difference time domain (FDTD) or MoMs). In the latter case asymptotic methods corresponding to geometrical optics (GO) and its extension allow for a ray-based representation of radio wave propagation. A further, quite known option is physical optics (PO), which is something in between GO and full electromagnetic formulations.

With respect to statistical and empirical channel models, deterministic channel characterisation allows for a more thorough insight into the propagation phenomenon. Nevertheless, high complexity, heavy computational burden and performance degradation due to lack in databases accuracy represent drawbacks that still need to be improved.

9.2.1 Full-Wave Electromagnetic Models

Full-wave models can be classified as being frequency domain or time domain depending on the form of Maxwell's equations used. These equations are most commonly expressed in the form of coupled differential equations but can also be manipulated into the form of a coupled set of equations involving integrals over volumes and bounding surfaces. Hence approaches can be further classified as being differential equation or integral equation based and the section which follows reviews full-wave electromagnetic modelling work carried out in COST Action IC1004 under these two headings.

9.2.1.1 Integral equation-based methods

There are several integral equation formulations of scattering problems, but in each case the principle is the same. The physical problem of a source radiating in the vicinity of a scatterer or scatterers is replaced with one or more *equivalent problems* where the scatterers have been replaced by a collection of unknown sources (mostly currents but in some cases charge densities). These

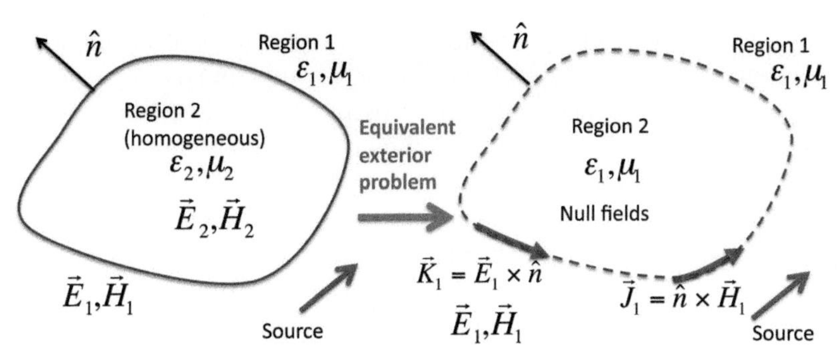

Figure 9.5 Equivalence principle for homogeneous scatterer. The configuration on the right produces the same fields exterior to the scatterer as the physical problem on the left. A similar arrangement is possible to recreate the fields in the scatterer's interior. Applying boundary conditions along the surface results in a set of coupled integral equations.

are assumed to radiate in an unbounded medium, allowing us to use potential integrals to write explicit expressions for the fields. Separate expressions can be thus obtained for fields exterior and interior to the scatterer. Applying boundary conditions along the scatterer surface results in coupled equations for the unknown sources. Figure 9.5 illustrates an example where the physical problem on the left is replaced with an equivalent problem on the right which can reproduce the fields exterior to the scatterer. Depending on the fields one works with it is possible to derive the electric field integral equation (EFIE), magnetic field integral equation (MFIE) or combined field integral equation (CFIE). In addition, one can derive volume integral equations in cases where one is considering inhomogeneous scatterers. In this case the unknown currents are distributed throughout the scatterer volume rather than residing on its surface. A simplified form of the surface EFIE equation holds for the case of perfectly conducting (PEC) scatterers as one only need consider a surface electric current (as the tangential electric field, and thus $\vec{K}$, is zero).

The MoMs [Har93] is used to discretise the integral equation. N basis functions are introduced to describe the unknown currents and charges and these are tested against N testing functions[1]. This results in a $N \times N$ linear system that can be solved using direct or iterative linear algebra methods. As one would expect increasing N generally results in a more accurate solution (albeit at the expense of greater computation times). Van Lil and De Bleser [VLDB14a, VLDB14b] examine the influence of the number and positioning

[1] One could use $M > N$ testing functions, in which case the resultant system would be solved in a least squares sense.

of the basis and testing functions on the accuracy of a method of moments (MoM) solution to a simple 2D problem involving computing the capacitance of a PEC strip. Using a higher density of basis functions at the edges where the charge density varies more rapidly is shown to give better results than a uniform distribution. Using a slightly higher number of testing functions than basis functions results in a rectangular system which can lead to improved accuracy, although this is sensitive to the precise numbers of functions chosen.

Another numerical issue that affects the surface EFIE is the so-called low-frequency breakdown whereby the solution becomes unstable as one approaches zero frequency. The instability is due to the explicit inclusion of charge continuity via the Lorenz gauge which removes the surface charge density as an independent unknown but introduces a $\frac{1}{\omega}$ term which results in a singularity at DC. An alternative is to include the surface charge density as a separate unknown in addition to the surface currents. Such an approach was considered for the case of PEC scatterers in De Bleser et al. [DBVL$^+$11a] and extended to the case of scattering from dielectric scatterers in De Bleser et al. [DBVL$^+$11c]. The extension necessitates the introduction of both electric and magnetic surface charge[2] and the paper outlines the relevant details needed to numerically compute singular integrals etc. Good agreement is obtained for scattered fields at low frequencies indicating the stability of the method. The accuracy of this stabilised method was examined in De Bleser et al. [DBVL$^+$12a, DBVL$^+$11d] by applying it to more complex problems, such as scattering from wind turbine blades, cubes with sharp and rounded edges as well as multi-layered bodies. Further scrutiny is carried out in De Bleser et al. [DBVL$^+$12b, DBVL$^+$11b] wherein quantitative verification against a Mie series solution is provided as is qualitative verification as to whether the boundary conditions are properly satisfied etc. Good agreement is observed at low frequencies as well as a small mesh-dependent error as frequency increases. Further technical detail is provided relating to singularity removal (for the self terms of the impedance matrix) as well the choice of testing and basis functions (including the need to test both tangential and normal components).

While the combined charge and current formulation improves the stability of the method it increases the number of unknowns which has an adverse affect on the computation time. An alternative formulation, referred to as the split formulation, is given in De Bleser et al. [DBVL$^+$13] and [DBVL$^+$14]. In this approach the surface charge is eliminated as an unknown, as in the standard EFIE, but this time without introducing the $\frac{1}{\omega}$ dependency which is

[2]Magnetic charge, while unphysical, is a useful and valid mathematical abstraction.

responsible for the low-frequency instability. The resultant approach is faster and remains stable at low frequencies.

Another key challenge in applying moment–method solutions is their associated computational burden. Discretising an integral equation with N basis and testing functions results in a dense linear system of size $N \times N$. Standard iterative solution using a Krylov solver scheme such as GMRES has computational complexity $\mathcal{O}(N^2)$ due to the necessity to perform at least one matrix vector multiplication at each iteration although various acceleration methods such as the Multilevel Fast Multipole Method (MLFMA; [SLC97]) and spectral acceleration [CJ00] can reduce this burden to something approaching $\mathcal{O}(N \log N)$ which can speed up analysis significantly. Spectral acceleration is used in conjunction with the forward backward method by Brennan et al. [BTM+13] and [BT12] to analyse the fields scattered from randomly rough surfaces. As communication frequencies increase the small random perturbations on a surface become more prominent with respect to the wavelength and the consideration of non-specular or diffuse reflection becomes more significant. The forward backward method is an alternative iteration scheme which sequentially computes the fields along the rough scatterer profile firstly away from the source and then back towards it, implicitly computing all multiple scattering events in a natural fashion as it does so. While less robust than Krylov solvers it has been shown to yield a convergent solution in fewer iterations when applied to certain classes of problem. The work mentioned above includes a novel numerical check to identify when the iteration solution update (i.e. the difference between two successive solution estimates) has broadly converged to a single direction (in the solution space) and, if this is the case, takes an optimised step in this direction. This simple expedient can significantly reduce the number of iterations needed. The method is applicable to TE and TM polarisation and can accurately compute quantities such as the normalised bistatic-scattering coefficient more rapidly than competing full-wave techniques.

More aggressive approximations are required when developing full-wave path loss models for large-scale propagation problems. The fast far field approximation (FAFFA) can yield greater efficiencies at the expense of introducing slightly higher, albeit acceptable and controllable, error and has proven useful for creating 2D propagation models, notably the tabulated interaction method (TIM) which optimises this process for propagation over undulating rural terrain profiles. Extension of such acceleration techniques to more complex propagation problems such as propagation in the vertical plane over the rooftops of buildings poses several challenges such as the effect of sharp edges and the enhanced levels of multiple backward and forward scattered fields. Brennan et al. [BTD11] and [BT13] address this problem in two ways. First, the FAFFA can be used to reduce the cost of each matrix vector multiplication. Second, a block forward backward method can be used to iterate the solution. The block method is similar to the forward–backward method discussed above but instead conceptually breaks the profile into connected strips (each containing many discretisations) and marches a solution from strip to strip (rather than from discretisation to discretisation) thereby working with blocks of the impedance matrix rather than individual entries. In practice, a hybrid method is used, using blocks of varying sizes—this is referred to as the hybrid forward backward method (HFBM). Results published in Brennan and Trinh [BT14b] (Table 9.1) indicate greater accuracy when predicting path loss for three urban routes in Munich, Germany when compared to SDMs Figure 9.6 for an example).

Table 9.1 Mean (η) and standard deviation (σ) of error between slope diffraction method (SDP), HFBM-FS (forward scattering) and measurements for points far from source such that vertical plane propagation is dominant

	METRO 200		METRO 201		METRO 202	
	η	σ	η	σ	η	σ
HFBM-FS	–1.2	7.3	4.26	7.95	0.57	9.7
SDP (Knife edges)	14.8	8.1	15.6	8.24	14.3	10.1
SDP (Wedges)	–1.1	8.0	–	–	–3.8	10.8

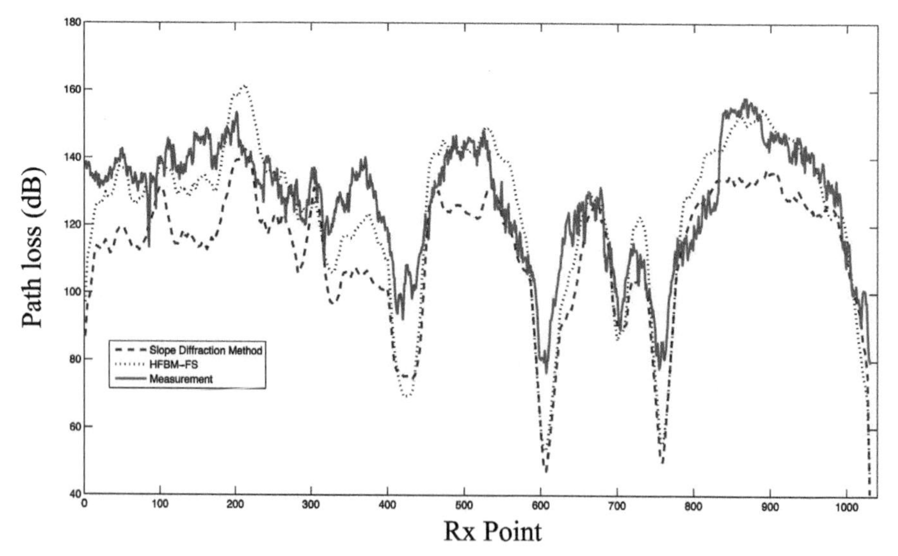

Figure 9.6 Predicted versus measured fields at for the commonly used METRO 202 route in Munich, Germany. Slope diffracted fields assume knife edge buildings.

Another MoM-based 2D path loss model is described in Brennan and Trinh [BK14] and Kavanagh et al. [KPCB15] for the case of indoor propagation. In this work, a volume EFIE (V-EFIE) is used. This requires the discretisation of the entire interior of the building but this is compensated by the resultant ability the use the fast fourier transform (FFT) to reduce the cost of each iteration. In addition the papers outline how the unknowns in free-space, which make up the vast majority of the unknowns, can be effectively removed from the iterative process, as they do not influence the fields elsewhere and are only introduced in order to ensure a regular grid which facilitates use of the FFT. Removing the influence of these non-interacting unknowns further reduces the number of iterations required. The model can include various material types (wood, glass, etc.) and initial validation shows that the model produces shadowing whose characteristics are similar to commonly used statistical models. The model is currently 2D but work is ongoing to extend it to 3D and to develop a wide-band version which can predict other aspects of the radio channel other than path loss.

9.2.1.2 Differential equation-based methods

An alternative starting point for the development of full-wave models is to use Maxwell's equations in differential form. From this a number of models are possible, such as those based on the application of finite elements or finite differencing. The potential of two time domain models for indoor propagation modelling was discussed in Virk et al. [VVW13]. Recognising the need for accurate models to facilitate femto cell design and cognitive radio deployment the paper firstly gives a brief overview of the FDTD method. In particular, the Yee cell discretisation and time-stepping approach is described. The paper also discusses the time domain ParFlow algorithm. This is based on Lattice Boltzmann methods and is related to the transmission line method(TLM).

A multi-resolution frequency domain version of the ParFlow method (MR-FDPF) is applied in Luo et al. [LLV$^+$12a] and [LLV$^+$12b]. Due to computational constraints the simulation is 2D but this is compensated for by way of calibration against measurement which informs the tuning of material parameter values (including an attenuation coefficient for air). The work cited above is specifically concerned with modelling shadow fading and starts by separating the various propagation mechanisms. A simple path loss model is assumed

$$\mathrm{PL}(d) = L(d) + X_\sigma + F, \tag{9.1}$$

where X_σ represents shadow fading while F represents fast fading. The mean path loss $L(d)$ can be further written as

$$L(d) = L_0 + 10n\log_{10} d, \tag{9.2}$$

The document describes how a local mean path loss (comprising mean path loss and shadow fading) can be predicted by averaging power values computed using the ParFlow model over areas of 3.8 $\lambda \times$ 3.8 λ to average out the effects of F. Simple curve fitting techniques can then be used to estimate L_0 and n. The model is verified against data collected by Stanford University in a 16 m × 34 m office space. The fitted path loss exponent n was 1.59, in close agreement to the value of n computed from the measured data (Eq. 1.44). The extracted simulated shadow fading was shown to fit a normal distribution with σ = 5.87 dB (see Figure 9.7) while the value fitted from the measured data was σ = 7.66 dB.

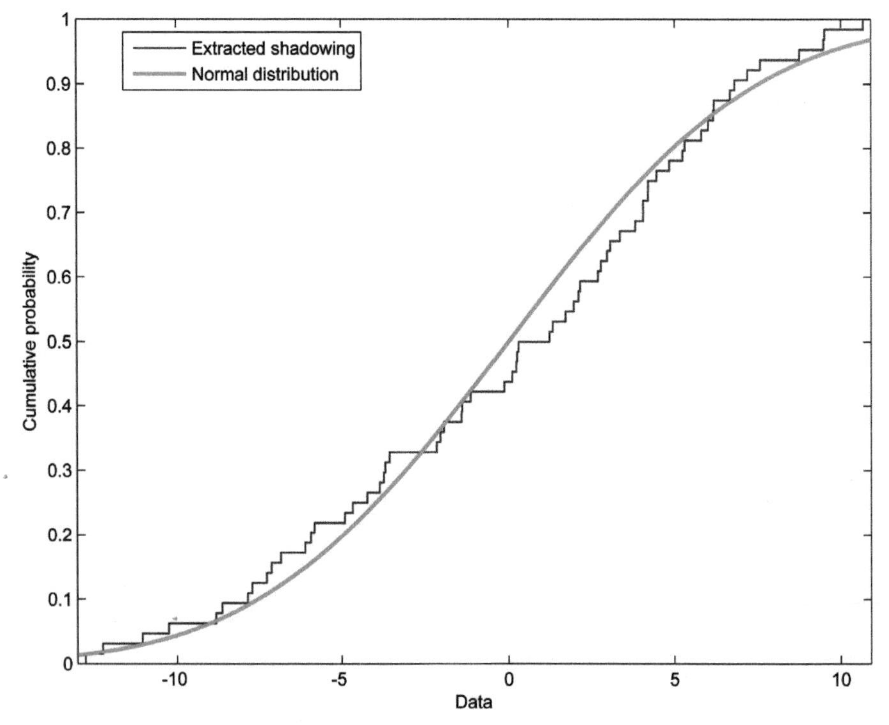

Figure 9.7 Cumulative distribution function of the shadow fading computed using MR-FDPF model (*blue*) versus fitted normal distribution (*green*).

9.2.2 Ray Tracing (RT) Models for Radio Channel Characterisation

Ray-Tracing models are based on the geometrical theory of propagation (GTP), that is an extension of GO to radio frequencies, i.e., not limited to optical frequencies. In order to account for diffraction effects, which can be important at radio frequencies but are completely neglected by GO, GTP includes the so-called uniform theory of diffraction (UTD) [MPM55]. Like GO, GTP is based on the *ray* concept: multipath propagation of electromagnetic waves in real environment is represented by means of rays, that spring out from the TX and may reach the RX after some *interactions* with the objects inside the propagation scenario. Beside standard interactions like specular reflection, transmission and diffraction, RT models have been recently extended to diffuse scattering, which can be important for a reliable multidimensional characterisation of the radio channel [DEFV+07, DDGW03, MQO11, VDEMO13].

Rays trajectories are commonly tracked according to two, different techniques, namely *images RT* and *ray launching* method [IY02].

Since a ray-based description of radio wave propagation somehow corresponds to mental, intuitive visualisation of the actual propagation phenomenon, RT models represent very helpful study and teaching tools. On the contrary, they have not still achieved widespread application to wireless systems design and deployment problems, mainly because of the heavy computational burden and the unavailability of detailed and reliable environment description databases. Moreover, radio system planning has traditionally aimed at pursuing a satisfactory coverage level, and simple, ready-to-use models such as the Hata model [Hat80] have commonly represented a satisfactory and fast solution to achieve the necessary narrowband, path loss predictions.

In the near future things will probably change. As already discussed, RT models are naturally fit to simulate multipath propagation and therefore, to provide the extensive, multidimensional channel characterisation that is necessary to effectively design and deploy the incoming wideband, multigigabit radio systems. Besides, the ongoing idea of accomodating 5G radio networks in the millimeter-wave bands [RSZ+13] may also spur a larger resort to RT simulations, since the smaller the wavelength the more acceptable the ray-optic approximation and the more accurate the RT results. Finally, the computational effort required by RT simulations will be reduced by the increase in computing capacity, and the access to detailed digitised maps of outdoor and even indoor environments will be also easier and simpler in the next future. RT models are therefore gaining increasing consideration and their use can be expected to become more common for both research and commercial purposes.

9.2.2.1 RT-assisted multipath tracking

Since the experimental characterisation of radiowave propagation is often expensive and time consuming, many effort have been devoted to the development of reliable field prediction models. In this prospect, once the propagation tool has been validated and/or tuned by means of measured data achieved in some reference cases, it should be used *instead of* measurement to assist the radio network planning phases.

In spite of this rather common view, a different option is also possible, aiming at exploiting prediction capabilities *together with* measurement to provide a thorough characterisation of the radio channel. This is done in Meifang et al. [MST11], where a ray launching tool is combined with channel measurements to identify and track the most likely multipath components. In order to increase efficiency, rays are launched from both TX and RX within a cone centred around the measured angles of departure and arrival; a rays matching procedure is then carried out, i.e., common interactions points are searched among rays launched from TX and RX, in order to get complete rays path. Each ray is then associated with a specific multipath component achieved from measurement (basically comparing the propagation delays). Physical scatterers can be visualised onto the map of the environment, and the interaction points with similar delays and rays intensity can be grouped into clusters, which represent a key concept of geometric–stochastic propagation models.

Another method to find the corresponding multipath contributions between measurement and RT simulation is described in [GvDH11] and consists of the following 5-steps procedure: (i) *clustering*: both simulation and prediction are clustered, i.e., multipath components with similar directions of arrival and delays are grouped together; (ii) *calibration*: different conditions between simulation and measurements are eliminated, e.g., the arbitrary start time of propagation delay in measurements; (iii) *feature generation*: each cluster is associated with a single value of delay, azimuth and elevation, as the average of such values among the multipath components belonging to the cluster; (iv) *matching*: each simulated cluster is associated with a measured cluster (v) *evaluation*: check if the matched clusters; and actually come from the same object in the environment.

In some cases, a measured multipath component may have several matched candidates from theRT tool, as well no matching is found in other cases. This is due to the unavoidable inaccuracies afflicting the RT simulation, e.g., some lack of details in the environment databases, where small objects belonging to the environmental clutter (such as lampposts) are not included but can play an important role from a propagation perspective.

9.2.2.2 New expedients for RT prediction improvement

Although RT application to radiowave propagation modelling dates back to the early nineties, its full potential is still partly unexpressed. A strong effort has been, therefore, devoted to the enhancement of RT performance, aiming at improving the prediction capabilities and/or reducing computation time. To this regard, a sophisticated but effective method is presented in Gan et al. [GMK+13]; the study basically aims at modelling the indoor multipath propagation channel for a mobile terminal moving over a distance smaller than the wavelength λ. In order to compute the time-variant channel frequency response, a standard RT approach would require to track all the propagation paths at different time samples. The required overall complexity can be reduced through a two-step procedure: (i) since the covered distance is lower than λ, the channel is assumed wide sense stationary, i.e., RT can be performed just at the starting position, (ii) each path is projected on a subspace spanned by two-dimensional discrete prolate spheroidal (DPS) sequences [CKZ+10]. The effectiveness of the DPS-based solution in terms of computational complexity reduction is shown in Figure 9.8.

An important source of inaccuracy affecting RT simulators is represented by the rough, oversimplified description of the electromagnetic properties of the materials. Electromagnetic parameters such as the electrical permittivity and conductivity are known only for a restricted set of materials and over limited frequency intervals. In many cases, real materials are a heterogeneous mixture of different and sometimes unknown components, and an electromagnetic characterisation is not available at all. These uncertainties of course reduce the actual accuracy of a RT prediction. A solution to this problem is discussed in Navaroo [Nav13], where the basic idea is to tune the material parameters to optimal values using some calibration measurements as a reference target.

9.2.3 DS Physical Modelling

It is well known that GTP can reliably model propagation in presence of smooth and homogeneous walls and surfaces, where multipath actually consist just of specular, transmitted and diffracted coherent contributions. Nevertheless, propagation in a real scenario is commonly affected also by the so-called

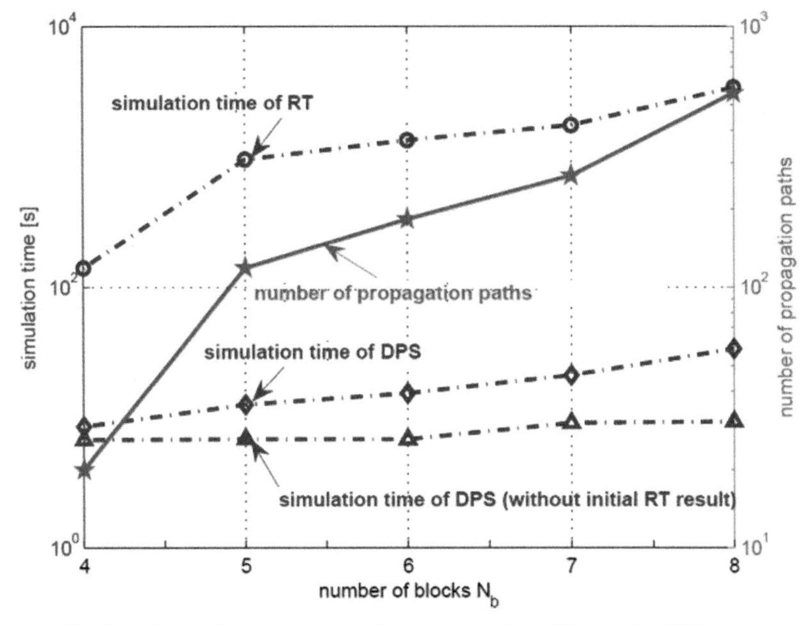

Figure 9.8 Simulation time comparison between standard RT and the DPS approach. N_b represents the number of objects inside the scenario.

DS (also known as dense multipath components (DMCs)), that is the result of the superimposition of a large number of incoherent micro-contributions, i.e., with unknown, random phase value an polarisation state. In urban and indoor environment, DS may originate from strong surface irregularities (e.g., windows frames, balconies, rain pipes, etc.) and volume inhomogeneities (e.g., cables and pipes buried inside walls, compound building materials, etc.). Vegetation also represents an important source of scattering, which can be produced by the trees canopy as well as by trunks and branches.

9.2.3.1 Scattering in urban and indoor environment

Owing to its importance for an effective modelling of the multi-dispersive properties of the radio channel, DS from building walls has been recently widely investigated. In Minghini et al. [MDDEV14], the signals scattered by some smooth and rough sandstone slabs placed in an anechoic chamber have been collected and compared in the [2–10] GHz band for different incidence angles of the impinging wave. Results highlight that the field is spatially spread over a wide angular range in the case of rough surface at the expense of the specular reflected component that is attenuated if compared to the smooth slab.

Several models have been then developed to take DS into account, like the effective roughness (ER) model, which has been recently proposed and can be easily embedded into RT conventional tools. According to the ER approach, each wall/object surface is subdivided into small tiles, and the power impinging on each tile is partly reflected specularly and partly scattered according to a proper scattering coefficient S and a specific scattering pattern [DEFV+07]. Both the *S* value and the width of the scattering lobe should be tuned based on measurement data collected in some reference cases in order to empirically account for the overall scattering effect, since surface irregularities and volume inhomogeneities are usually not included in the input database and a strictly deterministic approach to DS is, therefore, often unpracticable.

The ER model is adopted by Mani et al. [MQO11], where it is embedded in an RT tool to assess the contribution of DS to angular spreads through the comparison between measurement and RT prediction in a laboratory and in an office environment. Measurement have been carried out with a channel sounder working at 3.6 GHZ with a bandwidth of 200 MHz; the receiving antenna was composed of a dual-polarised uniform linear array (ULA), whereas the TX was equipped with a tri-polarised antenna, which was then moved to create a virtual uniform cubic array. Comparison results show that RT with scattering reduces the angle spread prediction error with respect to simulations limited to specular components only. The root mean square error (RMSE) is decreased by 14% for azimuth of departure (AAoD) and 39% for elevation of departure (EoD) in laboratory scenario. Similar results have been achieved for the office scenario.

Although DS is basically an incoherent phenomenon, its depolarisation properties might be not as random as the phase, i.e., the polarisation of the scattered wave could be partly deterministic. The polarimetric properties of diffuse power scattered off building walls is therefore investigated in Vitucci et al. [VMDE+11] by means of broadband measurements carried out at 3.8 GHz along the facades of a rural building and of an office building by means of ULAs of dual-polarised patch antennas at both the link ends. RT simulations including DS according to the ER approach are exploited to extract the DMC from measurements. The specular multipath components, the DMC and the full received signal are then compared in terms of cross-polarisation discrimination (XPD) defined as:

$$\mathrm{XPD} = 10\log_{10}\left[\frac{P_{\mathrm{co}}}{P_{\mathrm{xp}}}\right] \tag{9.3}$$

being P_{co} and P_{xp} the powers received in co- and in cross-polarisation, respectively. Results show that the coherent part of the power generally overestimates the measured XPD, thus highlighting that DS is a key aspect for an accurate prediction of the polarisation behaviour of the channel.

In order to extend the ER model to the polarisation domain, an additional parameter $K_{xpol} \in [0, 1]$ is introduced into the model [VDEM+13]. According to Equation (9.4), K_{xpol} sets the amount of power transferred into the orthogonal polarisation state after a scattering interaction:

$$\mathbf{e}_s = \sqrt{1 - K_{xpol}} \left\| \mathbf{e}_s \right\| e^{j\chi 1} \, \hat{\mathbf{i}}_{co} + \sqrt{K_{xpol}} \left\| \mathbf{e}_s \right\| \hat{\mathbf{i}}_{xp}, \tag{9.4}$$

where $\hat{\mathbf{i}}_{co}$ is a unit vector representing the polarisation of the scattered field if it were a non-depolarising interaction like specular reflection, and $\hat{\mathbf{i}}_{xp}$ is the unit vector orthogonal to $\hat{\mathbf{i}}_{co.}$Comparison between measurement and simulations is exploited in Vitucci et al. [VDEM+13] to tune the K_{xpol} values in different environments; the best-fit K_{xpol} are summarised in Table 9.2 and show that the depolarisation degree of the DMC seems to increase with the complexity of the environment.

In Tian et al. [TDEV+14], a deterministic model for DS is proposed in the framework of the graph theory. If multiple TXs and RXs are considered together with a spatial distribution of scatterers corresponding to the actual shape and position of the objects inside the environment, the channel transfer function is computed through an analytical expression involving several matrices accounting for propagation in line-of-sight (LOS) conditions, from TXs to scatterers, from scatterers to RXs, and among scatterers respectively. The non-zero matrices entries are expressed according to the ER model assuming a Lambertian scattering pattern. It's worth noticing that the matrices to be computed are independent of the number of "bounces" of propagation among the scatterers, i.e., the computation cost is not affected by the number of allowed scattering interactions. The model performance is assessed by

Table 9.2 Best-fit K_{xpol} values in different scenarios

Environment Type	K_{xpol}
Rural building	0.05
Office building	0.2
Campus scenario	0.3
Street-canyon scenario	0.3
Indoor office scenario	0.5
Indoor lab scenario	0.55

comparing the simulated power delay profile (PDP) with RT predictions and real channel measurement data (Figure 9.9). With respect to a previous work of Pedersen and Fleury [PF07], where the propagation graph is set up assuming a random distribution of the scatterers, the deterministic approach generates the PDPs with more realistic decaying slopes. Results are also in good agreement with a RT prediction, provided that the same lambertian model is implemented.

Based on the observation and the analyses of previous studies, DS from real building walls is broken down in Ait-Ighil et al. [AIPFL+12] into three different propagation phenomena, namely specular reflection, backscattering and incoherent scattering (Figure 9.10). Based on a fast formulation of PO [Bal89], a pre-calculated form of GO and five different types of rough surfaces [RBSK70], the three scattering contributions can be computed and combined into the three component models (3CMs), which has been checked against MoM as a reference tool for a complex, isolated building. Results show that the total scattered power calculated using the 3CM approach are in good agreement with the MoM prediction (note that other simulations have been performed using an asymptotic tool and differences with respect to the MoM-based results are similar).

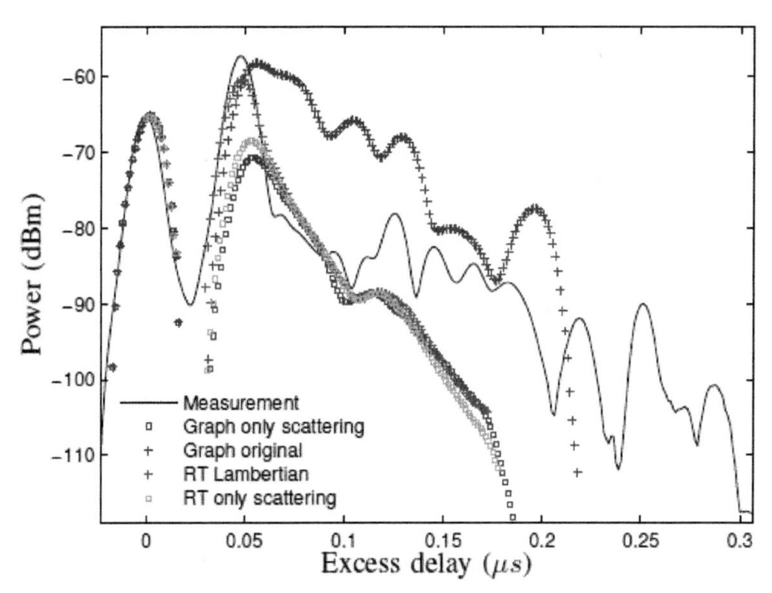

Figure 9.9 Comparison between measurement and different prediction models: deterministic graph model (Graph only scattering), stochastic graph model (Graph original), RT with lambertian scattering pattern (RT lambertian), RT limited to only lambertian scattering (RT only scattering).

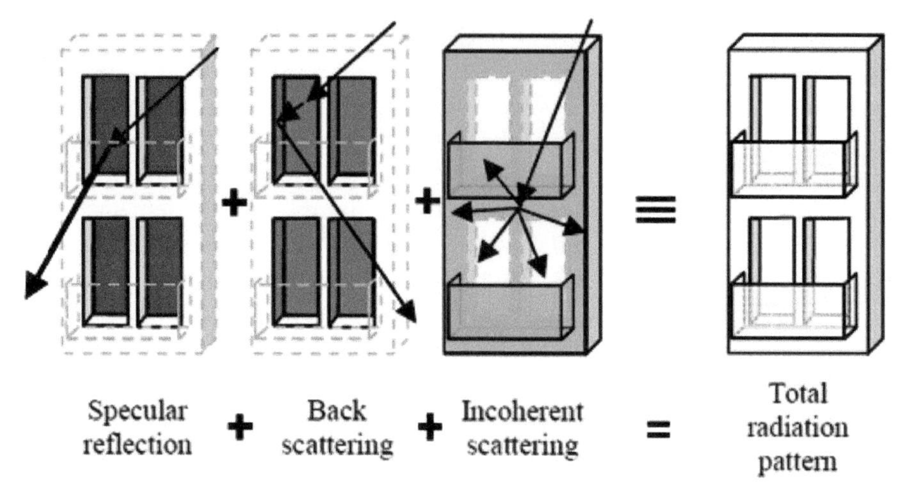

Figure 9.10 Decomposition of the scattering pattern of a complex building.

9.2.3.2 Scattering from vegetation

An electromagnetic wave propagating through a vegetated area undergoes a twofold effect: in addition to the attenuation of the main, forward coherent component, power is also scattered in all the directions (incoherent component). Such effects can be of primary importance in propagation within rural, suburban, or open urban areas.

In Mani and Oestges [MO12], scattering from tree branches is modelled according to a RT approach. Tree branches are modelled as cylinders and a digital representation of a tree is created by means of a 3D-fractal generator. Every time a ray impinges on a tree branch, scattered rays are generated in other-than specular directions. The scattered field is assumed totally depolarised and its overall amplitude is equal to the reflected contribution reduced by a proper coefficient dependent on the angle between the scattered and the specular directions and according to a cosine, linear or exponential law. Comparison with measurement carried out over 15 receiving outdoor locations in a campus scenario shows that the RMSE of the received power decreases from 6.5 to 3.7 dB if scattering from trees is included into the prediction model.

Both the attenuation and the scattering effects due to vegetation are computed in Torrico et al. [TLU13] based on the radiative transport theory applied to a tree trunks dominated environment. The TX and the RX are assumed at the same height and the trunks are represented as identical, circular lossy dielectric cylinders. The intensity of both the coherent and the incoherent contributions to the total received field is computed by means of the 2-D transport equation. According to the result shown in Figure 9.11, the coherent

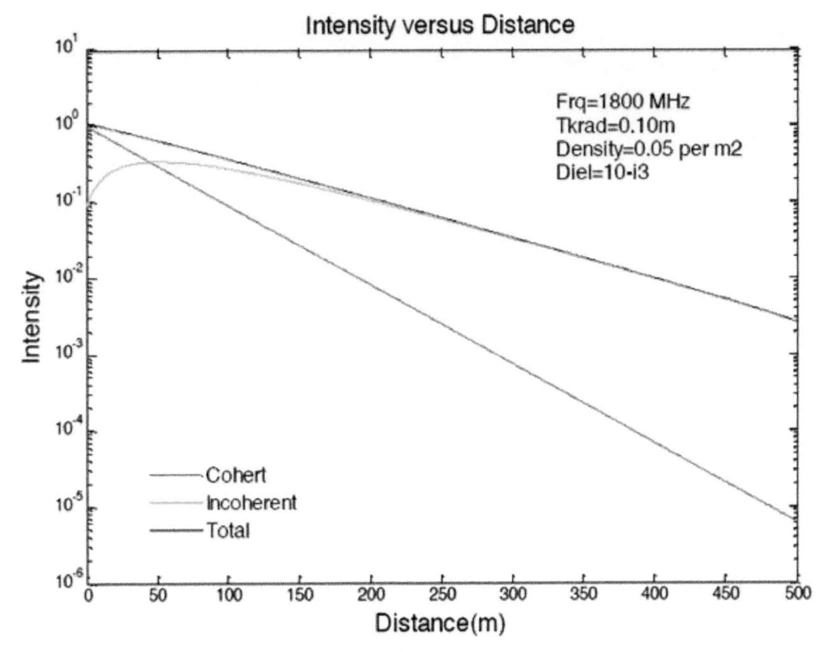

Figure 9.11 Intensities of the received signal contributions versus distance in a trunk dominated environment.

field is dominant only at low penetration distance into the forest layer, where the forward path is affected by few trees; at higher distance multiple scattering effect sets up as the leading propagation mechanism. It's worth mentioning that this behaviour is less evident at lower frequencies and/or for smaller trunks size, i.e., the incoherent waves become increasingly important as frequency and the trunk radius increase.

In Chee et al. [CTK13] the COST-231 Walfisch–Ikegami urban model is extended to vegetated residential environment. Assuming the transmitting antenna elevated above the average rooftop height and separated from the RX by a row of houses and tree canopies, propagation mainly occurs in the vertical plane over the rooftops and through the canopies, and then from the rooftop down to the street level. The theory of Foldy–Lax [Fol45, Lax51] is then exploited to embed scattering and attenuation effects due to the presence of trees into the formulation proposed in the COST-231 Walfisch–Ikegami model for over-rooftop and roof-to-street propagation. At the end, the impact of vegetation on the final, analytical expression for the overall propagation loss is accounted through two main additional parameters, namely the width of the canopy (ω_T) and its specific attenuation (γ). The model performance are investigated in Chee et al. [CBZ$^+$12], where predictions with and without trees are compared with several cross-seasons measurements carried out at 800, 2300, and 3500 MHz. Results show that the extension to vegetation effects yields significant improvements in moset cases over the COST-231 Walfisch–Ikegami model. Furthermore, the change in the propagation loss due to the seasonal variation in the degree of foliage can be somehow tracked by the model, provided that the canopy-specific attenuation value is properly set for the different seasons.

9.2.4 Other Issues Related to Physical Channel Modelling

Other activities related to deterministic channel modelling have been carried out within the COST IC1004 Action and are addressed in this section. The study in Zentner and Katalinić. [ZK13] and Mataga et al. [MZK13] basically aims at obtaining some scheme for preserving the accuracy of the RT approach, but reducing the the computational effort and the complexity.

The proposed solution is based on the concept of "*ray entity*", that is a group of rays undergoing the same type of propagation interactions, in the same order and on the same object. The space region where the ray entity is present defines its *visibility region.* According to GO basic theory, different rays reflecting on the same walls, i.e., belonging to the same ray entity, appear as originated from a single, fixed virtual transmitter (VTX), whose position corresponds to the multiple-reflected image of the actual TX with respect to the reflecting walls. RT simulations can be, therefore, run with a limited number of RXs, thus reducing the computational burden; then, the results within each visibility region can be interpolated to a higher resolution degree exploiting the ray entity concept at a negligible increase of the computational effort. This opportunity can be also extended to RT simulations including diffraction and DS.

The different strategies to handle VTX position in diffracted rays are investigated in Zentner et al. [ZKD12] with reference to the mobile multipath environment, where Doppler shifts need to be estimated in order to assess the channel temporal variability. With reference to single diffraction for the sake of simplicity, the virtual source is often assumed on the corner edges of buildings, i.e., coincident with the diffraction interaction point. A second option for the VTX position is a point in the ray's direction of arrival, as seen by the user, at a distance corresponding to the total, unfolded path length from the TX to the user. Although they may appear equivalent, the two solutions lead to different values for the Doppler shift related to propagation along the diffracted ray. In particular, a wrong evaluation is achieved when assuming the virtual source on the diffracting edge.

Diffraction problems are also addressed in Martinez-Ingles et al. [MRMP+12], where an hybrid UTD-PO formulation for the evaluation of multiple-diffraction effects over an array of obstacles is proposed and validated against measurement carried out at 62 GHz. With respect to previous studies, mainly related to the urban propagation context where obstacles (buildings) are usually modelled as knife edges [XB92] or wedges [JLR02], rectangular shapes are considered for the diffracting elements, which could be more appropriate in many cases. The solid agreement between predicted and measured values of the excess attenuation with respect to free space shown in Figure 9.12 corroborates the effectiveness of the UTD-PO solution for the study of wireless communication in the millimetre-wave band. Moreover, since the array of rectangular blocks used during the measurement campaign can be assumed to some extent a scaled-model of an urban environment, the proposed UTD-PO formulation can be also fit to model urban radiowave propagation.

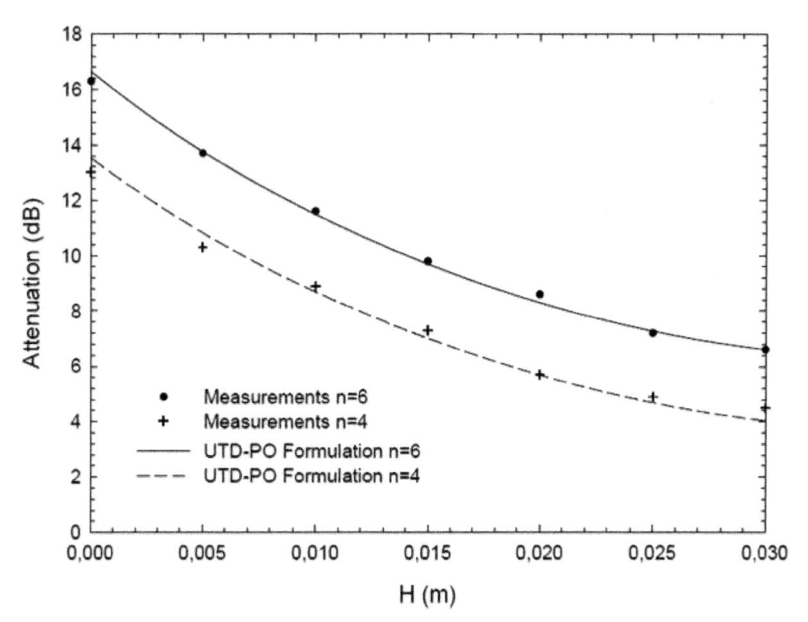

Figure 9.12 Comparison of the UTD-PO prediction with measurements at 62 GHz. H is the relative height of the TX with respect to the array of rectangular obstacles and n is the number of obstacles. Hard/vertical polarisation is considered.

9.3 Analytical Modelling Approaches

Opposite to the physical modelling techniques in the previous section, the present section deals with so-called analytical models. In this terminology, the notion of an "analytical" model carries essentially the meaning of a "non-physical" model. The common approach is to deliberately abandon the physics of electromagnetic wave propagation and to rely instead on conceptually and mathematically convenient simplifications. Thus, the goal is not to accurately reflect the physical processes taking place in the radio channel. Rather, the goal is to design (very) simplified models capable of reproducing and predicting dedicated physical phenomena which can be measured in practice. Modelling designs of this spirit are in the literature often encountered as so-called statistical or stochastic radio channel models.

9.3.1 A Statistical Model of Signal Strength Variability

The statistical model in Giménez [GGGC12] serves as a first illustrative example of an analytical model. Received signal strength is a frequency-dependent quantity in general, and the aim has been to design a model capable of revealing potential gains of time–frequency slicing techniques in digital terrestrial television applications. According to the proposed model, the difference in received signal strength at frequencies f_1 and f_2 depends on two distinct terms: A first term (reflecting propagation losses) parameterised to change deterministically with f_1 and f_2 and a second term modelled as a random variable with a standard deviation parameterised to change as a function of the frequency separation f_2–f_1. Overall, this analytical model includes in its design a total of four parameters whose numerical values are fitted/estimated from measurement data using non-linear optimisation techniques.

9.3.2 Stochastic Models in Two Narrowband Studies

In a study concerned with secure key generation in wireless communications, Mazloum et al. [MMS14] proposed a so-called disc of scatterers-based channel model. The main idea is to exploit (for secrecy purposes) the decorrelation of complex-valued channel gains observed by two different terminals separated at distance d in a narrowband channel. One terminal is the intended RX while the other is an eavesdropper. The physical cause for decorrelation of channel gains versus separation distance is not sought to be explained or justified. To the contrary, decorrelation serves as a premise and is "constructed" in a mathematically convenient way using a channel gain model in which statistically independent omnidirectional point-scatterers are placed uniformly at random on a disc centred around the intended RX. Individually phased and attenuated contributions from the different scatterers are summed up to yield the complex channel gain at a certain geographical location, yielding increasingly decorrelated gains when separated at larger distances. Using this stochastic channel model, different scenarios are then analysed in terms of secrecy, depending on what type of additional knowledge the eavesdropper has about the propagation environment.

Along similar lines as above, and still mainly in the context of narrowband radio channels, Yin et al. [YYC+13] provides a characterisation of channel polarisation-(an)isotropy. A stochastic model of the channel impulse response is formulated in terms of the so-called bidirection-delay-Doppler spread function of the propagation channel between individually polarised transmit and receive antennas. In the narrowband case and under the assumption that the channel is polarisation-isotropic, analytical expressions are derived for the K factor and the delay spread as functions of the TX-RX polarisations. Both quantities are found to be bivariate quadratic in, and to vary continuously with, the TX-RX polarisations $(p_1, p_2) \in (0, \pi) \times (0, \pi)$. From indoor channel measurements it is however found that the theoretical results are violated, i.e., the properties derived under the assumption that the channel is polarisation-isotropic are not supported by the measurement data.

9.3.3 Combining Partial Models into a Unified Propagation Model Using Signal Flow Graphs

In Vainikainen et al. [VVH13] a framework has been proposed for how to combine various partial models into a single (unified) propagation model, e.g., combining signal strength variability and polarisation anisotropy. The proposed framework can be seen as a network of individual propagation elements. These individual elements are combined to form signal flow graphs as known from, e.g., control systems. Thus, the main idea is to use small and simple building blocks to form branches of a signal flow graph describing complicated multi-dimensional connections between TXs and RXs in mobile communications or navigation. A crucial part in the creation of these signal flow graphs is the elimination of propagation elements that do not contribute significantly to the overall power.

9.3.4 Point Processes in Analytical Channel Modelling

In Jakobsen et al. [JPF14a], the theory of spatial point processes has been employed to revisit a particular class of time-variant stochastic radio channel models. Specific for all models in this class is that individual multi-path components are emerging and vanishing in a temporal birth-death alike manner. The suggested point process perspective has proven to be analytically advantageous, among others by circumventing enumeration issues of traditional modelling approaches. Specifically, traditional integer-indexed sums has been replaced by equivalent expressions indexed by points from spatial point processes. In essence, this allows for keeping track of individual path components by use of the same stochastic mechanism which is also generating the temporal birth-death behaviour of the channel. Under facilitating assumptions the time-variant transfer function of the channel is then shown to be wide-sense stationary in both time and frequency (despite the birth-death behaviour of the individual multipath components). The practical importance of being able to analytically characterise such temporal birth-death channel models is clearly evidenced since key parameters enter explicitly in measurable quantities such as the PDP.

Besides its analytical virtues, the point process perspective is also useful for purposes involving computer simulation. In [JPF14b] it is demonstrated how heuristic and approximate simulation guidelines from earlier channel modelling literature can be replaced by exact (i.e. non-approximate) equivalents. In particular, theoretical properties of Poisson point processes play a fundamental role in obtaining exact schemes for initialisation and time-discretisation.

9.3.5 Statistical Models of Antennas: An Overview

Since the transmit and receive antennas are part of the radio channel it seems out of balance to ignore the inherent stochastic nature (imperfect characteristics) of antennas, while spending vast efforts in investigating and modelling radio wave propagation by itself. Statistical modelling approaches are traditionally uncommon in antenna designs, but this need not necessarily be the case in the future since the current trend is that antennas become more and more complex and hence intuition and experience renders less useful in practice. Quite a lot of scientific work have recently been carried out in the area of statistical antenna modelling. This is evident from the overview provided in [Sib13] along with the list of references therein. One major challenge is argued to be with demonstrating an overall benefit of using statistical modelling approaches compared to current antenna modelling traditions. It is suggested that the most direct way of getting statistical antenna models to be used in practice is to incorporate them, possibly combined with propagation models, into wireless communication standards.

9.4 Parameter Estimation Techniques

The estimation of channel parameters plays an essential role in bridging the more practical oriented channel sounding task and channel modelling. Channel parameters to be derived from the gathered data sets are well connected to the channel models discussed within the subsequent sections. Within this section three main contributions in the field of parameter estimation can be distinguished:

1. Estimation of basic parameters as fading distributions, spatial degrees of freedom or from the Delta-K model.
2. Estimation of specular and dense multipath parameters which are subject to a defined data structure based on the understanding of the propagation phenomenas.
3. Estimation of multipath clusters and their parameters based on the results of high-resolution multipath parameter estimations.

9.4.1 Estimation of Basic Channel Parameter

Among various channel parameters the fundamental ones are the path loss and fading statistics. Both parameters are important for a proper system design since the outage performance depends on it. Fading basically arises from the superposition of multiple propagation paths (fast fading) and from the interaction of obstacles (shadow fading) as buildings, trees, and so on. The analysis results and interpretation of it depends on the proper definition of analysis methods. A traditional way to analyse the path loss and mitigate the fading is to average over a certain region in space/time conform to the wide-sense stationarity assumption. In Bühler et al. [BZG14] this approach is questioned since the arithmetic mean in the sense of averaging is not robust against outliers and noisy samples. It is proposed to consider the median (e.g., at 50% percentiles) as a more robust estimator. Furthermore the authors proposed not to ignore samples close to or below the noise floor. The results at their selected examples showed that the estimation of the fading statistics based on the two proposals is more robust and reliable.

Another fundamental channel parameter or characteristics is the spatial degree-of-freedom (DoF) since it allows insight to the maximum number of antenna elements in a given volume which optimally support the spatial diversity and multiplexing. In Haneda et al. [HKD$^+$13] an estimation method of the spatial DoF is presented. The proposed concept is based on the spherical wavemode expansion of a radiated electromagnetic field in a MISO configuration and has been verified inside an anechoic chamber before measurements have been performed in an realistic indoor lab scenario. The results show that the spatial DoF is increasing if the size of volume (aperture) and/or the carrier frequency increases. Whereby under LoS condition the increasing volume is not that effective as under NLoS or OLoS. Finally based on the solid angle an antenna-independent metric of the multipath diversity has been introduced. Together with the antenna aperture size the solid angle gives an upper bound of the spatial DoF within the wireless propagation channel.

A different view point into the stochastic radio channel modelling is approached by Jakobsen [Jak14]. The discussion is based on Suzuki's Delta-K model which was proposed to render with a flexible stochastic mechanism the observed grouping (clustering) of multipath components, whereby the renewal Poisson point processes plays an important role. Based on the explicit expression of the log-likelihood function derived in [Jak14] a joint maximum-likelihood estimator for the three parameters of the Delta-K model could be derived. The authors argue the results are promising in the perspective that based on the analytical description and available tools the research on stochastic radio channel modelling could attract again more interest.

9.4.2 Estimation of Specular and Dense Multipath Parameters

It is commonly understood that high-resolution multipath parameter estimation is based on a structured data model comprising two conceptional different contributions: SC and DMCs. The reliability of the estimated parameters depends on the accuracy of the considered data model, used antenna arrays as well as the estimation method. Within that scope several aspects have been discussed.

In Käske [K̈11], the influence of spectral windowing on the estimation results of the DMC is studied. It is known that window functions play an important role in signal processing and estimation, where in practical applications typically only a limited number of data samples are available. In particular in the case of high signal power/high dynamic range of the data the estimation of weaker contributions can be affected by sidelobes of the window function. It is shown in Käske [Kï1] that using a Hann window instead of a rectangular one increases the estimation accuracy of the noise variance, while the accuracy of the peak value of the DMC is slightly decreased.

Another interesting aspect in high-resolution multipath parameter estimation is the usage of the complex beam pattern. Both contributions [ST13b] and [ST13a] address this, where in Skoblikov and Thomä [ST13b], the impact of a selected elevation cut; and in Skoblikov and Thomä [ST13a], the polarimetric quaternion effective aperture function are discussed. If a uniform linear antenna array (ULA) is considered during the channel sounding and subsequent high-resolution parameter estimation step it is important to know that the estimation results are limited by the geometry of the array: methods based on ULAs (oriented in azimuth) are not able to estimate parameters in the elevation domain. Hence, a fixed elevation angle is often used assuming that all multipath components impinge on the array under that angle. However in practical situations this can not be meet. To discuss the consequences the ambiguity cone is introduced in Skoblikov and Thomä [ST13b]. It basically comprises the effect that multiple azimuth and elevation pairs/combinations will lead to the same phase pattern (measurement result) along a ULA. Subsequently, the estimation results get biased and the variance of the multipath model parameter is increased. In the case that the multipath components occur only in azimuth with common elevation (other cases are also possible, if the ULA is appropriate oriented) the selection of the corresponding elevation cut is necessary. If the selected cut differs from the common propagation elevation again the estimation results in azimuth undergo a bias and increased variance. Both lead to an increased non-realistic spread of the multipath propagation in the spatial domain.

For reliable high-resolution parameter estimation based on channel sounding data sets with dedicated experimental antenna arrays the full polarimetric complex beam pattern of the antenna arrays is required [ST13a]. Typically, the antenna manifold is measured for a finite grid only. For high-resolution estimation of the spatial multipath parameters a precise pattern interpolation is therefore required. A widely used approach is the effective aperture distribution function (EADF) which provides a compact description with low complexity and computational effort for pattern interpolation as well as for calculation of their derivatives (which are necessary for gradient-based estimation approaches as RIMAX [Kï1, ST13b, ST13a, TGJ+12]). However, the EADF is not directly designed for polarisation. In Skoblikov and Thomä [ST13a] a polarimetric extension to the Quaternion EADF (QEADF) is introduced, where both polarisations are described within one mathematical form and the manifold interpolation is done jointly for both.

The performance of different high-resolution multipath parameter estimation algorithms is evaluated in Tanghe et al. [TGJ+12]. Whereby three algorithms: ESPRIT, SAGE and RIMAX are validated in terms of their robustness under channels including different DMC contributions. While the RIMAX accounts for both SC and DMC parts of the wireless channel ESPRIT and SAGE account only for the SC part. The results for different power ratio's between the SC and DMC part show that estimators which do not consider the DMC part in their data model lead to significant increased estimation errors. In a more extensive study [LKT12], a detailed insight into the influence of inaccurate and incomplete data models on the multipath estimation results is given. It is strongly recommend to include both parts SC and DMC into the data model.

Another interesting approach to investigate the reliability of the high-resolution parameter estimation is discussed in Sommerkorn et al. [SKST13]. Estimators like RIMAX [Kï1, ST13b,TGJ+12, LKT12]) provide the variance of each SC for each propagation dimension. These information can be used to analyse the impact of the estimator on system level performance predictions, e.g., on the MIMO channel capacity. This allows for a different view point on the performance uncertainty of the given estimator. The proposed method consists of 4 main steps: multipath estimation by RIMAX, generation of randomly distorted sets of multipath parameter based on the derived estimation variances, channel synthesis to create the channel impulse responses for the final performance evaluation step. It is important to note that the RIMAX estimation can be interpreted as an antenna de-embedding and the channel synthesis as an antenna embedding, whereby the de-embedding is focused on the dedicated measurement antenna arrays and in contrast to that the embedding is rather flexible in considering more practical antenna arrays. Therefore, the performance evaluation is performed based on the embedding of the measurement antenna arrays and on a typical base station antenna array from Kathrein. In particular for the later array with very narrow beam width in elevation it was of interest to understand if small changes at the direction of departures will create small or large variation at the MIMO performance. The results in Sommerkorn et al. [SKST13] are based on a channel sounding data set from an urban macrocell scenario at Berlin, Germany, and in Käske et al. [KSST14] for the same class of scenario a different data set from Cologne, Germany, has been considered. Both contributions highlight that the estimators variance only weakly affects the

MIMO capacity with slightly larger influence for the more practical-oriented antenna array.

In conventional high-resolution multipath estimators [Kï1, ST13b, TGJ+12, LKT12], it is assumed that the signal waveform of the TX is known. However, in Häfner and Thomä. [HT13], an approach is introduced to estimate the radio channel parameters for unkown TX. Whereby the challenging task is to derive an approach which jointly allows to estimate the channel impulse response (aka blind channel estimation) and to estimate the multipath parameters. Two different solutions, the constrained maximum likelihood and the channel cross-correlation cost function, could be derived which are only dependent on the channel parameters. For minimising the cost functions the Levenberg–Marquardt algorithm was applied. Monte-Carlo simulation-based-results showed that both concepts are nicely applicable.

9.4.3 Estimation of Clusters in Multipath Propagation

Well-known channel models as from COST273/2100/IC1004 or SCM/E or the WINNER family assume that the specular multipath components arrive/depart in clusters. Components with similar properties in the different parameter domains are grouped together—as a cluster. The estimation of such clusters as a post processing step of the above discussed high-resolution multipath parameter estimation is a challenging task since standard approaches as e.g., K-means require a reliable initialisation as well as estimation of the model order (number of clusters). The model order estimation can be considered as pre- or post-processing step, whereat in the latter usually many different clustering results are validated by so-called cluster validation indices (CVI). The contribution from Schneider et al. [SIH+14] studies the robustness and performance of different clustering algorithms based on synthetic channel data sets generated by the WINNER channel model. Within the proposed evaluation framework all clustering approaches have to process the same data sets which allows for the derivation of comparable performances. The WINNER model is taken since its data model is similar to the one considered by common high-resolution parameter estimators such as RIMAX and moreover it is easy to adjust it for different number of clusters and cluster sizes. The spatial dimension of a cluster can be controlled by the inner cluster spread in elevation as well as azimuth, whereby for simplicity only the inner azimuth spread of arrival was changed. In Schneider et al. [SIH+14] two different clustering concepts are compared: the K-power-means and a new hierarchical Multi-Reference Detection of Maximum Separation (MR-DMS).

Within both approaches the selection of the final clustering result (i.e., number of clusters) is based on the fusion of different CVIs. The statistical results showed that the hierarchical MR-DMS outperforms the K-power-means in terms of the probability to estimate the correct number of clusters provided by the WINNER model.

The relation between physical clusters and clusters obtained by high resolution parameter estimation is investigate in Zhu and Tufvesson. [ZT12]. This research work is challenging since the multipath parameter estimators are not directly providing any information on the number of interaction points which the multipath undergoes between the TX and RX. The method used in Zhu and Tufvesson. [ZT12] combines the cluster identification based on a ray launching tool (including a detailed map of the environment) and the clustering results based on the K-power-means. The results show that single bounce (single interaction point) clusters can easily related to physical clusters within the environment. Whereas the physical clusters in the case of multi-bounces are not easy to identify. It is interesting to note that single physical interaction points (scatterer) can simultaneously be part of a single- as well as a multi-bounce cluster. This is in contrast to the COST2100 channel model, where each type of cluster is treated separately. Furthermore, the cluster visibility region, cluster spreads and the cluster selection factor are derived for the considered scenario.

9.5 COST 2100 Channel Model Updates

The COST2100 model is nowadays a well known Geometry-Based Stochastic Channel Model (GSCM). As it was defined in the the previous COST action, COST 2100, there have been further parameterisations of the model for larger range of scenarios and functionality, such as dense multipath, has been added to the publicly available source code. There have also been suggestions for extensions, ideas for further improvements of the model itself or for GSCMs in general. In this section we start discussing some specific considerations on GSCMs and the we review the most important updates to the COST 2100 channel model and channel model implementation.

9.5.1 Scenario Distance

When defining propagation scenarios it is often convenient to refer to the environment where the system is supposed to be used, e.g., indoor office, outdoor-to-indoor, suburban, rural macro cell etc. Often, those propagation scenarios are further specified by the expected mobility, whether it is a bad or "typical" case, LoS, NLoS etc. However, from a modelling point of view, such an approach may lead to an over-specification of the scenarios, and there migth be cases that from a modelling point of view are very similar, despite their different names. In [NSK15], an objective method to characterise the similarity between large scale parameters of different scenarios and between different measurements in the same scenario was developed. As a similarity metric the authors used the mean Kullback–Laibler divergence, that measures the distance between two distributions. In the special case of two multivariate normal distributions, which is a common case in contemporary channel models such as WINNER and COST2100, the divergence can be calculated as

$$D_{\mathrm{KL}}(P\|Q) = \frac{1}{2\log_e 2}\left[\log_e \frac{\det(\sum_Q)}{\det(\sum_P)} + tr\left(\sum_Q^{-1}\sum_P\right) + (\mu_Q - \mu_P)^T \sum_Q^{-1}(\mu_Q - \mu_P) - k\right], \tag{9.5}$$

where $\sum$ is the $k \times k$ covariance matrix and μ is the mean, respectively, of the multivariate normal distributions P and Q. However, this divergence is not symmetric and therefore the *mean* Kullback–Leibler divergence is used,

$$D_{KL}(P\|Q) = \frac{1}{2}(D_{\mathrm{KL}}(P\|Q) + D_{\mathrm{KL}}(Q\|P)). \tag{9.6}$$

By using the divergence it was concluded that the grouping of the 18 (sub-) scenarios in the WINNER model not always correspond to the smallest distance. With the approach it is possible to, in an objective manner, combine propagation scenarios that from a large scale parameter perspective anyway are quite similar to each other. Similarly, the method can be used as a measure of show close the large scale parameters from a measurement are to the reference scenarios.

9.5.2 Polarisation Characterisation

Polarisation is today included in the major GSCMs like COST 2100 and WINNER models. Typically, the two polarisation states, horizontal (H) and vertical (V), are modelled as independent and thus with independent phases. If slanted polarisations are used, e.g., $\pm 45°e$, the standard cross polarisation ratios alone do not provide information about the ratio between co- and cross polarisation reception; It is important to consider the phase relation between the H and V components as well. The behaviour of polarisation states for slanted polarisation was investigated in [KyK13]. It was concluded that for outdoor suburban channels it seems that the polarisation state for $\pm 45°$ transmission is preserved quite well, whereas circular polarisation has a tendency to have more of an elliptic shape at the RX. For indoor channels, it seems that the polarisation states are quite well preserved for the LoS component, but for the NLoS components the relative phase between the two polarisation states seems to have a more uniform distribution, but it is not completely random. Hence, the WINNER-modelling approach with uniform phase distribution for the NLoS paths and zero phase difference for the LoS path may provide a reasonable approximation, but further work is needed in the area.

9.5.3 Vitual Multi-Link Measurements versus True Multi-Link Measurements

Multi-link communication has become more and more important over the last years, and hence there is a need to capture and characterise the multi-link behaviour in a proper way in measurement campaigns in order to extract, e.g., the joint distributions between different large scale parameters or at least the correlation between them. Ideally, synchronous multi-link measurements should be performed and used for the characterisation, but due to practical limitations this is not always possible. Hence, for static environments repeated measurements along the same route, but with different base station positions, are sometimes used for characterising the multi-link behaviour. The natural question, which was examined in [DFT14b] is though whether both those measurement approaches give the same statistical distributions of the large-scale parameters. For the small-scale behaviour and instantaneous values it is clear that the results of the two approaches give different values, but for the studied static semi-urban microcell scenario the deviation between the results based on the two approaches differed only marginally. No significant difference was found for the multi-link parameters obtained from the synchronous measurements and the repeated measurements, which supports the results in [NKS11].

9.5.4 Cluster Multiplexing

In [DRZ15], a new multiplexing concept was studied, where the aim was to find the probability of having only one user connected to specific cluster, or equivalently a dedicated visibility region for that specific user. The calculated value gives and upper limit of the probability of successfully using directional-based

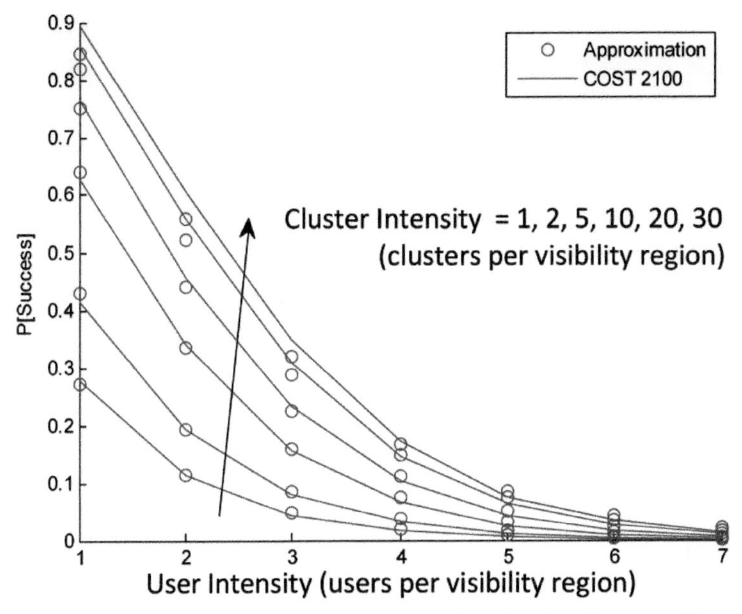

Figure 9.13 The probability of successful cluster multiplexing, i.e., that a specific user has a dedicated cluster.

multiplexing, e.g., beamforming, to communicate with the user of interest give no other collaboration. The main limiting factor for successful reception is the product of the area of the visibility region and the intensity of the users, which in some sense is quite obvious. The probability of successful cluster based multiplexing in a system is close zero when this product is larger than 7.

9.5.5 Physical Clustering

The clusters can often be connected to physical objects in the environment, this is especially true for single bounce clusters. Larger objects in the environment often give rise to a number of MPCs that have similar delay and angular parameters, e.g., a building often have a number of scattering points and therefore contributes with several MCPs to the impulse response. In Zhu et al. [ZHKT14], the concept of physical clusters are introduced and the properties of them are analysed. The physical clusters are extracted based on a measurement-based ray launching tool [ZST12], where the measured estimated MPCs are tracked and visualised in a 3D-map, see Figure 9.14. In this way the interaction points and the interacting objects can be identified and clustering can be performed based on the physical location of the interacting

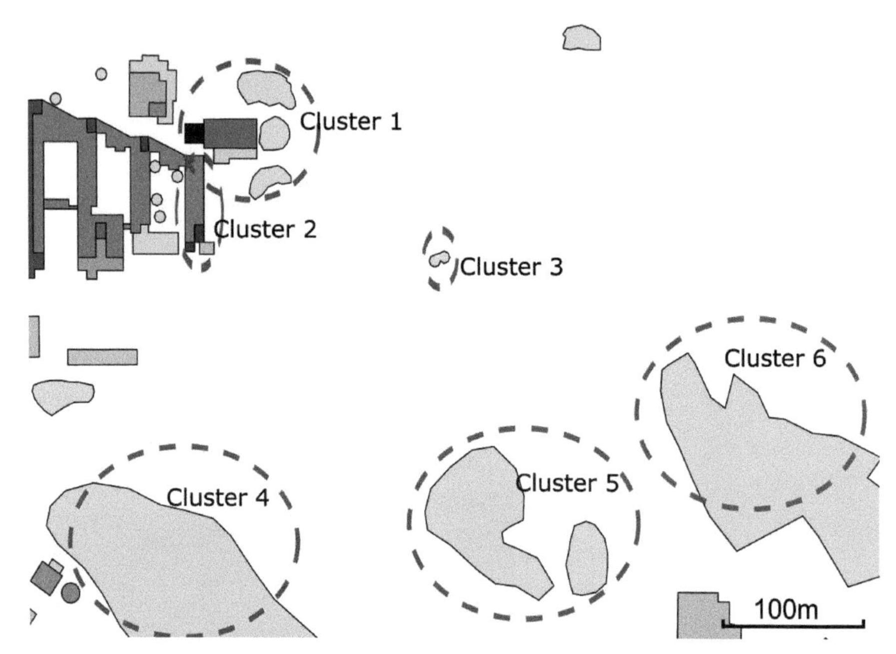

Figure 9.14 Six physical clusters in a sub-urban scenario.

points. The physical clusters show somewhat longer life time (i.e., larger visibility regions) compared to the parmeter-based clusters. They can constitute single-bounce and multiple bounce clusters at the same time, which is not the case for the carameter based clusters. In channel simulations, it was seen that it is somewhat easier to control the delay spread when using physical clusters instead of parameter-based clusters, but the extraction of physical clusters is quite complicated and a more sophisticated physical cluster algorithm is needed in the future. For multi link simulations, the physical cluster approach might be an attractive way forward as the extraction of common clusters to a large extent is based on a physical interpretation of the cluster locations. Parameters for physcial clusters for an sub-urban scenario at 300 MHz and for an urban scenario at 5.3 GHz are given in [Zhu14].

9.5.6 Dense Multipath Component Add On

The propagation of Energy in the radio channel is seen to have two parts: specular components, typically coming from reflections from-large physical objects, and DMCs, that represent the part of the received signal that can not be resolved in angle and delay. In the COST 2100 model, DMC is represented by many weak Multipath Components (MPCs), each cluster of specular components has a corresponding cluster of DMC, with the same cluster centroid, but with a somewhat larger angular spread and delay spread. In VVirk et al. [VHW14] the implementation of the DMC add on for the COST 2100 model is briefly introduced. The DMC add on is a parallell function to the function generating specular components, the overall structure can be seen in Figure 9.15

9.5.7 Massive MIMO and Distributed MIMO

Massive MIMO is an emerging technology that could enable a significant improvement in data TP, link reliability, and energy efficiency. The key to massive MIMO is to have a large number of antennas at the BS, serving multiple users equipped with rather few antennas, typically one or two. If the antenna array at the BS is physically large, it is important to consider possible variations in the channel characteristics over the array, for example

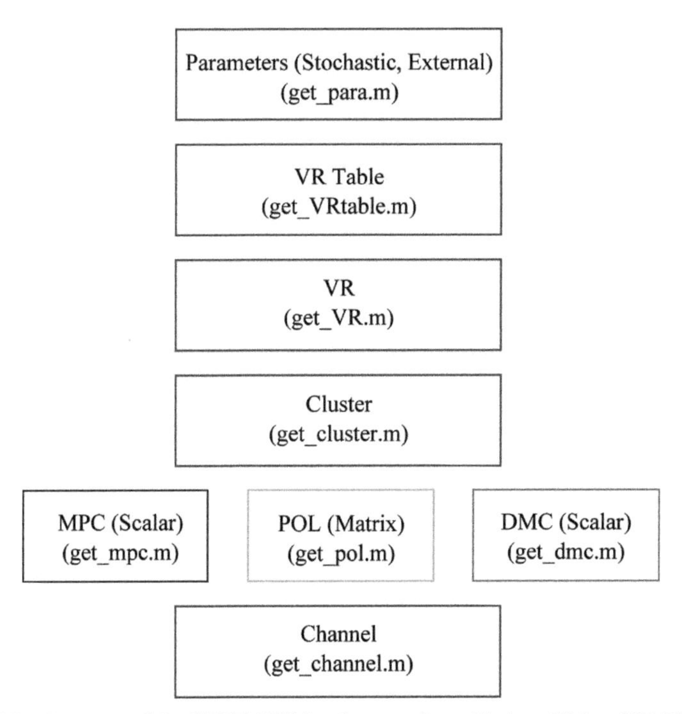

Figure 9.15 Structure of the COST 2100 implementation with the additional DMC function.

with respect to angular spread, delay spread, and large-scale fading, and include such variations in the models in an appropriate way. The COST 2100 channel modelling framework has the advantage that the scatterers in the clusters have a explicit position in the simulation area, hence the effect of spherical wavefronts are implicitly taken care of in the model, which is not the case in, e.g., the WINNER approach. In [Gao16], an extension of the COST 2100 model for massive MIMO was proposed where the concept of visibility regions is applied to the BS side as well. Conventionally, visibility regions for the clusters are applied only on the MS side. By using this concept at the BS side it is possible to let the different BS antennas "see" different clusters, even though they are co-located. By allowing a linearly decreasing or increasing gain function for each visibility region, it is possible to capture and model varying cluster power levels over larger array structures, a property that has been seen in measurements with physically large arrays. A schematic illustration of the BS visibility region concept is shown in Figure 9.16.

Parameters for the BS visibility regions have been extracted from a massive MIMO measurement campaign in Lund, Sweden in a sub-urban NLoS scenario. For each user the total number of clusters that can be seen

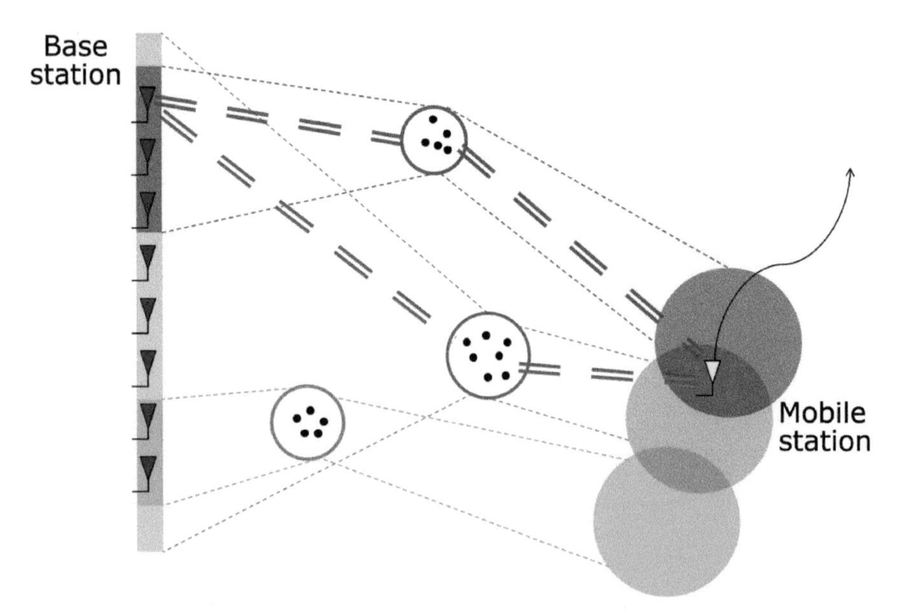

Figure 9.16 Extension of the concept of cluster visibility regions to the base station side. Each cluster has two types of visibility regions: MS-VR and BS-VR.

over the physically large array, in this case a 7.4-m long linear array, can be approximated to follow a negative binomial distribution, the median total number of clusters per user over the array is 10 (r = 2.43 and σ = 0.16 for the negative binomial distribution). The length of the BS visibility regions is modelled as log-normal with median 0.7 m and a standard deviation for the log value of 2. Finally, the slope of the linear (in the dB domain) gain function is modelled as Gaussian with zero mean and standard deviation of 0.9 dB/m for longer visibility regions (above 2 m), whereas the are modelled without a slope for shorter visibility regions.

The influence of the distance between the antennas at the BS side for a massive MIMO or somewhat distributed MIMO setup was investigated in [DFT15]. Especially two-modelling approaches to reflect the fact that the correlation of large scale parameters tend to decrease with increased antenna separation were investigated when a cluster that is created from the same physical object is used for communication. In the first approach, the decorrelation effect was modelled using a power profile for the visibility region, similar to the gain function in [FGD+15], to have an increasing or decreasing power contribution from the different parts of the physical cluster. In the second approach, an observation window that let the BS "see" different parts of the physical cluster was applied. Both methods found to capture the basic behaviour of decreased correlation for increased BS antenna distance and can be used to describe the effects seen in the measurements, the first approach is more consistent with the COST 2100 modelling approach, but further measurements and analysis are needed to validate the approaches.

The high-spatial resolution that is achieved in massive MIMO makes it a good candidate for some tricky scenarios where the used density is high and often under LoS conditions, e.g., during sports events in a stadium or during a concert in an open field or at an arena. To investigate whether massive MIMO has the potential to resolve and separate closely spaced users in LoS conditions, a measurement campaign was conducted and analysed in [FGD+15]. As opposed to the majority of massive MIMO measurement campaigns the measurements were fully synchronous with eight users confined in a circle with a 5-m diameter. It was concluded that the singular value spread in general was low, which is an indication that also the weakest user gets reasonable communication possibilities. The sum-rates achieved through massive MIMO and linear precoding (zero forcing or matched filter precoding) was at the levels of 70–80% of the optimum dirty paper coding capacity, whereas conventional

MIMO schemes fell much shorter attaining 18–34% of the capacity. Also in terms of fairness, it seemed that all users were allocated power while maximising the sum-rates, which is an indication that all the users experienced favourable propagation conditions. All in all, the conclusion was that it seems that massive MIMO is able to reasonably well spatially separate the closely spaced users and provide simultaneous communication possibilities also in the considered scenario, which conventionally is seen as tricky.

10

Innovations in Antenna Systems for Communications

**Chapter Editors: B. K. Lau, Z. Miers, A. Tatomirescu,
S. Caporal Del Barrio, P. Bahramzy, P. Gentner, H. Li and G. Lasser**

10.1 Introduction

As wireless communications continue to grow in importance by providing exciting new services beyond the conventional voice and text-only communications, antenna systems must evolve in order to keep up with ever increasing requirements and technical challenges. One prominent example is the demand for high data rates in mobile communications, due to data intensive applications such as interactive gaming and video streaming. Apart from the challenge of implementing multiple antennas in physically small terminals to support multiple-input multiple-output (MIMO)-enabled data rate enhancement, suitable measures are also needed to ensure that the antenna performance does not degrade due to proximity of the user and variability in the channel conditions. Moreover, the rapid adoption of machine-to-machine (M2M) communications and internet-of-things (IoT) concept also provides ample opportunities for innovations in antenna systems, including the use of smart antennas in RFID technology to enable smart wireless sensor nodes for tyre pressure monitoring.

In this context, Chapter 10 deals with recent advances and innovations in antenna systems for communications. Section 10.2 first examines the design of multiple antenna systems, focusing on the optimisation of antenna performance using decoupling techniques. Apart from designing MIMO terminal antennas, where moderate isolation (e.g., 10–15 dB) is sufficient, the section also presents new results on antenna decoupling for applications that require very high transmit-to-receive isolation (e.g., 80 dB for full duplex operation). As a promising new approach to provide efficient MIMO antennas, the growing field of antenna design using the theory of characteristic modes (TCMs) is summarised. Then, Section 10.3 explores the topic of smart reconfigurable

Cooperative Radio Communications for Green Smart Environments, 383–422.

antennas, where antennas are designed to be frequency reconfigurable to fulfill extreme bandwidth requirements of long-term evolution (LTE) systems. The emphasis is on its application to architectures advocating very high transmit-to-receive isolation, since such architectures favour the use of antennas with narrow instantaneous bandwidth. More ambitious reconfigurable antenna structures that can be harmonised to the immediate environment including user and propagation channel are also introduced. In Section 10.4, the attention is turned towards RFID and sensor antenna innovations, with different antenna design and modelling issues discussed for a variety of applications, with operations in high frequency (HF), ultra high frequency (UHF), and ultra-wideband (UWB) bands.

Following the recent innovations in antenna design for various applications, Section 10.5 deals with the holistic characterisation and measurement of antenna performance. Topics covered include latest advances in performance metrics, modelling of the user-channel effects and antenna evaluation in realistic scenarios. One important message to convey in this section is that, ultimately, smart innovations in antenna systems are not meaningful if they cannot be proven to provide real performance benefits in real life operations. Lastly, Section 10.6 is dedicated to a discussion of antenna measurement techniques, especially on the measurement of antenna patterns. Although measurement of antenna parameters is not a new topic, many challenges remain in obtaining good quality measurements, including the effect of feed cable on small antenna measurement, accurate pattern measurement of millimeter wave (mmW) frequencies and distortion of antenna pattern measurement due to support structures.

10.2 MIMO Antenna Innovations

High transmission rates of next generation communication systems require significant attention to individual antenna element design. Implementing multiple antennas in both the base stations and user terminals is key to increasing the channel capacity without sacrificing additional frequency spectrum and transmit power. In theory designing both base station antennas as well as mobile terminals is well researched and documented. However, practical design constraints prevent traditional antenna implementations from performing well in MIMO systems. The compactness of today's terminals is one such constraint that complicates the design of MIMO antennas.

In an effort to solve some of the problems associated with implementing multiple antennas in mobile terminals, several areas of terminal antenna design are actively being researched. The most challenging application of MIMO for mobile devices is given by the electrically small nature of mobile phones in the lower end of the frequency spectrum used in LTE. Different decoupling and decorrelating methods are discussed and evaluated further, ranging from the more straightforward antenna placement optimisation to some more complex methods such as the self-interference cancellation. Key design and implementation challenges are discussed with the focus on MIMO performance of the antenna system. Another approach to solving problems caused by compact MIMO terminals is through the use of TCMs. Characteristic modes allow a designer to determine all the orthogonal modes a structure is able to produce and determine where and how to feed these modes. This allows for optimal MIMO antenna placement as well as physical insights into how to feed each orthogonal mode [LLH11].

10.2.1 MIMO Antenna Decoupling and Optimisation

From a system perspective, the design of MIMO antennas on an electrically small ground plane is a complex optimisation problem due to the fact that the system's capacity depends on power balance, level, and correlation in each of the communication links which are a function of antenna pattern and channel properties [DFPS09]. The placement of the separate antenna elements has been investigated using multiple monopoles and inverted-F antenna (IFA) antennas on a card type ground plane [OC04, ANW11] for the higher frequencies used in LTE. A more general and systematic approach is presented in Karimkashi et al. [KKK11] where infinitesimal small dipoles are used to model the interactions between the antenna elements and the ground plane offering a fast optimisation process to enhance MIMO performance. However, the lower frequency bands of LTE are still not not addressed. Since the shape of gain pattern for electrically small antennas (ESAs) is very similar to a dipole pattern, a simplified metric using the efficiency and antenna correlation is presented for evaluating MIMO performance in Derneryd et al. [DSSW11]. This metric, also know as the multiplexing efficiency [TLY11], provides an efficient cost function for the multi-objective optimisation presented in Derneryd et al. [DSSW11] of the design for a dual band dual-antenna system which supports the 750–800 MHz and 2.5–2.7 GHz frequency bands.

Depending on the electrical size of the device, in some cases the optimal placement for the antenna elements of a MIMO antenna does not guarantee acceptable MIMO performance due to the high coupling between antennas. Thus, methods to cancel the coupling have been investigated in the past. These can be split into two main families, one containing methods to isolate the output ports of the antennas and the other focusing on decoupling the antennas.

Implementations from the first family can be seen as early as 1976 in Andersen and Rasmussen [AR76] followed by different embodiments of this concept as parasitic scatterers [LA12], a neutralisation line [CLD$^+$08], using a feeding network under the form of hybrid couplers, decoupling networks or lumped elements [TPK$^+$11, BYP09, CWC08]. The main drawback of this solution is the loss of bandwidth and efficiency depending on the level of initial coupling between the antennas [LAKM06]. The effect of having a smaller bandwidth can be compensated by introducing a tunable decoupling mechanism as shown in Tatomirescu et al. [TPFP11]. A simple tunable capacitor can extend the operating frequency of the neutralisation line by up to a factor of five. A lumped component feeding network is illustrated as well. However, the number of components and the high insertion loss attributed to them makes this solution undesirable.

For a given geometry, the concept of introducing an extra coupling path to cancel the initial coupling between radiators can be implemented using a several types of scatterers, as illustrated in Andersen and Rasmussen and Lau and Andersen [AR76, LA12]. In Zhang et al. [ZLSH11], two planar inverted-F antenna (PIFA) elements designed for 2.4 GHz are well isolated using a T-shaped slot in the ground plane placed in between the two antennas. The slot enables the design to be very efficient and compact array for a MIMO Wi-Fi application.

There is an asymmetry in demand for uplink data rate versus that of downlink data rates, with higher user interest in download than for upload. Therefore, the classical arrangement in current smart phones is to use MIMO only during downlink communications and utilise only one antenna for uplink communications. In Tatomirescu et al. [TBP13], a scenario with two diversity antennas is investigated. The two dual band folded monopoles are collocated at one end of the PCB to save space in the mobile phone. Because the antennas cover only the receive part of the spectrum, they can be made very compact, and using tunable matching networks, they are able to cover several bands. Although the matching network has only a tunable capacitor and two fixed inductors, matching is achieved over the whole tuning interval for both bands. The low-band antenna correlation is controlled by a simple tunable capacitor in between the elements which controls the backscattered signal thus achieving pattern diversity and a tunable correlation coefficient. Simulations with and without the user in isotropic and anisotropic environments confirm that the initial free space loss is compensated by a lower degradation due to user effect.

In Tian and Lau [TL12], it was shown that six degrees-of-freedom (DOF) can be achieved with a compact co-located array. They are the result of angle and polarisation diversities, as in the case of an ideal E/M dipole array. Further confirmation is obtained when a larger array is studied. Space diversity is found to contribute significantly to the larger array's effective degree-of-freedom (EDOF) performance. This finding indicates that the space diversity should be exploited whenever possible in array design, in order to complement angle and polarisation diversities and maximise overall EDOF performance, as shown in Figure 10.1. However, the proposed design is only intended to be proof of concept. Implementing the concept practically in the form factor of mobile terminal remains a significant challenge.

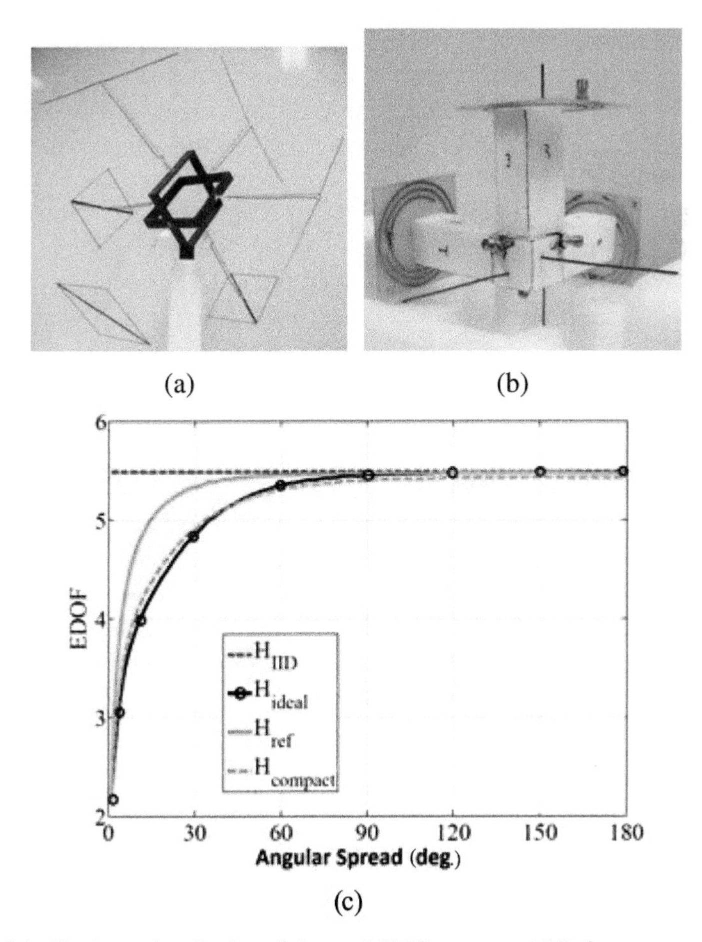

Figure 10.1 Design and evaluation of six-port MIMO antennas: (a) Reference transmit array, (b) compact receive array, (c) EDOF of 6 × 6 independent and identically distributed (IID) channel **H** and 6 × 6 channels with ideal, reference, and compact arrays for different angular spreads.

There are some examples in the literature of a topology with both balanced and unbalanced antenna combinations using a fixed [HHG11, IKA+11] or tunable [TAP12] balun. This type of systems has been demonstrated to obtain good isolation. One of the notable drawbacks of this approach is the high Q of the balanced element and the cutback required to minimise the ground plane shorting effect. The narrow-band (NB) operation of the balanced elements can be used for spatial filtering configurations where the Receiver (Rx) and Transmitter (Tx) chains have a dedicated tunable antennas which acts as a duplex filter; this is illustrated in Bahramzy et al. [BSP14]. The work in Bahramzy et al. [BSP14] also investigated the isolation of three antennas in a MIMO mobile handset at 700 MHz, where apart from the isolated Tx and Rx antennas, an additional Rx antenna is added for downlink MIMO. Due to the electrically small structure of the terminal, achieving decoupling among three antennas is a major challenge. Further results on this topic were reported in Bahramzy et al. [BSP15], where decoupling was achieved through the use of a loop antenna and two regular IFAs. This is due to the loop's inherently different electromagnetic properties relative to those of IFAs. One potential practical limitation is that the loop antenna requires some ground clearance.

Recently, full-duplex (FD) systems have attracted a lot of interest in the research community for their ability to allow a radio terminal to transmit and receive simultaneously on the same frequency band. FD systems can potentially double the data rates that can be achieved in half-duplex systems. However, it is very difficult to achieve this in practice due to implementation challenges. Usually, the self-interference (SI) cancellation techniques that were proposed in the prior works use a combination of analog-circuit and digital techniques, or exploit propagation properties [JCK+11, DS10, BMK13]. In Snow et al., Khandani et al., and Aryafar et al. [SFC11, Kha13, AKS+12], the focus is on using the transmit beamforming technique for interference cancellation. Most of these works consider the use of SI cancellation techniques mainly for a single-input single-output (SISO) system [ATTP13]. A natural follow-up is the transmit beamforming approach and an investigation of the feasibility and implementation limitation of the method for a compact MIMO-FD antenna system. In Foroozanfard et al. [FTF+14], an antenna system consisting of six patch elements on the top and bottom layers of a three-layer substrate is proposed, where the transmit antennas are located on the two sides of the substrate and Rx antennas are located in the centre. By arranging the transmit antennas symmetrically from the receive antennas, simple beamforming weights can be applied at the transmit elements to cancel out the SI at the central receiving elements. Antenna polarisation diversity and spatial diversity is employed to achieve higher isolation between different ports. The results show a significant amount of cancellation within the range of 60–75 dB. A different antenna setup is investigated in Foroozanfard et al. [FdCP14], where three dual-polarised patch antennas are mounted on the ground plane and radiate in the same direction. Furthermore, the investigation on the effect of the implementation accuracy on the system performance came to the conclusion that passive transmit beamforming is very sensitive to antenna misalignment. However, to remove the sensitivity of the antenna misplacement, active beamforming can be used to adapt the system to the impairments and also any reflection from the environment. This method can be a potential solution for achieving high-SI cancellation where the price is having redundant antennas.

10.2.1.1 Other antennas for high data rate systems

MIMO systems have been adopted in many mobile communication systems. However, the current throughput of MIMO systems will not permanently accommodate the data rates needed by individual users and operators. In an effort to support the growing data rates and to understand the limitations of both MIMO and massive MIMO systems, the characterisation of multiple links is becoming increasingly more important. In order to measure multiple links accurately, a three-dimensional array which is capable of measuring both polarisations is necessary. In Müller et al. [MKH+14] this type of array was built, tested, and optimised for high-resolution parameter estimation. Other efforts have focused not on the channel but on combining UWB and MIMO for significantly increasing total throughput. Achieving this type of system is difficult as each pair of antennas in the UWB MIMO system must maintain a low envelope correlation coefficient (ECC) across the entire frequency band. In Zhang et al. [ZLSH12], a 3.15- to 5.15-GHz two element MIMO universal serial bus (USB) dongle was developed which maintained an ECC of below 0.1 across the entire band. If data rate trends continue, massive MIMO UWB systems may become the standard for high data rate mobile communication systems. In these systems the channel must be well characterised in order to obtain the benefits of both systems.

10.2.2 MIMO Terminal Antennas with Characteristic Modes

The challenges of implementing efficient antennas within small mobile terminals have dramatically increased due to the adoption of MIMO technologies. MIMO systems require implementing multiple antenna elements that are packed into a single electrically small volume. Designing antennas in this type of systems becomes challenging due to the close proximity of radiating elements which will strongly couple to one another.

Numerous techniques have been proposed in the literature to mitigate coupling between antenna elements but relatively little attention has been given to the problem of chassis induced coupling, TCM provides solutions to this problem. TCM is based on the generalised eigenvalue equation $XJ_n = \lambda\ RJ_n$, which is derived by optimising the far-field radiation of a given structure. This equation is often solved using the method of moments (MoM) impedance matrix (i.e., $Z = R + jX$) and when solved provides the currents J_n for each orthogonal far-field which a structure is capable of radiating [HM71].

Initial work done through designing MIMO antennas based on TCM proved it is possible to understand the fundamental characteristics of a mobile chassis and why it is difficult to obtain both wide bandwidth and low correlation among multiple elements. The fundamental characteristic modes of a candy bar style mobile phone can be solved for using TCM and are shown (black) in Figure 10.2(a) [LML13]. For this type of chassis there is only one resonant mode below 1.5 GHz ($\lambda_{\mathrm{Fundamental}}$) which is referred to as the fundamental mode. It should be noted that a mode is resonant when it has an eigen value of zero, if the eigenvalue is negative or positive this indicates that the mode is reactive. The more capacitive or inductive a mode is, the higher quality factor the mode will maintain, and thus when properly fed, will have a maximum bandwidth as set by the quality factor.

Characteristic modes can be utilised to reduce the mutual coupling between multiple antennas on a mobile chassis. An example of this concept was described in Li et al. [LLH11], where it was shown that a single antenna on a chassis can be used to exploit the fundamental chassis mode, whereas a second antenna can be designed to avoid chassis excitation. This same idea was further expanded upon in Li et al. [LLYH12] by co-locating an electric antenna with a couple fed magnetic antenna. This enabled having two fully decoupled antennas while adhering to industry accepted antenna size constraints. The magnetic antenna, which did not couple to the fundamental mode, was fed using a magnetic loop consisting of two half square rings, while the fundamental mode was fed using a standard electric monopole antenna. In this same work it was shown that the TCM-based design is capable of providing nearly optimal MIMO performance. This analysis has been done by comparing the capacity of the TCM designed antenna against both the ideal capacity and the capacity of a typical monopole/PIFA antenna configuration. The channel capacity for the three cases were calculated and presented in Li et al. [LLYH12].

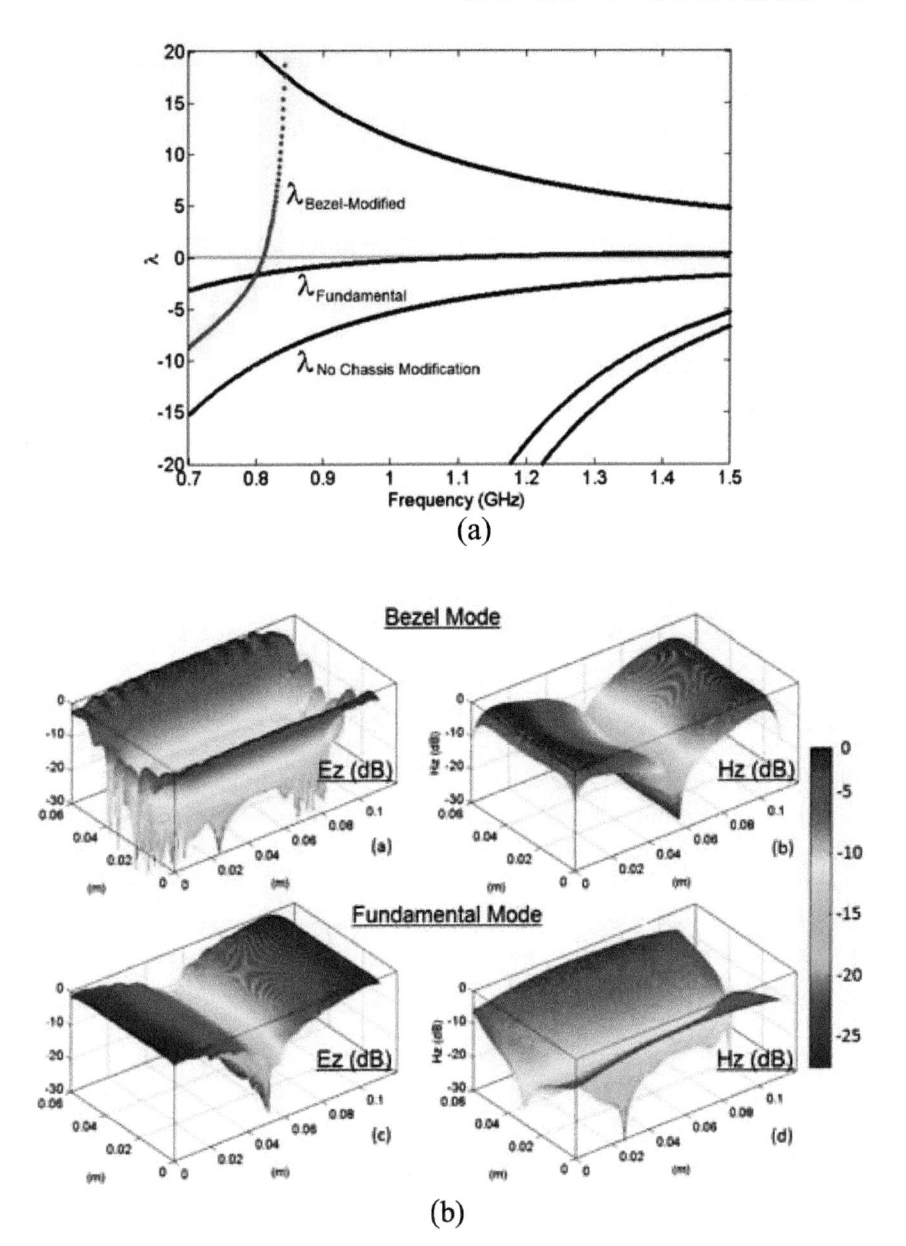

Figure 10.2 (a) First five characteristic modes of flat chassis (*black*), modified bezel mode (*red*). (b) Near-fields of fundamental and bezel modes.

In Li et al. [LLYH12], it was shown that performance increase can be obtained when using TCM as a way to decouple antennas. However, this led to reduced bandwidth and the designed system could not be used for LTE operation, unless the narrowband magnetic antenna can be tuned on demand. Nevertheless, Li et al. [LML13] and Bouezzeddine et al. [BKS13] describe a new design methodology which is based on utilising TCM to develop new modes in a structure which then can be excited. This design approach relies on manipulating the chassis so that more than one characteristic mode is resonant at the frequency of interest. To illustrate this design concept, the authors of Li et al. [LML13] added a metallic bezel to the terminal casing. When this bezel structure is analysed using TCM a new mode becomes resonant below 1 GHz as is shown in Figure 10.2(a) in red, where λ_{Bezel} is the new bezel mode, $\lambda_{\text{Fundamental}}$ is the fundamental mode, and $\lambda_{\text{No Chassis Modification}}$ is no longer present. This new mode is orthogonal to the fundamental chassis mode and as such when the far-field ECC of the two modes is calculated the result is 0. In the companion work of Miers et al. [MLL13] it was shown that through the use of TCM near-fields and currents it is possible to pinpoint possible feeding locations which can excite one mode without coupling into the other mode. The bezel mode was fed using the characteristic reactive near-fields seen in Figure 10.2(b). On either of the short edges of the structure the bezel mode has high-magnetic field strength and the fundamental mode has high-electric field strength. If an electric and magnetic feed elements are placed at this location the magnetic element will feed the bezel mode while the electric element will feed the fundamental mode.

It is possible to use TCM to manipulate a structure to support multiple resonances, as done in Miers et al. [MLL14]. This is applied by cross correlating the area around the feeds with the currents and near-fields of higher frequency modes. Using this information it is possible to determine what modifications to a structure can produce uncorrelated resonances at multiple frequency bands. In the structure described in Miers et al. [MLL14], the authors accomplished this by introducing a small HF resonant structure to create a new mode which the low-frequency feeds excite at higher frequencies.

The application of the TCM is not restricted to small wireless devices but can be extended to larger structures equipped with more than two antennas. The system considered in Bouezzeddine et al. [BKS13] and Bouezzeddine and Schroeder [BS14] was designed to be used as a four-port MIMO customer premises equipment (CPE) antenna system for operation in TV white space bands. Characteristic mode analysis of the chassis of the device was applied to determine location and types of couplers for excitation of four mutually orthogonal superpositions of characteristic modes. A four-port antenna system with high isolation was obtained.

Designing efficient antennas through the use of TCM is not confined to the types of systems described above, many other systems have been designed using this method including, but not limited to, vehicular [IF14], mmW [BVV+14], and dielectric resonant [AEM14] antennas.

10.3 Reconfigurable and Channel-Aware Antennas

With their increasing functionality, mobile phones are embedding better screens, better cameras, larger batteries and more antennas, among others. In order to keep the portability of such devices, a high degree of integration is needed. Section 10.3.1 discusses frequency-reconfigurability as a miniaturisation technique. Moreover, the effects on the antenna total efficiency are investigated. However, when assessing the performance of multi-antenna systems, the efficiency alone cannot provide enough insight into the handset performance. Section 10.3.2 relates the antenna characteristics to the propagation conditions, via the channel capacity. Channel-aware antennas are considered, they evaluate the channel and accommodate to it in order to enhance the capacity.

10.3.1 Frequency-Reconfigurable Antennas

Chipset miniaturisation has seen a large success over the last years. However, antenna volume is ruled by fundamental laws that relate size, efficiency, and bandwidth. To support the latest mobile communication standards, LTE and LTE-Advanced (LTE-A), antennas need to operate in frequency bands ranging from 698 MHz to 2.690 GHz. Tunable antennas are a very promising way to reduce antenna volume while enhancing its operating bandwidth. Tunable antennas use a tunable component in order to reconfigure their resonance frequency. These antennas exhibit an instantaneous narrow bandwidth, that can be reconfigured to a wide range of frequency, thus resulting in an effectively wide bandwidth.

10.3.1.1 Loss mechanism of frequency-reconfigurable antennas

A system using tunable antennas offers a better interference rejection due to the NB nature of the system, leading to relaxed specifications on channel select filters, e.g., SAW filters. Fornetti et al. [FSHB12] uses antenna tuning to increase the handset efficiency and sensitivity. Hence the reliability of the wireless data link and the battery life are enhanced and the pressure on the infrastructure is decreased. A tunable antenna has been built using digitally tunable capacitors (DTCs) provided by Peregrine semiconductors to address LTE bands at 1.7 GHz. These designs exhibited a large amount of loss. However, it must be noted that the loss of performance introduced by the tunable antenna may be acceptable at system level and lead to an overall increase in transceiver performance, hence bringing economical advantages.

The work by Del Barrio et al. [DBPFP12] focuses on the impact of the tuner loss on the efficiency of NB antennas, depending on the location and capacitance of the tuner. The results indicate that different combinations of added capacitance and tuner location can lead to the same resonance frequency, but different losses and required capacitance range. Measurements show a loss difference of 1.2 dB between different combinations of location and capacitors, at the same resonance frequency.

Del Barrio et al. [DBMP14] uses a capacitively-loaded handset antenna and a MEMS tunable capacitor from WiSpry [WiS]. This paper highlights the importance of co-designing the tuner and the antenna to optimise radiated performance. A prototype is shown in Figure 10.3(a), where the antenna volume is 0.5 cc. Its tunability in the low-LTE-bands is shown in Figure 10.3(b). The antenna is highly sensitive to the insertion loss of the tuner, due to its high antenna quality factor. The measured total efficiency decreases as the antenna is tuned towards lower frequencies. It has been noted that the tuner loss is not the only source of loss. The metal loss due to non-perfect conductor (i.e., copper) is significant for a high-quality factor antenna. Thermal loss is inherent to the antenna manufacturing and draws a limit to the achievable performance of tuned antennas.

Simulations of two planar handset antenna designs addressing the low-bands of LTE are presented in Tatomirescu et al. [TAP13]. They are frequency-reconfigurable but differ by their quality factor (Q). The tunability is enabled

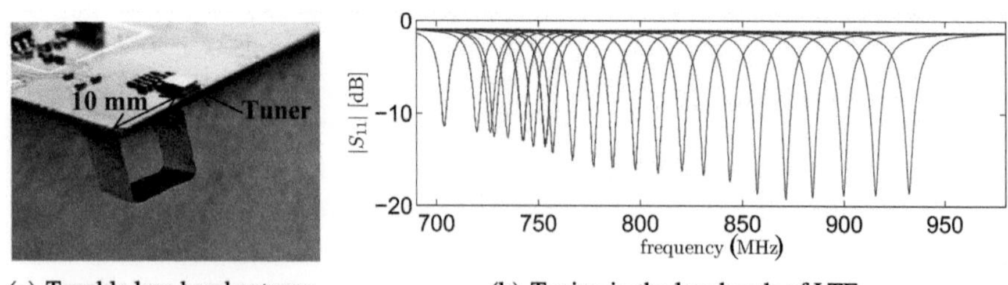

(a) Tunable low-band antenna. (b) Tuning in the low bands of LTE.

Figure 10.3 Frequency-reconfigurable antenna for LTE mobile phones [DBMP14].

using a MEMS tunable capacitor. For MIMO purposes, the antenna design is duplicated and placed on the same board. Five different placements of the secondary antenna are investigated. The results show that isolation between antennas is improved by using NB elements. Moreover, antenna placement on opposite sides of the PCB achieves a high level of isolation, even when the antennas are tuned to 700 MHz. In spite of the lower radiation efficiency that high-Q antennas present, their overall performance is better than the low-Q antennas, due to their inherently reduced coupling.

Due to the popularity of tunable antennas, the demand for manufacturing cost reduction is stronger than ever. Hence, Buskgaard et al. [BTDBP15] looks into manufacturing of antennas using a silver-based conductive ink on a plastic foil, which is a very cost efficient way of producing antennas, because it is possible to mass produce antennas onto a big roll of foil. For low-Q antennas the process looks feasible. However, high-Q antennas, because of the high currents, are extra sensitive to the sheet resistance of the metal used. The measurements show that the efficiency is much worse for the printed high-Q antennas than for a copper reference antenna.

10.3.1.2 NB tunable Tx and Rx antennas for a duplexer-less front-end

Conventional FD radio communication systems require that the radio Tx is active at the same time as the radio Rx. The increasing number of frequency bands require either a bank of NB filters with a switch or agile duplexers. While practical agile duplexers are not available, a bank of narrow-band filters with a switch is bulky and incurs switching loss. A novel front-end architecture, addressing the increasing number of LTE bands as well as multiple standards, is shown in Figure 10.4(a). NB antennas are used in order to provide sufficient rejection between the Tx and Rx. The investigation by Bahramzy et al. [BJSP14a] compares loaded antennas, capacitively and inductively. The inductive kind exhibits a smaller antenna volume for a given Q, where the inductance tuning is done with an LC circuit. A complete front-end [BOM+14] has been designed and fabricated to demonstrate the performance of the proposed architecture, where each component, including tunable antennas from Bahramzy et al. [BJSP14b] (Figures 10.4(b,c)), are designed specifically to fulfill the system requirements.

An alternative method for obtaining duplex function is described in Laughlin et al. [LZB+15]. This technique is about exploiting electrical balance (EB) in hybrid junctions to connect the Tx and Rx chains to a shared antenna, while providing high isolation between them. High Tx–Rx isolation can only

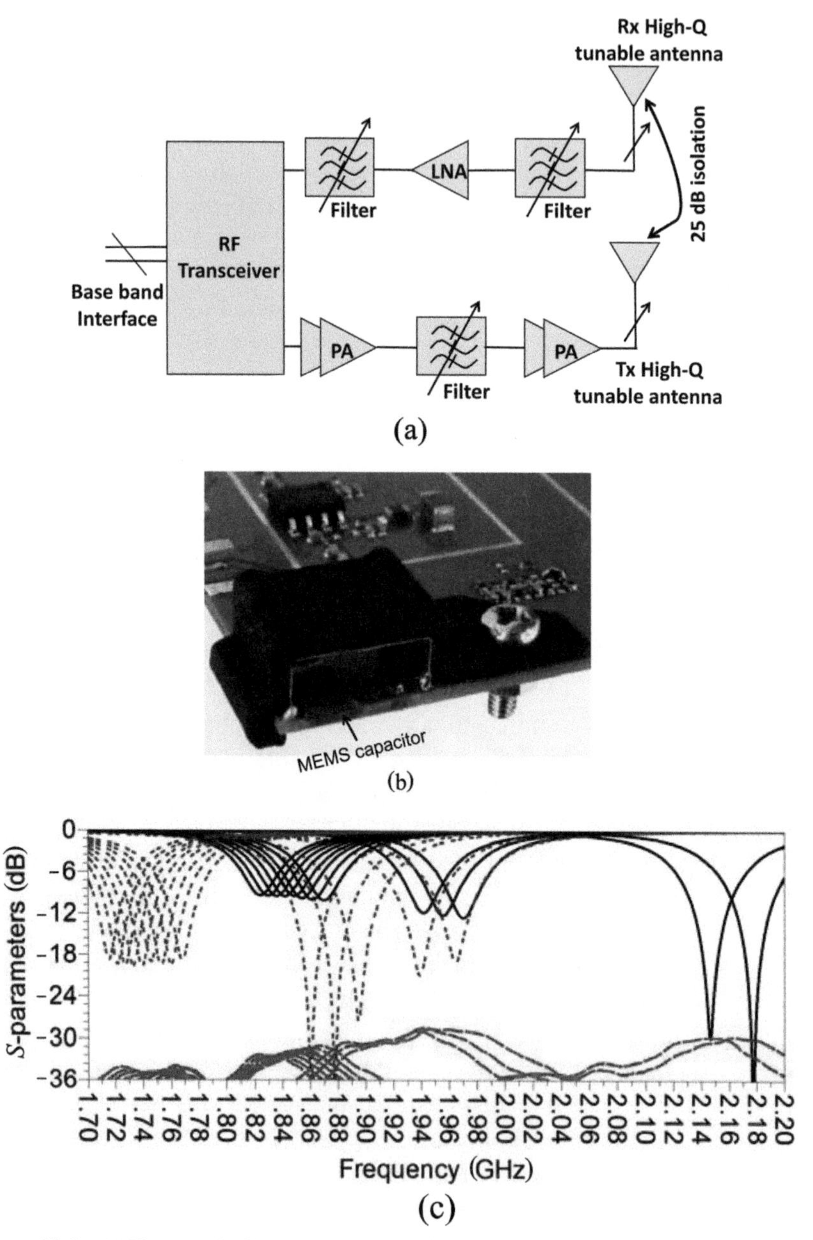

Figure 10.4 (a) Front-end with no duplex filter, (b) close-up of the antenna, and (c) measured scattering parameters of Tx and Rx antennas.

be acheived when the antenna impedance is closely matched by the balancing impedance, and the typically divergent nature of the antenna as well as the balancing impedances limits the isolation bandwidth of the EB duplexer. EB duplexing alone cannot provide the neccesary level of analog cancellation required for full duplex operation. However, when combined with active Analog Cancellation (AC), more than 80 dB Tx–Rx isolation is achievable over a 20 MHz bandwidth.

For LTE bands below 1 GHz, mobile phone antenna performance strongly depends on the ground plane size. Tatomirescu and Pedersen [TP14] shows how NB tunable antennas, due to their high-Q nature and thereby confined near-fields, can be somewhat ground plane size independent. The proposed tunable PCB antenna comprises two elements, occupies an area of $10 \times 30\ \mathrm{mm}^2$ and satisfies requirements of carrier aggregation combinations through NB dual-resonator design. With PCB size variation from $130 \times 65\ \mathrm{mm}^2$ to $70 \times 50\ \mathrm{mm}^2$, it is noted that impedance match and bandwidth stays reasonably stable.

10.3.2 Channel-Aware Antenna Design

For MIMO operation, LTE requires multiple antennas to be deployed at both the Tx and Rx. This requirement provides significant benefits in terms of higher data rates and better link reliability, but it also imposes significant challenges on antenna designers: (i) multi-antenna elements distributed within a compact volume increase the likelihood of the antenna elements being detuned by a user; (ii) the MIMO performance of multi-antennas (e.g., channel capacity) does not only depend on the received power, but also on correlation, which is a function of the antenna patterns and propagation channel characteristics.

10.3.2.1 User effect compensation for terminals with adaptive impedance matching

Adaptive impedance matching (AIM) can provide large performance enhancements in the presence of users, compared to free space operation. Apart from compensating for the detuning of an antenna caused by user proximity, adaptive matching also introduces a degree of frequency-reconfigurability for covering an increasing number of operating bands.

A dual antenna system is investigated in Plicanic et al. [PVTL11] and Vasilev et al. [VPL13], where the potential MIMO capacity gain from adaptive matching is examined in Plicanic et al. [PVTL11]. The S parameters were measured in an indoor office environment, with the terminal held by phantom hands. Ideal adaptive matching circuits were then added in post-processing. From the capacity performances with and without matching in the NLoS scenario at 0.825 GHz, it is observed that optimal matching for the two hand case enables a large capacity enhancement of 44% at 50% outage probability relative to no matching, which is mainly due to increased received power. The potential of AIM in the 0.8 GHz band based on two fabricated MIMO terminal prototypes (A and B) of significantly different antenna properties is also investigated [VPL13]. Prototype A has good impedance matching but poor isolation and NB behaviour, whereas prototype B has poor matching but high isolation and wide bandwidth. Assuming a reference signal-to-noise-ratio (SNR) of 20 dB, the capacity gain for prototype A is very high in the two-hand case, suggesting that AIM can significantly improve the capacity performance in cases where there is severe antenna mismatch. The performances of three dual-antenna terminals with adaptive matching were evaluated under four different user scenarios [VFL13]. It was observed that the prototype with the smallest antenna bandwidth and largest isolation can benefit the most from adaptive matching in the presence of a user (i.e., 24% capacity gain), which is because the adaptive matching networks effectively compensate for user-induced detuning.

The benefits of AIM, in compensating for performance degradation from propagation channel and user effects is studied in Vasilev et al. [VPTL13], based on field measurements. A MIMO terminal prototype equipped with either two ideal matching networks or two mechanical tuners was measured in three user scenarios and two propagation environments, see Figure 10.5. It was found that, due to the relative position between the hand and the terminal

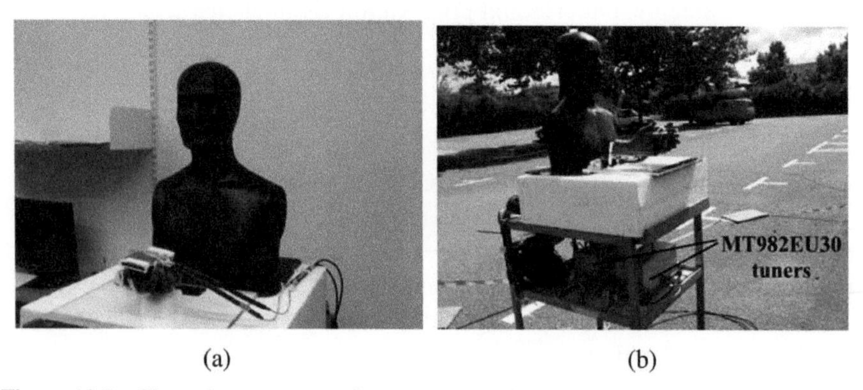

Figure 10.5 Channel measurement Rx setup: (a) Indoor scenario with one-hand grip and (b) outdoor scenario with two-hand grip, with MT982EU30 tuners shown [VPTL13].

antennas, the mismatch and therefore capacity gains in the one-hand cases were significantly lower than those in the two-hand cases. Further, since the user effects are less severe in the high band, the capacity gain is only up to 8% for these cases.

10.3.2.2 Channel adaptation to enhance capacity and localisation

UWB detection is an attractive solution for low-cost radar and localisation in the far field, while medical imaging has used UWB in the radiating near field. An understudied use of UWB is the possibility to apply detection methods in the sensors' reactive near field, where initial studies have been carried out in Ma et al. [MBH13]. The detection of eggs in a smart fridge is chosen as a case study. Exhaustive measurements have been carried out using an array of UWB sensors illustrated in Figure 10.6 where the filtering effect due to the presence of an egg can be quantified regardless of position by evaluating the maximum group delay of the UWB impulse or using a newly derived correlation coefficient when an egg is on top of the sensor. It is assumed the egg box is used and its presence above the sensors is known. The presence of clutter in the fridge such as the tomato in Figure 10.6 is also considered. Such concepts can be reliably implemented in smart container and packaging applications, where the sensors are invisible when fabricated with graphene or carbon nanomaterials.

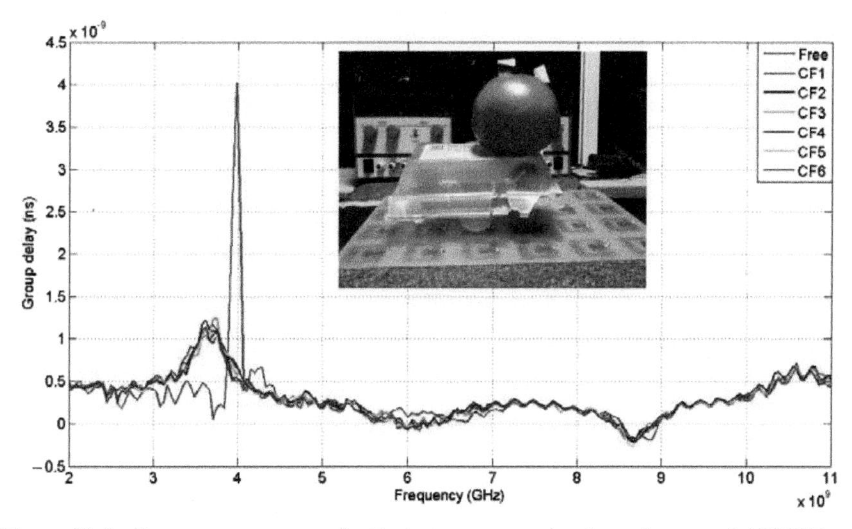

Figure 10.6 Frequency responses for the test case scenario of reactive near field UWB egg detection, when an egg is placed over a sensor [MBH13].

MIMO systems require a high degree of complexity due to the necessity of multiple radio-frequency (RF) chains. An adaptive reconfiguration of basis patterns in both Tx and Rx creates the possibility of reducing the transceiver complexity and provides the ability to design very small MIMO capable devices. In Vasileiou et al. [VMTK13], electronically steerable passive array radiators were used for beamspace MIMO systems. Previous work used these arrays with a specific set of basis patterns, in order to obtain a spatial multiplexing scheme for multiple data streams. This ensures orthogonal transmission of multiple data streams, though only in ideal channel conditions. The work in Vasileiou et al. [VMTK13] considers the variations of the channel and proposes to adaptively select the most effective basis pattern for the Tx and Rx sides. The goal is to diagonalise the beamspace channel matrix and ultimately to maximise the system capacity. This is done by developing channel-aware basis pattern calculations, which are based on SVD factorisation. The main contribution of the work is to propose a solution that works under realistic conditions. The work outperforms the current state-of-the art techniques, which do not take into account the channel, as can be seen in Figure 10.7.

Author Narbudowicz et al. [NAH15] propose a compact pattern-reconfigurable antenna design for femto-cell base station application. The advantages of the design are two fold: the direction of the beam can be continuously steered within 360° using phase shifting and the steering can

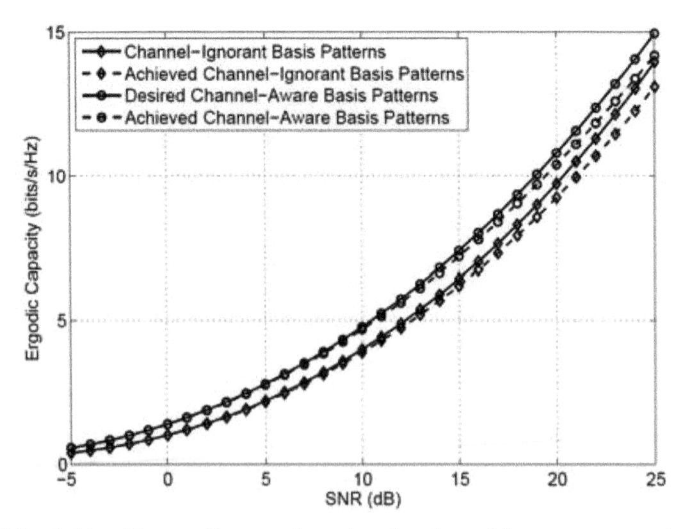

Figure 10.7 Achievable ergodic capacity using the channel-ignorant versus the channel-aware basis patterns [VMTK13].

be controlled by the modulation scheme. Consequently, the design allows independent steering for each channel. Applications of the proposed antenna tuning concept can be extended to wireless local area network (WLAN) access points, sensor networks or even enhanced localisation support.

The work by Li and Lau [LMCL15] present a pattern-reconfigurable antenna design for mobile phone application. The investigation is carried out on the low bands of the 4G spectrum, which is a real challenge since the chassis is the dominating radiator at these frequencies. The design is based on the characteristic mode analysis of a bezel-loaded chassis and uses PIN diodes to switch between the states. Simulations show patterns providing an envelope correlation below 0.2 within the investigated frequencies.

In Glazunov and Zhang [GZ11], the interactions between the propagation channel and the antennas are discussed in a MIMO context. A unified approach to characterise such interactions is proposed, using both the spherical vector wave expansion of the MIMO channel and the scattering matrix of the antennas. This method provides a better insight into the behaviour of the radio propagation channel, and helps characterising throughput and channel correlation, among others. The goal of the work is to obtain uncorrelated antenna radiation patterns, in order to achieve the specifications of LTE systems. The contribution shows that the spherical vector wave coefficients of the transmit antenna are the eigenvectors of the multi-mode correlation matrices at the transmitting antenna. The same conclusion is valid for the receiving antenna.

A previously developed synthesis method for NB antennas is extended to UWB antennas by Sit et al. [SRL+12]. The method maximises the capacity by tuning the patterns of the MIMO antenna system. This proves to be challenging since the channel characteristics change over the large UWB spectrum, hence affecting also the radiation pattern and the orthogonalisation process. The synthesis method aids in reducing the number of antenna elements needed in the final antenna system. In this paper, the resulting optimal radiation pattern for an indoor scenario is shown and the capacity analysis is also presented.

10.4 RFID and Sensor Antenna Innovations

This section highlights novel ideas in the research area of RF Identification. The trends for future applications are towards smart antenna research, where more functionality is introduced in the RFID chip cards as well as in the antennas for RFID readers. The miniaturisation of antennas is the key driver for the research on tags and wireless sensor nodes.

10.4.1 Antennas for RFID Readers and Chip Cards

A smart antenna at the reader side of an RFID system is presented in Lasser et al. [LMPM15]. The horizontally polarised, azimuth switched-beam antenna is designed for novel RFID-based tyre pressure monitoring systems in vehicles. The intended mounting position is below the body floor pan, which usually constitutes a large conducting object. To maintain the efficiency and pattern properties of the switched-beam antenna, the authors of Lasser et al. [LMPM15] use a dual-band frequency selective surface (FSS) manufactured from printed circuit boards mounted between the antenna and the body floor pan (Figure 10.8). This surface is designed to reflect a plane wave without phase shift and, therefore, enables efficient operation of a close-by antenna. An efficiency penalty of just 0.8 dB is measured at 868 MHz for a small FSS panel of 0.87 λ_0 edge lengths closely spaced at 0.08 λ_0.

A smart active UHF RFID tag is proposed for the use in a warehouse scenario in Dufek [Duf12]. Equipped with an adaptive antenna array, the tag is able to steer its beam towards a reader station using phase shifters embedded in the substrate. This system is considered advantageous for example in warehouses, where the to be tracked goods are moving, or if interferences are present due to multipath propagation. In addition to that, the communication range can be extended.

For designing contact-less chip cards and HF tags for various application scenarios, a good model of both the coil antenna and the RFID chip are necessary to match both components perfectly for an efficient power and data transfer. In Gvozdenovic et al. [GPM14] a planar spiral inductor synthesis method is discussed, which generates physical dimensions of an HF RFID

Figure 10.8 Photograph of the switched-beam antenna mounted atop the FSS using foam spacers for a spacing of 28 mm [LMPM15].

antenna according to specified equivalent circuit parameters. A numerical model for a spiral coil based on the partial element equivalent circuit (PEEC) method is combined with a non-linear optimisation algorithm to optimise and synthesise coils. The PEEC engine is implemented in Matlab and the inductance values of the synthesised circular spiral coils are comparable with the simulation results obtained from commercially available 3D EM simulators as well as with measurements [GMP$^+$13]. In Gvozdenovic et al. [GMM14] the non-linear behaviour of an RFID chip is evaluated by measurement and described by a simplified harmonic model. A manual for designing HF RFID tags is created which includes a procedure for measuring the impedances of coil and chip, and simple fitted formulas for quick calculation of coil parameters in order to achieve required resonance frequency and bandwidth.

10.4.2 On-Chip Antenna Design for Miniature RFID Tags

To decrease the size of an RFID tag dramatically (some mm^2) and to reduce the overall manufacturing costs, the metallisation layers of a cost efficient 130 nm CMOS process are utilised for antennas operating below 10 GHz. An UWB-pulsed data stream is considered in the uplink from the tag to a reader once the tag is powered and controlled via an UHF signal, forming an asymmetric or hybrid communication scheme. The addition of UWB into a tag introduces a substantially increased data rate, while compatibility to standard UHF reader is kept. Towards such a system, several compact on-chip antennas (OCAs) with a size of 1 mm^2 such as loops, dipoles and monopoles on a CMOS substrate are designed, manufactured and analysed to investigate the different communication links [GSM13].

A CMOS testbed, consisting of a circuitry and an OCA on the same die, has been manufactured [GWHM11] to investigate the UWB communication. This is considered critical, because the large Q-factor of ESAs limits the maximum achievable bandwidth. The circuitry features a voltage controlled oscillator, selectable data sources, a bandwidth adjustable glitch generator for a pulse amplitude modulator and a power amplifier. A horn antenna was placed 60 mm above the device, which was able to capture the wideband PAM signal. In Gvozdenovic et al. [GGM11], an antenna array is proposed to increase the gain at the reader station and to roughly localise the tag. Depending on the centre frequency and the embedded OCA type a large radiated bandwidth is measured, which can be explained that the efficiency of the OCAs is low, due to the small size of the antenna and the losses introduced by the silicon substrate and, therefore, the Q-factor is lowered as well. The measured bitrate is up to 100 Mb/s. With this technique a detailed characterisation of the radiated field of the manufactured on-chip antenna types is presented in Gentner and Mecklenbräuker [GM11]. In Gvozdenovic et al. [GGM12], a near-field channel from the actively powered UWB RFID tag with an OCA is measured and a simple channel model is proposed. The measurement and characterisation of a tiny tag with OCA is best done without any mechanical connection (bondwires). This is necessary to avoid any mutual coupling between the measurement feed connection, which can blur the measurement results by an angular tilt of the radiation pattern [GASM12].

A miniature UHF/UWB hybrid silicon RFID tag is presented in Gentner et al. [GLS+13]. This system-on-a-chip (SoC), shown in Figure 10.9, features two OCAs for an asymmetric communication. By inductive coupling, energy is received by a loop OCA and stored in a buffer capacitor on-chip. The tag receives control commands via the UHF link established with the loop, and transmits a 10 μs long hardcoded 4-PPM stream with the monopole OCA. This data stream, with a datarate of 117Mb/s, has been successfully captured and analysed for a measurement distance of 5 mm.

10.4.3 ESAs for Wireless Sensors

The design of a circular inverted-F antenna (CIFA) by Kakoyiannis and Constantinou [KC12b, Kak14] demonstrated that the exploitation of the circumscribing sphere of a wireless sensor to the greatest possible extent is the most effective miniaturisation technique, even for planar/printed (2-D)

Figure 10.9 Micro-photograph of the on-chip antennas of an UHF/UWB hybrid silicon RFID tag. This tag has a size of 3.5 mm × 1 mm.

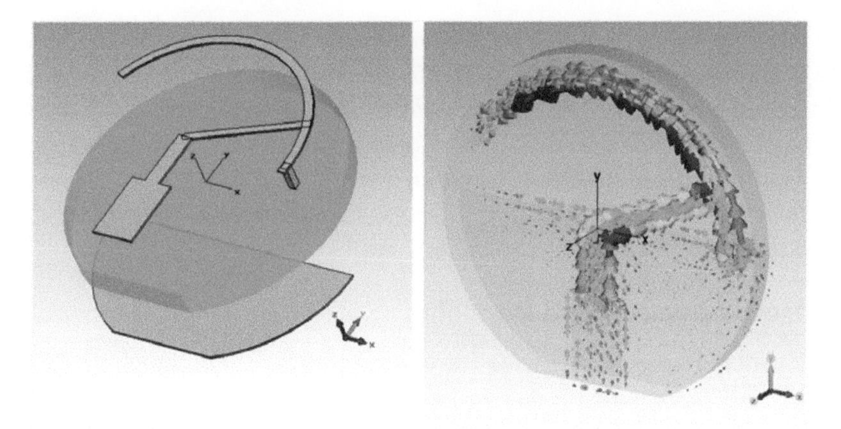

Figure 10.10 Layout and current density of a CIFA antenna. Picture published in Kakoyiannis and Constantinou [KC12b].

antennas. By shaping the radiating part of the current distribution into a circle close to the boundary, an ESA of size $ka = 0.5$ rad is obtained at 2.5 GHz with a 3.5% BW and a 90% efficiency (see Figure 10.10). The antenna is based on a minimal design space (two angles and a radius) and features a wide tuning range by way of modifying one or both angles. Based on first principles, the system model of the CIFA naturally leads to an upper bound on the directivity of ESAs. The upper bound can be tight or loose, depending on the shape of the current distribution. If efficiency proves to be high, then this bound also becomes a tight upper bound on the gain of ESAs. Finally, based on reasonable assumptions regarding the equivalent circuit at the input to the CIFA, a minimum transmission frequency for ever-shrinking sensor nodes can be derived. A forecast model predicts that Smart Dust nodes need to enter deep into the mmW region (approx. at 42 GHz), if they were to use microwaves for wireless communication.

10.5 Modelling and Characterisation of Antenna Systems

As an important component of wireless communication, an antenna is never independent from its surroundings, i.e., the propagation channel. Modelling and characterising antennas in a systematical way is very important, taking into account system performance and channel influence. In Section 10.5.1, multiplexing efficiency is introduced as a system metric for convenient evaluation of MIMO antenna performance. After that a simple method to measure the signal correlation of MIMO antennas is illustrated, followed by statistical antenna pattern analysis and pattern interpolation for use in channel modelling and emulation. The antenna system is analysed in real conditions in Section 10.5.2, including the factors of mobile devices, the mobility of the terminal and the real propagation scenario. The discussion is then continued into the interaction between users and antennas in Section 10.5.3, where hand, head and body are all taken into account.

10.5.1 Methodologies and Modelling

MIMO technology has been widely introduced in wireless communications due to its ability to increase the data rate linearly with the number of antennas, without any additional expense in transmit power or spectrum. To characterise the system performance of a MIMO link, several figures-of-merit (FOM) have been proposed, such as channel capacity and diversity gain, which take into account both the antenna and propagation effects. However, those figures cannot provide intuitive information to antenna designers regarding the impact of individual antenna design parameters. To address the effect of antenna parameters on MIMO performance, multiplexing efficiency was introduced [TLY11], which is a power-based metric.

Multiplexing efficiency is defined as the power penalty of a non-ideal antenna system in achieving a given capacity, compared with an ideal antenna system with 100% total antenna efficiencies and zero correlation among the antennas. Assuming that there is no correlation at the Tx and the receive antennas being the MIMO antennas under test, the multiplexing efficiency for a $M \times M$ MIMO system at high SNR can be approximated by

$$\eta_{\mathrm{mux}} = \left(\prod_{i=1}^{M} \eta_i \right)^{\frac{1}{M}} \det(\bar{\mathbf{R}})^{\frac{1}{M}} \tag{10.1}$$

where η_i is the total efficiency of the ith antenna port, and $\bar{\mathbf{R}}$ is a normalised correlation matrix whose diagonal elements are 1 and off-diagonal element denotes the complex correlation coefficient between the 3D radiation patterns of the antenna ports. In Equation (10.1), the efficiency imbalance and correlation between the antennas are translated to an equivalent power loss in decibel (dB). Based on the separate contributions from the antenna efficiency and the correlation, multiplexing efficiency can also be used to clearly explain the total impact of the user on MIMO terminals [TLY12]. This insight allows antenna designers to effectively quantify and address key parameters in designing MIMO terminal antennas for typical usage scenarios.

As shown in Equation (10.1), signal correlation among the antennas is a critical parameter for the MIMO performance. The conventional way of measuring the correlation is time-consuming, costly and needs specialised measurement facilities [Cla96], which requires information of phase and polarisation. To simplify the correlation measurement, a method based on equivalent circuits was introduced in Li et al. [LLLH13] for estimating correlation coefficients in lossy antenna arrays accurately. The idea is described as a cascade network model in Figure 10.11. A lossy dual-antenna array is equivalent to a network consisting of a lossless circuit in series with loss resistors, which represent the conduction and dielectric losses of the antennas. The value of the resistance r_{loss} can be calculated with the dual antenna circuit model with the knowledge of antenna efficiency. With the information of S parameters of the lossy array and the resistance, the correlation coefficient of the lossy array can then be calculated from the S parameters of the equivalent lossless circuit using the closed form expression from [BRC03]. The method requires only S parameters and antenna radiation efficiencies, with the latter being easier to acquire than radiation patterns. Detailed examples on how to estimate the correlation for dipole antennas were provided in Li et al. [LLLH13], and the method can be extended to calculate correlation for more than two elements.

To design high-performance MIMO antennas with low correlations, orthogonal or quasi-orthogonal radiation patterns are required. Fortunately, TCM provides a convenient tool for MIMO antenna design, as it enables orthogonal radiation patterns to be excited in a given antenna structure. However, eigenmodes solved from the MoM impedance matrix [HM72] are not always maintained across the frequency points, as some modes can

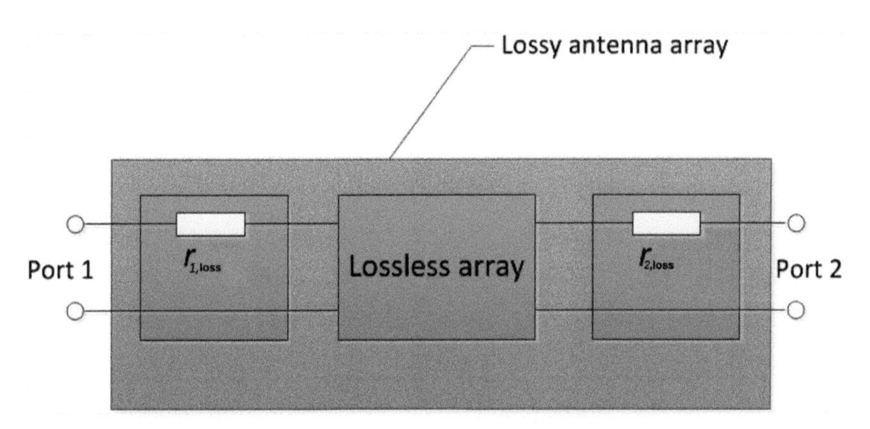

Figure 10.11 Cascade network of a dual lossy antenna array.

become unstable and cease to exist, whereas other modes can appear and become stable without having any relation to previous modes [CHHE11]. An improved method to track the mode based on far field patterns was presented in Miers and Lau [ML14]. This method is based on the fact that the far fields remain relatively stable with respect to variations in structural complexity and frequency. Since the eigen-far-fields are orthogonal to one another, every individual mode is unique and has no correlation to any other mode. By applying the standard multi-antenna ECC equation [VTK10] to modes at different frequencies, the modes with high correlation at different frequencies are successfully tracked as the same mode.

The antenna in real channel realisation is important for practical reasons. The antenna-channel joint modelling of multiple antenna terminals in a wireless communications context was carried out in Sibille [Sib15]. The research is based on the generation of a finite set of data, which represents the terminals characteristics and exploits the existing propagation channel models. In the paper, low-pass filtered antenna patterns are used to exploit the angular structure of the propagation scenario, and complex spatial correlation stemming from the angular dispersion of the propagation was accounted by using the full set of discrete DoA. In this way, the model is simplified with a limited set of parameters. For channel estimation purpose, the complex response of the embedded antenna must be known for all angles of incidence. As the antenna pattern is measured at discrete angles, interpolation is needed. A simple and fast method to characterise, compress and embed antenna patterns is introduced in Kotterman and Landmann [KL12] using effective aperture distribution function (EADF). A critical step in the method is to make the measured pattern periodic, so that the EADF can be applied. For measurements in spherical coordinates, the azimuth direction is periodic by nature, and periodicity in elevation is achieved by concatenating measurements from two opposing meridians, as were the measurements taken along one great circle on the measurement sphere. By performing an inverse 2-D discrete fourier transform (DFT) on the EADF for those particular values of azimuth and co-elevation pairs, a compact representation of complex antenna patterns are achieved after interpolation.

To describe the variability in the power gain patterns of a set of antennas, such as tag-like planar dipoles, a statistical model is a suitable way, which can present data in a simplified model that could be easily included in standards and in radio channel simulators. The statistical modelling of a UWB tag antenna describing the variability of its frequency averaged radiation patterns was described in Mhanna and Sibille [MS11]. The modelling is based on Fourier series expansion in the angular domain. In the first stage, the model achieves a compression rate of more than 81 on a per antenna basis, while having a relative error smaller than 1%. In a second step, the statistical properties of the model coefficients have been extracted, allowing to further reduce the amount of parameters. Finally, the statistical model coefficients were further modelled, resulting in a final number of 32 parameters to describe the whole set of tag antenna radiation patterns.

10.5.2 Characterisation in Realistic Conditions

To achieve the full benefit that MIMO technology has promised theoretically, besides good antenna designs, a series of conditions need to be satisfied in both the propagation environment and the devices [VMS09]. A joint evaluation of MIMO antennas with the actual device and the working environment has been performed in Sánchez-Heredia et al. [SPTG13]. Two MIMO antennas, labelled as 'good' and 'bad' according to their MIMO properties and radiation characteristics in the chamber, were used for comparison. The evaluated was performed for two phones [a recently released phone A (good phone) and a first generation LTE phone B (bad phone)] whose antennas were replaced by the good or bad MIMO antennas as well as two scenarios, i.e., Urban Micro and Urban Macro. The throughput of the system is shown in Figure 10.12. For phone A, good antenna shows a consistent improvement of about 2 dB for both scenarios over the bad antenna. For phone B, the good and the bad antenna give almost the same throughput performance for the Urban Macro-cell model, whereas for the Urban Micro-cell model, the bad antenna even performs slightly better than the good one. Thus, a bad LTE device can actually mask the benefits that a good antenna design provides, or even making a bad antenna look better than a good design.

Due to the importance of the scenarios, channel characterisation is of high interest for MIMO systems. For short range communications [LN10], near field channel characterisation is necessary since the classical assumptions of IID Rayleigh do not apply anymore. A channel for a 2×2 MIMO system in near field communication is measured in Gvozdenovic et al. [GTB12]. Three antenna sets including bicone antennas, monopole and loop pair, and dipole and slot pair are utilised to obtain the channel matrix, with which channel capacity and eigenvalues of the channels were calculated. The main results are summarised in Table 10.1. The dipole/slot antenna set behaves best out of three pairs, i.e, low-pattern correlation results in increase of spectral efficiency.

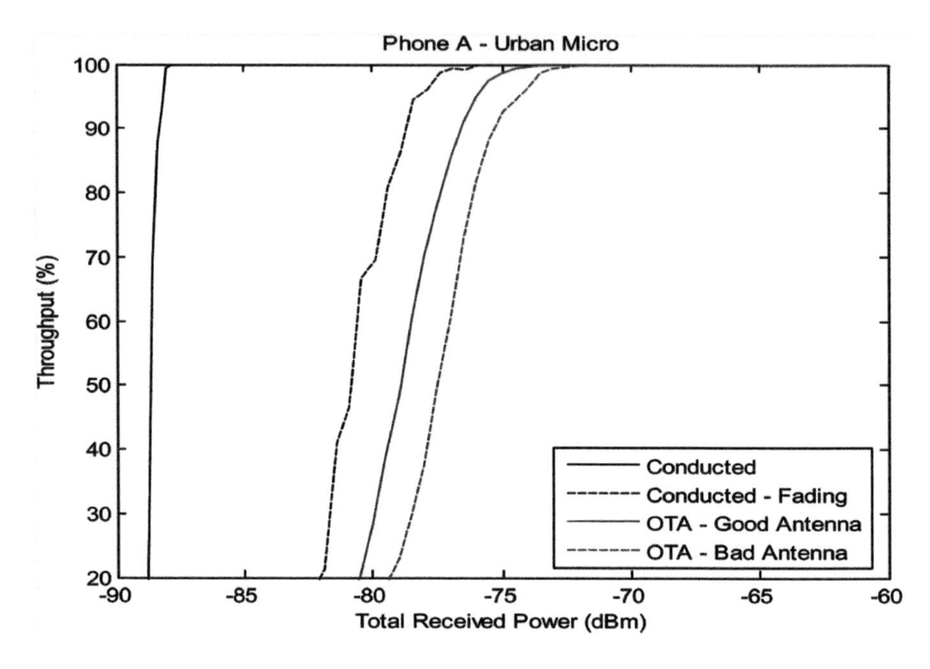
Phone A - Urban Micro
Throughput (%)
100
90
80
70
60
50
40
30
20
-90
-85
-80
-75
-70
-65
-60
Total Received Power (dBm)
Conducted
Conducted - Fading
OTA - Good Antenna
OTA - Bad Antenna

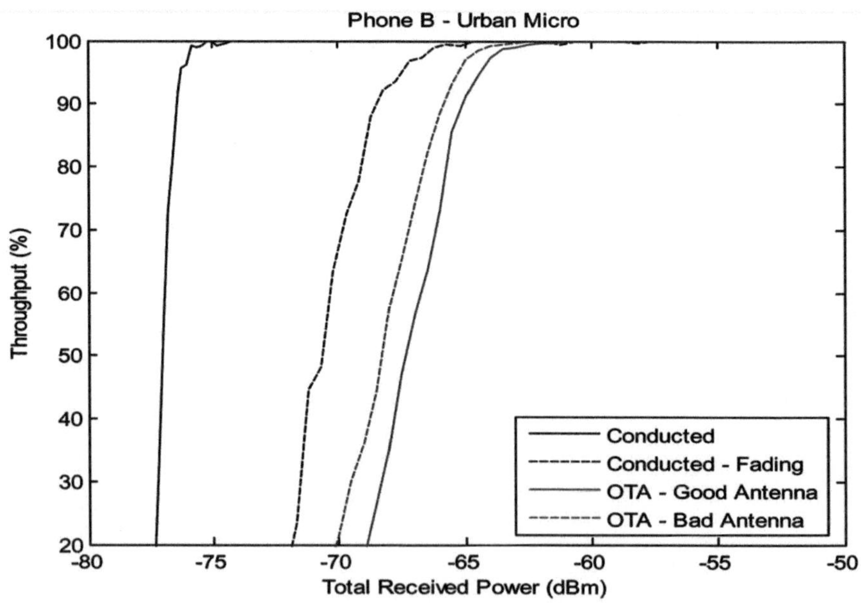
Phone B - Urban Micro
Throughput (%)
100
90
80
70
60
50
40
30
20
-80
-75
-70
-65
-60
-55
-50
Total Received Power (dBm)
Conducted
Conducted - Fading
OTA - Good Antenna
OTA - Bad Antenna

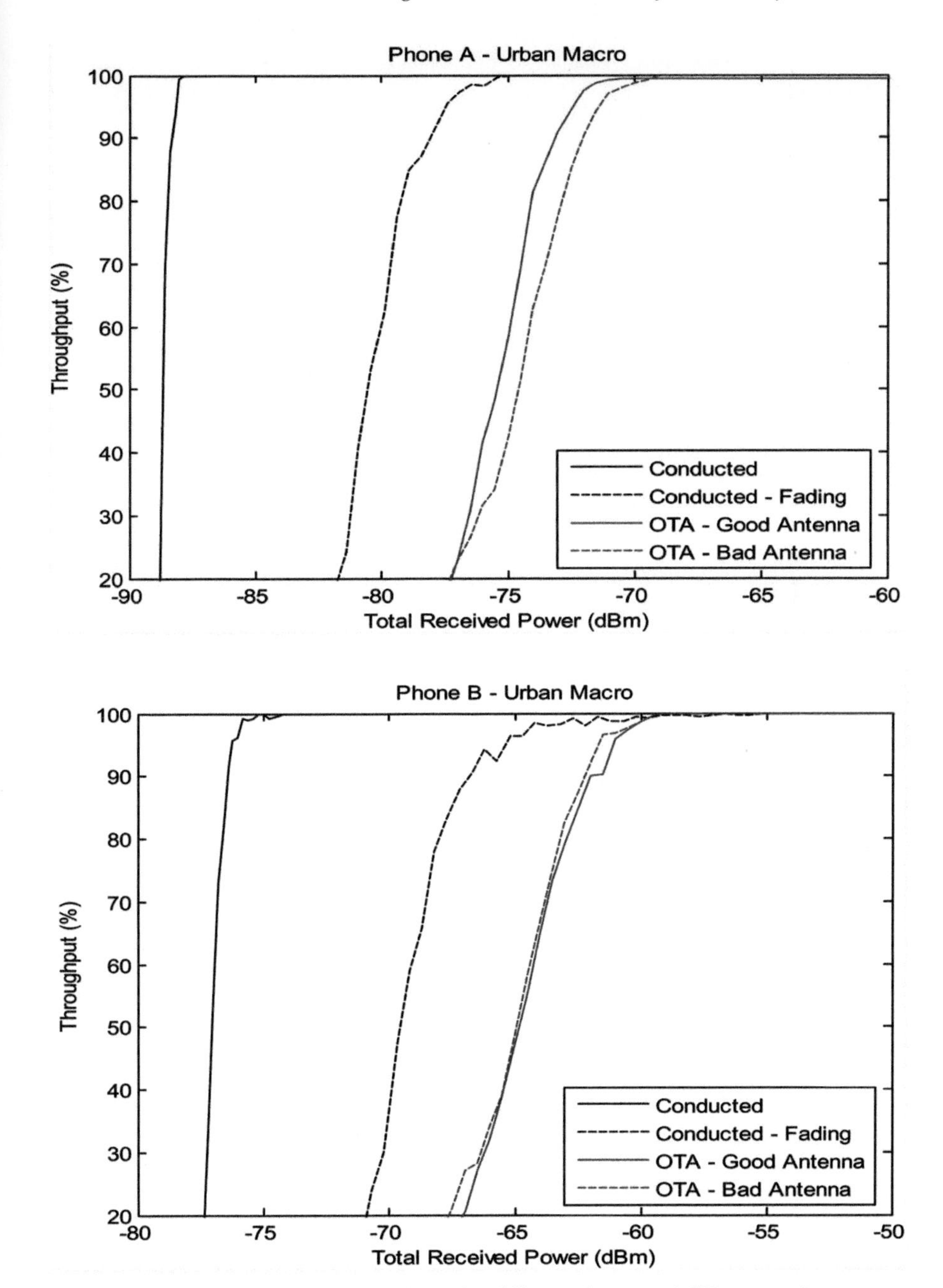

Figure 10.12 Measured LTE through put for different phones and different environments.

Table 10.1 Channel for different antenna sets

Antenna Set	ρ	d(m)	η(bit/s/Hz)	$\lambda_2 - \lambda_1$ (dB)
Bicone co-pol	1	0.7	6	15
Mono/loop 1	0.97	0.7	4	23
Dipole/slot 1	0.1	0.7	8	5

For wideband communication systems, group delay variation versus frequency is an essential issue which can cause distortion and degradation in the signals [Kwo06]. Recently, it was raised in Bahramzy and Pedersen [BP13] that group delay can also become a concern for high-Q antennas, due to its big and steep phase change over frequency. The results showed that the group delay of a low-Q antenna (dipole antennas) is around 1.3 ns, whereas a high-Q antenna (patch antenna) has group delay of around 22 ns. Therefore, it is important to measure the group delay for high-Q antennas in order make sure that its variations are within the specifications of the radio system design.

Interaction with nearby object and device mobility can also change the antenna performance as well as other components on the devices, such as EB duplexer, which can provide high Tx–Rx isolation over wide bandwidths [AGL13]. The variation in antenna impedance changes the performance of the duplexer, in terms of bandwidth and achievable isolation. When in proximity of the users, there is variation of up to 25 dB in mean isolation across a 20 MHz bandwidth [LB14], as indicated by the circuit simulations incorporating measured antenna S11 data. Interestingly, objects in the near field, especially user interaction, can improve the isolation of the duplexer. This is due to the reduction in radiated power from the antenna, which reduces the reflections from objects at further distances. Significant variation of mean Tx–Rx isolation was also observed when the device is moving, with higher device speeds causing more rapid variations [LHB+14]. The performance variation also depends on the operational environment and the dynamic balancing network. The environment with higher reflection (e.g., lab) causes significantly more variation in Tx–Rx isolation than the environment with lower reflection (e.g., cafe), as shown by Figure 10.13. When the balancing network is not dynamically updated, duplexer performance is reduced by up to 12 dB compared to an ideally balanced duplexer.

Regarding real integrated mobile devices, radio performance of popular smart-phones as well as the more classical phones has been investigated with head and hand phantoms [TP13, PT14]. Total isotropic sensitivity (TIS) is used as the criteria to compare the phones. It has been observed that the

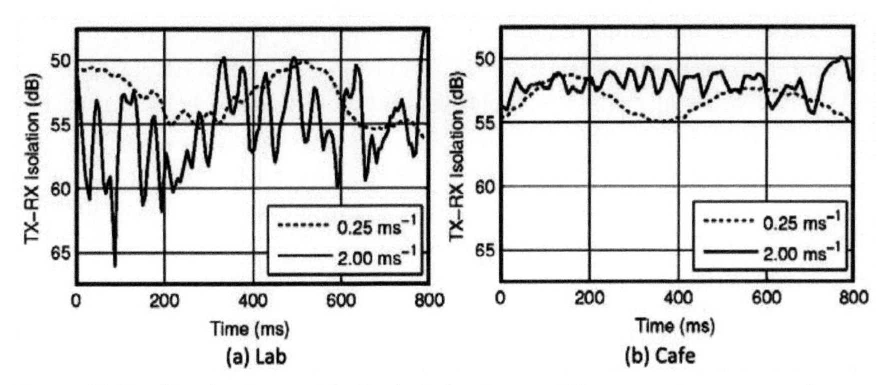

Figure 10.13 Simulated mean Tx–Rx isolation in two different environments and at two different device speeds.

newer generation thin smartphones perform worse than the classical phones, and there is even a loss in performance between generation of the same brand of smartphones. For example, the latest generation of the Galaxy and iPhone suffers 5 dB loss with respect to the last generation, which means that a doubling of the mobile base stations is required to provide the same coverage. Comparing different brands of phones, the Sony phones have very good performance among the ones used for this study, whereas the iPhones and most of the Samsung phones dominate the bottom of the performance list. Generally, the spread of TIS for most phones in all bands is within 6 dB, and half of them is within 4 dB.

10.5.3 User Effect on Antenna Systems

User effect is an important topic for terminal antennas, as it changes the antenna performance dramatically, or the user even becomes part of the antenna. The user's hand and head can lead to impedance mismatch and also greatly reduce the radiation efficiency of the antenna through absorption, deteriorating the communication performance. Figuring out how the antenna is affected by the users can provide useful information for the adaptive impedance matching of the antenna system.

The absorption loss and the mismatch level depend on the antenna type, user hand and head size and the grip style. Pelosi et al. [PFP12] compares the user effect on high-Q and low-Q antennas. The reference antenna is a PIFA, with its Q factor modified by reducing the height of the PIFA (high-Q antenna) and adding a capacitor between the PIFA and the ground plane (tunable high-Q antenna). Both data mode and talk mode were investigated, with two-hand grips, i.e., holding the phone tight or loose. It was observed that high-Q antennas only detuned slightly for both hand grips, so that the mismatch loss is below 1 dB. Nevertheless, the total losses are not improved because the absorption losses in the high-Q antennas are higher than in the low-Q antenna. The tunable high-Q antenna has a similar behaviour, i.e., the detuning is reduced by 95%, yet the absorption loss is larger. This indicates that the three antenna models have quasi-equal total losses for all simulation environments.

The influence of the index finger's position on different antenna sets have been studied in Buskgaard et al. [BTB$^+$14b]. Simulation are carried out for both single antenna and MIMO antennas in the talking mode. For the single antenna study, antennas with different bandwidth (NB or wide band (WB)) are either top mounted (TM) or bottom mounted (BM) on the mobile handset. As shown by Figure 10.14, the total loss is about 18 dB for TM and 11 dB for BM antennas and slightly worse for WB than for NB antennas. For the MIMO antenna cases, one main antenna is placed either at the top or at the bottom, and a diversity antennas is on the side of the mobile phone. It turns out that the diversity antenna is detuned significantly, whereas the main antenna is less affected by the user though it still suffers up to 6.4 dB of mismatch loss.

In real life, the natural grip position is quite related to the size of the mobile phone. For a slim mobile phone, the index finger normally has a limited reach,

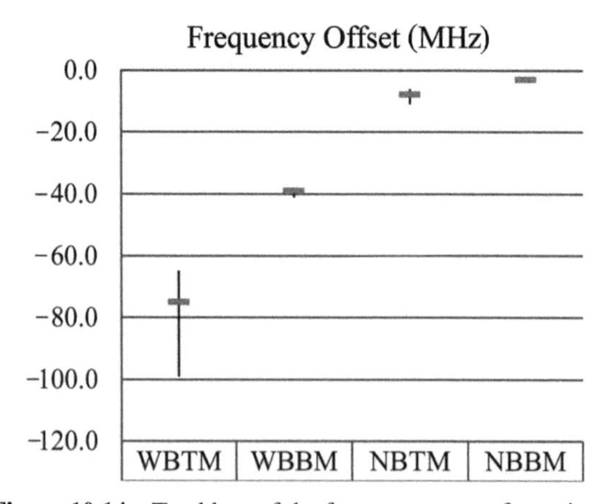

Figure 10.14 Total loss of the four antenna configurations.

i.e., it is actually shifted down from the top antenna. In this case, the factors that make NB and WB antennas perform differently with users were studied in Buskgaard et al. [BTB+14a]. Those factors includes power absorbed in the hand, head, plastic cover, index finger and the finger tip. As the mobile phone industry progresses, the mobile phones become larger and thinner [RE12]. For wider mobile phones, the index finger can again touch the end of phone, as in the case of Buskgaard et al. [BTB+14b].

Besides the antenna itself, the decoupling network connected to antennas also interact with the users, which was considered in Tatomirescu et al. [TPFP12]. When the antennas with a lumped decoupling network is in the proximity of the user, the performance of the decoupling network is degraded considerably due to the limited bandwidth of the antenna system. To counteract this degradation, either a tunable decoupling network must be implemented or the antenna placement and distance from the user must be optimised. It is also found that a distributed decoupling network is more robust than a lumped one against the user presence. Nevertheless, if the antenna position is not optimised considering the user, the decoupling network performance is still affected considerably.

Aside from the antenna performance, the exposure of the body to a RF electromagnetic field is also important for the sake of safety. To evaluate the exposure, the metric of specific absorption rate (SAR) is used, which is defined as the power absorbed per mass of tissue. The SAR performance of MIMO-enabled mobile handsets for the body-worn case was evaluated and compared for different antenna locations and types in Li et al. [LTDL14]. The results showed that stand-alone SAR and average simultaneous SAR give the same trend for SAR values. For antenna location, different from the intuition, the co-located and separated antenna setups have similar SAR values. For antenna type, the PIFA gives a lower SAR value in general than the monopole antenna. However, the smaller bandwidth of PIFA also needs to be taken into account since it may be severely detuned when it is placed close to the body. In practice, it is challenging to measure the simultaneous SAR when multi-antennas are transmitting due to the need to account for power allocation and relative phase among the antennas. A new metric time averaged simultaneous peak SAR (TASPS) is defined in Li and Lau [LL15] to circumvent simultaneous SAR evaluation, taking advantage of the random and fast-varying channel-dependent MIMO precoding. It is shown that TASPS will satisfy the SAR limit by default, as long as the stand-alone SAR of every individual antenna satisfies the limit. Hence, only stand-alone SAR measurement is needed.

10.6 Antenna Measurement Techniques

10.6.1 Measurement Hardware-Induced Pattern Distortions

The art of antenna measurements evolves with emerging antenna technologies, frequencies and mathematical methods. One trend in communications is the shift to higher frequencies which demands low- and medium-gain antennas at mmW frequencies. An example is the patch antenna presented in [HHS12] designed for the 60 GHz WLAN band, manufactured on a liquid crystal polymer (LCP) substrate. To rule out permittivity uncertainties as a cause of discrepancies between simulation and measurement results, the substrate panel carrying the antennas was additionally covered with 35 microstrip ring resonators. Using the measured dimensions of the manufactured resonators, the average permittivity of the panel was found to be $\varepsilon_r = 3.102$ with a standard deviation of $\sigma = 0.0143$. This permittivity is then used to simulate the antenna again, which still shows a discrepancy as indicated in Figure 10.15(a). In fact, the discrepancy is caused by manufacturing tolerances of the lithographic process which were measured as $\pm 30\ \mu\text{m}$ at maximum using a microscope. Further discrepancies were observed on the measured pattern, whose source was clearly identified as the probe used to connect the microstrip line on the substrate with the coaxial signal source of the measurement system (Figure 10.16(b)). This issue is further discussed in Renier et al. [RvDHH13, RvDHH14b] where the radiative, reflective and obstructive properties of the probe are examined (Figure 10.15(b)). The probe poses an additional source of radiation, which creates the interference ripple indicated in the green curve in Figure 10.15(b). Covering the probe with foam absorbers reduces the ripple (blue curve) but the obstruction for $\theta < -30°$ remains. The issue is solved by a bent probe as reported in Renier et al. [RvDHH14a]). The disturbing effects of the bent probe on the radiation pattern measurements are negligible.

Pattern discrepancies and ripple do not only occur at mmW frequencies, but also arise when low-gain antennas are measured on a θ over φ scanner with limited ($-160° < \theta < 160°$) scan range, as is reported in Lasser and Mecklenbräuker [LM14]. In this survey the 14 measurements of the same AUT at different mounting positions are evaluated. The goal is to find the required Rohacell foam spacer height between the antenna under test (AUT) and the dielectric support structure to receive consistent pattern measurements, using the measurement setup shown in Figure 10.17(a). However, the measurement series reveals that increasing the spacer length itself does not improve the ripple in the θ domain. Instead, the θ ripple is increased when the AUT

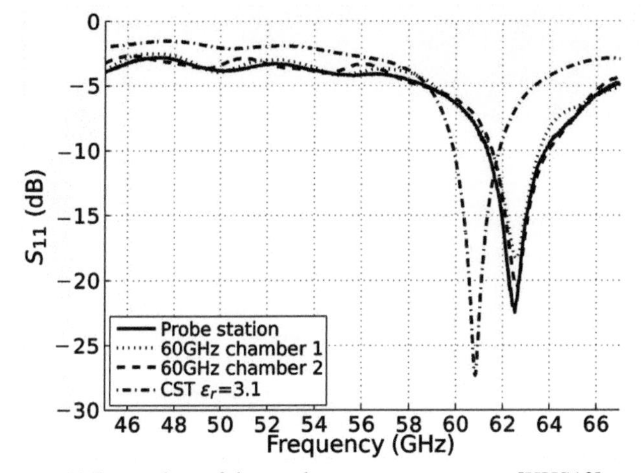

(a) Return loss of the patch antenna, compare to [HHS12].

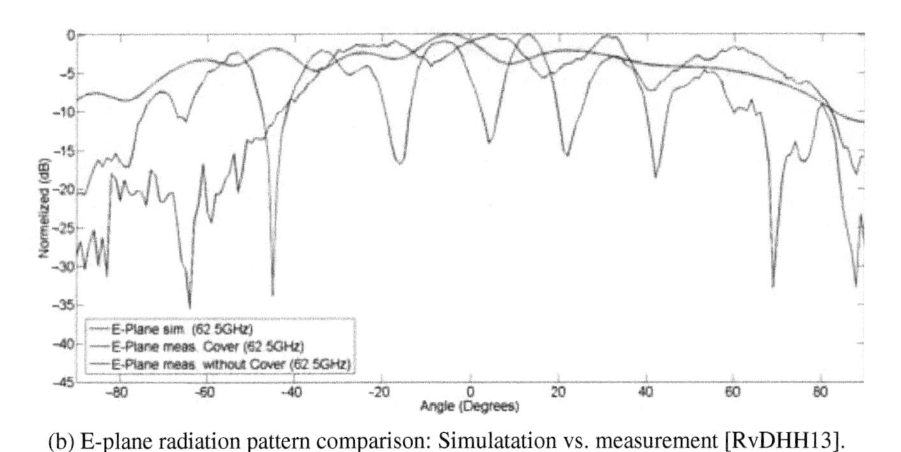

(b) E-plane radiation pattern comparison: Simulatation vs. measurement [RvDHH13].

Figure 10.15 Return loss and pattern measurement comparison of the patch antenna manufactured on LCP substrate.

is not mounted in the scanner's coordinate system centre, due to a larger resulting maximum radial extend (MRE) and less filtering in the spherical domain. Further investigations [LLM14b] reveal the θ near-field truncation as one error source. In the near-field to far-field transform the truncation acts as a step function on the measurement data which produces a Gibb's phenomenon in the far-field data. This truncation effect is extenuated when the AUT's main lobe points towards the ceiling and the truncation occurs in areas of low radiation from the AUT. To enable different orientations without

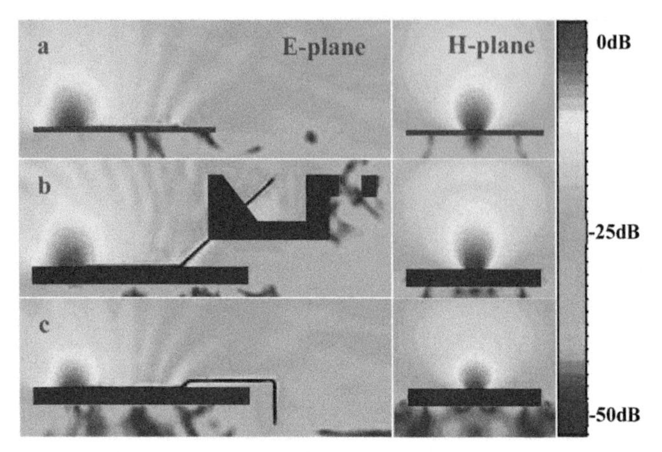

Figure 10.16 Power density comparison for: (a) ideal antenna, (b) antenna using classical probe feed, and (c) antenna fed by bent probe [RvDHH14a].

additional errors due to the feed cable, the AUT is fed using a small battery powered oscillator as described in Lasser et al. [LLM14a]. With this method the comparison shown in Figure 10.17(b) is produced, where the measurement data in orientations B and C are converted to the coordinate system A for comparison. The least ripple is produced in orientation B where the AUT's main beam points toward the ceiling of the chamber. The final solution to this measurement problem is reported in Lasser et al. [LLM15]. Using the same hardware setup (Figure 10.17(a)) and mounting the AUT offset in the z-axis, a measurement with a large MRE is obtained. The transformed far-field data is then phase shifted to mathematically move the AUT in the coordinate centre. This shifted far-field measurement data is now filtered in the spherical wave domain and again transformed into the far-field. The resulting ripple-free measurements are presented in Lasser et al. [LMPM15].

10.6.2 Efficiency Measurements for Small AUTs

Instead of using a small oscillator, a fibreoptic system as presented in Yanakiev et al. [YØCP10, YNCP12] can be used to feed an AUT without cable impairments. This is especially important to obtain accurate gain and pattern measurements for electrically small AUTs and omnidirectional radiators. The setup presented in Figure 10.18(a) is designed for a receiving AUT suitable for long range transmissions while maintaining a high SNR. The receiving AUT concept is chosen because no power hungry power amplifier is required in the small (40 $\times$ 40 $\times$ 10 mm) battery-driven unit connected to the AUT.

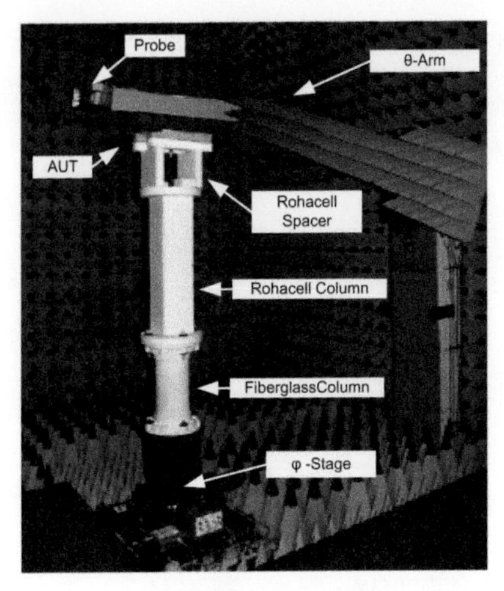

(a) Measurement setup.

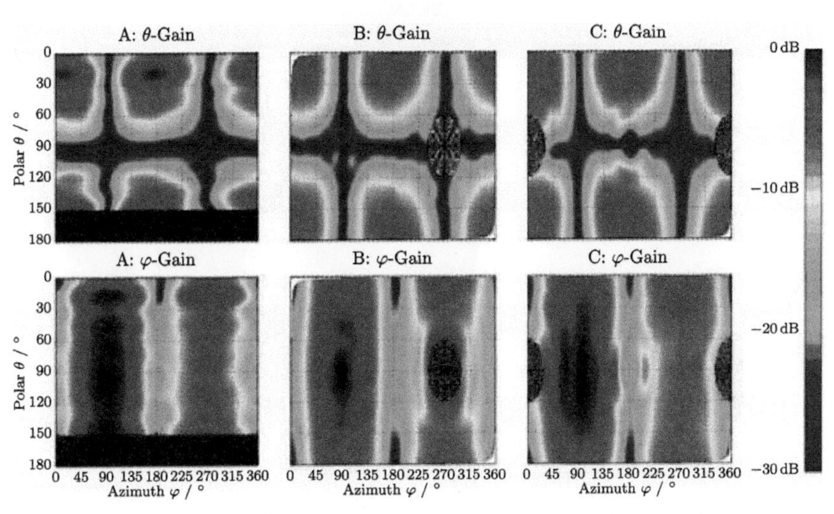

(b) comparison of gain plots in three mounting configurations, tranformed to coordinate system A.

Figure 10.17 Measurement setup and gain plot comparison of same AUT in three mounting configurations, transformed to coordinate system A [LLM14b].

This setup supports a high-power Tx on the other side of the channel, while the signal received at the AUT is amplified, directly modulates a laser diode, and is converted back to an electrical signal at the stationary equipment with an overall gain of 10.1 dB at 776 MHz. At this frequency the overall dynamic range of the presented optical feed system is 39 dB.

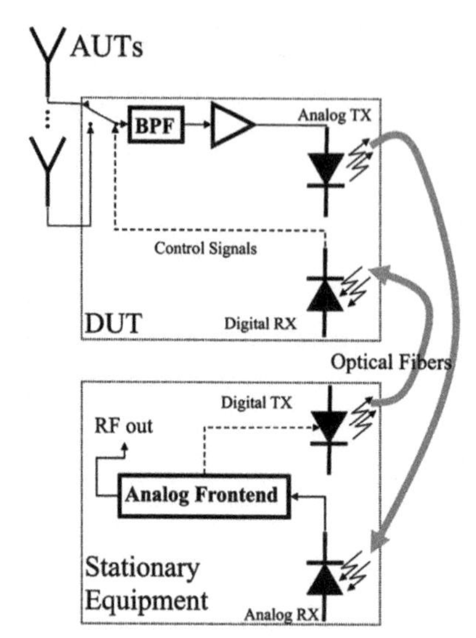

(a) Block diagram for the long range receiving AUT configuration [YØCP10].

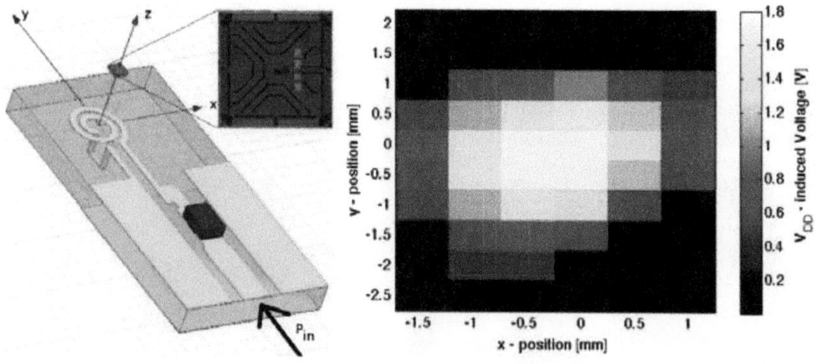

(b) Contactless evaluation of the induced voltage of a tiny RFID tag with on-chip antennna [GHSM12b].

Figure 10.18 Two different approaches for measuring AUT efficiencies.

While the presented optical solution is very compact and, therefore, supports many ESAs, extremely small AUTs such as on-chip devices still cannot be characterised with such a solution. A different approach is presented in [GHSM12b, GHSM12a]: For the development of a single chip RFID tag with an on-chip antenna measuring $1 \times 1\mathrm{mm}^2$, a contactless evaluation method is required to analyse the on-chip antenna and the rectifier for the setup's efficiency. The circuitry of the developed evaluation platform is designed so that the modulation frequency of the signal backscattered from the chip is proportional to the induced voltage V_{DD} of the rectifier which is delivered from the on-chip antenna. For the actual measurement the chip is moved in the $x-y$ plane at a constant height above the excitation coil (see left side of Figure 10.18(b)). At each point of the measurement grid, the modulation frequency is measured and subsequently V_{DD} is calculated. The result can be seen in the right side of Figure 10.18(b), where the chip was in a height $z = 0.5$ mm above the excitation coil. In this $x-y$ plane a maximum induced voltage of 1.7 V is measured, where the input power at the excitation coil is $P_{\mathrm{in}} = 8.6$ dBm.

Besides this method to measure the performance of an on-chip antenna and the connected the rectifier based on voltage mapping, a more conventional approach is to enclose the AUT in a cavity – called Wheeler cap – and measure the altered return loss at the antenna port. Comparing this value to the free space return loss, the efficiency of the AUT is obtained. However, for fixed cavity geometries, the bandwidth of the Wheeler method is limited, due to cavity resonances of the Wheeler caps. In Kakoyiannis and Constantinou [KC12a], an extensive survey on the achievable bandwidth of this method is presented, given the dimensions of the AUT. For rectangular cavities operated below cut-off, flat AUTs with dimensions of $0.6\ \lambda_0 \times 0.3\ \lambda_0$ are measurable. Using the intermodal spectrum, the AUT geometry may reach $0.78\ \lambda_0 \times 0.39\ \lambda_0$, achieving 33% of measurement bandwidth. Using spherical Wheeler caps, bandwidths up to 84% are obtained for small AUTs.

11

MIMO OTA Testing

Chapter Editors: Wim Kotterman and Gert F. Pederesen,
Section Editors: Istvan Szini, Wei Fan, Moray Rumney,
Christoph Gagern, Werner L. Schroeder and Per H. Lehne

11.1 Topical Working Group on Multiple-Input Multiple-Output (MIMO) Over-The-Air (OTA)

The main goal of the Topical Working Group on MIMO OTA is to gather all the relevant research across the Working Groups in the IC1004 Action for backing-up choices to be made in standardisation on technologies for OTA testing of multi-antenna devices. As no standards are conceived in European cooperation in science and technology (COST) IC1004, discussions are generally held in an easier atmosphere than in standardisation bodies. Contributions to a broader understanding of OTA testing of multi-antenna systems and its implications are welcomed as much as investigations of particular technologies or concepts. Such contributions come from industry and academia. Compared to earlier work in, for instance in COST Action 2100, the focus has shifted from RF performance (the present OTA standard) to overall device performance as seen by the user, without regarding any specific hardware/subsystem performance. This also means not primarily finding out why a certain terminal in a particular radio environment behaves the way it does, the focus is on how it performs w.r.t. to exchanging information (data, speech, images). The impetus comes from, among others, mobile service providers that want to rank UE for their portfolio. The targeted application of MIMO OTA in standardisation is the conformance testing cycle, currently targeting RF performance only and not production testing. In this Chapter, contributions over the project duration are documented and resumed in a coherent way.

11.1.1 The Organisation of This Chapter

Originally, four methods for OTA testing of MIMO terminals were under discussion for standardisation and an appreciable part of the research within

Cooperative Radio Communications for Green Smart Environments, 423–468.

the Topical Working Group is devoted to these particular methods which are multi-probe anechoic, reverberation, two-stage, and decomposition. At the time of writing this book, 3rd generation partnership project (3GPP) is continuing to develop standards for three of the methods but has stopped work on the decomposition method. Of the three remaining methods, international association for the wireless telecommunications industry (CTIA) has selected two for the first draft of its MIMO OTA test plan and is considering the third method for a future release. However, independently from any decisions taken by 3GPP and CTIA, results for all the methods are presented here as part of the research within COST IC1004.

In addition, research into MIMO OTA in a broader sense was also undertaken. Therefore, after this introduction we will first give some general underlying concepts of the present state of the art of MIMO OTA and will then describe the concepts behind the four proposed technologies. In the succeeding sections, this structure will be repeated by first presenting research results that are relevant for all OTA technologies and then successively treating contributions to each of the particular technologies. A summary and outlook conclude this chapter.

11.2 OTA Lab Testing: Models and Assumptions

11.2.1 General Considerations of OTA Testing

The main reason for OTA-testing is the interaction between device antenna and EM environment. Simply put, the more directive the wavefields produced by the environment, the more the directivity of the antennas matters. As preceding COST Actions have gone a long way modelling the non-isotropic directionality of the mobile radio channel, its polarisation state, and its time variance, accounting for the effects these properties have on reception through the antennas of a specific terminal is a logical consequence. The first standard on single-input single-output (SISO) OTA concentrated on capturing the power received or transmitted by a system through its antennas and RF subsystems [3GPP12, CTIA14], as the amount of transmitted/received power determines the achievable throughput (TP). For MIMO systems, it is of importance to know how much power is received in or transmitted into the available independent channels as provided by the radio environment. Scattering richness of the channel determines how many of these links exist, with their number indicated by the channel rank. How much of this channel rank is available to the transmission system depends on both the antenna array constellation, the respective radiation patterns of the individual antenna elements, and the signal-to-noise ratio (SNR). Also, orientation of the device antennas with respect to the environment plays an additional role through their directivity, as mentioned above. Because device orientation generally will differ from user to user, measurements have to be performed with different device orientations in order to properly describe these orientation influences. The channel models prescribed in standardisation are adaptations of 3GPP spatial channel model extended (SCME) [BHdG+05] effectively modelling two-dimensional (2-D) incident fields with static angular distributions, that are run in "drops", i.e., without temporal evolutions of large-scale effects.

11.2.2 Anechoic Chamber or Multi-Probe Method

This method is the one that attempts to physically recreate the radio environment of the device under test (DuT) by emulating the important features of the incident wave fields. For this, the DuT is encircled in an anechoic chamber by a set of OTA antennas, typically in an annular array. The individual antennas are excited with signals that are jointly optimised to produce, by superposition, a wave-field that represents the radio environment. A characteristic of this method is that the size of the test area with good field quality (the "test zone" or "sweet spot") scales linearly with the number of OTA antennas and wavelength, see Section 11.4. As the annular set-up intrinsically is 2-D, projections onto the azimuthal plane are necessary when emulating three-dimensional (3-D) environments, which is still realistic under perfect power control [LGP+13], as long as both polarisations are emulated. Pirkl and Remley [PR12] reached similar conclusions without mentioning power control.

Advantages of multi-probe method are:

- No access to antenna ports is needed, as no conducted tests are needed.
- The method allows for emulating virtually any radio environment, not only the ones modelled in standardisation.
- Spatial channel characteristics, e.g., for MIMO operation, can be emulated with great accuracy.

Disadvantages are:

- Its cost, especially when emulating 3-D channel models. An anechoic chamber is needed, as are channel emulators (CEs) (with sufficient interconnectivity) for every separate antenna element, i.e. for testing a 2×2 setup in a dual-polarised 8-antenna ring (16 antenna ports) 16 dual-input CE are needed. Therefore, the size of the test zone cannot easily be made large.
- 2-D set-ups with annular OTA arrays tend to amplitude decay over the test zone.

- Of the proposed wave synthesis strategies, coherent synthesis demands coherent operation of all the CE channels with minimal drift. This, too, is expensive and creates a lot of overhead in terms of calibration. Pre-faded synthesis (PFS) is suited best for Non-line-of-sight (NLoS) situations, see Section 11.4.

11.2.3 Two-Stage Method

The philosophy behind the two-stage method is that communication hardware works on signals, not on fields. Therefore, imposing the correct signals at the device receivers will result in correct measurement. Consequently, cabled connections to the terminal are possible as long as the combined effect of the transmitting antenna array, the propagation channel, and the receiving antenna array is accurately emulated. For this, the antenna patterns of the DuT are measured in both polarisations and over 3-D (stage one). Then, the antenna patterns are embedded in the channel that is emulated, over cable connections (stage two). A recent enhancement is a radiated two-stage (RTS) method that avoids cabled connections thud relieving some of the disadvantages of the method, see Section 11.6.

The conducted two-stage method has the following advantages:

- No dedicated room (anechoic or reverberant chamber) needed for the second-stage throughput measurements. However, the device radiation patterns need to be determined in an anechoic antenna measurement chamber.
- Less emulation hardware resources needed than for wave-field synthesis; for instance, emulation of a 2×2 MIMO transmission will require emulating four independent links, meaning four CE channels, whereas a wave-field synthesis with eight dual-polarised antennas requires 16 dual-input CEs, see Section 11.2.2.
- No need for wave-field emulation to be restricted to 2-D.

The disadvantages are:

- In the conducted version, access to antenna ports is needed and the termination impedance mismatch must be considered.
- Devices with adaptive antennas cannot be tested properly, because of the variability of the patterns. This also applies to the radiated version.
- When connected by cable, it is not possible to measure device's radiated desensitisation caused by signals leaking from the DUT transmit antennas back into the DUT receiver as the antennas are disconnected by the cabled connections, see Section 11.6.3. The radiated two-stage method resolves this disadvantage.

11.2.4 Reverberant Chamber Method

As the name of the method already indicates, a reverberant chamber is used, creating a rich scattered wave field with large angular spreads. The fields are homogenised by continuously changing the geometric properties of the reflecting surfaces, by use of so-called stirrers and turn table rotations over time, see Section 11.5. These devices are historically mainly used for measuring power output irrespective of antenna directivity, by storing the energy in the reverberant field and spreading it angularly. Along similar lines, reception sensitivity can be measured too. As such, the use of reverberation chambers (RCs) is standardised as one of methods used for the SISO OTA test.

Advantages of the RC method are:

- No access to antenna ports is needed, as no conducted tests are needed.
- After installation (which includes properly loading the chamber, i.e., tuning the reverberation time), there is no need for extensive calibration.
- The size of the chamber can be made smaller than the anechoic chambers typically used for the anechoic chamber method.

Disadvantages are:

- Many important aspects of the emulated wave-field are fixed.
  - The fading profile typically is a Rayleigh distribution, related to the vast number of scattered field components, combined with the large angular spread. Other distributions are, therefore, difficult to generate.
  - Spatial correlation cannot be tuned because the angular distribution is random based on the stirrer positions.
  - The cross-polarisation power ratio (XPR) of the emulated fields is very close to 0 dB, meaning total depolarisation. This removes any effect of, e.g., polarisation diversity of the base station and DuT antennas.
  - However, in a variation of the RC method, some temporal characteristics can be added by the use of CEs:
    * Additional delay spread on top of that determined by the reverberation time of the chamber.
    * Additional Doppler profiles to those of fixed shape with relatively small spread determined by the rotation speed of the stirrers.
- Instantaneous angular distributions of the field are not isotropic, but are assumed isotropic only after sufficient averaging over time. Averaging

of measured TP has had considerable discussion due to the strongly non-linear behaviour of the measured TP with respect to received power.

- The evolution of the angular distribution over time is not typical of a cellular mobile radio channel. As a result, smart antenna adaptivity is not likely to develop its full potential.

11.2.5 Decomposition/Two-Channel Method

This method evolved during the project duration. At its start as the two-channel method, it deviated from the other methods described here as it considered component testing rather than testing device performance, in this way more or less continuing the line of thought that produced the first OTA standard. The main idea was to test OTA only the antenna with RF front-end, without any fading. All other components can be tested conducted. Later, accommodating standardisation's requirements for a full end-to-end characterisation instead of a component test, and under fading conditions, a connected test was added to the test suite and the name of method changed to "decomposition method", see Section 11.6.

A clear advantage of the method is:

- The two-channel method principally does not use CEs. The decomposition method, though, requires two single-input CEs for two channels, three for three channels, etc.

Disadvantages of the method are:

- An anechoic chamber is needed with a two-way mechanical positioner. For future higher-order constellations, multiple-way positioners could be required.
- Influences of the spatial characteristics of incoming fields are only available in a non-linear fashion as with the RC method, i.e., through (in this case, optionally 3-D angularly weighted) averaging over measured TPs per combination of two incidence angles.
- The conducted test added afterwards does not embed antenna directivity in the channel (in contrast to the two-stage method), implicitly assuming some generic (presumably omnidirectional) pattern.

11.3 General Research Topics

At the start of TWG MIMO OTA mid 2011, standardisation and certification groups such as 3GPP and Cellular Telecommunication and Internet

Association (CTIA) were heavily engaged in MIMO OTA. The TWG MIMO OTA chose to support those efforts consolidating the expertise of its members on topics important for properly evaluating MIMO OTA devices.

The initial scope adopted by standardisation and certification groups was based on tests intended to determine what constitutes a good versus a bad performing MIMO-capable user equipment (UE). As the project progressed, it became clear that just determining what constitutes a good or a bad performing MIMO UE is quite challenging, and, therefore, MIMO OTA performance would require more than one test condition in order to properly assess MIMO-capable devices across their entire performance range.

The challenges associated with MIMO OTA performance evaluation differ substantially from those the wireless communications industry faced when defining a measurement methodology for SISO-radiated performance measurements [CTIA14]. Because a SISO receiver (Rx) does not require a special propagation environment, an LoS radio path within an anechoic chamber is employed. MIMO, on the other hand, requires a spatially-diverse radio channel in order to deliver maximum performance, resulting in more than one MIMO OTA candidate methodology being proposed.

Given the diverse nature of the methodologies available for assessing MIMO OTA performance, it became evident that the fundamental aspects of the OTA measurement should be scientifically validated. In addition, result comparison between labs would also be required. The comparison of measurement techniques is nothing new. Over the years, the industry has established round-robin test efforts designed to identify the strengths and weaknesses of proposed measurement methodologies, while at the same time attempting to quantify device performance in such a way that results can be directly compared between labs and measurement techniques. However, because MIMO performance is related to the spatial–temporal aspects of the operating environment, special techniques were required. In response to this need, COST IC1004 collaborated with those groups validating fundamental concepts and defining measurement techniques. These research topics, relevant to all MIMO OTA test methodologies were defined as follows:

- Definition of MIMO reference antennas
- Fundamental limitations of test environments
- The expected data TP value for a real mobile device
- Definition of figure of merit (FoM) post-processing
- Definition and characterisation of measurement campaigns

11.3.1 Definition of MIMO Reference Antennas

Understanding that the industry was overlooking fundamental aspects of good academic and engineering practices the MIMO reference antennas as shown in Figure 11.1 were proposed. Initially solving the issue surrounding the unknown antenna radiated performance in all devices under test, the MIMO reference antennas shown in Szini et al. [SPDBF12, SPSF12, SYP14] was proposed as a solution to this problem, therefore, eliminating this variable from the list of unknowns that affect the MIMO OTA measurement campaign outcome. Those antennas were proposed in mid 2011 during the first face-to-face meeting of the recently formed MIMO OTA sub group (of CTIA) (MOSG) and initial designs in the COST IC1004 action, and were adopted during two additional measurement campaigns organised by CTIA and supported by 3GPP RAN4 MIMO OTA *ad hoc*.

11.3.2 Fundamental Limitations of Test Environments

A number of common channel models including 3GPP SCME and wireless world initiative new radio (WINNER) interpret the cross-polarisation coupled power (as described by the XPR) as power added to the co-polarised powers [SPSF12]. Thus, if the co-polarised V–V or H–H signals were normalised to unit power, the cross-polarised terms would represent scattered or reflected power originating from the other polarisation and modelled as power added to the co-polarised unit power driving the value above unity. Thus, receiving

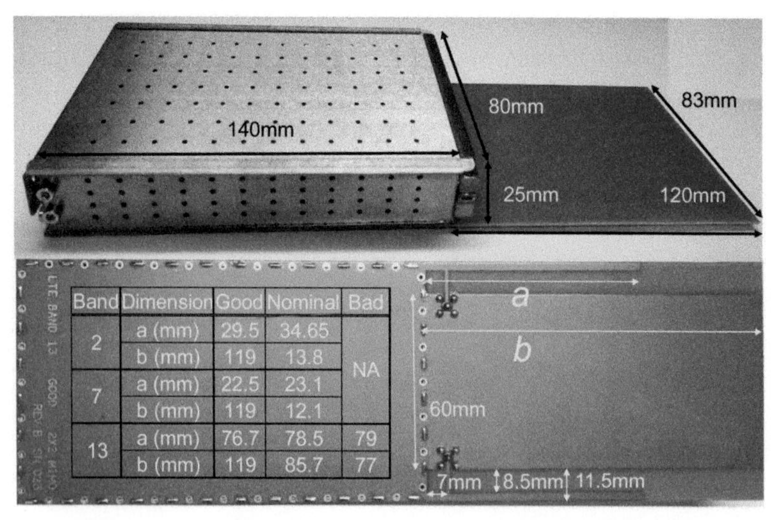

Figure 11.1 MIMO reference antennas and respective dimensions.

in one polarisation benefits from transmitting in both polarisations due to the scattering in the environment. In a system simulation, the cross-polarised powers vary due to the SCME and WINNER phases selected on each drop. However, for link level evaluation, such as in a MIMO OTA test environment, the random processes are replaced by a fixed set of parameters including the polarisation phases in order to generate reproducible channel conditions. The following simple power normalisation produces unit power in a Rx test volume:

$$\frac{P_V}{P_V + P_H} + \frac{P_H}{P_V + P_H} = 1, \tag{11.1}$$

with P_V, P_H the powers in the vertical and horizontal polarisation, respectively. When multiple taps are defined, the power in each polarisation is normalised per tap and the powers per tap are then weighted according to the power delay profile (PDP) to arrive at the total normalised power. With this normalisation, a constant unit power will be presented to the DuT, while having a variety of spatial, temporal, correlation, XPR, and polarisation properties as defined by the given channel model. When different channel models are selected, they will use these parameters only at a normalised power level.

Manufacturers of cellular handsets most commonly adopted radiation pattern diversity as MIMO antenna system design technique. The reference antennas described previously were, therefore, designed to emulate such designs. However, this does not mean that the OTA test methods should tune their tests optimally to these types of antenna design. Different perspectives do exist as shown in Szini et al. [SFR+14], where a MIMO antenna system based on pure polarisation diversity was presented. The objective was to bring to attention that conclusions based on limited sample of reference antennas cannot be extrapolated to a wide variety of MIMO antenna systems. It was demonstrated that fundamentally different test methodologies, regarding channel models and cross-polarisation definitions, evaluated identical devices quite differently. Especially, RC-based methods cannot discriminate between MIMO antenna systems based on pure polarisation diversity, due to the lack of cross-polarisation control (Figure 11.2). While achieving pure polarisation diversity is a challenge in low frequency and small form factor units like handsets, it is common in larger mobile devices such as tablets, laptops, machine-to-machine (M2M) devices, etc.

Traditionally, the channel models used in standardisation are exclusively 2-dimensional, i.e., only azimuthal angles are taken into account, with development of 3-D models under way. Although reducing a 3-D to a 2-D propagation environment is a clear simplification of reality, it does not need to have a relevant impact on the results. Simulation results, based on channel

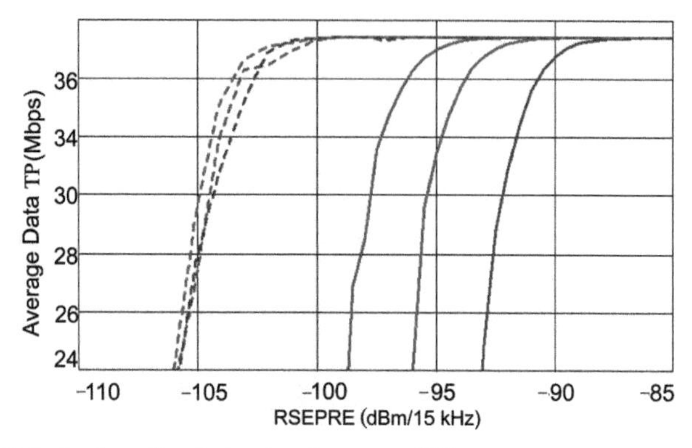

Figure 11.2 Polarisation discrimination between different antennas, measured in MPAC (solid lines) and RC (dotted). Red: Band 13 "Bad", green: Band 13 "Nominal", and blue: Band 13 "Good" antenna.

measurements in Ilmenau in which the MIMO Reference Antennas were embedded, [LGP+13], showed little loss of TP (less than 10% uplink), as long as the channel remains full-polarimetric and perfect power control is assumed (among others, the UE is located away from the cell edges). Further simplifying the channel environment, from 3-D dual-polarised to 2-D single-polarised, clear deviations in channel characteristics can be noticed (on average 10–15% additional loss, up to 50%). There are indications the rank of the channel may be impaired by the simplifications. This bears implications for the multi-probe anechoic chamber (MPAC) method with annular arrays or the use of 2-D cuts in the two stage methods, when dealing with real 3-D propagation data instead of with 2-D SCME models.

During work on device characterisation, it was discovered that operating the UE at high TP for extended periods of time created unreliable results [Jen11]. The issue was found to be related to a temperature rise within the UE, impacting the transceiver (TRx) IC. This is the reason why further study of MIMO OTA has been carried out using an uplink power of only –10 dBm. In real life use cases, cell edge reference sensitivity also coincides with the highest UE output power and so it should be understood that continued reliance on low UE output power is not fully representative of performance for devices whose performance degrades at high temperatures.

11.3.3 The Expected Data TP Value for a Real Mobile Device

With the MIMO antenna performance issue addressed, the next issue to be faced by the standardisation community was the validation of base station (BS) settings and channel model emulation in the test environment baseline realisation. These technical problems were the motivation for the follow-up work shown in Szini et al. [SPTI13]. The Absolute Data Throughput Framework was a method coined to establish a deterministic MIMO OTA figure of merit and stimulate the industry to proper defined the channel model pertinent to each proposed MIMO OTA test methodology as demonstrated in Figure 11.3. Adopting the MIMO reference antennas already accepted by the industry, the Absolute Data Throughput Framework compares the conducted data TP measurement (through the CE including the spatial and temporal characteristics of the defined channel model and the embedded complex radiation pattern of the MIMO reference antennas) with an over the air measurement using the same channel model, same DuT and same reference antennas (of which the complex radiation pattern was measured). In this way, the expectation of radiated channel model emulation is validated. This method was extensively used during the conclusion of the 3GPP RAN4 MIMO OTA work item, and was considered one of the fundamental criteria to validate MIMO OTA test methodologies.

11.3.4 Definition FoM Post-Processing

Although data TP had been agreed and defined as the fundamental MIMO OTA figure of merit, the way to post-process raw data continues to be a topic of discussion and constant investigation that must be addressed before the conclusion of MIMO OTA certification process.

Based on measurement in both an anechoic chamber and a reverberant chamber set-up, differences where noticed between calculating average TP versus power, typical for a reverberant chamber result, and calculating TP

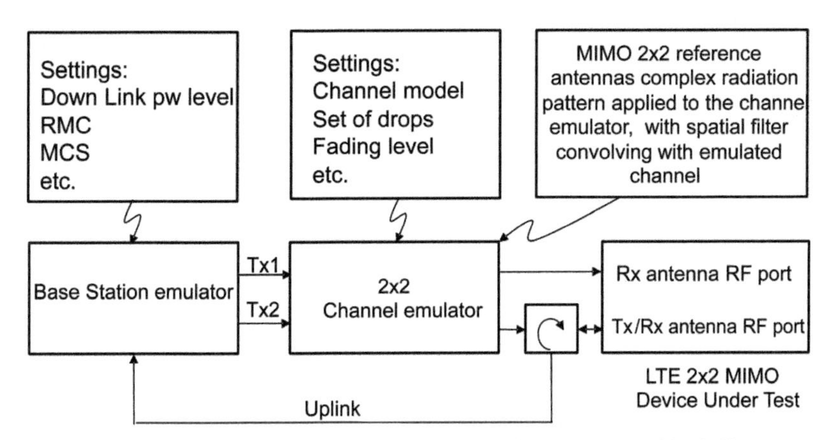

Figure 11.3 Absolute Data Throughput conducted measurement block diagram.

versus average power, with average power defined as the inverse of the average of the inverse used for sensitivity [Szi14b]. Since the shape of a single TP versus power curve is very different from that of the average of the TP across the total spread in power, the resultant curves differ considerably. The results indicate that a simple average over all data spread does not provide proper characterisation or discrimination of the device's radiated performance in user centric defined modes, such as hands free, navigation/data portrait, navigation/data/gaming landscape, or hotspot.

11.3.5 Definition and Characterisation of Measurement Campaigns

One of the relevant open issues in the MIMO OTA industry and standards nowadays is the definition of the test measurement uncertainty (MU), different test methodologies have different hardware and software requirements and unique system implementations. While some test methodologies require multiple CE ports, PAs, probe antennas, cables, connectors etc., other MIMO OTA methods are based on antenna system complex radiation patterns gathered in SISO anechoic chambers and conducted measurements, clearly a single MU value cannot capture both methods uncertainty properly.

A measurement campaign was started for investigating the root cause of MU in anechoic chamber multi-probe set-ups, in which three independent implementations of the same test methodology [FSF$^+$14] were compared. Two of these implementations, those of Aalborg University and Motorola Mobility were built from ground-up, where the third set-up of ETS-Lindgren was based on a commercially available installation. Details are given in Section 11.3.4.

Other measurement campaigns revealed consequences of strong correlation at the BS side for the SCME urban-macro (UMa) channel model, being approximately 0.95. A cross-polarised 45 slanted antenna at the BS is defined, assuming to represent the most common network deployment scenario. As the angles of departure are close to 90° (representing the end-fire direction of the array), the AoD spread is about 2°. Based on the foreshortening effect of the array elements, the horizontal component is also reduced in this model. Results from these measurements indicate that there is a significant improvement in DuT performance, approximately 6 dB, for the UMa model when the BS correlation is removed. It was later decided to preserve the BS antenna correlation effects so that the test conditions could distinguish device performance differences. The urban-micro (UMi) model is quite different, having the AoDs all near 0°, which results in a nearly balanced polarisation ratio and very low correlation between BS array elements. The DuT will perform better when using the UMi channel model due to these effects. It was also noted that the slope of the TP curve, averaged over the various device orientations, is affected by the variation in the individual TP curves. The larger the variations over the orientations, the shallower the slope of the average curve.

The emulation of a well-determined SNR during testing showed to be non-trivial. Four different methods of generating noise in the test environment were analysed [JKR13], from noise injection before channel faders to noise injection at OTA antenna feed points. As a result, a definition of SNR based on omnidirectional unfaded additive white Gaussian noise (AWGN) was proposed, to be used for the future evaluation of MIMO OTA test methods along with non-AWGN test cases. This definition of SNR minimises the correlation between the signal and noise without going as far as defining directional or time-variant noise which remain items for future study. The use of omnidirectional noise has since been adopted by CTIA and is under consideration by 3GPP. An analysis showed that in low-noise environments, antenna efficiency is dominant, but in high-noise environments, antenna correlation is far more important [JKR13].

Apart from noise, also interference needs to be considered, as the performance of long-term evolution long-term evolution (LTE) mobile terminals in cellular systems is limited by interference, e.g., the inter-cell interference. Open questions are which the characteristics of interference are in real environments, which characteristics are essential for radio link performance in an OTA measurement, and how to emulate interference realistically, especially in the MPAC set-ups. To answer these questions, background interference was measured, power levels were determined, and variations depending on the AoA at the mobile location [NFP13]. Background interference has been defined as signals and noise received within the band of a particular cellular system, excluding the signals originating from the system itself. A small series of initial exploratory measurements were performed with a spectrum analyser connected to a spherically scanning horn antenna. The measurements were done in different geographical locations, urban, sub-urban, and rural. Power distributions were successfully obtained within frequency bands where various systems are known to transmit. However, the median power levels were generally too close to the system noise floor around. To solve the problem of too high-noise floors in the analysis, future work will likely involve measurements selected frequency bands only.

Averaging TP results became a topic too as two different ways to derive average TP were proposed in industry. One way is to generate an average

TP curve across all 12 DuT rotations and then compute the reference signal energy per resource element (RS EPRE) value necessary to reach the 70 or 95% of the peak TP on this average curve. The other way is to compute the RS EPRE values (i.e., energy per resource element of the reference signal and is expressed in dBm/15 kHz) necessary to reach the 70 or 95% of the peak TP for each curve corresponding to the rotations and then calculate the average performance metric across all rotations. It was shown that the first method incorrectly estimates the FoM due to the non-linear relationship between the RS EPRE and TP, whereas the second method provides the correct average representation of the performance metric and also generates a useful visualisation of the data [IMU13, Iof14].

11.4 MPAC Method

The antenna design and the propagation channels are the two key parameters that together ultimately determine MIMO device performance [JW04]. As antennas are considered inherently in the OTA testing, it is important to also include realistic channel models for MIMO device performance evaluation. The MPAC set-up has attracted great research attention both from industry and academia due to its capability to emulate realistic multipath environments with controllable channel characteristics, making it a suitable method for testing terminals equipped with multiple antennas. This part is organised as follows. The MPAC set-up and the basic idea are introduced first. Then, channel emulation techniques, which are widely discussed and investigated in the literature, are described. And finally, a state-the-art of topics related to MPAC set-ups is presented.

11.4.1 Introduction

An illustration of the MPAC set-up is shown in Figure 11.4. The MPAC system often consists of a radio communication tester, a CE, a PA box, multiple probe antennas located around the DuT in an anechoic chamber. The radio communication tester is used to emulate the cellular network end of the link. The CE and the multiple probes are used to create desired spatial–temporal channels and intended interferences within the test area. The PA are used to adjust the signal to the desired power level. A network analyser is often used for channel validation investigations. As illustrated in Figure 11.4, the current set-up is focused on emulating realistic downlink channel models (i.e., communication from BS to mobile terminal), while the uplink is realised by a direct antenna and cable connection. As the testing is performed in the anechoic chamber, the generated multipath environment will be free from reflections inside the chamber and

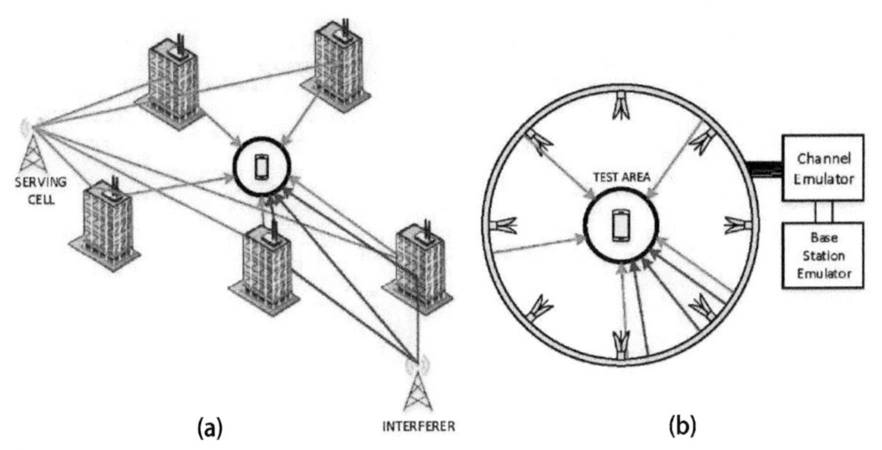

Figure 11.4 The multipath environment (a) and channel emulation in the MPAC set-ups (b).

external unwanted interferences. The testing is realistic as well, since the DuT is evaluated as it is used in the real network. The main disadvantage with the MPAC method is the cost of the set-up. The number of output ports of the CE is often limited, and, therefore, the number of probes utilised for synthesising the channel is limited, which would result in a test area with a limited size.

11.4.2 Radio Channel Emulation Techniques

One of the main technical challenges for OTA testing of MIMO capable devices is how to emulate the spatial channel models in the volume where the device is to be tested. The key idea of channel emulation is to ensure that the signals emitted from the probe antennas are properly controlled such that the emulated channels experienced by the DuT approximate the target channel models within the test area. In this section, different channel emulation techniques are revisited and summarised.

11.4.2.1 Prefaded signal synthesis

The PFS technique was proposed in Kyösti et al. [KJN12], and has been widely used in commercial CEs. With the PFS technique, fading signals, generated with the sum of sinusoid technique, are transmitted from each probe antenna. Each cluster is emulated by several probe antennas. Fading signals associated with the same cluster are independent and identically distributed. The emulated channel, which is a linear summation of contributions from the multiple probes, matches with the target channel in the temporal domain. For each cluster, the Rx side spatial characteristics are reconstructed by allocating appropriate power weights to the fading signals from the probes. The size of

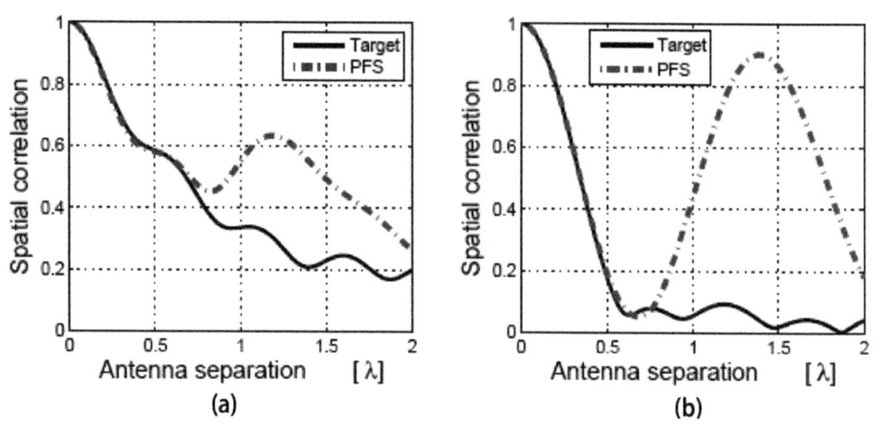

Figure 11.5 Target and emulated spatial correlation at the Rx side for the SCME urban macro channel model (a) and SCME urban micro channel model (b), with eight OTA probe antennas.

test area with acceptable accuracy is only determined by how well the Rx side spatial characteristics can be emulated, as temporal characteristics could be perfectly reproduced. An example of how well the emulated channel matches with the target channel in terms of spatial correlation at the Rx side is shown in Figure 11.5, where a test area of 0.7 wavelength diameter can be achieved with eight probe antennas. For channel models that consist of multiple clusters, each cluster is emulated independently. For dual-polarised channel models, vertical and horizontal polarisations are emulated independently. The effects of other channel characteristics, e.g., the transmitter (Tx) antenna array, channel spatial characteristics at the Tx side, are considered and modelled in the fading signals. Geometry-based stochastic channels (GBSCs) are often selected as the target channel models for the PFS technique. The PFS technique has gained its popularity due to its capability to emulate GBSC, with only probe power calibration required in the MPAC set-ups.

11.4.2.2 Plane wave synthesis

The basic idea of the plane wave synthesis (PWS) technique is that a static plane wave with an arbitrary impinging angle can be generated within a test area by allocating appropriate complex weights to the probe antennas on the OTA ring. Target plane wave is with a uniform power distribution and ideal linear phase front along the impinging direction within the test area. Different techniques have been proposed to obtain the complex weights, see, e.g., least square technique in Kyösti et al. [KJN12] Fan et al. [FCnN$^+$12] Kotterman et al. [KSLDG14], and trigonometric interpolation in Fan et al. [FNF$^+$13].

An example of the emulated field with eight probe antennas for a test area of 0.7 wavelength diameter is shown in Figure 11.6, where the target plane wave is with impinging angle 22.5° (i.e., from between two adjacent probes). Two ideas were proposed to create spatial–temporal channel models based on static plane waves. Each snapshot of a time-variant channel can be considered as static, and can be modelled by multiple static plane waves, each with a complex amplitude, angle-of-arrival (AoA) and polarisation. The PWS technique can then be applied to approximate each snapshot. Another idea is to emulate GBSC models. A cluster with a stationary power angular spectrum (PAS) can be discretised by a collection of plane waves, each with a specific AoA. Each plane wave can be approximated by the PWS technique. A Doppler shift can then be introduced to each static plane wave to enable time variant channels [KJN12]. With the first idea, arbitrary multipath environments (e.g., channels with time-varying AoA) can be reproduced. For the second idea, the reproduced channel is stationary with a fixed AoA, as the incoming power angel spectrum has a specific shape. The main disadvantage is that both phase and power calibration are required for the multiple probes, as complex weights have to be obtained. Otherwise in hardware requirements both the PFS and PWS methods are alike.

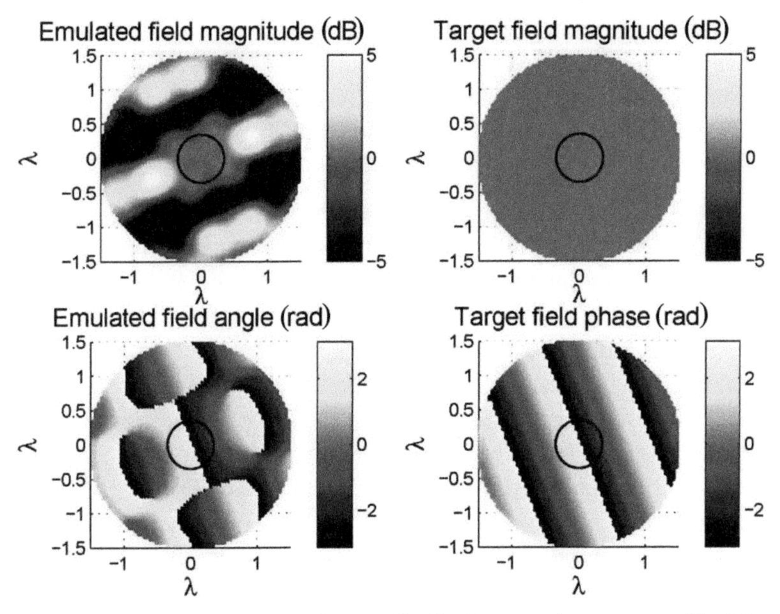

Figure 11.6 Emulated magnitude and phase distribution over the test area with eight probe antennas. *Black circle* denotes the test area.

11.4.3 The MPAC Set-up Design

The cost of the MPAC set-up depends directly on its design. Key aspects related to the MPAC design are physical dimensions of the OTA ring on which the probe antennas are located, number of required probe antennas, probe antenna design, probe configuration, MIMO OTA testing in small anechoic chambers, and the probe selection concept [FSNP13].

11.4.3.1 Chamber size

The physical dimension of a ring of antennas in an OTA set-up is limited by the size of the anechoic chamber. The physical dimensions are important in planning of the MPAC set-ups. It is important to understand up to which maximum size devices can be accurately measured for given dimensions of a ring of OTA antennas and a given range of frequencies. The physical dimension criteria of the MPAC set-ups based on field strength stability and phase stability across the test zone were often investigated [KH12]. Due to the limited distance between OTA probes and test area, the path loss is non-uniform and phase fronts are curved over the test area. Three classes of criteria for physical dimensions of the MPAC set-ups are considered. The criteria were

1. Non-uniform field strength caused by varying path loss, which may result in power imbalance between DuT antennas,
2. Phase variation caused by curved non-planar waves, which may result in correlation errors on DuT antennas, and
3. CTIA far field criteria. As error thresholds, 0.05 root mean square (RMS) correlation error and 0.5 dB average power imbalance were selected.

With the selected error thresholds, a maximum ratio of $r/R = 0.33$ was found from the power imbalance criterion and $r/R = 0.1$ from the correlation error criteria on the frequency range of 0.5–6 GHz, where r and R denote the test area radius and OTA ring radius, respectively.

OTA testing of MIMO capable terminals is often performed in large anechoic chambers, where planar waves impinging the test area are assumed. Furthermore, reflections from the chamber, and probe coupling are often considered negligible due to the large dimensions of the chamber. It is interesting to explore the possibility of performing MIMO OTA testing in a small anechoic chamber. It was concluded that 1.5 m distance between test antennas and DuT is generally sufficient [Mli11]. It was also investigated how to accommodate probe antennas used to synthesise clustered radio signal with 35° rms Laplacian distribution per the SCME standard. However, the proposed set-up is limited to a single spatial cluster with restricted AoA.

11.4.3.2 Probe configuration

Probe configuration is another topic related to the MPAC set-up design. Different 3-D (full sphere) and 2.5-D (three elevation rings) probe configurations are assessed with 3-D extended IMT-Advanced channel models in Kyösti and Khatun [KK13b]. The FoMs for assessing the probe configurations are the RMS error on spatial correlation function and the synthesis error. It is demonstrated that the emulation accuracy depends on both channel model and probe configuration. It is shown that a configuration with 16 dual-polarised probes could be sufficient for testing of terminals with diameter of 0.75 or even 1λ. In Kotterman et al. [KSLDG14], it is shown that the orientation of dual-polarised antenna elements has influence on the size of the test zone for 3-D electromagnetic wave field synthesis. The customary choice of polarisation directions along azimuth and elevation allows for the full angular range over which wave fields can be synthesised. But, this choice effectively reduces the number of available active radiators in case a single-polarised wave is to be synthesised with direction of incidence near maximum elevation. One option to maintain synthesis quality is driving both types of polarised elements, meaning also adding CEs that are an appreciable cost factor. However, not all applications need fully 3-D incident fields. For instance in outdoor cellular applications, a limited elevation range is not uncommon, avoiding the problematic angular region.

With respect to the number of probes, the flexibility of field emulation increases with the number of probes. However, probes become increasingly closely spaced, causing increased scattering from the neighbouring probes. In a 16-probe set-up, scattered fields were 25 dB below the main signal, in an eight-probes set-up 30 dB [BFKP14]. The consequences of scattering from neighbouring probes on field quality in the test zone need to be further investigated, in order to define a maximum acceptable level.

11.4.3.3 Probe design

One part of the MPAC set-up is OTA probe design. The antennas need to offer good polarisations properties and, at the same time, to be directive for creating variable radio channel conditions within the test zone. One option is to use narrow band transmitting antennas for every test frequency. This is not a very handy approach as huge banks of antennas are needed to cover different test frequencies. Another option is the use of wideband antennas to cover all necessary test bands. Several probe antennas are utilised in the MPAC set-ups, e.g., horn antenna, dipole, and Vivaldi antennas. In Sonkki et al. [SSEH^{+}15], a wideband dual-polarised cross-shaped Vivaldi antenna is presented. The antenna offers good polarisation properties over a wide frequency bandwidth with good impedance matching and very low mutual coupling between the antenna feeding ports.

11.4.3.4 Calibration

For the MPAC set-up, proper calibration of the system is required before the actual measurement. For the PFS technique, as the signals transmitted from the probes are power weighted, it is required to ensure identical path losses from the probes to the test area centre. For the PWS technique, complex weights are allocated to the probes, and hence it is required to ensure both identical path losses and phase lags from the probes to the test area. A calibration antenna, usually an electric or magnetic dipole, is placed instead of the DuT and connected to the vector network analyser (VNA). The main drawback is that for each calibration, the set-up has to be changed. In Fan et al. [FCnN+13], it is shown the main cause of the signal drifting over time is the active elements of the set-up, hence a specific calibration method focusing on those elements should be considered [CnFN+13]. By adding electronic switching units after the power amplifier (PA), a connection between the VNA, the CE, and the PA can be created. This way, the chamber and all the elements inside are bypassed and a calibration of the active elements can be done without physically changing the set-up [CnFN+13].

11.4.3.5 Test area size investigation

One of the key questions to be addressed is how large the test area can be supported with a limited number of probes. The test area is an area where the desired channel models can be accurately reproduced. The antenna separation on the DuT should be smaller than the test area size to ensure that the DuT is evaluated under the desired channel conditions.

Different FoM are proposed and analysed in the literature to determine the test area size for different channel emulation techniques. For the PWS technique, often field synthesis error $|E - \hat{E}|$ is selected as the FoM, where E and $\hat{E}$ represents the target and emulated field, respectively. $|V - \hat{V}|$ is suggested as the FoM, with V being the received voltage for the target plane wave and $\hat{V}$ being the received voltage for the emulated plane wave [FNF+13]. In this FoM, the DuT antenna pattern is included in the evaluation. Other FoMs could be adopted as well, e.g., spatial correlation, wave front direction accuracy, power flow/time-averaged Poynting vector, phase of the field vector elements, ellipticity, and group delay are under discussion. The question remains, though, which FoM describes the reaction of the DuT on the emulated field best, the answer likely being (radio) system dependent.

For the PFS technique, often the spatial correlation error at the Rx side $|\rho - \hat{\rho}|$ is selected, as it represents how well the emulated impinging PAS follows the target. In Fan et al. [FNF+13], the antenna correlation error $|\rho_a - \hat{\rho}_a|$ is proposed to determine test area size, where ρ_a is the correlation of the received signals at antenna output ports for the target impinging power angle spectrum and $\hat{\rho}_a$ is the similar correlation for the emulated impinging power angle spectrum. Fan et al. [FKNP16] investigated how well the capacity of the emulated channels matches with that of the target channel models. The test zone size depends on the probe configuration, carrier frequency, number of probes, channel emulation techniques, target channel models, acceptable error level, and DuT radiation patterns.

Intrinsic disadvantages of 2-D synthesis are amplitude drop-off within the test area and a small usable height. Coherent wave-field synthesis through the use of an annular antenna array suffers from remaining wave-front curvature normal to the plane of the 2-D array, and hence the usable test volume is limited to a thin disc [KLHT11]. Additionally, the amplitudes of the (approximately) cylindrical emulated waves drops with the inverse of the square root of distance and this decay over the test area is noticeable, especially when using a metric like EVM. As an alternative to real 3-D synthesis, the use of small sub-arrays is proposed, replacing the OTA antennas. Each of the sub-arrays locally generates wave-fronts with only curvature in the plane of the test area, which can be compensated by the field synthesis [Kot12]. Using three antennas per sub-array, with two wavelengths separation and passive power division and phasing, the test area size could be enlarged to approximately a sphere with less than 1° wavefront direction error and a total amplitude variation of 0.7 dB.

The influence of complex amplitude errors on the quality of synthesised wave fields was investigated in Kotterman [Kot13]. Simulation of the influence of errors in the excitation signals (i.e., the complex weights), with the aim to determine which accuracy is needed when including all errors from different sources like calibration, drift, mechanical vibration, phase noise, etc. It was noted that the influences of individual error distribution realisations were quite different, depending on whether the strongest excitation signals were impaired with larger or smaller random errors. Based on the simulation results, it is recommended to keep the maximum phase error span limited to $[-10°, +10°]$.

The test area size can also be expressed in terms of capacity emulation accuracy [FKNP16]. The investigation is based on the well accepted channel models in the standards for OTA testing of MIMO capable terminals, i.e., the SCME Umi and SCME Uma. The impact of spatial correlation at the Tx side, the channel model, and the spatial correlation at the Rx side on the capacity emulation accuracy was investigated. Simulation results show that the number of probes is irrelevant when the spatial correlation at the Tx side is in the high region (e.g., $\rho > 0.7$). Furthermore, when correlation at the Tx side is low, the spatial correlation accuracy is less critical with small correlation at the Rx side. The simulation results are supported by measurements in a practical set-up [FKNP16].

Attempts have been made to define the test area size in terms of TP, being the relevant FoM in standardisation [Szi14a, IY14], but further study is required.

11.4.4 Practical Channel Emulation

11.4.4.1 Measurement uncertainty

A mandatory step for evaluating MIMO devices in practical set-ups, is analysis of the sources of errors and uncertainties in the measurements. The uncertainty level can help to understand the level of confidence associated with testing results.

Some investigations on MU were reported in the literature, where some error sources were identified and analysed, as detailed below. However, the actual impact of the error levels on the testing results is still unclear. Quantifying the impact of errors on the important parameters, e.g., signal correlation accuracy, received voltage accuracy on the antenna, capacity and TP would be more interesting.

The probes are often assumed accurately placed on the OTA ring, however, probe placement error, e.g., probe orientation error and probe location mismatch error might exist in practical set-ups. Probe placement can introduce error in the system [FNCn+12, FNCn+13]. Probe orientation error might effectively modify the complex weights allocated to the probes and hence have an impact on the field synthesis accuracy. It was concluded that radial location errors are most critical, since the synthesised field for a radial error of a quarter of a wave length is no longer the plane wave-field with the target AoA. The impact of probe placement error on spatial correlation emulation was investigated in Fan et al. [FNCn+13], the emulated correlation depending on the power weight allocated to each of the probe and on the probe angular location. The simulation results show that the probe angular location error is critical for spatial correlation emulation. Note that both probe orientation errors and radial location errors can be compensated during the calibration of the set-up.

Measurement uncertainty levels for different labs, i.e., at Aalborg University (AAU), Motorola Mobility (MM), and in ETS-Lindgren (ETS) were investigated to show key aspects related to MPAC set-up design [FSF+14]. The MPAC set-up in AAU was equipped with an aluminium ring of 2 m and 16 dual-polarised horn antennas. Polystyrene placed on top of the turntable was used to support the DuT. Cables are connected to the DuT directly. The MM set-up was equipped with eight uniformly placed horn antennas on a OTA ring with 1.2 m radius. Choke and cartridge at various frequency bands were used to connect to the DuT. The set-up in the ETS was equipped with 16 dual-polarised Vivaldi antennas on a ring of radius 2 m. Ferrite-loaded cables are used to connect to the DuT. An illustration of the MPAC set-ups are shown in Figure 11.7. The main testing items of the MU investigations include, e.g., dipole radiation pattern measurements, turntable stability, CE stability, system frequency response, power coupling between probes, reflection level inside the chamber, and field synthesis. It was concluded that cable effect will distort the radiation pattern of the DuT and hence affect the results of the measurements. By the use of a choke/cartridge or ferrite-loaded cable, the cable effect can be minimised. Field synthesis measurements demonstrated the improved results with chokes/cartridges and ferrite-loaded cables. The polystyrene, used to support the DuT in the AAU set-up, introduces mechanical instability after movement. Non-flat frequency response of the OTA system can be introduced by the CE, termination of the cables (probe antenna) and mismatch between the components. Good agreement between the measured plane wave and the target plane wave both for the vertical and horizontal polarisations is obtained in the MM and ETS set-up. Sources of errors and uncertainties and probe coupling levels between neighbouring probes were addressed in Fan et al. [FCnN+13] Barrio et al. [BFKP14] as well.

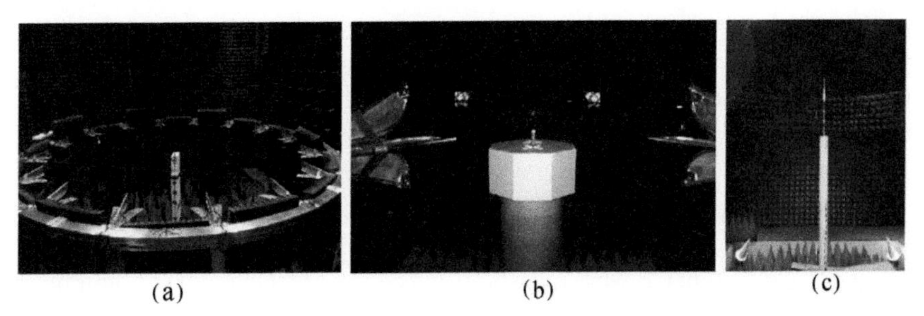

(a) (b) (c)

Figure 11.7 Three different MPAC set-ups, that of AAU (a), of Motorola Mobility (b), and of ETS-Lindgren (c).

A fundamental way of measurement system analysis is setting up assessment procedures under the Gage Repeatability and Reproducibility framework, of which an example for a specific MIMO OTA test set-up was given in Wu et al. [WCY+12]. The advice, however, is that laboratories yet do not rely on GRR alone for assessment of set-ups.

The probes are often in the near field, and, importantly, there may also be scattering from the neighbouring probes. It is anticipated that these near-field and scattering effects will increase the uncertainties in the multi-probe testing. Therefore, it would be desirable to have a way to compensate those effects to generate fields identical to a plane wave in the test zone. The work in Parveg et al. [PLK+12] proposed a calibration technique for partially compensating the near-field effects and scattering contributions from the neighbouring probes in 2-D MPAC system. The results show that both the near field effects and scattering contributions in the test zone can be partially compensated effects by using the proposed technique.

11.4.4.2 Validation of the emulated channel

The goal of the channel validation is to ensure that the created channels within the test area follow target channels in the practical set-ups, and hence, comparable testing results could be obtained among different laboratories. Validation of four domains of GBSC is required in 3GPP and CTIA, i.e., delay (PDP), temporal (temporal correlation or Doppler power spectrum), polarisation (cross-polarisation ratio), and spatial (spatial correlation or PAS) domains. The focus of PWS validation measurement was to check whether the measured complex field in the test area matches the target field.

For static PWS, good agreement between the measured and emulated field in the test area has been achieved for all the scenarios in Fan et al. [FCnN+12, FCnN+13, FSF+14]. Similar correspondence was observed for the simulated and measured plane wave field for all scenarios outside the test area. Analysis of the results made it possible to identify sources of inaccuracies like DuT placement errors and cable effects resulting from bending. Several aspects of the PFS were subject of investigation [FCnN+13, FCnA+13, WCY+13, SAG14]. An investigation of channel model validation in the MPAC set-up with a radius of 3.2 m and eight dualpolarisation probes was performed in Sun et al. [SAG14], where the characteristics of the channel environment emulated using different CE were measured and compared. Channel validation results for the single spatial cluster channel models are presented in Wu et al. [WCY+13].

When estimating the PAS of the emulated channel with the PFS technique, one should realise that the emulated PAS at the Rx side is discrete, characterised by the angular locations and power weights of the active probes [FNP14]. In practical set-ups, knowledge on how the channel is emulated in commercial CE is very limited. Therefore, estimation of the discrete PAS can be used to verify how well the target channel is implemented in the test area. Beam-forming techniques on measurements on virtual arrays are proposed. However, direction of arrival (DoA) and power estimates are prone to inaccuracy due to low spatial resolution and side lobes. In Fan et al. [FNP14], the MUSIC algorithm was chosen for its high resolution. The power estimates based on DoA estimates match well with the target in the measurements. To improve accuracy and robustness in elevation DoA estimation, the use of an (virtual) array with large aperture in elevation too is recommended.

11.4.4.3 Actual OTA measurements

Data TP has been selected as the FoM in MIMO OTA standards to rank MIMO capable terminals, as it reflects the end-user experience. The Inter-Lab OTA performance comparison testing campaign of CTIA started in 2012, where the focus was on comparing results of the same methods in different labs. Extensive measurement campaigns have been performed in different laboratories and numerous results have been reported [KHNK11, IMU13, Iof14, CnFN+13]. However, deviations in terms of TP in measurement results still exist among laboratories and explanations for the causes are not determined yet. There is a strong need to develop a TP simulation tool with reasonable accuracy, as it would give more insight into the test results and would help with eliminating systematic errors in measurements.

The TP performance of a commercial LTE mobile terminal, subjected to different channel models in practical 2-D and 3-D MPAC set-ups, was investigated in Kyösti et al. [KHNK11]. More specifically, GBSC models, e.g., IMT-Advanced, WINNER, SCME, and different single spatial cluster channel models were selected to evaluate the TP performance of the device. The DuT was evaluated with three different tilt angles. The measurement results indicate that the channel model has impact on TP performance. The 3-D channel model gives higher TP than the 2-D. Different multi-cluster models with the 2-D configuration have performance variation from 0.5 to 2.5 dB, while the single cluster model with different angular spread parameters with the 2-D configuration has more than 14 dB performance variation. It is also pointed out that DuT TP results over different tilt angles under the same channel model are different. Note that the DuT TP results over different tilt angles in RC are expected to be the same due to the isotropy of the channel.

11.4.4.4 Arbitrary spatial channel emulation

The MPAC method is known for its capability to physically synthesise arbitrary radio propagation environments under laboratory condition. 2-D GBSC, where the incoming power angular spectra of the channels are defined only on the azimuth plane, are targeted in PFS current set-ups. GBSC are generated based on sum-of sinusoids techniques, with each sinusoid characterised by its amplitude, Doppler frequency, and random initial phase. Although different in their synthesis approach, both synthesis methods are capable of equal performance, as shown by simulations [RBRH14, Kyö12], and have almost equal variation of ergodic capacity and equal time variance over random initialisations. Also, simulation and emulation are comparable for both methods times. Note that conclusions in Kyösti et al. [Kyö12] are valid in the single-polarised case only. When introducing a dual-polarised configuration, the matrix product of 2×2 random initial phase matrix and dual-polarised Tx (/Rx) antenna gain patterns will result in variant gains of rays, which will lead to a non-ergodic simulator [Obr13, RBRH14].

In general, three aspects of the radio field are to be considered in emulation, i.e., directivity or spatial correlation, polarisation properties, and 3-D field incidence (Section 11.2.7). With respect to the latter aspect, only 2-D standard channel models have been used in MPAC set-ups so far, as the channel models in standardisation are still 2-D. However, since long, from measurements is known that elevation spread cannot be ignored in many propagation environments [KLV+03]. In order to evaluate MIMO terminals in realistic environments in the lab, it would be desirable that 3-D radio channels can be accurately reproduced in MPAC set-ups. However, costs, in terms of the much greater number of CE needed, become a major issue when an appropriate 3-D probe configuration is required [KK13b, KSLDG14].

The discussions on channel models in MIMO OTA standards concentrate on SCME channel models, i.e., on Rayleigh fading channel models.

On one hand, attempts are made to simplify the structure of the SCME models, in order to save on emulation hardware. The basic idea is to simplify the SCME sub-paths and then to evaluate the uncertainty resulting from this simplification [Szi11a, Szi11b]. For this, data TPs for the same complex antenna radiation pattern are compared between SCME and sub-sets of modified SCME channel models. On the other hand, a strong need is felt to include Rician channel models as well for lab-testing in more realistic environments. A novel technique is proposed to model the Rician fading channel models in the MPAC [FKH+14], in which a LoS path with arbitrary incidence is possible and a NLoS component with arbitrary PAS shape can be modelled.

More specifically, the specular path is modelled using the PWS technique, with the scattering NLoS component modelled by the PFS technique. Simulation results showed that the reproduced Rician channels match very well with the target models, in terms of field envelope distribution, estimated K-factor, spatial correlation, and Doppler power spectrum. The emulated spatial correlation follows the target curve well up to 0.71 λ distance and deviates after that. Replaying ray tracing simulated channels in the MPAC set-ups was brought up in Llorent et al. [LFP15]. MIMO OTA performance testing requires devices to be tested under realistic channel conditions. Standard channel models such as SCME or WINNER aim at modelling environments that are generic, representing defined general channel conditions, e.g., urban, suburban, rural, or indoor environments. As an alternative, replaying field measurements or using ray tracing models would result into more realistic models since they are site-specific. Ray tracing simulations of an urban environment with LoS and NLoS conditions are used in Llorent et al. [LFP15] to obtain the complex amplitudes of rays that subsequently are to be emulated in a MPAC set-up using PWS. An evaluation of simulated fields promised high accuracy both for an arbitrary ray and for the total received field.

11.4.5 Other Applications

The main driver for MIMO OTA research up to now has been the radiated performance of small cellular mobile UE, like handsets and laptops. But, there are many more radio systems that depend on interaction with the EM environment in which they operate, not all necessarily radio communication systems. Often they share with cellular systems that they operate in Multi-User environments with the resulting (directional) interference. In those cases, OTA testing can be well applied, as its essence is emulating the (system-) relevant properties of the system's radio environments. In this context, the expression virtual electromagnetic environment (VEE) has been coined [SKL+13]. Radio systems not primarily intended for communication are, e.g., radio location and positioning systems that observe their environment mainly from the directional/angular spectrum point of view. Then, installed performance can only be tested OTA. The same is true for cognitive radio (CR) systems whose operational environment is characterised by the (time-variant) interference that normally also shows directionality [SKL+13]. Note that truly emulating interference is likely to be the next step in MIMO OTA for cellular mobile UE too. Intelligent transportation systems (ITS) like ITS-G5 with their road/user-safety relevance are thought to be interference-prone when massive deployment is reached. With ITS G5 installed on cars,

OTA installations necessarily become big and TU Ilmenau had to build a separate, large MPAC facility for vehicular OTA applications [HBK+15], in connection with C2X-research, which will have real-time connections with other testbeds on the campus. The goal is creating a large VEE for communication with an operational vehicle, virtually driven by a human driver through a defined, virtual, traffic environment while subjected to generic traffic scenarios. Coherent synthesis in the MPAC is unachievable, though, as the test objects, cars, have largest dimensions of the order of 100 wavelengths at 6 GHz, for example. Therefore, simpler approaches are taken elsewhere too, as in Nilsson et al. [NAH+13].

11.5 RC Method

The RC can be used for OTA measurements [WB10, Che14b]. A typical measurement set-up is depicted in Figure 11.8. Detailed descriptions on the theory and operation of RCs can be found in Hill [Hil09].

Any lossy objects present in the chamber, including the building material of the chamber itself, antennas, and microwave absorbers, will load the RC cavity. Thus, when the RC is excited by an antenna in the chamber, it decays exponentially [DDDLD08]. This decay is usually described using the RMS delay spread (DS). The RMS DS can be decreased by adding lossy material inside the chamber. The size of RCs used for OTA testing typically have an inherent RMS DS of 200 ns without any added microwave absorbers.

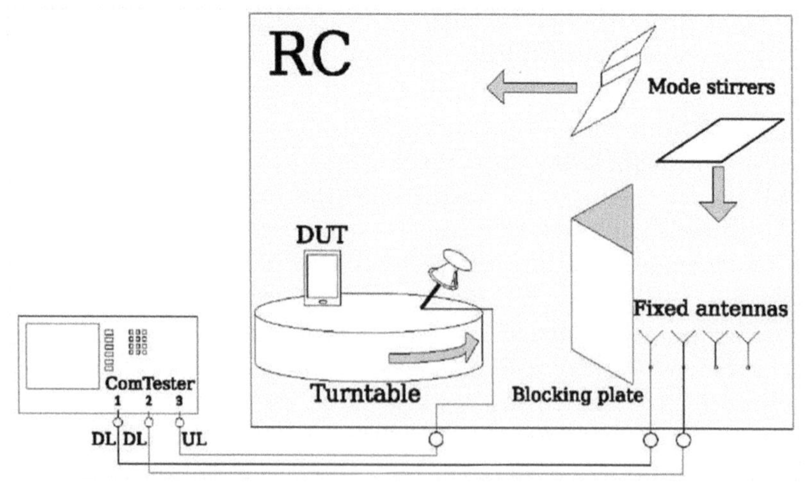

Figure 11.8 Typical set-up of an OTA measurement using an RC.

A decrease in RMS DS corresponds to an increase in coherence bandwidth. The spatial receiving characteristics are cumulative isotropic, meaning that isotropy is achieved only after completing the full measurement sequence. The field in each mode stirring position is not isotropic. Channel properties of RC are discussedin Skårbratt et al. [SLR15] and Kildal et al. [KCO+12].

11.5.1 RC as An OTA Measurement Environment Extended Using a CE

For advanced multi-antenna Rxs the RC can be complemented by a CE to provide testing in more complex channels. This set-up is depicted in Figure 11.9. The receiving spatial properties of the RC are not affected by the addition of a CE. The temporal properties of the measurement set-up can be further controlled by the addition of a CE. DS profile, fading statistics, and Doppler spread can be modified. Also the BS antenna correlation, as seen by the DuT Rx, can be changed by modifying the BS correlation using the CE. The channel properties of the RC test set-ups are further elaborated in Skårbratt et al. [SLR15].

The total downlink MIMO channel experienced by the communication system when using both an RC and a CE can be described by

$$\mathbf{r} = \mathbf{H}_{\mathrm{RC}}\mathbf{H}_{\mathrm{CE}}\mathbf{s} + \mathbf{n}, \tag{11.2}$$

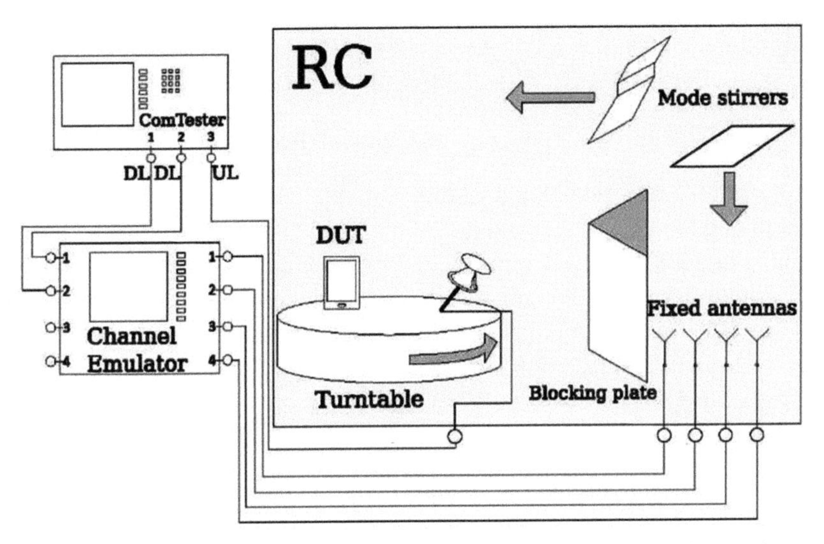

Figure 11.9 Typical set-up of an OTA measurement system using an RC complemented with a CE.

where $\mathbf{H}_{RC}$ represent the channel matrix of the RC and the DuT and $\mathbf{H}_{CE}$ the channel matrix of the CE and the communication tester; **s** and **n** are the signal and additive noise vector, respectively. The RC creates a Rayleigh faded environment due to the stirrers moving in the chamber. Often Rayleigh fading is also enabled in the CE. Due to the cascading of these two according to Equation (11.2), the DuT Rx experiences a double-Rayleigh faded signal. This can be mitigated by using multiple antennas, independently faded between the CE and the RC. This increases the richness of the MIMO channel and makes it behave closer to regular Rayleigh fading (see Skårbratt et al. [SLR15]).

11.5.2 Common RC Channel Realisations

The most commonly used channel models for the RC+CE set-up are the short delay low correlation (SDLC) an long delay high correlation (LDHC) channel models, see 3GPP [3GPP14b], even though other channel models can be realised as well (see for example Skårbratt et al. [SRL15]). These are based on the SCME UMi and UMa channel models [BHdG^{+}05], but modified to be realisable in an RC environment with the average isotropicAoA.

As a complement to these, the National Institute of Standards and Technology (NIST) model exists, described in 3GPP [3GPP14b]. The NIST model does not require a CE to be realisable in an RC, which can be favourable for some applications due to the lower complexity of this test set-up. As an example, this test set-up does not require phase calibration for stable measurements, which has been shown to be a significant source of uncertainty for methodologies including a CE. More detailed discussions about the NIST model canbe found in Matolak et al. and Remley and Kaslon [MRH09, RK13].

11.5.3 LTE Measurements in the RC Test Set-ups

The testing of LTE UE revealed a number of separate issues, related to the LTE system and the test methodology.

When testing LTE devices with a CE augmenting the RC, it is important that the output phase from the communication tester is calibrated to the CE. The problem is illustrated in Figure 11.10, where different phase offsets cause the TP to vary significantly [SLR15]. However, as it relates to the definition of the channel models (Section 11.2.10), this is not only an issue for the RC method.

The UE total isotropic sensitivity (TIS) depends on the fading conditions. It might be appropriate to define a new TIS measurement to handle the wide-band signal and different MIMO technologies incorporated in the LTE standard. No conclusions were drawn on how this procedure should be defined.

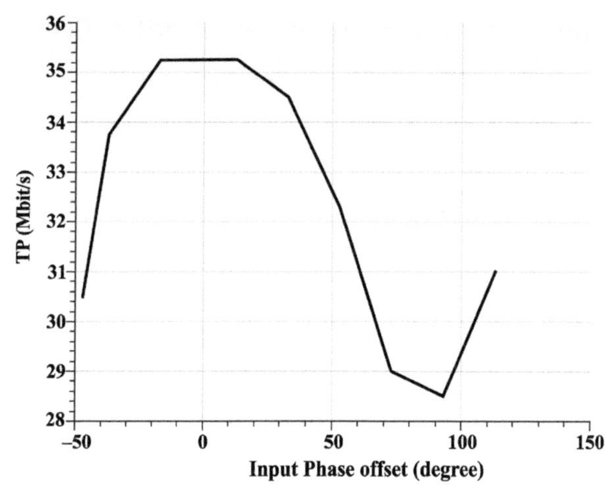

Figure 11.10 TP variation as a function of input phase offset for a DuT with LDHC as channel model.

For example, CTIA is considering to replace the existing TIS test with transmission mode 2 (TM2) TP testing, since this is considered a more realistic test due to the included fading properties of the channel models used for the testing. This is further discussed in Arsalane [Ars13].

Regarding ranking LTE UE based on TP, it was found that SDLC, LDHC, and NIST more or less agree on UE ranking when using the 50% TP level [PF11b]. Repeatability and reproducibility are reported to be within 0.5 dB of the measurements when using an RC.

During the testing of laptop LTE modems, influences of the host laptop on the dongle-under test were noticed [PF11a]. In order to prevent such influences from affecting the testing, a laptop phantom is required to yield repeatable and accurate results. Such a laptop phantom was later developed and standardised in 3GPP [3GPP12].

The use of adaptive modulation for OTA testing LTE UEs was investigated. It was found that in an RC, adaptive modulation gives the same device ranking as using the fixed modulation and coding scheme (MCS) [SRL15]. However, while good devices rank similarly when using adaptive modulation, the bad device performed even worse using adaptive modulation compared to the fixed MCS.

11.6 Two-Stage Test Method

For an overview of the two-stage MIMO OTA test method, refer to Sections 11.2.3 or 6.3.1 of 3GPP [3GPP14b].

11.6.1 UE Antenna Pattern Measurements Proof of Concept

In order to avoid cumbersome and possibly biased measurements of UE antenna patterns at the UE antenna ports by the use of external equipment, the definition of a standardised UE-internal measurement routine was proposed, the UE antenna test function (ATF). As proof of concept of the ATF, patterns of reference dipoles measured using the ATF and by the traditional passive approach were compared [KJZ11a]. The relative accuracy and linearity of UE-measured antenna amplitude and phase was seen to be <0.1 dB from –30 to –60 dBm and 1° at –50 dBm. The channel capacity resulting from the antenna patterns measured by passive and active ATF methods were similar with $<2\%$ difference in channel capacity at 25 dB SNR, thus proving the principle of the two-stage test method. A similar procedure was performed on real (hence unknown) antennas of two commercial universal serial bus (USB) LTE dongles [KJZ11b]. One of the dongles was modified to enable traditional passive antenna pattern measurement and for the other, unmodified, device of the same type the active ATF approach was used. TP was measured (using the conducted second stage approach) to show consistency. A channel capacity simulation based on manufacturer-provided theoretical antenna patterns was also performed to cross check with the TP test results. It was shown that the two-stage MIMO OTA test method can rank the antenna performance correctly. The measured test results aligned with the antenna channel capacity simulation results once the differences in conducted performance between the dongles were taken into account.

11.6.2 Study of 2-D versus 3-D Device Evaluation

To investigate the differences in MIMO OTA performance using 2-D and 3D evaluation fields, the variation in performance of a UE in a 2-D field at different elevation angles was determined [JZK12]. As the elevation angles were independent of azimuth, the 2-D incident fields were defined on conical surfaces. The analysis found an 8 dB variation in performance. Furthermore, a model was proposed to use the capability of the two-stage test method to emulate 3-D fields and it was shown that the performance from a single 3-D field equalled the average performance from 10 (conical) 2-D cuts as shown in Figure 11.11. The conclusion was that UE orientation relative to the field is important and that a single 2-D cut is not sufficient to determine total performance [JZK12]. Further analysis of reference antenna performance over different 2-D elevations was carried out with band 13 reference antennas [Jin12a]. This showed a smaller variation of 5 dB than the device with real antennas used in the study of Jing et al. [JZK12].

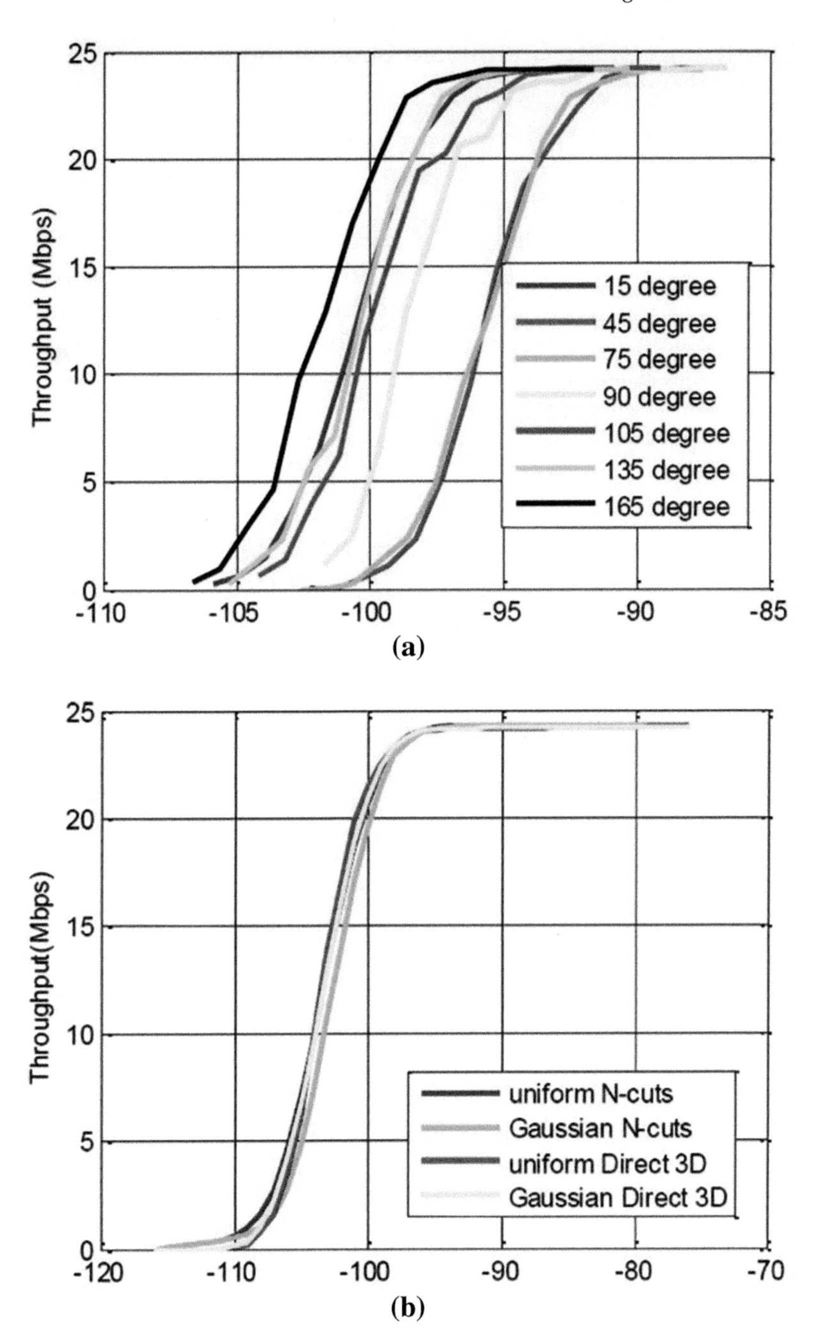

Figure 11.11 Variation in UE performance for different 2-D cuts and comparison of averaging ten 2-D cuts with a single 3-D measurement using the two-stage method. Along the abscissae, received power in 15 kHz bandwidth [dBm].

11.6.3 Limitations of the Conducted Second Stage with UE Desensitisation

A limitation of the two-stage method using the conducted second stage is that UE self-desensitisation is not measured since the UE antennas are disconnected at the temporary antenna connector. One method to overcome this limitation the use of UE-based noise estimation (Iot, [3GPP15]) was studied [Jin12b]. This analysis showed that it was possible to very accurately measure UE self-interference with reference signal received quality (RSRQ) measurements. This measurement of Iot can then be added back into the conducted second stage signal TP measurements to get the same overall results as would be seen using a fully radiated approach. However, a better solution to the limitation of the conducted second stage was later developed by using a radiated second stage, as described in Rumney et al. [RKJZ15], see Section 11.5.4.

11.6.4 Introduction of the Radiated Second Stage

An alternative approach to that taken in Jing [Jin12b] to correctly measure device desensitisation was developed, known as the radiated two-stage method and described in [RKJZ15]. This development means that the radiated desense is now fully covered by the two-stage method, overcoming the limitations of the conducted second stage approach. A further advantage of the radiated second stage is that the calibration method used for the radiated second stage means that the absolute accuracy of the UE measurements from the first stage pattern measurements does not contribute to the overall accuracy of the two-stage method. The only requirement on the UE is that the ATF measurements used to build the antenna pattern are monotonic over a give power and phase range. When monotonicity is fulfiled, the test system can fully validate, and if necessary linearise the measurements against test signals of known accuracy.

11.6.5 Formal Definition of the Two-Stage ATF

The formal definition of the ATF measurements was specified in 3GPP TR 36.978 [3GPP14a] and described in Rumney et al. [RKJZ15]. The TR defines two ATF measurements, reference signal antenna power (RSAP) and reference signal antenna relative phase (RSARP). In addition, a layer-3 signalling protocol is defined enabling the test system to query the UE antenna attributes without relying on proprietary UE interfaces as has been the case to date. The ATF message definition includes two important aspects for future flexibility, firstly, the number of UE receive antennas is a reported parameter enabling up to 8 RSAP and RSARP results to be reported, and second, the carrier number is specified in the request message making the ATF extendible to the use of arbitrary numbers of channels for carrier aggregation.

11.7 Two-Channel/Decomposition Method

11.7.1 Introduction

In this Section, 2 × 2 down-link (DL) MIMO OTA testing is addressed from the point of view of commercial testing of UE. Consequently, it strives for MIMO OTA metrics that can unambiguously be related to physical attributes of a DuT and for measurement procedures that are simple and of high reproducibility. The name "Decomposition Method" relates to the fact that the proposed radiated measurements focus on performance and spatial properties of UE antennas. Other properties of the UE can also be measured in conducted testing. "Two-Channel Method" is motivated by the fact that the relevant physical attributes of a DuT are characterised in measurement set-ups where two data streams from the evolved Node-B (eNB) are mapped to two measurement antennas (probes). In a later stage, this measurement method was enhanced to a test plan named "Decomposition Method", comprising a part that focuses on performance and spatial properties of UE antennas and a part that focuses on Rx performance under conditions of a fading channel, as described in the second paragraph of this subsection.

The most comprehensive summary of ideas behind the two-channel method, its theoretical foundation, and its development up to the year 2012 can be found in Feng et al. [FSvG$^+$12, Fen13, BvGT$^+$11, FJS11].

The basic set-up can be seen in Figure 11.12(a). Two dual-polarised test antennas can be set to arbitrary elevation angles in a plane around the DuT. A turntable allows changing the azimuth of the DuT. The test antennas are fed from an eNB emulator via a switching unit that allows selecting freely the polarisations of the test antennas. Each selected azimuth and elevation setting together with the chosen polarisations is named a "constellation". While varying the DL power level, the observed TP is recorded. For the up-link (UL), an independent communication antenna is used.

11.7.2 Two-Channel Method

This approach focuses on characterisation of the performance of UE antenna systems that can only be tested OTA. It nevertheless qualifies the DuT as a whole. Selecting a channel model out from an infinite choice hardly is reproducing reality. Following established engineering practice, the goal is, therefore, to isolate the very properties of the DuT. Second, state-of-the-art

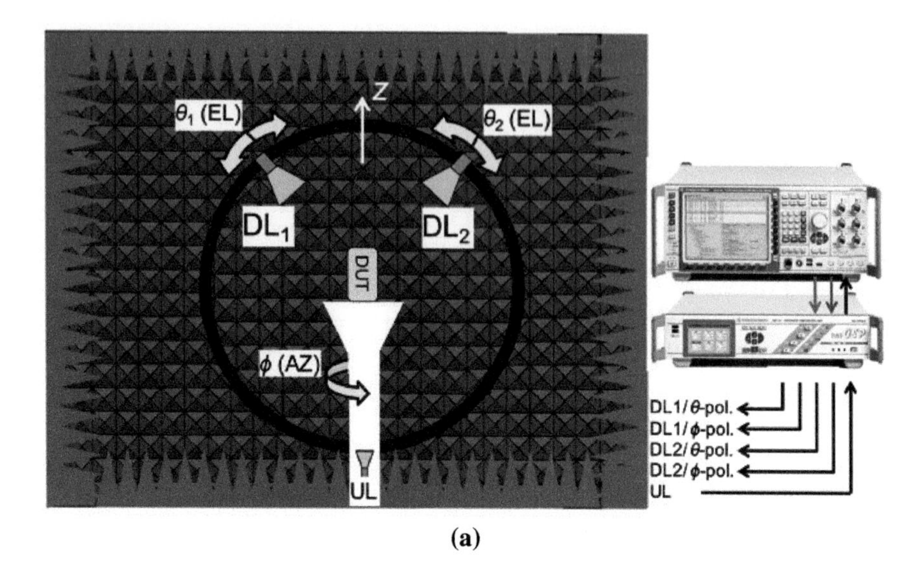

(a)

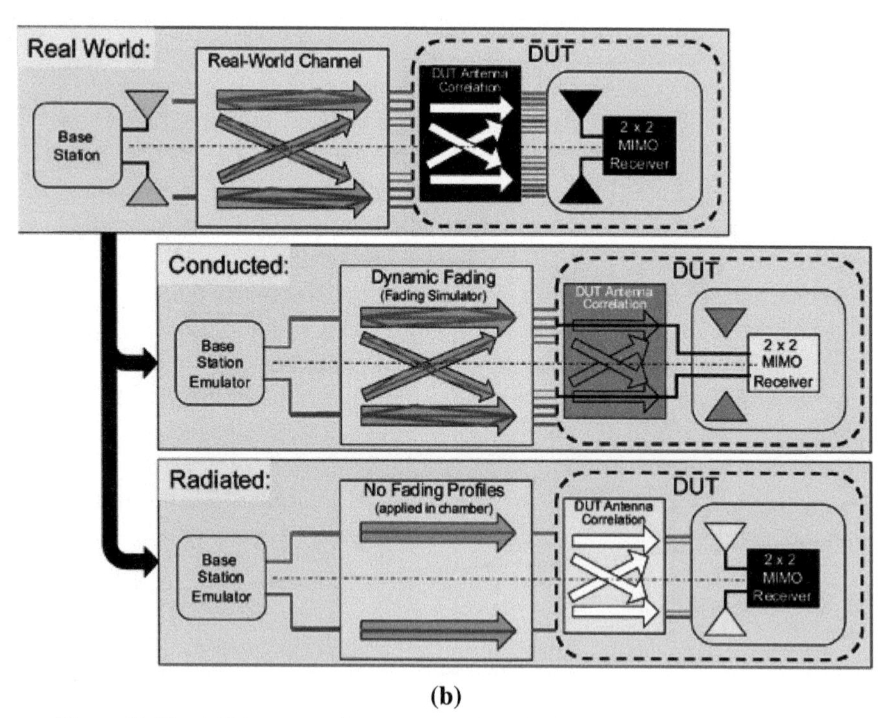

(b)

Figure 11.12 (a) Test set-up for two-channel method; (b) Decomposition elements.

mobile communication standards such as LTE are highly adaptive. The MIMO TM in particular is adaptively switched between, e.g., DL transmit diversity (TD), open-loop spatial multiplexing (OL-SM) and closed-loop spatial multiplexing (CL-SM) based on current channel state information. Likewise, the MCS is permanently adapted to the current channel conditions. It is, therefore, not meaningful to subject UE to arbitrary channel conditions unless an eNB emulator is also employed that fully supports the adaptive features of the standard. Still then, the adaptation rules used by the eNB emulator, that are not standardised, would enter into the UE test result. These and further fundamental aspects of MIMO OTA testing were discussed in Schroeder and Feng [SF11, STFvG13].

The two-channel method, therefore, builds on two complementary test cases and associated metrics that are in agreement with requirements for the DL TD and the OL-SM MIMO TMs, respectively, and with the use of fixed reference channels (FRCs). A detailed overview of the overall test plan was given in Böhler et al. [BvGT$^+$11].

The first test case evaluates isotropic sensitivity in DL TD mode for a noise-limited scenario. The test set-up is similar to a conventional TIS measurement with the difference that the two orthogonal copies of the DL transmit signal are mapped to the two polarisations of a dual-polarised horn antenna. Results are reported in terms of a cumulative distribution function (CDF) of sensitivity over all AoAs realised by the probe. As shown in Feng et al. [FSvG$^+$12, Fen13], the test evaluates the impact of the UE antenna system on first order channel statistics and includes Rx sensitivity as well as self-interference. It is noteworthy that this test fully qualifies diversity performance independently from the number of UE antennas and that the significance of the result could not be improved by adding additional test antennas carrying further de-correlated or faded copies of the DL signals. Detailed analyses were presented in Feng et al. [FSA$^+$11, FSK12].

The second test case evaluates SM performance in the high SNR regime. It is applied in OL-SM TM (for which FRCs are defined). The test set-up is extended by a second dual-polarised horn antenna whose angular position relative to the first can be varied independently. Outage power levels relative to a given block-error-rate (BLER) for a high-order MCS are recorded over a set of constellations. As shown in [FSvG$^+$12, Fen13], the test fully qualifies the SM performance of a UE with two antennas mainly influenced by the antenna correlation. The complementary cumulative distribution function (CCDF) curves for a given DL power level show clear differences for the various combinations of incoming polarisations.

A RAN4 round robin campaign with LTE USB modems allowed to do more tests with different constellations [BvGT+11]. For example, in order to compare more easily with the results of other methodologies, a subset of geometrical constellations representing a 2-D plane was analysed.

The two-channel method is characterised by its simple set-up. In a comparison exercise, asmart phone was tested in similar ways in a large reference chamber, in a compact test chamber (R-Line), and in a desktop anechoic chamber (DST200). With some limitations in the constellations that can be used, each of the three environments correlated well with the others.

11.7.3 Decomposition Method

The set of tests in the decomposition method consists of a conducted measurement without channel impairment, a conducted measurement with channel impairment, and a radiated measurement using the 144 constellations mentioned earlier. The channel models used for the channel impairment are based on the SCME models UMa and UMi, with the exception that the spatial aspects are not included. The channel impairment is applied in a conducted test where no spatial information is used. The constellations used in the radiated test comprise sets of different elevation, azimuth, and polarisation settings for each of the two test antennas. Figure 11.12(b) shows the elements of the decomposition method, Figure 11.13 indicates the hardware set-ups.

The conducted test with channel impairment assesses primarily the MIMO Rx. Tests with UMi or with UMa channel models give quite different results. The radiated test without channel impairment, on the other hand, evaluates primarily the MIMO performance of the antenna system.

CTIA organised another Round Robin test in which two smart phones operating in bands 7 and 13 were tested. In addition, the downlink signals carried an additional noise contribution for some of the tests, resulting incurves of TP versus SNR.

The results of the decomposition method for band 13 show a variety of interesting aspects. In a first test campaign, the elevation and azimuth angles were taken from a grid with 30° spacing, as in earlier testing. As can be seen in Figure 11.14(a), the three different reference antenna systems can clearly be distinguished. More background information and additional results can be found in Rohde and Schwarz [Sch12a, Sch12b].

Another selection of 128 constellations, based on phyllotaxy ("growth pattern constellations") with an optimised distribution, was used in a subsequent measurement. No additional noise was injected. Figure 11.14(b) shows the final result obtained with the decomposition method for UMi channel models.

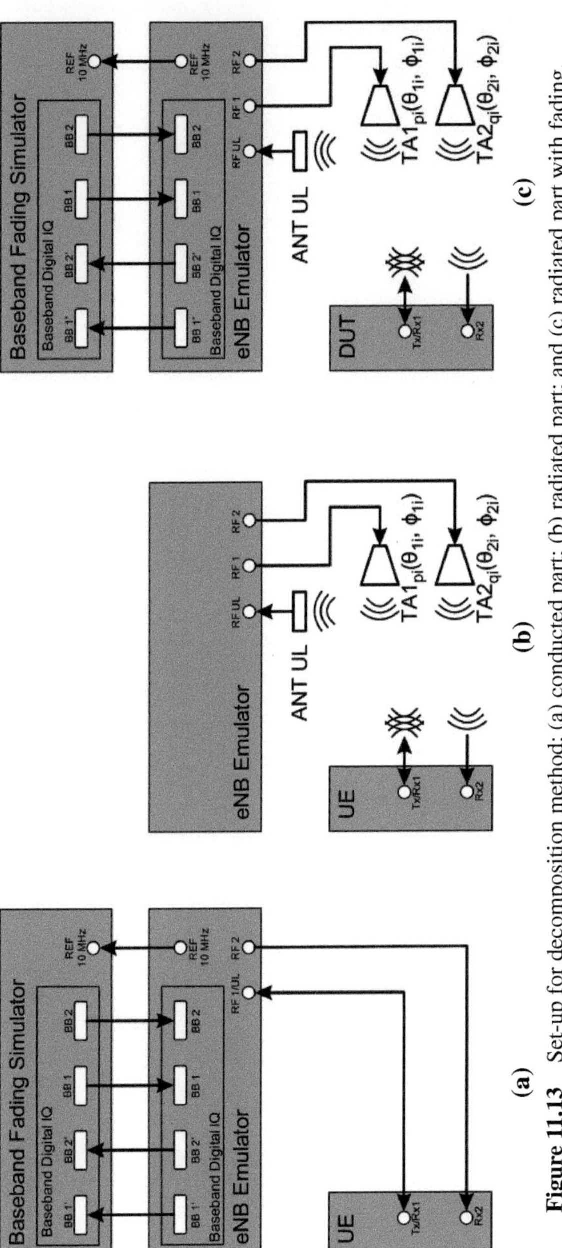

Figure 11.13 Set-up for decomposition method; (a) conducted part; (b) radiated part; and (c) radiated part with fading.

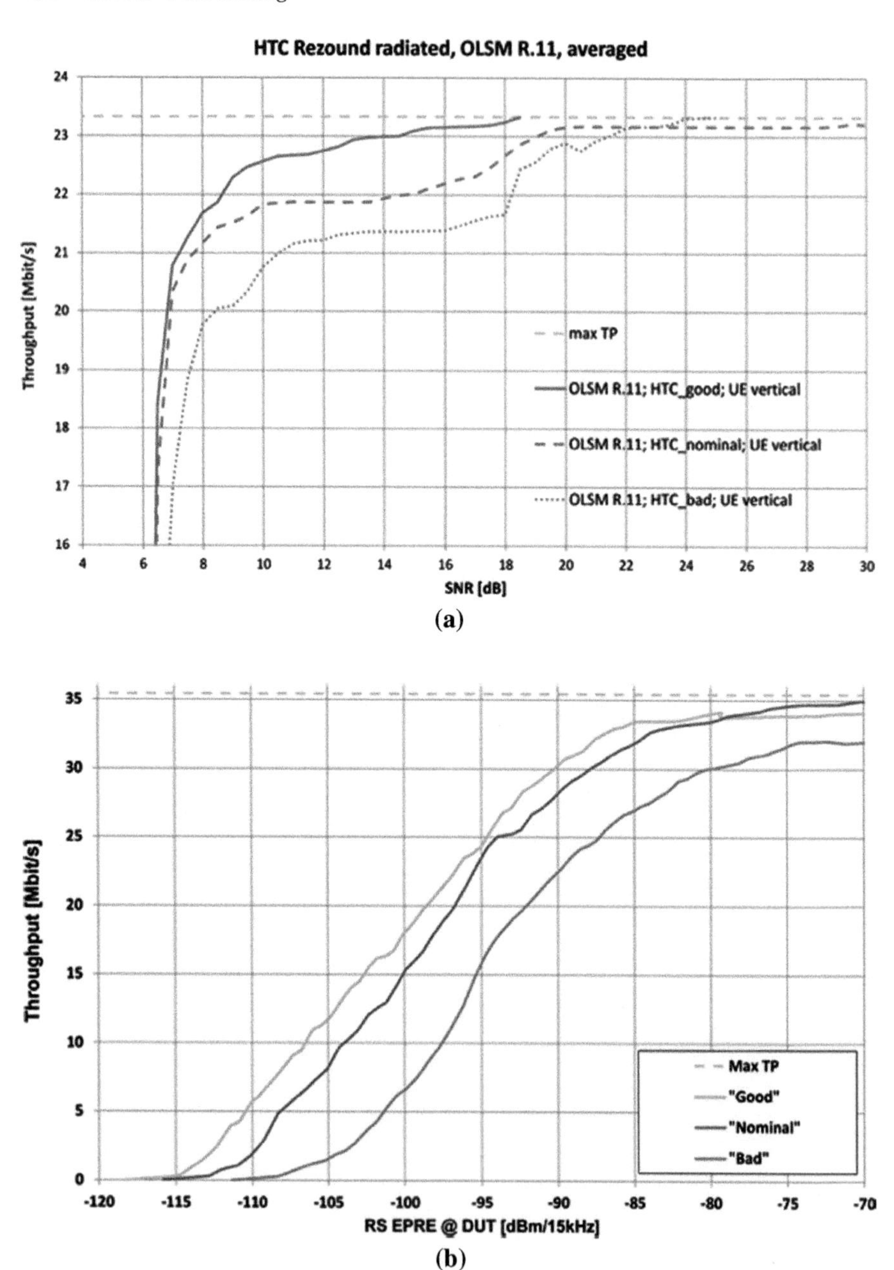

Figure 11.14 (a) Radiated TP, OL-SM R.11, averaged over 144 constellations; (b) Decomposed TP curves for UMi channel model.

The TP curves for UMa channel models are identical in shape, but with a shift of 2dB towards lower sensitivity.

A further extension of the decomposition method comprised, instead of applying the faded environment in a conducted test, moving this step to a radiated test as well [TvG14]. Figure 11.13(b) depicts the extension into faded radiated measurements. This way, the faded measurement can also be applied to a UE that does not carry any external connector. The agreement of results using either method was very good.

In order to underline the validity of the decomposition method, system level simulations using the simulation software SystemVue were performed for one of the CTIA Round Robin reference antennas as UE antenna. The results with different geometrical constellations show good agreement with measurements, especially when the condition number of the channel matrix is small [ATG+13].

11.8 OTA Field Testing

11.8.1 Evaluating Low-Cost Scanners as Channel Measurement Devices in LTE Networks

Radio channel measurements using specialised multi-dimensional channel sounders offer good resolution and high accuracy which is beneficial for exploratory research, however the down-side is their cost and complexity. A cheaper and quicker alternative is using commercial channel scanners typically used in drive-test campaigns by operators. Some of these scanners can optionally be enabled for low-layer channel sampling, and they typically support two antenna ports. By using the network BSs as Tx sources, the need for a dedicated Tx and test licences to utilise the frequencies are eliminated. This also means that tests done using scanners reflect real operating environments. The resolution, however, is limited to the system parameters of the standards.

An evaluation of scanners for channel measurements has been conducted with a laboratory set-up as well as a field test [KK13a, KKJ14]. The study was performed using an LTE signal on band 3 (1800MHz). In the lab, the bandwidth was 10 MHz at 1842.5 MHz, while in the field, the bandwidth was 20 MHz at 1815 MHz. In the lab, a communication tester and fading emulator with different reference channel models were used, as shown in Figure 11.15.

The field measurements were done on a live LTE network with a number of BSs covering a closed route in the Oulu Technopolis area in Finland. Tests were done using both two omni directional orthogonally polarised antennas separated by one wavelength and directive Vivaldi antennas [AZW08, SS11].

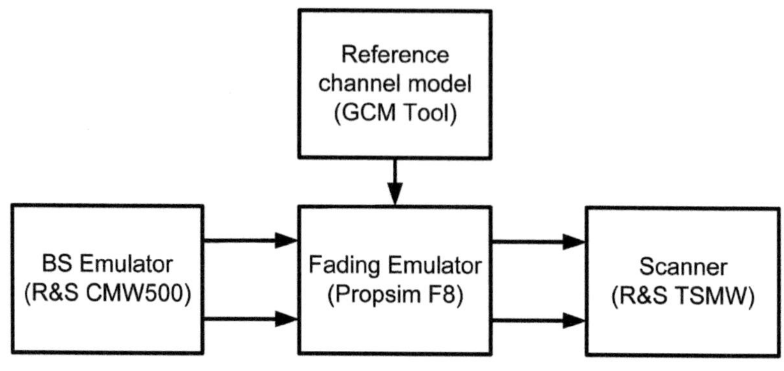

Figure 11.15 Set up of the laboratory measurements.

The data was analysed with respect to the frequency response, path loss, PDP, Ricean K-factor, Rx polarisation factor, and antenna correlation.

Path loss can be almost exactly estimated in the laboratory; however, in the field, this requires knowledge of the BS power. Still, the relative dynamic variation of path loss and shadowing can be measured. Similarly, only excess delays can be estimated in the PDP measurements, since the absolute delay is unknown. In measuring the PDP, estimations of fixed paths are quite accurate, less than 1 dB, however, dynamic delays are difficult to capture, most likely due to averaging over time and distance. The Ricean K-factor measurements show a good match as long as the reference K-factor is above 3 dB with a maximum deviation of roughly 4 dB. For lower K-factors the deviation increase, and in general dB-negative K-factors are difficult to estimate. Rx polarisation power ratio estimates follow the trend in the reference model, ranging from –16 to 11 dB. The scanner capability limits the estimation accuracy of the antenna correlation, mostly due to phase offsets between the elements of the frequency response matrix.

A comparison of the laboratory and field Rx power measurement was done using a field scanner measurement from a single BS link and an OTA radio channel emulation based on the same field measurement [KK13a]. Directional measurements include multiple cells, showing that the distribution of power in azimuth is not uniform. The variation is between 9 and 22 dB for nine different locations. The variation of the V/H XPR ratio is in the range of –8 and 13 dB and was measured directly using the two elements ($\pm 45°$ slant) of the directive Vivaldi antenna. The maximum excess delays of the PDP varied between 0.45 and 3.67 μs, which fits the 3GPP urban and macro channel models of 3GPP [3GPP14b]. As in the lab, the initial delay in the field cannot be estimated.

Scanner measurement and the corresponding analysis set-up can be utilised to create measurement based and site specific MIMO OTA channel emulations. By sampling a measurement route with a number of static locations, the extracted parameters can be utilised to generate fast fading inside an anechoic chamber with proper characteristics. The propagation parameters may be interpolated between locations or preserved constant. Overall Rx power level and PDP are obtained from the measurement with omni directional antennas. Angular power distributions and Rx XPR are captured with the directional antenna. Doppler spectra can be modelled with a synthetic model or Doppler shifts can be approximated based on the PAS and the selected velocity vector. All the necessary parameters for a measurement based MIMO OTA testing are available from the proposed set-up.

11.8.2 Measuring User-Induced Randomness for Smart Phones

An effect, that is not automatically included in OTA testing, is the randomness due to the user. Pure LoS and rich multipath (RIMP) environments are rarely present in real-life. Real-life environments will most likely show a mixture of LoS and NLoS conditions rather than one or the other. Further, introducing the user randomness means that the LoS component becomes 'random-LoS' due to the user [Kil13]. It means that the LoS experienced by a mobile terminal becomes completely random due to its random position and orientation with respect to the BS, as shown in Figure 11.16.

In real life, this randomness of the phone orientation is not known. One approach to estimate this is to use modern smart phones that all contain sensors providing information about the phones orientation in 3-D as well as sensing proximity to, e.g., head. Together with location information and different signal level and quality measurements, user-statistics could be produced. A smart phone application (app) can be be used to collect such sensor and radio

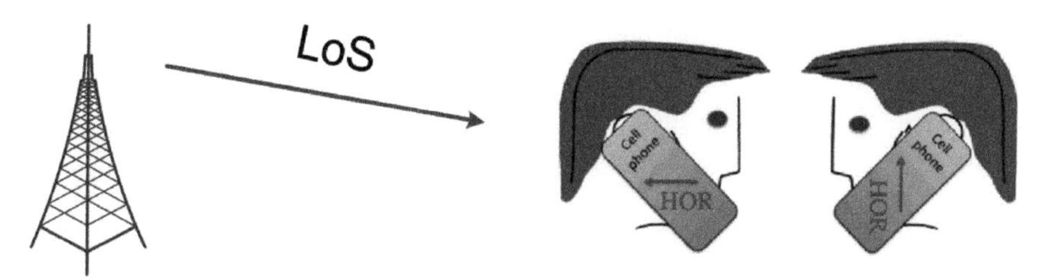

Figure 11.16 Illustration of random orientations of a wireless user device.

information data from a number of devices in daily use. Some of the available and relevant sensor data which can be easily retrieved on most smart phones are rotation, acceleration (linear and rotational), proximity, and location. Additionally, it is possible to retrieve some of the radio measurements, like received power and SNR.

Actually, more specific measurements would have been advantageous, especially more in-depth data on the RF link as defined by 3GPP. One example are measurement reports on channel state information related to MIMO for LTE, like the rank indicator (RI) and the precoding matrix indicator (PMI). However, this would require more in-depth access to the phone software than is possible using commercially available phones. By combining information from local sensors, network information, and signal quality, it is possible to find out whether user behaviour influences link quality and if so, how the correspondence is.

Data from a limited number of phones, collecting measurements over a period of more than 2 months, were analysed with respect to the phone orientation angles, pitch, roll, and azimuth [LMG+15]. The pitch and roll are tilt angles, while azimuth is the rotation relative to magnetic north.

One immediate observation is a high peak around 0 degree for the pitch and roll, which means the phone is lying screen up on a horizontal surface. This is to be expected since phone applications do a lot of background data traffic without user interaction, and in those cases, the phone is often lying on a table or another horizontal surface. More interesting is the behaviour in voice mode where the early test measurements in normal usage show trends of typical rotations of the phone in voice mode [LMG+15], however, much more data from a larger population are needed to draw any conclusions.

11.9 Discussion and Future Work

Even after the successful standardisation of OTA testing of SISO mobile terminals, the development of methods for multi-antenna devices proved to be a difficult one. Even though not all four proposed methods will make it into the standard, that does not mean the research effort into these methods is wasted, especially not while the respective research teams have contributed to the general understanding of the problems involved. Over the last 4 years, the process underwent a steep learning curve, in which first the influence of the antennas of the test objects had to be eliminated by the use of the CTIA reference antennas, then seasoned experts found themselves confronted with, among others, how to define the SNR of test signals or how to average measurement results. At the moment of writing, not all issues with the definition of the general measurement set-up are solved. Although looking forward to a successful completion of this standardisation process, most will agree it is just the first step. For instance, LTE, the most adaptive radio communication system in history, will be tested in stationary environments without any temporal evolution of large-scale effects and with its adaptation mechanisms switched off. Besides, the time-variant transmission channels are abstract models in two dimensions and the test set-up is that of a typical single-user link with infrastructure (interference represented by AWGN), at present only in DL.

Ahead lie enhancing the degree of realism of the test environments, with 3-D channels whose large-scale effects naturally evolve over time. Then, the operating point of test objects will be much closer to that in reality and technological advances like adaptive antenna systems or smart Rx algorithms/structures can prove their added value. Carrier aggregation will be a challenge for channel emulation, as will be multi-user environments and bi-directionality. Related to multi-user aspects are coexistence/interference issues between systems, of which operation of LTE-wireless local area networks (WLANs) in unlicensed bands at 0.7 and 2.6 GHz is one example. Another is coexistence between dedicated short-range communications (DSRC), ITS-G5, and WLAN at 5.8 and 5.9 GHz. Bi-directional multi-user environments are also important for other devices than those communicating with infrastructure, for instance, for adhoc network nodes in peer-to-peer communication (D2D, V2X) in which "uplink" and "down link" are meaningless terms. Furthermore, reactions to incoming messages are not necessarily directed towards the original sender(s). Just as challenging as these distributed communication partners is testing of devices with distributed antennas. It is expected that these subjects will become pertinent in 5G systems, that obviously will pose some other challenges. Certainly, there will be new physical layer and network concepts of which we should expect extremely wideband transmissions (apart from carrier aggregation) and migration into SHF and EHF bands, with many interworking issues. With the tight time schedule for 5G, it is foreseeable that testing of 5G BS will become pertinent in the next 4–5 years.

12

Future Trends and Recommendations

Chapter Editors: A. Sibille, B. K. Lau, C. Oestges, A. Burr, S. Ruiz and T. Jamsa

12.1 Introduction

Wireless technologies are constantly evolving and have done so for more than two decades, when only considering mobile communications, going from 1G to 4G and soon 5G. This strong evolution has been driven by meeting between the technology push that is the DNA of academia, research centres and industry, and the market pull resulting from the increasing greediness of consumers. The momentum is maintained not only by the younger ones, but also by the elder, who now see the digital technologies as an extension to their own body functionalities and are always depending more on these technologies in their everyday life. The Edholm law of bandwidth [Che04] predicts an exponential increase of the data rate, which has been verified and generates the same expectations in consumers as those prevailed in the decades between 1990 and 2010 for the computer's power in relation with Moore's law. However, in both cases, the laws are not magic theorems that are obeyed without doing anything; they require treasures of imagination and dedicated efforts of crowds of engineers and researchers to achieve their predictions.

We are not at the end of wireless technology development, far from that, as the penetration of wireless objects closer and closer to humans and more and more embedded environments is anticipated in the next decade and beyond. Then the question of how to do it arises. A major step should be crossed within the next 5 years through the elaboration of "5G" networks, but the increasing complexity and wide versatility expected from these networks generates much questioning about what 5G will be and how far it should be standardised? Beyond 5G there might be (or will certainly be?) further developments to better translate fundamental discoveries into practical solutions and improve

Cooperative Radio Communications for Green Smart Environments, 469–486.

system capabilities in parameter domains such as the energy consumption, the latency, the throughput, the reliability, the security, and others.

In this chapter, we modestly attempt to elaborate on the work presented in the previous chapters, based on the results obtained by the European COST IC1004 network of scientific experts, in order to risk a vision on the future challenges and developments in mobile and wireless communications, on a time scale of about 5 years. This is done below in the three main "pillars" of the success story of follow-up mobile communications COST Actions (COST 207, 231, 259, 273, 2100, and IC1004), which are antennas and channels, signal processing, and device networking.

12.2 Antennas and Channels

Channel modelling will likely continue to be of fundamental importance. A rich spectrum of radio channel measurements, characterisation, and modelling results is available in the 0.7- to 6-GHz band. For higher frequency bands (cm- and mm-waves), now being considered for some 5G implementations, there were studies made in previous projects [Ver12], but specific measurement campaigns and a global regulation of such spectrum for mobile services is still missing. The models available so far cover cellular access between base and mobile stations, vehicular communications, and device-on-body scenarios. A number of reference channel models are available, such as IEEE [Mal+09], WINNER [Kyo07], and COST2100 [LPQ+12].

Likewise, the impact of antennas "in free-space" on the radio channel characteristics has been thoroughly investigated between 0.7 and 6 GHz, leading to a number of design criteria for a "good antenna array" on a small device and antenna evaluation methods [Lau11]. However, this simplistic free space approach is no longer adequate in long-term evolution (LTE) and beyond systems that more critically depend on good antenna performance to deliver the specified system performance. This has led to recent requirements by major mobile operators for handset vendors to provide performance figures relating to realistic usage conditions. Moreover, new wireless devices are required to cover many more frequency bands (over 40 for LTE alone), as well as implementing beyond 2×2 MIMO, which are highly challenging due to compact design space.

More accurate radio channel models are, therefore, required, including: (i) "new" deployment scenarios, such as very high mobility (vehicular, drone), different human body postures and tissues, harsh physical environments, ultra-dense device deployment, very-short links, highly directional front/backhaul links, among others; (ii) Multiband and wideband channel modelling with carrier frequencies above UHF up to Terahertz; and (iii) 3D modelling with site-specificity.

New deployment scenarios in complex propagation median raise a problem of scale. In many heterogeneous environments, the size of the area to be modelled can indeed be very large and also contain a number of very complex objects (e.g., planes in an airport and human bodies in a room). This multi-scale aspect results in a methodological issue, each scale being usually modelled by a different method: on the one hand, stochastic and asymptotic physical models such as ray-tracing (used in earlier works by the authors) enable to model large-scale structures; the computational cost depends on the required level of accuracy and the determinism of the model (i.e., it is larger for ray-based tools than for stochastic models) but remains manageable thanks to simplified descriptions; on the other hand, full-wave electromagnetic models such as the method of moments (MoMs) and its fast extensions are able to calculate spatial fields in complex scenarios (in terms of shape, etc.) but the required meshing associated with these methods limits the size of the environment to be modelled. Although both approaches presented above are appealing and could in theory be combined, this combination cannot be achieved by a mere superposition for two reasons: not only both approaches use different tools and assumptions, but also small- and large-scale interactions are mutually coupled. As a result, heterogeneous deployments often call for the development of hybrid models which correctly account for mutual coupling.

Multiband channels should extend the current model range well above 6 GHz, in particular towards the centimetre and millimetre wavelengths. Whereas, quite a few models exist for indoor environments, directional outdoor models are still largely missing, in particular those based on measurements (recent results are only based on ray-tracing simulations). Such measurements are challenging and would likely require new methods to be devised. Furthermore, many other aspects are also lacking, e.g., indoor-to-outdoor propagation, polarisation modelling, and channel dynamics caused by moving scatterers.

Classical directional channel modelling in the literature, including the 3rd generation partnership project (3GPP)-SCM models, represents the angular characteristics in the azimuthal domain [Tuf14]. This is related to the fact that capacity improvements by multiple-antenna systems are greatest when the angular spread is large, which usually occurs in the azimuth domain. In the elevation plane, the angular spread observed at the base station is usually restricted to a fairly narrow range. However, the emergence of massive MIMO systems has renewed the interest in using the elevation domain for beam-forming purposes, in particular in multi-user settings. Within COST IC1004, an extension of the COST 2100 model for massive MIMO systems has been proposed. The biggest gaps in existing data sets and 3D models are related to outdoor-to-indoor links [Tuf14]. As examples, the impact of the indoor floor plan or the horizontal location of the mobile terminal within a building is still yet to be modelled accurately.

Finally, the needed radio channels also require new measurement methods, such as multiband channel sounding for a wide frequency range including micro-/millimeter-wave bands, distributed massive MIMO channel sounding, and portable channel sounding for drones, sensors, body environments and near-field links. Similarly, new channel emulation techniques should cope with carrier aggregation, isolating the different environments of multiple users, and testing objects up to the size of cars. To this end, innovative metrics for emulation accuracy are required, which could be user-oriented (subjective), or application specific.

On the antenna system side, the time is ripe for a more comprehensive strategy in optimising the interactions of the antennas, not only between the antenna elements, but also between the antennas and their surroundings (Figure 12.1). In this context, there has been some recent progress in co-optimising the antennas and the device platform using the theory of characteristic modes (TCM), as was described in Section 10.2. However, significant opportunities exist to extend the design framework to take into account nearby objects as well as the propagation channel. Moreover,

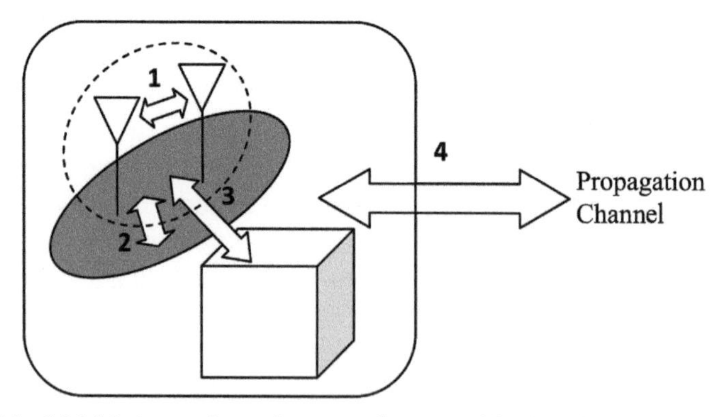

Figure 12.1 Multiple interactions of antenna elements with (*1*) each other and its surroundings, including (*2*) wireless device platform, (*3*) objects in proximity, and (*4*) propagation channel.

the non-stationary nature of the surroundings can be addressed using novel circuit technologies (e.g., low-loss MEMS circuits) to provide flexible and power efficient pattern-reconfigurable antennas to optimise near/far-field interactions with channels in any given deployment scenarios, under the constraints of size and disturbances in the vicinity (Figure 12.1). Existing work on pattern reconfigurable antennas are largely confined to canonical structures in free space operation, neglecting the device platform, nearby objects or even channel properties.

Apart from designing optimised antenna solutions, the properties of the antennas and objects in proximity that critically influence the physical channel must be derived in order to provide realistic channels for link and system evaluations. Prior work in joint antenna–channel modelling focused on traditional antenna solutions and statistical characterisation using large sample populations of possible antenna configurations, orientations, etc. The impact of flexible and ambient-optimised antenna systems must be accurately represented in order to facilitate overall system design.

12.3 New Challenges for the Fifth-Generation Physical Layer

COST IC1004 has focused on many of the issues that arise in communication networks for "green, smart environments". One of the most significant of these is interference, where there has been significant research in working group 2 (WG2) on both interference modelling and interference management approaches, including interference alignment. Another is advanced MIMO techniques, which have the potential to reduce the effect of interference, as well as to increase link capacity. Advanced iteratively decodable error correction codes have been investigated: again, these have potential to improve robustness to interference and to increase capacity. And in the context of "green" communications, important work on energy efficiency has been carried out. However, much more remains to be done. In particular, we are now well into the development of the fifth generation (5G) of wireless communication, and it is clear that many new challenges will arise from this. In this context, COST IC1004 with its emphasis on wireless communication in a range of different environments and applications was very well positioned to provide the basis of this development, since 5G is expected to fulfil precisely this function of an "inclusive" wireless communication system, covering a much wider range of environments and applications than previous generations. This also leads to a much broader set of key performance indicators (KPIs), for which very challenging targets have been set.

These KPIs and their target values include the "traditional" ones we have been used to in the fourth generation (4G), and especially both greatly increased peak data rates (increased to 20 Gbps) and increased capacity density (10 Tbps/km^2), plus increased spectral efficiency and user mobility (up to 500 km/h), but also new ones such as connection density increased to 1 million/km^2 and latency reduced to 1 ms. A challenging target for increased energy efficiency (100 times compared with 4G) has also been set. The new targets represent new applications, especially internet of things (IoT) applications, including critical system control applications, and provide particular challenges in the design of 5G systems.

These new applications, and also requirements such as increased capacity density also have implications for the architecture of 5G networks, which in turn have implications for the physical layer. This will require further extension of a theme that has already been well-established in WG2 (and especially in SWG2.1) of COST IC1004, as well as appearing in the title of the Action—that of *cooperation*. This turns out to be fundamental to achieving the required performance, and calls for a completely new paradigm for the role of the physical layer in wireless networks, which we call the *network-aware physical layer*. Rather than regarding the physical layer only as providing the link between two network nodes, with signals from other transmitting nodes treated as interference, a node is aware of its position in the network, which enables it to process all signals it receives appropriately, as well as to generate appropriate signals. We address some of the KPIs and their implications, and some possible research directions arising from them in what follows:

- **Peak data rate up to 20 Gbps**

This could in principle be achieved in a number of ways: (i) by increasing modulation order; (ii) by increasing the number of antennas, at both base station and at the terminal; or (iii) by increasing the bandwidth. Point (i) has limitations in practice, and cannot provide the extent of the increase required; point (ii) is also limited by the number of antennas that can reasonably be accommodated on a terminal, which leaves point (iii). The extent of the data rate required is such that several GHz of bandwidth may be called for – which would be very difficult to find at frequencies below 6 GHz, and hence had led to calls for the development of millimeter-wave (mm-wave) transmission. The problem with this is the inherent path loss increase at these frequencies. To balance this, the gain available from an antenna of given size also increases proportionately. Note, however, that this gain cannot be realised on non-line-of-sight channels, at least not simply by reducing beam width, because of the multipath spread which occurs on such channels. To realise the gain requires antennas at both transmitter and receiver that can adapt to the multipath environment, which may in turn require complex channel estimation and adaptation algorithms. A promising approach which may simplify this might be the use of passive beam-forming technologies, for example based on the Butler matrix.

Also, increased diffraction loss may make mm-wave links less reliable. It is likely, therefore, that "backup" links at lower frequencies may be needed. Because of the fragmentation of spectrum below 6 GHz, carrier aggregation may be required in these bands, and since relatively small spectrum slots may be dominated by the guard bands required for OFDM side lobes, technologies such as filter bank multi-carrier (FBMC) have been proposed to improve frequency localisation by filtering side lobes, albeit at the cost of reducing orthogonality between sub-carriers, and (inevitably) of degrading time localisation.

- **Capacity density 10 Tbps/km^2**

While increasing peak data rate might be expected also to increase capacity density, we should note that the targeted peak data rate increase over 4G is 20 times, while the capacity density increase is 100 times. Moreover, it may be desirable to provide capacity density increase even for users who do not need the full data rate, and may operate in the bands below 6 GHz. This will help to fulfil another KPI: user-experienced data rate of 1 Gbps.

Two main technologies have been proposed to fulfil this target: *small cells* and *massive MIMO*. The small cell approach continues the previous trend of increasing the density of cell sites—but also takes it further, such that the number of access points may exceed the number of terminals served. Massive MIMO, on the other hand, increases the number of antennas at a cell site (possibly to 100 or more), again so that there may be several times more antennas than terminals served. It then uses multi-user MIMO techniques on a massive scale to serve an increased number of users in the cell. This relies to a large extent on the richness of multipath in the channel, and the evaluation of its performance at the physical layer poses new challenges for channel modelling as well as many aspects of physical layer design. Of course both technologies have the effect of increasing the density of infrastructure antennas—the difference is the distribution of the antennas. There are also proposals for *distributed massive MIMO*, in which the antennas are distributed to a number of cell sites across the cell. Cloud radio access network (C-RAN) takes this to the extreme, by cocentrating the baseband processing functions of a large number of cells in one central baseband unit (BBU) and replacing access points with remote radio heads (RRHs) which forward sampled versions of the received signals to the BBU over what is now known as the *fronthaul* network. This has the benefit over the conventional small cell approach of enabling base station cooperation (as in cooperative multipoint (COMP)) over a wide area. To all intents and purposes the concept of the cell is abolished, and with it those disadvantaged cell-edge users, and the entire service area can use the same resources via a number of antenna sites which together act like a single MIMO base station.

The main problem with C-RAN, as is rapidly being realised by operators and manufacturers when they seek to implement it, is the load on the fronthaul, which can easily exceed 50 Gbps even for a relatively modest antenna site (because of the signal bandwidth operating over multiple antennas with dual polarisation), and may be tens or even hundreds of times the user data rate. Even if optical fibre is available, there are substantial costs in providing this bandwidth, and in many cases wireless may be the only medium available for fronthaul. There is thus a huge challenge for the 5G physical layer both in implementing the required fronthaul bandwidth, and also in evaluating the effect on access segment performance—since an imperfect fronthaul network will distort the signals forwarded and degrade access network performance. One approach, which has already been developed in COST IC1004, is the application of wireless physical layer network coding (WPNC) which provides benefits similar to CoMP with a front/backhaul load (in this case it is difficult to define which term applies) that is no greater than the total user data rate.

- **Latency 1 ms**

A dramatic reduction in the latency/delay limit has been proposed in order to enable a number of new applications, from the *tactile Internet* to safety–critical applications like vehicular and process control. This will, however, be an especially challenging target to meet, because it enforces changes in the way error correction coding has been implemented ever since Shannon. A latency limit implies a limit on code length, especially for machine-type communications (MTCs) where the data rate is low, and conventionally code designers have tried to approach the Shannon limit by increasing code length. This is one area in which the new targets may require a fundamental re-examination of the principles of communication theory. Perhaps new types of code will be required, not so much related to the random-like codes inspired by Shannon. This is also an additional challenge for C-RAN, as the fronthaul network inevitably adds further latency, which must be minimised.

- **Connection density 1 million/km^2**

MTCs in general are likely to involve very large numbers of devices which typically may be small and low power, requiring a similar density of access nodes to serve them. Because of the cost of individual backhauling, these in turn are likely to require in-band backhaul, in the form of a mesh network. Note also that this type of communication is likely to consist of very large numbers of very short data packets; moreover, with very low latency limits. Conventional mesh network paradigms will find it quite difficult to meet these requirements: this is another case where the new network-aware physical layer paradigm (mentioned above) is called for. For example for this application, the DIWINE project has proposed the *wireless cloud network*, in which the network becomes a multitude of small devices cooperating to relay data using the principles of WPNC. This minimises contention for resources, which is one of the major causes of latency in conventional wireless networks. WPNC is one example of how the conventional paradigm requiring orthogonal transmission of separate data streams can be circumvented. We have also seen that strict orthogonality in both time and frequency domain within the air interface may no longer be achievable. Part of the new paradigm is therefore coping with the loss of orthogonality, in such a way that the interaction between data streams provides diversity that enhances, rather than interference that degrades. This type of approach may inform the development of new air interfaces (or perhaps a single, but highly adaptive air interface) for the 5G physical layer.

12.4 Next Generation of Wireless Networks

Future 5G mobile ecosystem is oriented to ubiquitous ultra-broadband wireless connectivity, improving many metrics with respect current standards. What characterises a 5G Network is that it has to satisfy simultaneously extremely different requirements: it should be capable to deliver connections at “faster than thought” speeds (known also as “zero distance” connection) to offer ultra-high definition visual communications and immersive multimedia interactions, and simultaneously, massive IoT low-energy devices and applications with a large volume of data but at low data rates and being not sensitive to latency.

To accomplish this, 5G networks should, in addition to requirements mentioned in the previous section, provide:

- Massive connectivity
- Quality of experience (QoE) for the end user, independently of user’s location.

- Flexibility: an enormous, diverse and wide range of services and applications with different performance requirements.

5G radio access will be built upon both new radio access technologies (RATs) and evolved existing wireless technologies (LTE, high-speed packet access (HSPA), global system for mobile communications (GSM), and WiFi). A joint optimisation to efficiently use these radio resources providing on-demand resource and capacity wherever needed will lay on a combination of Cloud Architecture, software-defined networking (SDN), and network functions virtualisation (NFV) technologies. Additionally, Cognitive Network Management will be used, a smart network technology being the evolution of Self-Organising Network (SON) idea, which automatically learns from data demands and problems experienced on the network.

In particular, many breakthrough developments are required on different aspects of RATs involved in the network operating principles:

- New air interface and novel transmission techniques: multiple access and advanced waveform, coding, and modulation algorithms, as non-orthogonal multiple access (NOMA), FBMC, white space techniques, space-division multiple access (SDMA) with multi antenna pre-coding to serve multiple users in parallel while performing an adaptive control of interference, single frequency full duplex radio technologies, etc.
- New Traffic patterns to model machine-to-machine (M2M), human-to-human, and human-to-machine services.
- Algorithms to balance the centralised versus distributed control and execution of functions.
- Network slicing and user oriented radio resource management (RRM) algorithms.
- Energy Efficiency improvements through coordinated transmission, beam-forming, massive MIMO, new radio waveforms with less control overhead, shorter transmission ranges with ultra-small cells or device-to-device (D2D) communications, load adaptive, and context aware activation of additional resources, opportunistic transmission, etc.
- Dynamic spectrum access: efficient techniques to opportunistically use any portion of non-used available spectrum.
- Optimisation of wireless and optical backhauling.
- Definition of virtualised and cloud-based radio access infrastructure.
- Specific requirements for vehicular networks, advanced robotics, body centric communications, M2M, IoT, etc.

- Heterogeneous networking in terms of transmitted powers, frequency bands, bandwidths, antenna configuration, multi-hop architectures, duplex technologies, etc.
- heterogeneous networks (HetNets) in terms of seamless integration of all available technologies with an ultra-dense deployment of small cells, macro and micro cells, with different degrees of centralised/distributed networking. Research on seamless vertical handover, multi-technology load balancing, multi-operator roaming should be done.

12.5 From Ideas to Standards

12.5.1 Introduction

The road from a scientific concept to its implementation into a commercial system is often very long and obstructed by many pitfalls. Among them is the necessity to standardise, which implies the support of a sufficient number of actors, mainly from industry, agreeing to produce interoperable systems obeying common standards. In the area of wireless communications, the role of regulation is also extremely important, given that the spectrum is a natural public resource that can't be used freely. This is obviously related to interference between different users wanting to access the same resource defined by a frequency, a time instant, a geographical location and other parameters.

In Europe, the roles in regulation and standardisation are shared among several actors. The European Commission gives mandates to the European conference of postal and telecommunications administrations (CEPT) and especially to its electronic communications committee (ECC) in order to harmonise the use of radio spectrum in Europe and coordinate the views of European regulators in preparing the triennial world radio communication conference (WRC). This very important event, held every 3 or 4 years, can among others revise the Radio Regulations and any associated Frequency assignment and allotment Plans. Once the conditions of spectrum use are well established, the European telecommunications standards institute (ETSI) is able to be active in preparing standards for future technologies and systems employing the wireless resource. Actually ETSI signs memorandum of understanding with major bodies such as the international telecommunications union (ITU), CEPT and other such international organisations, so that the standardisation and regulation processes were jointly developed and converged efficiently. Major activities under ETSI are 3GPP, which proposes specifications for mobile communications systems (originally 3G and now encompassing all standards from 2G to 5G under European initiative), and M2M for M2M communications as part of the IoT.

12.5.2 Academia versus Industry Participation to Standardisation in the Context of a Network of Experts

Standardisation bodies commonly involve the main participation of large companies and SME/Mid-size companies, sometimes also prominent research centres (private/public), for which it is strategic to have their own technology included as far as possible in the standard. The motivation is much less for academia and the difficulty much higher, given that improvement of a standard comes after a large number of regular international meetings, the participation to which being costly and uneasy to valorise in an academic institution.

Still, gathering experts from academia and industry together toward a common goal, such as in IC1004 or other COST Actions, may be seen as a very powerful tool for achieving a consensual and technically/scientifically grounded proposal to standardisation bodies. In case sufficient symbiosis can be achieved by such a network, the benefits are mutual as the industry representative, participating in standardisation body meetings can but reference their proposal to the underlying joint work (giving it also more credibility), providing recognition to involved academic participants at a reduced overhead for them.

Given these preliminaries, the participation of academia and in particular of a COST Action can take place through a pre-standardisation approach, as follows:

- by producing a set of jointly achieved research results that pave the way to definition of a standard;
- by initiating new methods/new approaches that can be directed toward validating new kinds of standards yet non–existing;
- by encouraging relevant actors (mainly industry members or major research centers) to act in coordination in order to propose/influence standards in a way mostly issued from COST work;
- by making available (publicly or through an open and free membership) databases/software tools/reference scenarios, platforms... that can be exploited by standardisation bodies (and any contributor to these bodies) toward standardisation.

Among relevant contributions to standardisation, models have a recognised role in order to provide all actors who want to demonstrate the performance of some scheme a unified way to test this performance. In the past, much referenced work was carried out within the set of mobile communications COST Actions regarding the development of radio channel models, which are of paramount importance in the performance evaluation of physical layer schemes or network architecture principles. Radio channel models are indeed now, and have been for several decades (the GSM being a good example, based on COST 207/231 channel models), part of many published standards such as UMTS, IEEE 802.11n, IEEE 802.16, LTE to name just a few. COST IC1004 has pushed novel methods/approaches in terms of modelling of terminals antenna performance [Sib13], which may trigger future contributions to standards. Indeed such modelling, of statistical nature, may have numerous advantages in terms of realistic performance evaluation of physical layer schemes or of wireless networks, taking into account the high variability of terminals performance in use conditions [Sib14].

In the following, a vision on potential future contributions to standards is provided, focusing on 5G networks.

12.5.3 5G Timeline for Channel Models

5G is currently a "hot" topic and widely discussed among researchers and industry. However, 5G is not yet standardised, and there are different opinions about the 5G requirements and technologies. This section defines the timeline for 5G channel model standardisation as accurately as possible based on the currently available information.

Channel model is an essential tool for system evaluations—without channel model many 5G technologies cannot be evaluated. Understanding the timing of 5G channel model standardisation is important. Many good contributions are not taken into account because of wrong timing. The channel model should reflect reality, but shouldn't be too complex. There are several points for a 5G channel model:

- Frequency range: The channel model should cover frequency range from sub-1 GHz up to 100 GHz. Propagation is very different in the two extremes of that range. Therefore, frequency dependence of the channel model parameters should be investigated and a unified channel modelling methodology could be exploited to facilitate the implementation of system evaluation.
- Link type and deployment scenario: The variety of foreseen 5G link types and deployment scenarios (e.g., vehicle-to-vehicle (V2V), M2M, D2D, sensor networks, mobile base stations, ultra-dense networks, stadium, and festival cases) set additional requirements to channel models.
- More advanced MIMO features: Massive antenna arrays have much more accurate angular resolution than conventional diversity and MIMO antennas in 3G and 4G. It means the spatial characteristics of the channel need to be modelled very accurately.

It should be noted that some of the above points need to be considered jointly to fulfil the requirement of 5G system evaluation.

Recently, several research projects/organisations have shown their view on 5G development, and mentioned the requirement of channel models. The METIS project reports the "foundation" for the *beyond 2020* 5G mobile and wireless communication systems by providing the technical enablers needed to address the requirements foreseen for this time frame [METIS, FT15]. METIS vision of a future access is that information and sharing of data is available anywhere and anytime to anyone and anything. Next generation mobile networks (NGMN) 5G white paper says that propagation is not well understood for higher frequencies [NGMN15].

International mobile telecommunications (IMT)-2020 (5G) promotion group identified all-spectrum access as a key enabler for 5G system, and mentioned that more channel measurements on higher frequency are needed before well designing an all-spectrum access system working on both high- and low-frequency band [IMT15].

ITU and 3GPP described 5G schedule recently [ITU15, Flo15].

12.5.4 5G Schedules

Besides the channel modelling study, the industry is approaching 5G standardisation. The ITU and 3GPP schedule of 5G activities are summarised in this section, and a consideration on 5G channel modelling time plan is presented, based on the observed state-of-art of 5G channel model, and the ITU and 3GPP schedule.

12.5.4.1 ITU schedule

Recently, ITU radiocommunication sector (ITU-R) WP5D agreed the work plan for "IMT-2020" (i.e., the formal name of "5G" in ITU-R context) [ITU15, Flo15].

- Initial technology submission, deadline June 2019
- Detailed specification submission, deadline October 2020

ITU work plan for "IMT-2020" is shown in Figure 12.2. Channel models will be described in the evaluation criteria & method document during Feb 2016–June 2017 (highlighted in Figure 12.2).

12.5.4.2 3GPP schedule

To meet ITU's 5G schedule, 3GPP discussed the 5G schedule in its RAN#67 meeting.

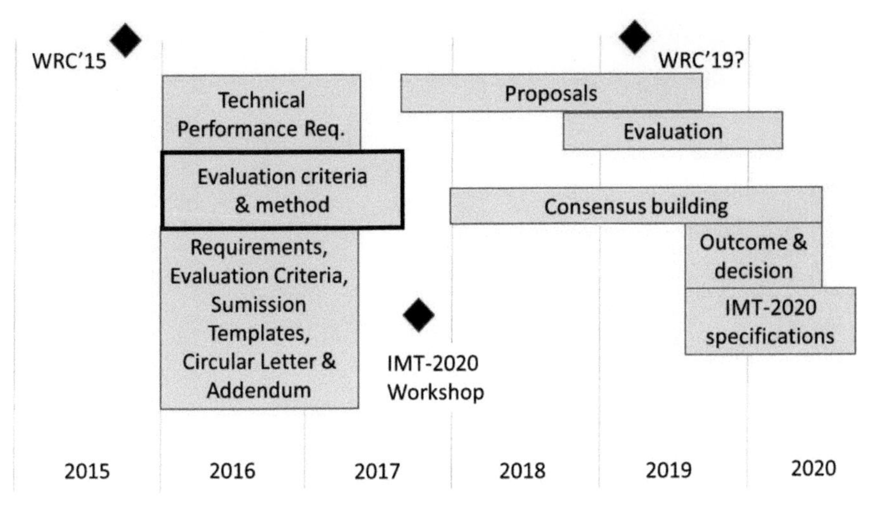

Figure 12.2 ITU-R 5G timeline.

It is assumed that 3GPP will submit a candidate technology for "IMT-2020". 3GPP RAN will lead the 3GPP submission process to ITU-R. 3GPP will import the relevant IMT-2020 requirements from ITU, and subsequently may add its own requirements on top of that. 3GPP assumes that a new radio technology and new system architecture are needed [Flo15]. However, the technical requirements and solutions are not yet exactly known.

The tentative timeline in 3GPP RAN is shown in Figure 12.3 [Flo15]. A "RAN workshop" on 5G requirements and scopes was held in the 3GPP RAN#69 meeting in September 2015. Then a study item (SI) dedicated to 5G channel modelling is likely to be approved (highlighted by black in Figure 12.3, 3GPP assumes that 5G work will also cover frequencies above 6 GHz). It is also mentioned that the first stage of this SI would be to identify the potential frequency band, which is vital information for channel modelling.

It is clear that the 5G channel modelling work would start in late 2015/early 2016 in 3GPP and ITU-R WP5D. WP5D will accomplish the work in June 2017, and requires the initial 5G proposal before June 2019. It is likely that 3GPP will submit its 5G proposal to ITU in early 2019, to meet the deadline of 5G proposal submission.

When considering the 5G channel model time schedule, it needs to take into account the state-of-the-art of channel modelling, and also the time plan of ITU and 3GPP. By the above observations, it seems appropriate to

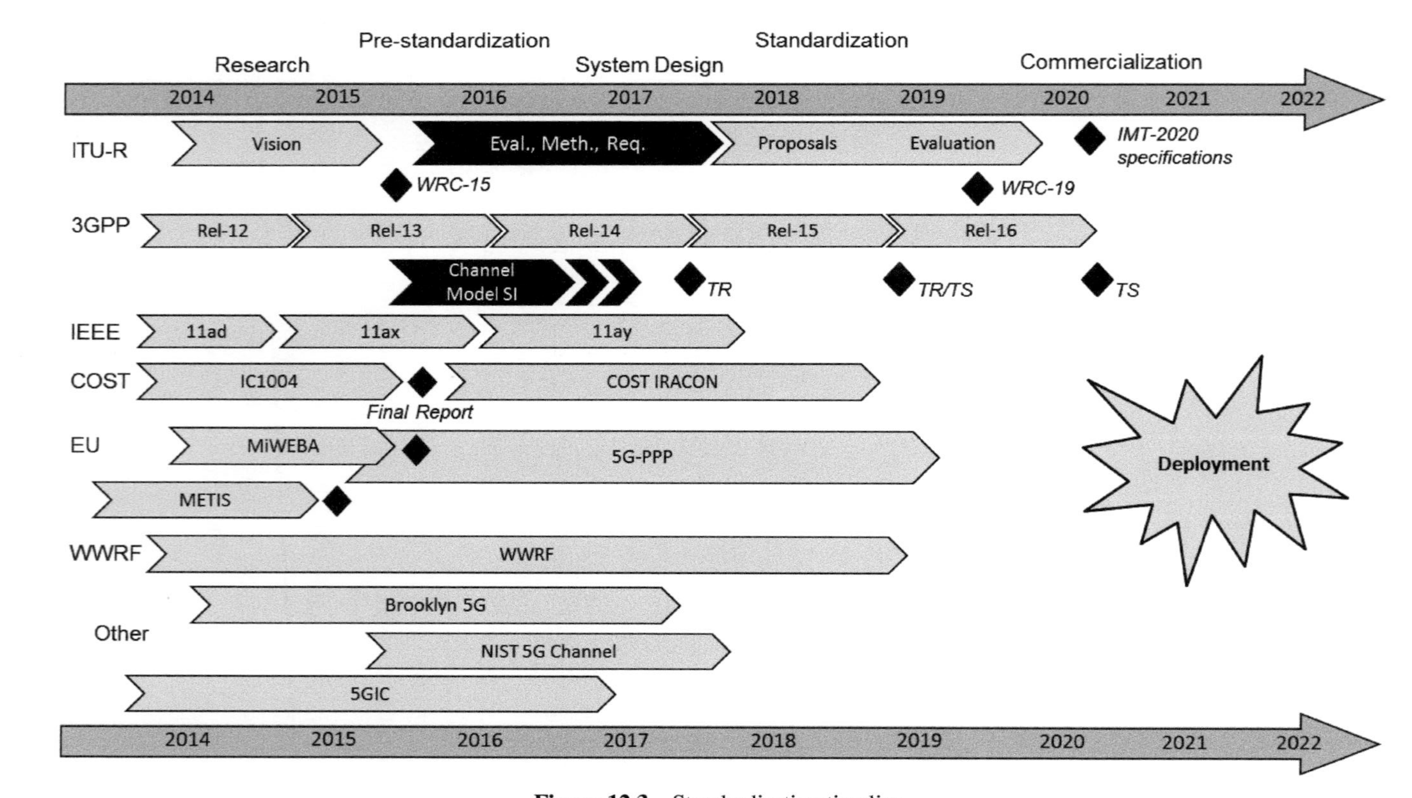

Figure 12.3 Standardisation timeline.
Note: This is a sketch only and is not accurate.

assume the accomplishment of the 5G channel modelling work in the end of 2016/beginning of 2017. On one hand, such a plan would fully meet the 5G standardisation timeline proposed both by ITU and 3GPP. On the other hand, it could guarantee a thorough and mature input to ITU and 3GPP. An earlier or aggressive schedule would be too rush, and may lead to misleading system design and deployment which raise the risk of 5G's grand success. A late or conservative schedule (e.g., accomplish in H2 2017) would be unacceptable for ITU, and would delay the 3GPP submission.

Figure 12.3 sketches the overall schedule of 5G channel model standardisation. METIS and COST IC1004 final reports could be useful inputs for next stage study that will appear in both 3GPP and ITU-R WP5D. It is proposed that experts of COST IC1004 continue the study of channel modelling by taking into account all the points mentioned in Section 12.1 (and to see if there is any others not mentioned), and contribute to 3GPP and ITU-R WP5D in their timeframe, as well as to collaborate with other organisations, e.g., NGMN, wireless world research forum (WWRF), 5G infrastructure public private partnership (5G PPP), etc. to polish this work.

12.5.5 Other Important Aspects

In addition to the mainstream 3GPP–ITU standardisation on IMT-2020, we should not forget other channel model standardisation activities such as IEEE Wi-Fi evolution, V2V, M2M, BAN, and sensor networks. They shall define their channel models and COST IC1004 members may consider contributing to them. Related topic is future test methodologies, e.g., OTA testing of adaptive antenna arrays, massive-MIMO, and vehicles.

12.5.6 Conclusions on Channel Model Standardisation

This section presented the state-of-the-art of 5G channel model studies, and the timeline for 5G channel model. It is shown that 3GPP and ITU-R WP5D will start the work in late 2015/early 2016, and ITU-R will finish the work in June 2017. Sufficient time for the 5G channel model study should be ensured, but it is also important to meet the standardisation time plan, e.g., it needs to be finished before H1 2017. Therefore it is proposed that experts of COST IC1004 continue the study of 5G channel model, and contribute to 3GPP and ITU-R WP5D in their timeframe. The following COST action (IRACON) should continue the channel modelling work, and could provide help for the members contributing to standardisation.

Bibliography

[3GP12] 3GPP. TS34.114 User Equipment (UE)/Mobile Station (MS) Over The Air (OTA) antenna performance; Conformance testing (release 11). Technical report, 3GPP, December 2012.

[3GP14a] 3GPP. TR 36.978 Evolved Universal Terrestrial Radio Access (E-UTRA) User Equipment (UE) antenna test function definition for two-stage Multiple Input Multiple Output (MIMO) Over The Air (OTA) test method (Release 13). Technical report, 3GPP, December 2014.

[3GP14b] 3GPP. TR37.977 universal terrestrial radio access (UTRA) and evolved universal terrestrial radio access (E-UTRA); verification of radiated multi-antenna reception performance of user equipment (UE). Technical report, 3GPP, March 2014.

[3GP15] 3GPP. TS 36.133 Evolved Universal Terrestrial Radio Access (E-UTRA); Requirements for support of radio resource management, Version 12.7.0. Technical report, 3GPP, April 2015.

[3GPP10] 3GPP. Further advancements for e-utra physical layer aspects. Technical report, 3rd Generation Partnership Project (3GPP), March 2010. TR 36.814.

[5GPPP] The 5G Infrastructure Public Private Partnership (online). Available at: https://5g-ppp.eu/

A

[AAM13] M. R. K. Az, K. Anwar, and T. Matsumoto. A joint RSS-DOA factor graphs based geo-location technique. Technical Report TD-06-084, Malaga, Spain, February 2013.

[AB07] T. C. Aysal and K. E. Barner. Generalized mean-median filtering for robust frequency-selective applications. *IEEE Transactions on Signal Processing*, 55(3):937–948, March 2007.

[Abb14] T. Abbas. *Measurement Based Channel Characterization and Modeling for Vehicle-to-Vehicle Communications.* PhD thesis, Department of Electrical and Information Technology, Faculty of Engineering, Lund University, 2014.

[ABSV13] M. Alasti, M. Barbi, K. Sayrafian, and R. Verdone. Inter-BAN interference evaluation and mitigation: a preliminary study. Technical Report TD(13)08060, Ghent, Belgium, September 2013.

[ACS09] A. J. Ali, S. L. Cotton, and W. G. Scanlon. Spatial diversity for off-body communications in an indoor populated environment at 5.8 GHz. In *Proceeding of 5th Loughborough Antennas Propagation Conference (LAPC2009)*, pp. 641–644, Loughborough, UK, 2009.

[ACD+12] J. G. Andrews, H. Claussen, M. Dohler, S. Rangan, and M. C. Reed. Femtocells: Past, present, and future. *IEEE Journal on Selected Areas in Communications*, 30(3):497–508, 2012.

[ACO15] M. Alodeh, S. Chatzinotas, B. Ottersten; "Constructive Multiuser Interference in Symbol Level Precoding for the MISO Downlink Channel," To appear in IEEE Transactions on signal processing, 2015. (Also available as TD(15)12002).

[AEM14] H. Alroughani, J. L. T. Ethier, and D. A. McNamara. Observations on computational outcomes for the characteristic modes of dielectric objects. In *IEEE Antennas and Propagation Society International Symposium (APSURSI)*, pp. 844–845, July 2014.

[AFM+13] P. Angueira, M. Fadda, J. Morgade, M. Murroni, Sancho R., M. Alexandru, and Popescu V. Co-Channel and Adjacent Channel Interferences for DVB-T2 over IEEE 802.22 WRAN. Technical Report TD(13)08043, Action COST IC1004, Cooperative Radio Communications for Green Smart Environments, Ghent, Belgium, September 2013.

[AFRJ14] J. Acevedo Flores, D. Robalo, and Velez F. J. Transmitted Power Formulation for the Implementation of Spectrum Aggregation in LTE-Advanced over 800 MHz and 2 GHz Frequency Bands. Technical Report TD-14-09079, Ferrara, Italy, February 2014.

[AFVC+14] J. Acevedo Flores, F. J. Velez, O. Cabral, D. Robalo, O. Holland, A. H. Aghvami, F. Meucci, A. Mihovsha, N.

R. Prasad, and R. Prasad. Cost/Revenue Performance in an IMT-Advanced Scenario with Spectrum Aggregation Over Non-Contiguous Frequency Bands. Proc. of International Conference on Telecommunications 2014 (ICT 2014), Lisboa, Portugal, 5–7 May 2014. (Also available as TD(14)09080).

[AGL13] S. H. Abdelhalem, P. S. Gudem, and L. E. Larson. Hybrid transformer-based tunable differential duplexer in a 90-nm CMOS process. *IEEE Transaction Microwaves Theory Technology*, 61:1316–1326, 2013.

[AGLV15] M. D. Abrignani, L. Giupponi, A. Lodi, and R. Verdone. Scheduling the 3GPP LTE Uplink over a *Dense* Heterogeneous Network. Technical Report TD-15-12014, Dublin, Ireland, January 2015.

[AGR05] B. Abdous, C. Genest, and B. Remillard. Dependence properties of meta-elliptical distributions. In *Statistical Modeling and Analysis for Complex Data Problems*, pp. 1–15, Springer, Berlin, 2005.

[AGT14] C. Gustafson, T. Abbas, D. Bolin and F. Tufvesson, "Statistical Modeling and Estimation of Censored Pathloss Data," in IEEE Wireless Communications Letters, vol. 4, no. 5, pp. 569–572, Oct. 2015. doi: 10.1109/LWC.2015.2463274. (Also available as TD-14-10084).

[AH13] R. Acedo-Hernández, M. Toril, S. Luna-Ramírez, I. de la Bandera, and N. Faour. Analysis of the impact of PCI planning on downlink throughput performance in LTE, Computer Networks, vol. 76, no. 1, pp. 42–54, 2015. (Also available as TD(13)06081).

[AHZ+12] B. Ai, R. He, Z. Zhong, K. Guan, B. Chen, P. Liu, and Y. Li. Radio wave propagation scene partitioning for high-speed rails. *International Journal of Antennas and Propagation*, 2012:7, 2012.

[AIPFL+12] M. Ait-Ighil, F. Perez-Fontan, J. Lemorton, C. Bourga, and M. Bousquet. Simple model for complex building scattering in urban environment based on a three macro propagation phenomena approach. Technical Report TD(12)04058, Lyon, France, May 2012.

[AK13a] S. J. Ambroziak and R. J. Katulski. An empirical propagation model for mobile radio links in container terminal environment. *IEEE Transactions on Vehicular Technology*, 62(9):4276–4287, 2013. (Also available as COST IC1004 TD(13)08002).

[AK13b] S. J. Ambroziak and R. J. Katulski. "Path Loss Modelling in the Untypical Outdoor Propagation Environments." In XXXI URSI General Assembly and Scientific Symposium (URSI GASS), Beijing, China, Aug. 16–23, 2014. (Also available as TD-08-002).

[AK14] S. J. Ambroziak and R. J. Katulski. "Statistical Tuning of Walfisch-Ikegami Model for the Untypical Environment." In 8th European Conference on Antennas and Propagation (EuCAP), Hague, Netherlands, Apr. 06–11, 2014. (Also available as TD-09-006).

[AKS+12] E. Aryafar, M. A. Khojastepour, K. Sundaresan, S. Rangarajan, and M. Chiang. Midu: enabling MIMO full duplex. In *18th Annual International Conference on Mobile Computing and Networking*, pp. 257–268. ACM, 2012.

[AKT11] T. Abbas, J. Karedal, and F. Tufvesson. "Measurement-based analysis: The effect of complementary antennas and diversity on vehicle-to-vehicle communication." IEEE Antennas and Wireless Propagation Letters, vol. 12, pp. 309–312, 2013. (Also available as TD-01-033).

[AKT13] T. Aoyagi, M. Kim, and J.-I. Takada. Characterization for a electrically small antenna in proximity to human body towards antenna de-embedding in body area network channel modelling. In *Proceeding of European Conference on Antenna nd Propagation (EuCAp2013)*, Gothenburg, Sweden, April 2013. (Also available as TD(13)06040).

[AKT14] T. Aoyagi, M. Kim, and J. Takada. Geometric modeling of shadowing rate for off-body propagation during human walking. Technical Report TD(14)11064, Krakow, Poland, September 2014.

[ALM15] G. Artner, R. Langwieser, and C. F. Mecklenbräuker. Measured antenna performance for different ground plane materials in vehicular applications. In *COST IC1004 12th MC & Scientific Meeting*, number TD(15)12030, January 2015.

[AM12] A. Alonso and C. F. Mecklenbräuker. Effects of co-located transmissions in the performance of DCC IEEE802.11p MAC and self-organizing TDMA for VANETs. In *COST IC1004 5th MC & Scientific Meeting*, number TD(12)05045, Bristol, United Kingdom, September 2012.

[AMAU14] N. Amiot, M. Mhedhbi, S. Avrillon, and B. Uguen. Pylayers. An indoor propagation tool for studying localization in WBAN context. In *IC1004*, Aalborg, Denmark, May 2014 (TD(14)10021).

[AMR07] F. W. Lindemans, N. F. De Rooiji and R. A. M. Receveur, Microsystem technologies for implantable applications. *Micromechanics and Micro-engineering*, 17:R50–R80, April 2007.

[AMS+14] S. J. Ambroziak, M. Mackowiak, J. Sadowski, J. K. Ryszard, and Correia L. M. Measurements of path loss in off-body channels in indoor environments. Technical Report TD(14)10007, Aalborg, Denmark, May 2014.

[ANK+14] T. Abbas, J. Nuckelt, T. Kürner, T. Zemen, C. F. Mecklenbräuker, and F. Tufvesson. "Simulation and measurement-based vehicle-to-vehicle channel characterization: Accuracy and constraint analysis." IEEE Transactions on Antennas and Propagation, vol. 63, no. 7, pp. 3208–3218, July 2015. (Also available as TD-07-059).

[ANP+07] J. B. Andersen, J. Ø. Nielsen, G. F. Pedersen, G. Bauch, and M. Herdin. Room electromagnetics. *IEEE Antennas and Propagation Magazine*, 49(2):27–33, 2007.

[ANW11] M. K. T. Al-Nuaimi and W. G. Whittow. Performance investigation of a dual element IFA array at 3 GHz for MIMO terminals. In *2011 Loughborough Antennas and Propagation Conference (LAPC)*, pp. 1–5, Nov 2011.

[APNG03] A. A. Paulraj, R. Nabar, and D. Gore. *Introduction to Space-Time Wireless Communications.* Cambridge University Press, Cambridge, UK, 2003.

[AR76] J. B. Andersen and H. Rasmussen. Decoupling and descattering networks for antennas. *IEEE Transactions on Antennas Propagation*, 24(6):841–846, 1976.

[ARD14] S. Arnesano, R. Rosini, and R. D'Errico. Study and implementation of space–time correlation for a dynamic on-body channel model. Technical Report TD-14-11040, Krakow, Poland, 2014.

[ARLM14] S. J. Ambroziak, J. K Ryszard, L. M. Correia, and M. Mackowiak. Measurement and analysis of the radio wave polarization in off-body communications in indoor environment. Technical Report TD(14)11015, Krakow, Poland, September 2014.

[Ars13] N. Arsalane. Total isotropic sensitivity (TIS) and Throughput Measurements For MIMO-LTE Terminals in Reverberant Cell. Technical Report TD(13)07008, 7th COST IC1004 Scientific Meeting, Ilmenau, Germany, May 2013.

[ASGL13] B. Azimdoost, H. R. Sadjadpour, and J. J. Garcia-Luna-Aceves. Capacity of wireless networks with social behavior. *IEEE Transactions on Wireless Communications* 12(1):60–69, 2013.

[ASPM13] A. Alonso, D. Smely, A. Paier, and C. F. Mecklenbräuker. Impact of traffic priorities and channel load on the DCC mechanism of the IEEE802.11p. In *COST IC1004 6th MC & Scientific Meeting*, number TD(13)06009, Malaga, Spain, February 2013.

[ASU+11] A. Alonso, K. Sjöberg, E. Uhlemann, E. G. Ström, and C. F. Mecklenbräuker. Challenging vehicular scenarios for self-organizing time division multiple access. In *COST IC1004 1st MC & Scientific Meeting*, number TD(11)01031, Lund, Sweden, June 2011.

[AT11] T. Aoyagi and J. Takada. Propagation simulation for body area netrorks by GMT. Technical Report TD-11-01061, Lund, Sweden, June 2011.

[AT12] K. Anwar and T. Matsumoto. Iterative spatial de-mapping for simultaneous full data exchange in three-way relaying channels. Technical Report TD-03-073, Barcelona, Spain, February 2012.

[ATG+13] B. Auinger, A. Tankielun, M. Gadringer, C. von Gagern, and W. Bösch. Introduction of a simulation approach for the Two-Channel Method using SystemVue. Technical Report TD(13)07049, 7th COST IC1004 Scientific Meeting, Ilmenau, Germany, May 2013.

[ATK12] T. Abbas, K. Sjöberg, J. Karedal, and F. Tufvesson. "A Measurement Based Shadow Fading Model for Vehicle-to-Vehicle Network Simulations." International Journal of Antennas and Propagation, vol. 2015, Article ID 190607, 2015. (Also available as TD-05-020).

[ATK14a] T. Aoyagi, J. Takada, and M. Kim. Directional characteristics of shadowing between on-body nodes and off-body antennas during human movement in WBAN. Technical Report TD(14)09020, Ferrara, Italy, February 2014.

[ATK14b] T. Aoyagi, J. I. Takada, and M. Kim. Characterization of a chip antenna for 2.4 GHz in proximity to dielectric material for antenna de-embedding of body area networks. In *Proceeding of 8th European Conference on Antennas and Propagation*, pp. 4128–4132, 2014. (Also available as TD(13)08040).

[ATK15] T. Aoyagi, J. Takada, and M. Kim. Geometric modeling of shadowing rate for off-body propagation during various human movements. Technical Report TD(15)12060, Dublin, Ireland, January 2015.

[ATTP13] O. Alrabadi, E. Tsakalaki, A. Tatomirescu, and G. F. Pedersen. Broadband decoupling via symmetric antenna topologies. Technical Report TD-13-06015, Malaga, Spain, February 2013.

[AZW08] T. Adamiuk, T. Zwick, and W. Wiesbeck. Dual-orthogonal polarized Vivaldi antenna for ultra wideband applications. In *17th International Conference on Microwaves, Radar and Wireless Communications (MIKON 2008)*, Wroclaw, Poland, May 2008.

B

[Bal89] C. A. Balanis. *Advanced Engineering Electromagnetics.* John Wiley & Sons, Hoboken, NJ, 1989.

[BAS+05] K. Brueninghaus, D. Astdlyt, T. Silzert, S. Visuri, A. Alexiou, S. Karger, and G. Seraji. Link performance models for system level simulations of broadband radio access systems. *Proceedings of PIMRC 2005 – IEEE 16th International Symposium on Personal, Indoor and Mobile Radio Communication*, pp. 2306–2311, 2005.

[BB12] A. Botos and V. Bota. Theoretical performance evaluation of a two-level rateless-FEC coding scheme. Technical Report TD-04-024, Lyon, France, May 2012.

[BB13] V. Bota and A. Botos. Performance evaluation of relay assisted cooperative transmission schemes using two level rateless-FEC coding. Technical Report TD-06-033, Malaga, Spain, February 2013.

[BBN+14] P. Bagot, M. Beach, A. Nix, J. McGeehan, and P. Moss. Adaptive Broadcast Techniques for Digital Terrestrial Television. Technical Report TD-09-041, Ferrara, Italy, February 2014.

[BBV14a] V. Bota, B. Bartha, and M. Varga. On the BLER and spectral efficiency performance of two-way relay channel algorithms on rayleigh-faded channels. Technical Report TD-11-019, Krakow, Poland, September 2014.

[BBV14b] N. Barroca, L. Borges, and F. J. Velez. Block Acknowledgment in IEEE 802.15.4 by Employing DSSS and CSS PHI layer. Technical Report TD(14)11018, Krakow, Poland, September 2014.

[BBVC14] N. Barroca, L. Borges, F. J. Velez, and P. Chatzimisios. IEEE 802.15.4 MAC Layer Performance Enhancement by employing RTS/CTS combined with Packet Concatenation," in Proc. of IEEE International Conference on Communications (ICC 2014) – Ad-hoc and Sensor Networking Symposium, Sidney, Australia, June 2014. (Also available as TD(13)09081).

[BCE+11] M. Barbiroli, C. Carciofi, P. Grazioso, D. Giuducci, V. Petrini, and G. Riva, "Planning criteria to improve energy efficiency of mobile radio systems," *IEEE – APS Topical Conference on Antennas and Propagation in Wireless Communications (APWC)*, September 2011.

[BCF+13] M. Barbiroli, C. Carciofi, F. Fuschini, P. Grazioso, and D. Guiducci. Methodologies and Tools for the Evaluation and Mitation of Mutual Interference between DVB and LTE. Technical Report TD(13)07024, Action COST IC1004, Cooperative Radio Communications for Green Smart Environments, Ilmenau, Germany, May 2013.

[BCGP12] M. Barbiroli, C. Carciofi, D. Guiducci, and V. Petrini. White spaces potentially available in italian scenarios based on the geo-location database approach. IEEE Symposium on new Frontiers in Dynamic Spectrum Access Networks (DySPAN), 2012, pp. 393–398, October 2012. (Also available as TD(12)05044).

[BDH+11] C. Buratti, R. DErrico, S. Huettinger, M. Maman, R. Martelli, F. Rosini, and R. Verdone. Design of a body area network for medical applications: the WiserBAN Project. Proceedings of the 4th International Symposium on Applied Sciences in Biomedical and Communication Technologies – ISABEL 2011. (Also available as TD(11)02020).

[BDZ13] D. Bajić, G. Dimić, and N. Zogović. A hybrid procedure with selective retransmission for aggregated packets of unequal length. *Proceedings of IEEE ICC 2013, Budapest, Hungary*, pp. 4078–4082, 9–13 June, 2013.

[BF14] A. Burr and Dong Fang. Linear block physical-layer network coding for multiple-user multiple-relay wireless networks. Technical Report TD09077, 2014.

[BFKP14] S. C. Barrio, O. Franek, G. Krenz, and G. F. Pedersen. Evaluation of reflections in a MIMO OTA test setup. In *IEEE-APS Topical Conference on Antennas and Propagation in Wireless Communications*, pp. 761–764, May 2014. (Also available as TD(14)10080).

[BFW14] A. Burr, D. Fang, and Y. Wang. Linear wireless physical layer network coding for multilayer relay networks. Technical Report TD10069, 2014.

[BGJ+13] M. Ballester, J. J. Giménez, T. Jansen, D. Rose, and N. Cardona. A multi-wall approach for the indoor propagation in lte femto-cell scenarios. In *IC1004*, Malága, Spain, February 2013 (TD(13)06073).

[BGW11] M. Beach, N. Gautam, and M. Webb. Virtual MIMO benefit analysis in a shared vehicular environment. In *COST IC1004 2nd MC & Scientific Meeting*, number TD(11)02038, Lisbon, Portugal, 2011.

[BHdG+05] D. S. Baum, J. Hansen, G. del Galdo, M. Milojevic, J. Salo, and P. Kyösti. An interim channel model for beyond-3G systems: extending the 3GPP spatial channel model (SCM). In *Vehicular Technology Conference, 2005. VTC 2005–Spring. 2005 IEEE 61st*, Vol. 5, pp. 3132–3136, May 2005.

[BJRK12] J. Baumgarten, T. Jansen, D. M. Rose, and T. Kürner. Power consumption model for gsm, umts and lte macro cells. *COST IC1004 5th Scientific Meeting*, pp. 1–6, September 2012.

[BJSP14a] P. Bahramzy, O. Jagielsi, S. Svendsen, and G. F. Pedersen. Compact agile antenna concept utilizing reconfigurable front end for wireless communications. *IEEE Transactions on Antennas Propagation*, 62(9):4554–4563, 2014. (Also available as TD-14-11042).

[BJSP14b] P. Bahramzy, O. Jagielsi, S. Svendsen, and G. F. Pedersen. Self-matched high-Q reconfigurable antenna concept for mobile terminals. *IET Science, Measurement and Technology*, 8(6):479–486, April 2014. (Also available as TD-14-11043).

[BJT+12] A. Bamba, W. Joseph, E. Tanghe, G. Vermeeren, and L. Martens. Theory for exposure prediction in an indoor environment due to UWB systems. In *COST IC1004 4th Scientific Meeting*, pp. 1–3, Lyon, France, May 2012.

[BJT+13] A. Bamba, W. Joseph, E. Tanghe, G. Vermeeren, and L. Martens. Circuit model for diffuse multipath and electromagnetic absorption prediction in rooms. *IEEE Transactions on Antennas and Propagation*, 61(6):3292–3301, 2013.

[BJVNVB+14] T. Bernabeu-Jimenez, A. Valero-Nogueira, F. Vico-Bondia, E. Antonino-Daviu, and M. Cabedo-Fabres. A 60-Ghz ltcc rectangular dielectric resonator antenna design with characteristic modes theory. In *IEEE Antennas and Propagation Society International Symposium (APSURSI)*, pp. 1928–1929, July 2014.

[BK14] C. Brennan and I. Kavanagh. A full wave indoor propagation model. In *COST IC1004 9th Management Committee and Scientific Meeting*, Technical Report TD(14)09076, Ferrara, Italy, February 2014.

[BKS13] M. Bouezzeddine, A. Krewski, and W. L. Schroeder. A concept study on MIMO antenna systems for small size CPE operating in the TV white spaces. Technical Report TD-13-07051, Ilmenau, Germany, May 2013.

[blu10] Specification of the bluetooth system version 4.0. Bluetooth *SIG*, June 2010.

[BMG+14] A. Bamba, M.-T. Martinez-Ingles, D. P. Gaillot, E. Tanghe, B. Hanssens, J.-M. Molina-Garcia-Pardo, M. Lienard, L. Martens and W. Joseph, "Experimental Investigation of Electromagnetic Reverberation Characteristics as a Function of UWB Frequencies" IEEE Antennas and Wireless Propagation Letters, pp. 859–862, vol. 14, 2015

[BMK13] D. Bharadia, E. McMilin, and S. Katti. Full duplex radios. In *Annual Conference of ACM Special Interest Group on Data Communication (SIGCOMM)*, pp. 375–386. ACM, 2013.

[BMM+14] J. Blumenstein, T. Mikulasek, R. Marsalek, A. Prokes, T. Zemen, and C. F. Mecklenbräuker. "In-Vehicle mm-Wave Channel Model and Measurement," IEEE 80th Vehicular Technology Conference (VTC Fall), Vancouver, BC, Sept. 14–17, 2014, pp. 1–5. (Also available as TD-11-012).

[BMPZ13] A. Bazzi, B. M. Masini, G. Pasolini, and A. Zanella. Routing algorithms for V2V and V2R communications in vehicular sensor networks. In *COST IC1004 6th MC & Scientific Meeting*, number TD(13)06024, Malaga, Spain, February 2013.

[BMS15] S. Beygi, U. Mitra, and E. G Ström. "Nested sparse approximation: Structured estimation of V2V channels using geometry-based stochastic channel model." IEEE Trans. Signal Process., vol. 63, no. 18, pp. 4940–4955, September 2015.

[BNB15] T. H. Barratt, A. Nix, and M. A. Beach. 60 GHz channel characterisation Òseen throughÓ on chip antenna arrays. *COST IC1004, Dublin, Ireland*, 2015.

[BOM+14] P. Bahramzy, P. Olesen, P. Madsen, J. Bojer, S. C. Del Barrio, A. Tatomirescu, A. Bundgaard, P. Morris, and G. F. Pedersen. A tunable RF front end with narrow-band antennas for mobile devices. Technical Report TD-14-11044, Krakow, Poland, September 2014.

[BP13] P. Bahramzy and G. F. Pedersen, "Group Delay of High-Q Antennas," Antennas and Propagation Society International Symposium (APSURSI), pp. 1208–1209, 2013. (Also available as TD(13)08018).

[BPV13] V. Bota, M. T. Pavel, and M. Varga. Reliability performance evaluation of distributed alamouti-based schemes. Technical Report TD-08-023, Gent, Belgium, September 2013.

[BRC03] S. Blanch, J. Romeu, and I. Corbella. Exact representation of antenna system diversity performance from input parameter description. *Electronics Letter*, 39(9): 705–707, 2003.

[Bro14] T. W. C. Brown. Measured impact of polarizsation on interference from television white space devices. Technical Report TD(14)10044, Action COST IC1004, Cooperative Radio Communications for Green Smart Environments, Aalborg, Denmark, May 2014.

[BS12] A. Burr and J. Sykora. Cooperative wireless network coding for uplink transmission on hierarchical wireless networks. International Symposium on Signals, Systems, and Electronics (ISSSE), 2012, vol., no., pp. 1, 5, 3–5 Oct. 2012. (Also available as TD(12)03047).

[BS14] M. Bouezzeddine and W. L. Schroeder. Parametric study on capacitive and inductive couplers for exciting characteristic modes on CPE. In *2014 8th European Conference on Antennas and Propagation (EuCAP)*, pp. 2731–2735, The Hague, Netherlands, April 2014.

[BSA15] M. Barbi, K. Sayrafian, and M. Alasti. Inter-BAN interference mitigation using uncoordinated transmission scheduling strategies. Technical Report TD(15)12035, Dublin, Ireland, January 2015.

[BSG+13] N. Barroca, H. M. Saraiva, P. T. Gouveia, J. Tavares, L. M. Borges, F. J. Velez, C. Loss, R. Salvado, P. Pinho, R. Goncalves, N. Borges Carvalho, R. Chavez-Santiago, and I. Balasingham. Antennas and Circuits for Ambient RF Energy Harvesting in Wireless Body Area Networks. Proceedings of PIMRC 2013 – The 24th IEEE International

Symposium on Personal, Indoor and Mobile Radio Communications, London, United Kingdom, Sep. 2013. (Also available as TD(13)08068).

[BSM+12] A. Boulis, D. Smith, D. Miniutti, L. Libman, and Y. Tselishchev. Challenges in body area networks for healthcare: the MAC. *IEEE Communication Magazine*, 50(5):100–106, 2012.

[BSP14] P. Bahramzy, S. Svendsen and G. F. Pedersen, "Isolation between three antennas at 700 MHz: for Handheld terminals," IET Microwaves, Antennas & Propagation, Sep. 2014. (Also available as TD(14)10026).

[BSP15] P. Bahramzy, S. Svendsen, and G. F. Pedersen. Mutual coupling investigation between a loop and two inverted-L antennas operating at 700 mhz. Technical Report TD-15-12041, Dublin, Ireland, January 2015.

[BSVT11] A. Böttcher, C. Schneider, P. Vary, and S. Thomä. "Dependency of the power and delay domain parameters on antenna height and distance in urban macro cell," in *Proceedings of the 5th European Conference on Antennas and Propagation (EUCAP 2011)*, pp. 1395–1399, IEEE, Piscataway, NJ, Rome, Italy, 11–15 April 2011. (Also available as TD(11)01018).

[BT12] C. Brennan and X. D. Trinh. Improved forward backward method with spectral acceleration for scattering from randomly rough lossy surfaces. Technical Report TD(12)03057, Barcelona, Spain, February 2012.

[BT13] C. Brennan and X. D. Trinh. Full wave modelling of wave propagation in urban areas. Technical Report TD(13)08047, Ghent, Belgium, September 2013.

[BT14a] T. W. C. Brown and A. C. H. Tay. On the benefits of polarization for fixed area television white space devices. *IEEE Transactions on Antennas and Propagation*, 62(3):1147–1156, 2014.

[BT14b] C. Brennan and X. D. Trinh. Full wave computation of path loss in urban areas. In *Proceeding on 8th European Conference of Antennas and Propagation, EuCAP 2014*, The Hague, The Netherlands, April 2014.

[BTB+14a] E. Buskgaard, A. Tatomirescu, S. Barrio, O. Franek, and G. Pedersen. In-depth study of the user effect on a mobile phone. Technical Report TD-14-10092, Aalborg, Denmark, May 2014.

[BTB+14b] E. Buskgaard, A. Tatomirescu, S. Barrio, O. Franek, and G. Pedersen. User studies with semi-flexible simulation hand. Technical Report IC1004 TD(14) 09061, Ferrara, Italy, February 2014.

[BTDBP15] E. Buskgaard, A. Tatomirescu, S. C. Del Barrio, and G. F. Pedersen. On the feasibility of a low cost printing technology for mobile phone antennas. Technical Report TD-15-12050, Dublin, Ireland, January 2015.

[BTD11] C. Brennan, X. D. Trinh, and J. Diskin. Recent advances in full wave and asymptotic propagation modelling. Technical Report TD(11)02062, Lisbon, Portugal, October 2011.

[BTM+13] C. Brennan, X. D. Trinh, M. Mullen, P. Bradley, and M. Condon. Improved forward backward method with spectral acceleration for scattering from randomly rough lossy surfaces. *IEEE Transactions on Antennas Propagation*, 61(7):3922–3926, 2013.

[Bur03] A. G. Burr. Capacity bounds and estimates for the finite scatterers MIMO wireless channel. *IEEE Journal on Selected Areas Communication*, 21(5):812–818, 2003.

[Bur11] A. Burr. MIMO truncated Shannon bound for system level capacity evaluation of MIMO wireless networks. In *COST IC1004 2nd MCM, Lisbon, Portugal*, (TD(11)02039), 2011.

[Bur13] A. Burr. Linear mapping functions in wireless network coding. Technical Report TD08038, 2013.

[Bur14a] A. Burr. Equivocation performance of linear wireless physical layer network coding. Technical Report TD09078, 2014.

[Bur14b] A. Burr. Linear physical layer network coding based on rings of Gaussian integers. Technical Report TD11041, 2014.

[BV12] Ch. Buratti and R. Verdone. P-CSMA: A Priority-Based CSMA Protocol for Multi-Hop Linear Wireless Networks. Technical Report TD-12-04045, Lyon, France, May 2012.

[BV13] N. Barroca, F. J. Velez. Block Acknowledgment Mechanisms for the optimization of channel use in Wireless Sensor Networks," in Proc. of PIMRC 2013 – The 24th IEEE International Symposium on Personal, Indoor and Mobile Radio Communications, London, United Kingdom, Sep. 2013. (Also available as TD(13)08057).

[BVF+11a] M. Boban, T. T. V. Vinhoza, M. Ferreira, J. Barros, and O. K. Tonguz. Impact of vehicles as obstacles in vehicular ad hoc networks. *IEEE Journal on Selected Areas in Communications*, 29(1):15–28, 2011.

[BVF+11b] N. Barroca, F. J. Velez, M. Ferro, L. Borges, and A. S. Lebres. Block Acknowledgment Mechanism for Wireless Sensor Networks. Technical Report TD-11-01001, Lund, Sweden, 2011.

[BvGT+11] E. Böhler, C. von Gagern, A. Tankielun, Y. Feng, and W. Schroeder. Measurements of Over-The-Air performance of MIMO UE. In *33rd Annual Symposium of the Antenna Measurement Techniques Association (AMTA 2011)*, Englewood, Colorado, October 2011. In parts also available as TD(11)1044, TD(11)1045, and TD(11)1046.

[BY09] N. C. Beaulieu and D. J. Young. Designing time-hopping ultrawide bandwidth receivers for multiuser interference environments. *Proceedings of the IEEE*, 97(2):255–284, 2009.

[BYP09] R. A. Bhatti, S. Yi, and S. O. Park. Compact antenna array with port decoupling for LTE-standardized mobile phones. *IEEE Antennas Wireless Propagation Letters*, 8:1430–1433, 2009.

[BZBO13] E. Bjornson, P. Zetterberg, M. Bengtsson, and B. Ottersten. Capacity limits and multiplexing gains of MIMO channels with transceiver impairments. *IEEE Communications Letters*, 17(l):91–94, 2013. (Also available as TD(12)05039).

[BZG14] C. Bühler, H. anad Mecklenbraeuker, R. Zelenka, and T. Gyoergyfalvay. On measuring fast fading signals for channel modelling. Technical Report TD-011-056, Krakow, Poland, September 2014.

C

[CAM+09] V. Chandrasekhar, J. G. Andrews, T. Muharemovict, Z. Shen, and A. Gatherer. Power control in two-tier femtocell networks. *IEEE Transactions on Wireless Communications* 8(8):4316–4328, 2009.

[CAV13] E. Chatziantoniou, B. Allen, and V. Velisavljevic. An hmm-based spectrum occupancy predictor for energy efficient cognitive radio. In *Personal Indoor and Mobile Radio Communications (PIMRC), 2013 IEEE 24th International Symposium on*, pp. 601–605. IEEE, 2013.

[CAZ+13] I. S. Comsa, M. Aydin, S. Zhang, P. Kuonen, and Wagen J. Insight into a Self Organising LTE-A Scheduler. Technical *Report* TD-13-07056, Ilmenau, Germany, May 2013.

[CB14] R. Cavallari and C. Buratti. On the performance of IEEE 802.15.6 CSMA/CA with priority for query-based traffic. In *Proceeding of European Conference on Networks and Communications, (EuCNC2014)*, pp. 1–5, June 2014. (Also available as TD(14)09051).

[CBZ+12] K. L. Chee, J. Baumgarten, P. Zahn, M. Rohner, M. Braun, S. A. Torrico, and T. Kürner. Radiowave propagation prediction in vegetated residential environments, part ii: Verification by measurements. Technical Report TD(12)05055, Bristol, UK, September 2012.

[CCB95] P. S. Chow, J. M. Cioffi, and J. A. C. Bingham. A practical discrete multitone transceiver loading algorithm for data transmission over spectrally shaped channels. *IEEE Transactions on. Communication*, 43(2/3/4):773–775, 1995.

[CCC11] Q. H. Chu, J.-M. Conrat, and J.-C. Cousin. Experimental Characterization and Modeling of Shadow Fading Correlation for Relaying Systems. Vehicular Technology Conference (VTC Fall), 2011 IEEE, vol., no., pp. 1, 5, 5–8 Sept. 2011. (Also available as TD(11) 02004).

[CCC12] L. Caeiro, F. Cardoso, and L. M. Correia. Adaptive allocation of virtual radio resources over heterogeneous wireless networks. In *Proc. of EW2012 – European Wireless 2012*, Poznan, Poland, April 2012. (Also available as TD(11)02018).

[CCC14a] L. Caeiro, F. Cardoso, and L. M. Correia. Addressing multiple virtual resources in the same geographical area. Technical Report TD-14-11027, COST IC1004, Krakow, Poland, September 2014.

[CCC14b] L. Caeiro, F. Cardoso, and L. M. Correia. Wireless access virtualisation: Physical versus virtual capacity. In *Proceedings of 5GU 2014 – 1st International Conference on 5G for Ubiquitous Connectivity*, Levi, Finland, November 2014. (Also available as TD(13)08036).

[CCI86] CCIR Recommendations, Propagation in non-ionized media, ITU, Geneva, 1986.

[CCLK12] J. Choi, B. Clerckx, L. Namyoon, and G. Kim. A new design of polar-cap differential codebook for temporally/spatially correlated MISO channels. *IEEE Transactons on Wireless Communication*, 11(2):703–711, 2012.

[CCS+06] N. Czink, P. Cera, J. Salo, E. Bonek, J. P. Nuutinen, and J. Ytalo. A framework for automatic clustering of parametric MIMO channel data including path powers. In *Proceedings of the VTC 2006 Fall – IEEE 64th Vehicular Technology Conference*, pp. 1–5, September 2006.

[CDG+14] A. Conti, D. Dardari, M. Guerra, L. Mucchi, and M. Z. Win. Experimental characterization of diversity navigation. *Systems Journal, IEEE*, 8(1):115–124, 2014.

[CDP12] V. Cipov, L. Dobos, and J. Papaj. ToA estimation enhancement for indoor MANET cooperative localization algorithms. In *5th COST IC1004 Management Committee Meeting*, Bristol, United Kingdom, September 2012 (TD(12)05003).

[CEP09a] CEPT. The identification of common and minimal (least restrictive) technical conditions for 790–862 mhz for the digital dividend in the european union. Technical Report 30, CEPT, 2009.

[CEP09b] CEPT. Technical considerations regarding harmonisation options for the digital dividend in the european union y frequency (channelling) arrangements for the 790–862 mhz band. Technical Report 31, CEPT, 2009.

[CF12] J.-M. Conrat and M. Focsa. Path loss models for LTE-advanced Urban relaying systems with antenna height correction factor. Technical Report TD(12) 05012, September 2012.

[CGT+15] S. Cheng, D. P. Gaillot, E. Tanghe, P. Laly, T. Demol, W. Joseph, L. Martens, and M. Lienard. Full-polarimetric distance-dependent model for large hall scenarios. In *COST IC1004 12th Scientific Meeting*, pp. 1–18, Dublin, Ireland, January 2015.

[CGV+11] M. Chen, S. Gonzalez, A. Vasilakos, H. Cao, and V. C. Leung. Body area networks: a survey. *Mobile Network Application*, 16(2), April 2011.

[CH14] C. Chaudet and Y. Haddad. Wireless Software Defined Networks: Challenges and Oppportunities, in IEEE International Conference on Microwaves, Communications, Antennas and Electronics (IEEE COMCAS), Tel-Aviv, Israel, Oct. 2013. (Also available as TD(14)09022).

[Che99] P.-C. Chen. A non-line-of-sight error mitigation algorithm in location estimation. In *WCNC*, pp. 316–320. IEEE, September 1999.

[Che14a] X. Chen. A stackelberg learning framework for energy efficiency in heterogeneous cellular networks. Temporary document, IC1004 COST action, May 2014.

[Che14b] X. Chen. Experimental investigation and modeling of the throughput of a 2 × 2 closed-loop MIMO system in a reverberation chamber. *IEEE Transactions on Antennas and Propagation*, 62(9):4832–4835, 2014.

[Cherr04] S. Cherry. Edholm's law of bandwidth. *IEEE spectrum*, 41(7), 2004.

[CHHE11] M. Capek, P. Hazdra, P. Hamouz, and J. Eichler. A method for tracking characteristic numbers and vectors. *Progress In Electromagnetics Research B*, 33:115–134, 2011.

[CHP08] H. Claussen, L. T. W. Ho, and F. Pivit. Effects of joint macrocell and residential picocell deployment on the network energy efficiency. In *Personal, Indoor and Mobile Radio Communications, IEEE 19th International Symposium on*, pp. 1–6, September 2008.

[CIJW13] J. C. Ikuno, T. Jansen, and T. Werthmann. Harmonization of lte system level simulators addressing synthetic simulation scenarios. Technical Report TD(13)06-042, Malaga, Spain, February 2013.

[CJ00] H. T. Chou and J. T. Johnson. Formulation of forward-backward method using novel spectral acceleration for the modeling of scattering from impedance rough surfaces. *IEEE Translation Geo-science and Remote Sensing*, 38(1):605–607, 2000.

[CJ08] V. R. Cadambe and S. A. Jafar. Interference alignment and degrees of freedom of the K-user interference channel. *IEEE Transactions on Information Theory*, 54(8):3425–3441, 2008.

[CJ11] R. Chandra and A. J. Johansson. Antennas and propagation for the ear-to-ear propagation channel for binaural hearing aids. Technical Report TD-11-01027, Lund, Sweden, June 2011.

[CJ13] R. Chandra and A. J. Johansson. Analytical on-body propagation models for channel around the body based on attenuation of creeping waves over an elliptical surface. Technical Report TD-13-06053, Malaga, Spain, February 2013.

[CK99] D. J. Cichon and T. Kürner. *Digital Mobile Radio Towards Future Generation Systems*, chapter Propagation prediction models. COST 231 Final Report, 1999.

[CKKT12] Y. Chang, Y. Konish, M. Kim, and J. Takada. Indoor wideband 8Œ12 mimo channel analysis in an exhibition hall with presence of people. In *IC1004*, Bristol, UK, September 2012 (TD(12)05030).

[CKY10] T. Chen, H. Kim, and Y. Yang. Energy efficiency metrics for green wireless communications. In *Wireless Communications and Signal Processing (WCSP), 2010 International Conference on*, pp. 1–6, October 2010.

[CKZ+10] N. Czink, F. Kaltenberger, L. Zhou, L. Bernado, T. Zemen, and X. Yin. Low-complexity geometry-based modeling of diffuse scattering. In *Proceeding of 4th European Conference on Antennas and Propagation, EuCAP 2010*, Barcelona, Spain, April 2010.

[CL12] P. Cataldi, M. Lee, and R. Cepeda. An overview of the cambridge white spaces trial. Technical Report TD(12)05058, Action COST IC1004, Cooperative Radio Communications for Green Smart Environments, Bristol, United Kingdom, September 2012.

[Cla96] R. Clarke. A statistical theory of mobile radio reception. *Bell System Technical Journal*, 47(6):957–1000, 1968.

[CLD+08] A. Chebihi, C. Luxey, A. Diallo, P. Le Thuc, and R. Staraj. A novel isolation technique for closely spaced PIFAs for UMTS mobile phones. *IEEE Antennas Wireless Propagation Letter*, 7:665–668, 2008.

[CLV98] D. Chizhik, J. Ling, and R. A. Valenzuela. The effect of electric Field polarization on indoor propagation. In *Proceeding of International Conference on Universal Personal Communications (ICUPC 1998)*, pp. 459–462, October 1998.

[CM13] J.-M. Conrat and I. Maaz. Path loss models in los conditions for relaying technique. Technical Report TD(13) 08015, September 2013.

[CM14] J.-M. Conrat and I. Maaz. Path loss Models in NLOS Conditions for relaying technique. Technical Report TD(14) 11005, September 2014.

[CM15] J.-M. Conrat and I. Maaz. Channel Model Validation for the Relay-Mobile Link in Microcell Environment. Technical Report TD(15) 12024, January 2015.

[CMG+12] L. S. Cardoso, A. Massouri, B. Guillon, P. Ferrand, F. Hutu, G. Villemaud, T. Risset, J.-M. Gorce. Cortexlab: A large scale testbed for physical layer in cognitive radio networks. In *4th Meeting of the Management Committee of COST*

IC1004 – Cooperative Radio Communications for Green Smart Environments, 2012.

[CMM+11] O. Cabral, F. Meucci, A. Mihovska, F. J. Velez, and N. R. Prasad. Integrated common radio resource management with spectrum aggregation over non-contiguous frequency bands. *Wireless Personal Communications*, 59(3):499–523, 2011.

[CMW14] G. W. K. Colman, S. D. Muruganathan, and T. J. Willink. Distributed interference alignment for mobile MIMO systems based on local CSI. *IEEE Communications Letters*, 18(7):1206–1209, 2014.

[CnFN+13] X. Carreño, W. Fan, J. Ø. Nielsen, J. Singh Ashta, G. F. Pedersen, and M. B. Knudsen. Test setup for anechoic room based MIMO OTA *testing* of LTE terminals. In *Antennas and Propagation (EuCAP), 2013 7th European Conference on*, pp. 1417–1420, April 2013.

[CO13] B. Clerckx and C. Oestges. *MIMO* Wireless *Networks*. Waltham, MA: Academic Press, *2nd edn* edition, 2013.

[Col13] G. W. K. Colman. Spatial hopping in MIMO systems for impeded signal reception by multi-element eaves droppers. *IEEE Wireless Communications Letters*, 2(6):647–650, 2013. (Also available as TD(13)07016).

[COS01] COST 255, Radiowave propagation modelling for new SatCom services at Ku band and above, European Commission, Brussels, 2001.

[COST231] E. Damosso, D. Cichon, and T. Kurner. COST 231 Final report: digital mobile radio towards future generation systems. Technical Report, http://www.lx.it.pt/cost231/ (Online), 1998.

[Cot15] S. Cotton. Characterization of on-body fading channels operating in anechoic and reverberant environments at 2.45 ghz. Technical Report TD-15-12070, Dublin, Ireland, February 2015.

[CS07] S. L. Cotton and W. G. Scanlon. Characterization and modeling of the indoor radio channel at 868 MHz for a mobile bodyworn wireless personal area network. *IEEE Antennas Wireless Propagation Letters*, 6:51–55, 2007.

[CS09] S. L. Cotton and W. G. Scanlon. Channel characterization for single- and multiple-antenna wearable systems used for indoor body-to-body communications. *IEEE Transactions on Antennas Propagation*, 57(4):980–990, 2009.

[CS13] R. Chvez-Santiago and I. Balasingham. Perspectives and challenges for the ultra wideband capsule endoscope. Technical Report TD(13)08052, Ghent, Belgium, September 2013.

[CS14a] R. Chvez-Santiago and I. Balasingham. Modeling the implant ultra wideband radio channel. Technical Report TD(14)09031, Ferrara, Italy, February 2014.

[CS14b] R. Chavez-Santiago, C. Garcia-Pardo, A. Fornes-Leal, A. Valles-Lluch, I. Balasingham, and N. Cardona. Ultra wideband propagation for in-body to in-body communications. Technical Report TD(14)11060, Krakow, Poland, September 2014.

[CSB10] N. Blumm, C. Song, Z. Qu, and A. L. Barabsi. Limits of predictability in human mobility. *IEEE Transactions on Communications*, 327(5968):1018–1021, 2010.

[CSEM11] P. Castiglione, O. Simeone, E. Erkip, and G. Matz. Energy-neutral source-channel coding with battery and backlog sise constraints. *COST IC1004 2nd Scientific Meeting*, p. 111, October 2011.

[CSO13a] S. Chatzinotas, S. K. Sharma, and B. Ottersten, "Asymptotic Analysis of Eigenvalue based Blind Spectrum Sensing", in Proc. IEEE Int. Conf. Acoustics, Speech and Signal Processing (ICASSP), Vancouver, Canada, May 2013, pp. 4464–4468. (Also available as IC1004 TD(13)07018).

[CSO13b] S. Chatzinotas, S. K. Sharma, and B. Ottersten, "Frequency packing for interference alignment-based cognitive dual satellite systems", in Proc. IEEE Vehicular Technology Conference (VTC)-fall 2013, Las Vegas, Nevada, Sept. 2013, pp. 1–7. (Also available as TD(13)08076).

[CSO13] S. Chatzinotas, S. K. Sharma, and B. Ottersten. Frequency packing for interference alignment-based cognitive dual satellite systems. In *IEEE 78th Vehicular Technology Conference (VTC Fall), 2013*, pp. 1–7, Sept 2013.

[CT13] S. Cicalò, V. Tralli, Cross-Layer Algorithms for Distortion-Fair Scalable Video Delivery over OFDMA Wireless Systems; IEEE Globecom 2012, Workshop on Quality of Experience for Multimedia Communications; Anheim, USA, Dec. 3–7, 2012, pp. 1287–1292. (Also available as TD(13)06064).

[CT14a] S. Cicalò and V. Tralli. Adaptive Resource Allocation with Proportional Rate Constraints for Uplink SC-FDMA Systems; IEEE COMMUNICATIONS LETTERS, vol 18, pp. 1419–1422, 2014. (Also available as TD(14)09027).

[CT14b] S. Cicalò and V. Tralli. Resource Allocation with Proportional Average Rate Constraints: Performance Evaluation in Uplink SC-FDMA Single Cell Scenario. Technical Report TD-14-11049, Krakow, Poland, September 2014.

[CTI14] CTIA. Test plan for wireless device over the-air performance. Technical report, CTIA MOSG, December 2014.

[CTK12] K. L. Chee, S. Torrico, and T. Kurner. Radiowave propagation prediction in vegetated residential environment part i: theoretical modeling. Technical Report TD(12) 04017, May 2012.

[CTK13] K. L. Chee, S. A. Torrico, and T. Kürner. Radiowave propagation prediction in vegetated residential environments. *IEEE Transaction on Vehicle Technology*, 62(2):486–499, 2013. (Also available as TD(12)04017).

[CU93] A. Cichocki and R. Unbehauen. *Neural networks for optimization and signal processing*. John Wiley & Sons Ltd. & B. G. Teubner, Stuttgart, 1993.

[CW11] G. W. K. Colman and T. J. Willink. Distributed MIMO interference alignment in practical wireless systems. In *XXXth URSI General Assembly and Scientific Symposium, 2011*, pp. 1–4, 2011.

[CWC08] S. C. Chen, Y. S. Wang, and S. J. Chung. A decoupling technique for increasing the port isolation between two strongly coupled antennas. *IEEE Transactions on Antennas Propagation*, 56(12):3650–3658, 2008.

[CWS+12] L. Clavier, G. Wei, Fr. Septier, G. Peters, and I. Nevat. Interference modelling and cooperative communications. Technical Report TD-03-074, Barcelona, Spain, February 2012.

[CWW12] G. W. K. Colman, M. Wang, and S. Watson. Direction of arrival estimation for MIMO systems employing constellation-based precoding. In *Proceedings of VTC 2012 Fall – IEEE 74st Vehicular-Technology Conference*, pp. 1–5, September 2012. (Also available as TD(13)06007).

[CYT+14] J. Chen, X. Yin, L. Tian, et al., "Measurement-Based LoS/NLoS Channel Modeling for Hot-Spot Urban Scenarios in UMTS Networks," International Journal of Antennas and Propagation, vol. 2014, Article ID 454976, 2014. (Also available as TD(14)11033).

[CZA+14] I. S. Comsa, S. Zhang, M. Aydin, J. Chen, P. Kuonen, and J.-F. Wagen. Adaptive proportional fair parametreization based lte scheduling using continuous actor-critic reinforcement learning. In *Global Communications Conference (GLOBECOM), 2014 IEEE*, pp. 4387–4393. IEEE, December 2014.

[CZB+10] L. M. Correia, D. Zeller, O. Blume, D. Ferling, Y. Jading, I. Gódor, G. Auer, and L. Van der Perre. Challenges and enabling technologies for energy aware mobile radio networks. *IEEE on Communications Magazine*, 48(11):66–72, 2010.

[CZHK09] T. Chen, H. Zhang, M. Höyhtyä, and M. D. Katz. Spectrum self-coexistence in cognitive wireless access networks. In *Global Telecommunications Conference, 2009. GLOBECOM 2009. IEEE*, pp. 1–6. IEEE, 2009.

[Czi07] Nicolai Czink. The Random-Cluster Model — A Stochastic MIMO Channel Model for Broadband Wireless Communication Systems of the 3rd Generation and Beyond. PhD thesis, Telecommunication Research Center Vienna (FTW), FTW ForschungszentrumTelekommunikation Wien GmbH, Tech Gate Vienna Donau-City-Strae 1/3. Stock, 1220 WIEN, ÖSTERREICH, Tel: +43/1/505 2830-0 F-99, Mail: of-fice@ftw.at., 2007.

D

[dAMC+11] L. M. del Apio, E. Mino, L. Cucala, O. Moreno, I. Berberana, and E. Torrecilla. Energy efficiency and performance in mobile networks deployments with femtocells. In *Proceedings of IEEE 22nd International Symposium on Personal Indoor and Mobile Radio Communications (PIMRC 2011)* – pp. 107–111, September 2011. (Also available as TD(12)03-007).

[DBB14a] G. Dimić, D. Bajić, and M. Beko. Regenerative relay: constraining interference vs. increasing energy- and spectral-efficiency. *COST IC1004 10th MCM, Aalborg, Denmark*, (TD(14)10048), 2014.

[DBB14b] G. Dimić, D. Bajić, and M. Beko. Relay type la in LTE-advanced: can it increase energy efficiency? In *Proceedings of the 31st URSI General Assembly and Scientific Symposium 2014, Beijing, China*, pp. C05.3, August 16–23, 2014.

[DBMP14] S. C. Del Barrio, A. Morris, and G. F. Pedersen. Antenna miniaturization with MEMS tunable capacitors: Techniques and trade-offs. *International Journal of Antennas and Propagation*, pp. 1–8, Aug 2014. (Also available as TD-14-10079).

[DBPFP12] S. C. Del Barrio, M. Pelosi, O. Franek, and G. F. Pedersen. On the currents magnitude of a tunable planar-inverted-f antenna for low-band frequencies. In *2012 6th European Conference on Antennas and Propagation (EUCAP)*, pp. 3173–3176, March 2012. (Also available as TD-11-02026).

[DBVL$^+$11a] J. De Bleser, E. Van Lil, A. Van de Capelle, "A Charge and Current' Formulation of the Electric Field Integral Equation", European Conference on Antennas and Propagation, pp. 1–5, Rome, Italy, 11–15 April 2011. (Also available as TD(11)02061).

[DBVL$^+$11b] J. De Bleser, E. Van Lil, A. Van de Capelle, "A comparison of implementations of a combined charge and current formulation of the method of moments", URSI General Assembly and Scientific Symposium, pp. 1–4, Istanbul, Turkey, 13–20 August 2011. (Also available as TD(12)05024).

[DBVL$^+$11c] J. W. De Bleser, E. Van Lil, and A. Van de Capelle. Extending the MoM charge and current formulation to dielectrics. Technical Report TD(11)02061, Lisbon, Portugal, October 2011.

[DBVL$^+$11d] J. De Bleser, E. Van Lil, A. Van de Capelle, "Stabilizing the method of moments for dielectrics using a combined charge and current formulation of the EFIE", Progress in Electromagnetics Research Symposium, pp. 450–454, Suzhou, China, September 12–16, 2011. (Also available as TD(12)03059).

[DBVLVdC12a] J. W. De Bleser, E. Van Lil, and A. Van de Capelle. Stable method of moment simulations of complex objects. Technical Report TD(12)03059, Barcelona, Spain, February 2012.

[DBVLVdC12b] J. W. De Bleser, E. Van Lil, and A. Van de Capelle. Verification of the charge and current MoM for mixed PEC and dielectrics. Technical Report TD(12)05024, Bristol, UK, September 2012.

[DBVLVdC13] J. W. De Bleser, E. Van Lil, and A. Van de Capelle. Decoupling the charge and current equations in a stabilized mom formulation. Technical Report TD(13)06071, Malaga, Spain, February 2013.

[DBVLVdCl4] J. W. De Bleser, E. Van Lil, and A. Van de Capelle. Split formulation of the charge and current integral equations for arbitrarily shaped dielectrics. *IEEE Transaction on Antennas Propagation*, 62(1):302–310, 2014.

[DDDLD08] O. Delangre, P. De Doncker, M. Lienard, and P. Degauque. Delay spread and coherence bandwidth in reverberation chamber. *Electronics Letters*, 44(5):328–329, 2008.

[DDGW03] D. Didascalou, M. Dottling, N. Geng, and W. Wiesbeck. An approach to include stochastic rough surface scattering into deterministic ray optical wave propagation modeling. *IEEE Transactions on Antennas Propagation*, 51(7):1508–1515, 2003.

[DEFEF07] V. Degli-Esposti, F. Fuschini, E. M. Vitucci, and G. Falciasecca. Measurement and modelling of scattering from buildings. *IEEE Transaction on Antennas Propagation*, 55(1):143–153, 2007.

[DEFV+07] V. Degli-Esposti, F. Fuschini, E. M. Vitucci, and G. Falciasecca. Measurement and Modelling of Scattering From Buildings. *IEEE Transactions on Antennas and Propagation*, 55(1):143–153, 2007.

[DEFV+15] V. Degli-Esposti, F. Fuschini, E. M. Vitucci, M. Barbiroli, M. Zoli, D. Dupleich, R. Müller, C. Schneider, and R. S. Thomä. *Polarimetric* analysis of mm-wave propagation for advanced beamforming applications. In *Proceedings of the 9th European Conference on Antennas and Propagation* (EuCAP 2015), Lisbon, Portugal, 12–17 April 2015.

[DEGd+04] V. Degli-Esposti, D. Guiducci, A. de'Marsi, P. Azzi, and F. Fuschini. An advanced field prediction model including diffuse scattering. *IEEE Transactions on Antennas and Propagation*, 52(7):1717–1728, July 2004.

[Dey66] J. Deygout. Multiple knife-edge diffraction of microwaves. IEEE Transactions on Antennas and Propagation, 14: 480–489, 1966.

[DEY09] T. Dissanayake, K. P. Esselle, and M. R. Yuce. Dielectric loaded impedance matching for wideband implanted antenna. *IEEE Transactions Microwave Theory*, 57(10):2480–2487, October 2009.

[DFM+14] D. Dupleich, F. Fuschini, R. Müller, E. M. Vitucci, C. Schneider, V. Degli-Esposti, and R. Thomä. Directional characterization of the 60 GHz indoor-office channel. In *Proceedings of the 31th URSI General Assembly and Scientific Symposium*, Beijing, China, August 2014.

[DFPS09] A. Derneryd, J. Friden, P. Persson, and A. Stjernman. Performance of closely spaced multiple antennas for terminal applications. In *3rd European Conference on Antennas and Propagation (EuCAP)*, pp. 1612–1616, March 2009.

[DFT13] G. Dahman, Jose Flordelis, and Fredrik Tufvesson. On the cross-correlation properties of large-scale fading in distributed antenna systems. In COST IC1004 Technical Report TD-08-045, September 2013.

[DFT14a] G. Dahman, J. Flordelis, and F. Tufvesson. Analysis of multi-link large-scale parameters in suburban micro-cellular environment (COST IC1004) Technical Report TD-09-059, May 2014.

[DFT14b] G. Dahman and J. Flordelis and F. Tufvesson. *Distributed Antenna Systems: Synchronous Measurements versus Repeated Measurements.* 2014 February, Ferrara, Italy, TD-09-059.

[DFT15] G. Dahman, J. Flordelis, and F. Tufvesson. Experimental evaluation of the effect of BS antenna inter-element spacing on MU-MIMO separation. In *2015 IEEE International Conference on Communications (ICC)*, pp. 1685–1690, 2015. doi: 10.1109/ICC.2015.7248567. (Also available as TD-12-074).

[DGC13] A. Navarro, D. Guevara, J. lopez, and N. Cardona. A discussion about measurement-based ray-tracing models calibration in urban environments. In *IC1004 6th MC and Scientific Meeting*, Malaga, 2013.

[DGCG14] V. Dimanche, A. Goupil, L. Clavier, and G. Gellé. On detection method for soft iterative decoding in the presence of impulsive interference. *IEEE Communic Letter*, 18(6):945–948, 2014.

[Die13] J. Diez. Cooperative cognitive radio performance in indoor environments. In *8th Meeting of the Management Committee of COST IC1004 – Cooperative Radio Communications for Green Smart Environments*, Technical Report TD-13-08037, Ghent, Belgium, September 2013.

[DJL$^+$13] M. Deruyck, W. Joseph, B. Lannoo, D. Colle, and L. Martens. Designing energy-efficient wireless access networks: Lte and lte-advanced. *IEEE Internet Computing*, 17(5):39–45, 2013.

[DJT$^+$13] M. Deruyck, W. Joseph, E. Tanghe, D. Plets, and L. Martens. Designing green wireless access networks: optimising towards power consumption versus exposure of human beings. *COST IC1004 8th Scientific Meeting*, pp. 1–5, September 2013.

[DJTM14] M. Deruyck, W. Joseph, E. Tanghe, and L. Martens. Reducing the power consumption in lte-advanced wireless access networks by a capacity based deployment tool. *Radio Science*, 49(9):777–787, 2014.

[DLR14] Deutsches Zentrum fürLuft-und Raumfahrt (DLR). Institute of Transportation Systems – SUM F. Kaltenberger, A. Byiringiro, G. Arvanitakis, R. Ghaddab, D. Nussbaum, R. Knopp, M. Bernineau, Y. Cocheril, Yann, and H. Philippe. Broadband wireless channel measurements for high speed trains. ICC 2015, IEEE International Conference on Communications, 8–12 June 2015, London, United Kingdom. (Also available as TD(14)11017) O – Simulation of Urban MObility, December 2014.

[DMW+11] A. Damnjanovic, J. Montojo, Y. Wei, T. Ji, T. Luo, M. Vajapeyam, T. Yoo, O. Song, and D. Malladi. A survey on 3gpp heterogeneous networks. *IEEE Wireless Communications*, 18(3):10–21, June 2011.

[DRM11] R. D'Errico, R. Rosini, and M. Maman. A performance evaluation of cooperative schemes for on-body area networks based on measured time-variant channels. In *Proceeding of IEEE International Conference on Communications (ICC 2011)*, pp. 1–5, Kyoto, Japan, June 2011.

[Dru07] S. Drude. Requirements and application scenarios for body area networks. In *Mobile and Wireless Communications Summit, 2007. 16th IST*, pp. 1–5, July 2007.

[DRW+09] K. Doppler, M. Rinne, C. Wijting, C. B. Ribeiro, and K. Hugl. Device-to-device communication as an underlay to lte-advanced networks. *IEEE Communications Magazine*, 47(12):42–49, 2009.

[DRZ15] G. Dahman, F. Rusek, M. Zhu, and F. Tufvesson. On the Probability of Non-Shared Multipath Clusters in Cellular Networks *IEEE Wireless Communications Letters,* 2015, 4(2), 161–164, doi: 10.1109/LWC.2015.2388531

[DS10] M. Duarte and A. Sabharwal. Full-duplex wireless communications using off-the-shelf radios: Feasibility and first results. In *44th Asilomar Conference on Signals, Systems and Computers (ASILOMAR)*, pp. 1558–1562. IEEE, 2010.

[DSSW11] A. Derneryd, A. Stjernman, H. Strandberg, and B. Wastberg. Multi-objective optimization of MIMO antennas for dual-band user devices. In *IEEE International Symposium on Antennas and Propagation (APSURSI)*, pp. 2449–2452, July 2011.

[DTJM12] M. Deruyck, E. Tanghe, W. Joseph, and L. Martens. Energy efficiency of 802.11n versus LTE advanced femtocell networks based on a 3D deployment tool. *COST IC1004 4th Scientific Meeting*, pp. 1–7, Technical Report TD(12)04-015, Lyon, France, May 2012.

[DTP+15] M. Deruyck, E. Tanghe, D. Plets, W. Joseph, and L. Martens. Optimising wireless access networks towards both power consumption and electromagnetic exposure of human beings. *COST IC1004 12th Scientific Meeting*, pp. 1–15, January 2015.

[Duf12] I. Dufek. An increase of the communication range and a spatial selectivity of the active UHF RFID technology by using special types of antennas. Technical Report TD-12-05032, Bristol, UK, September 2012.

[DVVL+13] J. De Bleser, E. Van Lil, A. Van de Capelle, "Split Formulation of the Charge and Current Integral Equations for Arbitrarily-Shaped Dielectrics", IEEE Transactions on Antennas and Propagation, vol. AP-61, no. 1, pp. 302–310, *January* 2014. (Also available as TD(13)06071).

[DZB11] G. Dimić, N. Zogović, and D. Bajić. Energy efficiency maximization by power control and packet length adaptation under resource constraints in wireless (sensor) networks. *COST IC1004 1st scientific meeting*, pp. 1–24, June 2011.

[DZB12a] G. Dimić, N. Zogović, D. Bajić, "Energy Efficiency of Supportive Relay with a Novel Wireless Transmitter Power Consumption Model," International Conference on Selected Topics in Mobile & Wireless Networking – iCOST'2012, Avignon, France, 3–4 July 2012. (Also available as TD(12)04032).

[DZB12b] G. Dimić, N. Zogović, and D. Bajić. A wireless transceiver power consumption model and two-hop vs. single-hop energy efficiency ratio. *Future Network and Mobile Summit 2012, Berlin, Germany*, pp. 1–8, July 4–6, 2012.

[DZB13] G. Dimić, N. Zogović, and D. Bajić. Supportive relay with heterogeneous transceivers: quantification of energy efficiency improvement. *IEEE ICC 2013 – Energy Efficiency in Wireless Networks and Wireless Networks for Energy Efficiency Workshop, Budapest, Hungary*, pp. 458–462, June 9, 2013.

E

[EAH12] O. El Ayach and R. W. Heath. Grassmannian differential limited feedback for interference alignment. *IEEE Transactions on Signal Processing*, 60(12):6481–6494, 2012.

[EKM12] L. Ekiz, O. Klemp, and C. F. Mecklenbruker. Performance Evaluation of Automotive Qualified LTE MIMO Antennas. In *COST IC1004 5th MC & Scientific Meeting*, Technical Report TD(12)05071, Bristol, United Kingdom, September 2012.

[EM02] Y. Ephraim and N. Merhav. Hidden markov processes. *IEEE Transactions on Information Theory*, 48(6):1518–1569, June 2002.

[Eng03] A. Engelhart. *Vector detection with moderate complexity.* PhD thesis, University of Ulm, Institute of Information Technology, Ulm, Germany, 2003.

[EPKM13] L. Ekiz, T. Patelczyk, O. Klemp, and C. F. Mecklenbräuker. Compensation of vehicle-specific antenna radome effects at 5.9 GHz. In *IEEE Annual Conference of the Industrial Electronics Society (IECON)*, pp. 6880–6884, November 2013. (Also available as TD(13)07046).

[EPKM14] L. Ekiz, A. Posselt, O. Klemp, and C. F. Mecklenbräuker. Assessment of design methodologies for vehicular 802.11p antenna systems. In *COST IC1004 12th MC & Scientific Meeting*, number TD(14)10046, May 2014.

[EPL+13] L. Ekiz, A. Posselt, B. K. Lau, O. Klemp, and C. F. Mecklenbräuker. System level assessment of vehicular MIMO antennas in 4G LTE live networks. In *COST IC1004 8th MC & Scientific Meeting*, number TD(13)08029, Ghent, Belgium, September 2013.

[ETJ13] M. Lienard L. Martens E. Tanghe, D. Gaillot and W. Joseph. Experimental analysis of dense multipath components in an industrial environment. In *COST IC1004 7th* Scientific *Meeting*, pp. 1–11, Ilmenau, Germany, May 2013.

[ETJ14] M. Lienard L. Martens E. Tanghe, D. P. Gaillot and W. Joseph. Experimental analysis of dense multipath components in an industrial environment. *IEEE Transactions on Antennas and Propagation*, 62(7):3797–3805, 2014.

[ETL+02] A. Engelhart, W. G. Teich, J. Lindner, G. Jeney, S. Imre, and L. Pap. A survey of multiuser/multisubchannel detection schemes based on recurrent neural networks. *Wireless*

Communications and Mobile Computing, Special Issue on Advances in 3G Wireless Networks, 2(3):269–284, 2002.

[EVB15] E. M. Vitucci, V. Degli-Esposti, F. Fuschini, M. Barbiroli, J. N. Wu et al., "Ray Tracing RF Field Prediction: an Unforgiving Validation", International Journal of Antennas and Propagation, Vol. 2015.

[EVC11] E. M. Vitucci, F. Mani, V. Degli-Esposti, C. Oestges, "Polarimetric Properties of Diffuse Scattering from Building Walls: Experimental Parameterization of a Ray-Tracing Model", IEEE Transactions on Antennas and Propagation, Vol. 60 No. 6, pp. 2961–2969, June 2012.

[EVE14] E. M. Vitucci, V. Degli-Esposti, and F. Fuschini. Ray tracing simulation of the radio channel time and angle-dispersion in large indoor environments. In *Proceedings of the 8th European Conference on Antennas and Propagation,* (EuCAP 2014), The Hague, The Netherlands, 6–11 April 2014.

[Eyr30] C. F. Eyring. Reverberation time in 'dead' rooms. *Journal of the Acoustical Society of America*, 1(2A):217–241, 1930.

F

[Fallgren2015] M. Fallgren, B. Timus, Eds. Future radio access scenarios, requirements and KPIs, Deliverable D1.1, V1.0, ICT-317669, METIS project, 1st May 2013.

[FB12] D. Fang and A. Burr., Rotationally Invariant Coded Modulation for physical layer network coding in two-way relay fading channel. European Wireless, 2012. EW. 18th European Wireless Conference, vol., no., pp. 1, 6, 18–20 April 2012. (Also available as TD(12)05065).

[FC12] L. S. Ferreira and L. M. Correia. Radio Resource Management in Self-Organised Multi-Radio Wireless Mesh Networks. Technical Report TD-12-03012, Barcelona, Spain, February 2012.

[FC13] L. S. Ferreira and L. M. Correia. Efficient and Fair Radio Resources Allocation for Spontaneous Multi-Radio Wireless Mesh Networks. Technical Report TD-13-08016, Ghent, Belgium, September 2013.

[FCnA+13] W. Fan, X. Carreño, J. S. Ashta, J. Ø. Nielsen, G. F. Pedersen, and M. B. Knudsen. Channel verification results for the SCME models in a multi-probe based MIMO OTA setup. In *Vehicular Technology Conference (VTC Fall), 2013 IEEE 78th*, pp. 1–5, September 2013.

[FCnN+12] W. Fan, X. Carreño, J. Ø. Nielsen, K. Olesen, M. B. Knudsen, and G. F. Pedersen. Measurement verification of plane wave synthesis technique based on multi-probe MIMO-OTA setup. In *Vehicular Technology Conference (VTC Fall), 2012 IEEE*, pp. 1–5, September 2012.

[FCnN+13] W. Fan, X. Carreño, J. Ø. Nielsen, J. S. Ashta, G. F. Pedersen, and M. B. Knudsen. Verification of emulated channels in multi-probe based MIMO OTA testing setup. In *Antennas and Propagation (EuCAP), 2013 7th European Conference on*, pp. 97–101, April 2013.

[FdCP14] E. Foroozanfard, E. de Carvalho, and G. F. Pedersen. Antenna isolation technique for a MIMO full-duplex system. Technical Report TD-14-10087, Aalborg, Denmark, May 2014.

[Fen13] Y. Feng. *Over-the-Air (OTA) Measurement Method for MIMO-enabled Mobile Terminals*. PhD thesis, Universität Duisburg-Essen, 2013.

[FFK02] H.-B. Fang, K.-T. Fang, and S. Kotz. The metaelliptical distributions with given marginals. *Journal of Multivariate Analysis*, 82(1):1–16, 2002.

[FGD+15] J. Flordelis, X. Gao, G. Dahman, F. Rusek, O. Edfors, and F. Tufvesson. Spatial separation of closely-space users in measured massive multi-user MIMO channels. In *Proceedings of the ICC 2015 – IEEE International Conference Communication*, pp. 1441–1446, June 2015. (Also available as TD(15)12089).

[FGGP+14a] M. Fuentes, E. Garro, C. Garcia-Pardo, D. Gomez-Barquero, and N. Cardona. Coexistence of LTE and DTT Services in the First and Second Digital Dividends. Planning Studies and Potential Approaches. Technical Report TD(14)10034, Action COST IC1004, Cooperative Radio Communications for Green Smart Environments, Aalborg, Denmark, May 2014.

[FGGP+14b] M. Fuentes, E. Garro, C. Garcia-Pardo, D. Gomez-Barquero, and N. Cardona. Coexistence of LTE and DVB Technologies Digital Dividend. A Contribution to Transmission Parameters Optimization. Technical Report TD(14)09042, Action COST IC1004, Cooperative Radio Communications for Green Smart Environments, Ferrara, Italy, February 2014.

[FJS11] Y. Feng, J. Jonas, and W. L. Schroeder. Discussion of statistical metrics for MIMO OTA performance based on empirical results. In *5th European Conference on Antennas and Propagation (EuCAP 2011)*, Rome, Italy, April 2011.

[FK13] R. Fabian and P. Kulakowski. Geographic routing in realistic radio propagation conditions. Technical Report TD-13-06016, Málaga, Spain, February 2013.

[FKH+14] W. Fan, P. Kyösti, L. Hentilä, J. Ø. Nielsen, and G. F. Pedersen. Rician channel modeling for multi-probe anechoic chamber setups. *IEEE Antennas and Wireless Propagation Letters*, (99):1761–1764, 2014.

[FKNPed] W. Fan, P. Kyösti, J. Ø. Nielsen, and G. F. Pedersen. Wideband MIMO channel capacity analysis in multi-probe anechoic chamber setups. *IEEE Transactions on Antennas and Propagation*, To be published.

[Flore] D. Flore. Getting ready for '5G', 3GPP RP-150056, Shanghai, China, March 9–12, 2015.

[FLZ+10] S. Fu, K. Lu, T. Zhang, Y. Qian, and H.-H. Chen. Cooperative wireless networks based on physical layer network coding. *IEEE Wireless Communication*, 17(6):86–95, 2010.

[FMV+13] M. Fadda, M. Murroni, V. Popescu, J. Morgade, R. Sancho, P. Angueira, and M. Alex. Protection ratios of broadcast services in the presence of cognitive transmissions. Technical Report TD(13)07019, Action COST IC1004, Cooperative Radio Communications for Green Smart Environments, Ilmenau, Germany, February 2013.

[FMW12] M. Froehle, P. Meissner, and K. Witrisal. Tracking of UWB multipath components using probability hypothesis density filters. In *Proc. ICUWB 2012 – IEEE International Conference on Ultra-Wideband*, Syracuse, NY, September 2012. (Also available as TD(12)04003).

[FNCn+12] W. Fan, J. Ø. Nielsen, X. Carreño, O. Franek, M. B. Knudsen, and G. F. Pedersen. Impact of probe placement error on MIMO OTA test zone performance. In *Antennas and Propagation Conference (LAPC), 2012 Loughborough*, pp. 1–4, November 2012.

[FNCn+13] W. Fan, J. Ø. Nielsen, X. Carreño, J. S. Ashta, G. F. Pedersen, and M. B. Knudsen. Impact of system non-idealities on spatial correlation emulation in a multi-probe based MIMO OTA setup. In *Antennas and Propagation (EuCAP), 2013 7th European Conference on*, pp. 1663–1667, April 2013.

[FNF+13] W. Fan, J. Ø. Nielsen, O. Franek, X. Carreño, J. S. Ashta, M. B. Knudsen, and G. F. Pedersen. Antenna pattern impact on MIMO OTA testing. *IEEE Transactions on Antennas and Propagation*, 61(11):5714–5723, 2013.

[FNKS02] A. R. Forouzan, M. Nasiri-Kenari, and J. A. Salehi. Performance analysis of time hopping spread-spectrum multiple access system: uncoded and coded schemes. *IEEE Transaction Wireless Communication*, 1(4):671–681, 2002.

[FNP14] W. Fan, J. Ø. Nielsen, and G. F. Pedersen. Estimating power angular spectra in multi-probe setups for terminal testing. In 2014 *IEEE Antennas and Propagation Society International Symposium (APSURSI)*, pp. 707–708, July 2014.

[Fol45] L. L. Foldy. The multiple scattering of waves. i. general theory of isotropic scattering by randomly distributed scatterers. *Physical Review*, 67(3–4):107–119, 1945.

[FPH+14] L. S. Ferreira, D. Pichon, A. Hatefi, A. Gomes, D. Dimitrova, T. Braun, G. Karagiannis, M. Karimzadeh, M. Branco, and L. M. Correia. An architecture to offer cloud-based radio access network as a service. In *EUCNC – European Conference on Networks and Communications*, Bologna, Italy, June 2014. (Also available as TD(14) 09003).

[FS13] J. A. Fernández Segovia. Self-Planning of PUSCH Parameters on LTE Networks. Technical Report TD-13-06063, Málaga, Spain, February 2013.

[FSA+11] Y. Feng, W. L. Schroeder, R. Acharkaoui, W. Richter, A. Tankielun, and C. von Gagern. Straightforward MIMO OTA characterization and statistical metrics: Applied to LTE devices. Technical Report TD(11)02064, 2nd COST IC1004 Scientific Meeting, Lisbon, Portugal, October 2011.

[FSF+14] W. Fan, I. Szini, M. D. Foegelle, J. Ø. Nielsen, and G. F. Pedersen. Measurement uncertainty investigation in the multi-probe OTA setups. In *2014 8th European Conference on Antennas and Propagation (EuCAP)*, pp. 1068–1072, April 2014.

[FSHB12] F. Fornetti, K. Stevens, G. Hilton, and M. Beach. Design and performance characterisation of tuneable magnetic loop antennas. Technical Report TD-12-04023, Lyon, France, May 2012.

[FSK12] Y. Feng, W. L. Schroeder, and T. Kaiser. Straightforward MIMO OTA characterization and statistical metrics for

LTE devices. In *6th European Conference on Antennas and Propagation (EuCAP 2012)*, Prague, Czech Republic, March 2012.

[FSNP13] W. Fan, I. Szini, J. Ø. Nielsen, and G. F. Pedersen. Channel spatial correlation reconstruction in flexible multiprobe setups. *IEEE Antennas and Wireless Propagation Letters*, 12:1724–1727, 2013.

[FSvG+12] Y. Feng, W. L. Schroeder, C. von Gagern, A. Tankielun, and T. Kaiser. Metrics and methods for evaluation of over-the-air performance of MIMO user equipment. *IJAP*, 2012. doi:10.1155/2012/598620.

[FTF+14] E. Foroozanfard, E. Tsakalaki, O. Franek, E. de Carvalho, and G. F. Pedersen. Impact of antenna cancellation on a MIMO full-duplex system. Technical Report TD-14-09060, Ferrara, Italy, February 2014.

[FTH+99] B. H. Fleury, M. Tschudin, R. Heddergott, D. Dahlhaus, and K. I. Pedersen. Channel Parameter Estimation in Mobile Radio Environments Using the SAGE Algorithm. *IEEE Journal on Selected Areas in Communications*, 17(3):434–450, 1999.

[FV12] J. M. Ferro and F. J. Velez. Cost-based optimisation of the routing for wireless sensor networks in the presence of mobility. *COST IC1004 3rd Scientific Meeting*, February 2012.

G

[Gao16] X. Gao. Massive MIMO in real propagation environments. 2016, Lund University, Sweden. http://lup.lub.lu.se/ record/8567483/file/8569798.pdf

[GASM12] P. K. Gentner, A. Adalan, A. L. Scholtz, and C. F. Mecklenbräuker. Impact analysis of silicon and bondwires on an on-chip antenna. In *European Conference on Antennas and Propagation*, pp. 3168–3172, Prague, Czech Republic, March 2012. (Also available as TD-11-02058).

[GB14] J. Grosinger and W. Bosch. Measurement-based statistical evaluation of on-body backscatter rfid systems. Technical Report TD-14-10058, Ferrara, Italy, February 2014.

[GB94] R. Gahleitner and E. Bonek. Radio wave penetration into urban buildings in small cells and micro cells. In *Proceedings of the IEEE Vehicular Technology Conference (VTC1994)*, pp. 887–891, Budapest, Stockholm, May 1994.

[GBA12] R. K. Ganti, F. Baccelli, and J. G. Andrews. Series expansion for interference in wireless networks. *IEEE Transactions on Information Theory*, 58(4):2194–2205, 2012.

[GBS08] A. A. Goulianos, T. W. C. Brown, and S. Stavrou. A novel path-loss model for UWB off-body propagation. In *Proceeding of VTC 2008 Spring – IEEE 67th Vehicular Technology Conference*, pp. 450–454, Marina Bay, Singapore, 2008.

[GC11] T. A. Correia Goncalves and L. M. Correia. Energy efficient solutions based on beamforming for umts and lte. *COST IC1004 2nd Scientific Meeting*, pp. 1–9, October 2011.

[GC12a] T. A. C. Goncalves and L. Correia. Energy efficient solutions based on beamforming for UMTS and LTE. *COST IC1004 2nd MCM, Lisbon, Portugal*, (TD(11)02074), 2012.

[GC12b] Wei Gu and L. Clavier. Decoding metric study for turbo codes in very impulsive environment. *IEEE Communications Letters*, 16(2):256–258, 2012.

[GCA+10] H. E. Ghannudi, L. Clavier, N. Azzaoui, F. Septier, and P.-A. Rolland. α-Stable interference modeling and cauchy receiver for an ir-UWB *ad hoc* network. *IEEE Transactions on Communication*, 58:1748–1757, 2010.

[GCS14] S. Gherekhloo, A. Chaaban, and A. Sezgin. The generalized degrees of freedom of the interference relay channel with strong interference. 51st Annual Allerton Conference on Communication, Control, and Computing (Allerton), Oct 2013. (Also available as TD(14)09030).

[GDK06] N. Guney, H. Deliç, and M. Koca. Robust detection of ultrawideband signals in non-gaussian noise. *IEEE Transactions on Microwave Theory and Techniques*, 54(4): 1724–1730, 2006.

[GE11] N. Gvozdenovic and M. Eric. Localization of users in multiuser MB OFDM UWB systems based on TDOA principle. In *2nd COST IC1004 Management Committee Meeting*, Lisbon, Portugal, October 2011 (TD(11)02057).

[GEAT10] K. Gulati, B. L. Evans, J. G. Andrews, and K. R. Tinsley. Statistics of co-channel interference in a field of Poisson and Poisson–Poisson clustered interferers. *IEEE Transactions on Signal Processing*, 58(12):6207–6222, 2010.

[GERT11] X. Gao, O. Edfors, F. Rusek, and F. Tufvesson. Linear pre-coding performance in measured very-large MIMO channels. In *Proceedings of the Vehicular Technology Conference (VTC Fall) 2011 IEEE*, pp. 1–5. IEEE, 2011. (Also available as TD(10)01047).

[GES12] K. Gulati, B. L. Evans, and S. Srikanteswara. Joint temporal statistics of interference in decentralized wireless networks. *IEEE Transactions on Signal Processing*, 60(12):6713–6718, 2012.

[GGGBC12] J. J. Giménez, D. Gozálvez, D. Gómez-Barquero, and N. Cardona. Statistical model of signal strength imbalance between RF channels in DTT network. Electronics Letters, 48(12):731–732, 2012. (Also available as TD(04)049).

[GGM11] N. Gvozdenovic, P. K. Gentner, and C. Mecklenbräuker. Antenna array for the reader of an ultra-wideband identification tag with on-chip antenna. In 2011 Loughborogh *Antennas* & Propagation Conference, Loughborough, United Kingdom, November 2011. (Also available at TD(11)01032).

[GGM12] N. Gvozdenovic, P. K. Gentner, and C. Mecklenbräuker. Near-field channel measurements for UWB RFID with on-chip antenna. In iWAT 2012, pages 28–31, Tucson, Arizona, USA, March 2012. (Also available at TD(11)01032).

[GGRL13] D. Gonzalez G, M. Garcia-Lozano, S. Ruiz Boque, and Dong Seop Lee. Optimization of soft frequency reuse for irregular LTE macrocellular networks. *IEEE Transactions on Wireless Communications*, 12(5):2410–2423, 2013.

[GGRO11] D. González G., M. Garcia-Lozano, Ruiz S., and Olmos J. Impact of Downlink Signalling Capacity Constraints on the Provision of QoS in LTE. Technical Report TD-11-02041, Lisbon, Portugal, October 2011.

[GGRO12a] D. G. González, M. Garcia-Lozano, S. Ruiz, and J. Olmos. Novel ICIC-Oriented Channel State Information Feedback Scheme for Aperiodic Reporting in LTE. Technical Report TD-12-03038, Barcelona, Spain, February 2012.

[GGRO12b] D. G. Gonzlez, M. Garcia-Lozano, S. Ruiz, and J. Olmos. On the Need for Dynamic Downlink Intercell Interference Coordination for Realistic LTE Deployments. *Wireless Communications and Mobile Computing*, 2012. John Wiley & Sons, Ltd, New York doi: 10.1002/wcm.2191.

[GHP+12] H. Giddens, G. Hilto, D. Paul, and J. McGeehan. Electricall small, wearable, magnetically coupled loop antenna for on-body communications. Technical Report TD(12)05070, Bristol, UK, September 2012.

[GHSM12a] P. K. Gentner, G. Hofer, A. L. Scholtz, and C. F. Mecklenbräuker. Accurate Measurement of Power Transfer to an RFID Tag with On-Chip Antenna. In *Progress In Electromagnetics Research Symposium*, pp. 227–230, Moscow, Russia, 2012.

[GHSM12b] P. K. Gentner, G. Hofer, A. L. Scholtz, and C. F. Mecklenbräuker. A contact-less evaluation method on the powering of an active RFID tag with on-chip antenna. Technical Report TD-12-03054, Barcelona, Spain, 2012.

[GHWT14] C. Gustafson, K. Haneda, S. Wyne, and F. Tufvesson. On mm-wave multipath clustering and channel modeling. *IEEE Transactions on Antennas and Propagation*, 62(3):1445–1455, March 2014. (Also available as TD(13)06062).

[GJRC12] J. J. Giménez, T. Jansen, D. Rose, and N. Cardona. Indoor-to-outdoor measurement campaign for the development of a propagation model for femtocell environments. In *IC1004*, Bristol, UK, September 2012 (TD(14)05063).

[GLG11] V. Garcia, N. Lebedev, and J. Gorce. Model Predictive Control for Smooth Distributed Power Adaptation. Technical Report TD-11-02029, Lisbon, Portugal, October 2011.

[GLS+13] P. K. Gentner, R. Langwieser, A. L. Scholtz, G. Hofer, and C. F. Mecklenbräuker. A UHF/UWB hybrid silicon RFID tag with on-chip antennas. *EURASIP Journal on Embedded Systems*, 12:12, 2013. (Also available as TD-12-05073).

[GM04] S. Goh and D. Mandic. A complex-valued rtrl algorithm forrecurrent neural networks. *Neural Computation*, 16(12):2699–2713, 2004.

[GM11] P. K. Gentner and C. F. Mecklenbräuker. Analysis of the influence of silicon and bondwires to an on-chip antenna. Technical Report TD-11-02058, Lisbon, Portugal, October 2011.

[GMK+13] M. Gan, F. Mani, F. Kaltenberger, C. Oestges and T. Zemen, “A ray tracing algorithm using the discrete prolate spheroidal subspace”, in IEEE International Conference on Communications (ICC), Budapest, Hungary, June 2013. (Also available as TD(13)07020).

[GMM14] N. Gvozdenovic, L. W. Mayer, and C. F. Mecklenbräuker. Measurement of harmonic distortions and impedance of HF RFID chips. In 8th European Conference on Antennas and Propagation, euCap 2014, 2014. (Also available as TD(14)10005).

[GMP+13] N. Gvozdenovic, L. W. Mayer, R. Prestros, C. F. Mecklenbräuker, and A. L. Scholtz. PEEC modeling of circular spiral coils. In European Microwave Conference, 2013. (Also available as TD(13)06038).

[GPC+12] W. Gu, G. Peters, L. Clavier, F. Septier, and I. Nevat. Receiver study for cooperative communications in convolved additive alpha-stable interference plus gaussian thermal noise. In *Proceedings of 9th International Symposium on Wireless Communication Systems, ISWCS 2012*, Paris, France, August 2012.

[GPM14] N. Gvozdenovic, R. Prestros, and C. Mecklenbräuker. HF RFID spiral inductor synthesis and optimization. In 2014 International Symposium on Wireless Personal Multimedia Communications (WPMC), Sydney, Australia, September 2014. (Also available as TD(15)12008).

[Gro13] P. Große. A hybrid channel model based on WINNER for vehicle-to-x application. In *COST IC1004 7th MC & Scientific Meeting*, number TD(13)07040, 2013.

[GSM13] P. K. Gentner, A. L. Scholtz, and C. F. Mecklenbräuker. Compact antennas for UHF-actuated UWB-RFID tags. Technical Report TD-13-07050, Ilmenau, Germany, May 2013.

[GSML+14] M. Gan, G. Steinböck, P. Meissner, E. Leitinger, T. Zemen, K. Witrisal, and T. Pedersen. Experimental investigation of the characteristics of the electromagnetic reverberation in the UWB bands. In *COST IC1004 10th Scientific Meeting*, pp. 1–6, Aalborg, Denmark, May 2014.

[GT15] C. Gustafson and F. Tufvesson. Polarimetric propagation channel characterization at 60 ghz with realistic shadowing. In *COSTIC1004, Dublin, Ireland*, 2015.

[GTB12] N. Gvozdenovic, W. Thompson, and M Beach. Short range ultra-wideband multiple input multiple output channel measurements, WCNC 2013, Shanghai, China, April 2013. (Also available as TD(12)04011).

[GTB+12] N. Gvozdenovic, W. Thompson, M. A. Beach, C. Mecklenbräuker, G. Hilton. Short range ultra-wideband multiple input multiple output channel measurements. In WCNC 2013, Shanghai, China, April 2013. (Also available as TD(12)04011 and TD(12)05018).

[GTER12] X. Gao, F. Tufvesson, O. Edfors, and F. Rusek. Channel behavior for very-large MIMO systems — initial characterization. In COST IC1004, September 2012.

[GTER13] X. Gao, F. Tufvesson, O. Edfors, and F. Rusek. Measured propagation characteristics for very-large MIMO at 2.6 GHz. In *Proceedings of the ASILOMAR 2013 – 46th Conference on Signals, Systems and Computers*, pp. 295–299, November 2013. (Also available as TD(13)06060).

[Gud91] M. Gudmundson. Correlation model for shadow fading in mobile radio systems. *Electronics letters*, 27(23):2145–2146, 1991.

[GvDH11] P. Guo, A. R. van Dommele, and M. H. A. J. Herben. Method to match waves of ray-tracing simulations with 3-D high-resolution propagation measurements, Proceedings of the 6th European Conference on Antennas and Propagation (EuCAP2012), Prague, Czech Republic, 26–30 March 2012, pp. 3351–3355. (Also available as TD(11)02023).

[GWHM11] P. K. Gentner, M. Wiesflecker, G. Hofer, and C. F. Mecklenbräuker. Bandwidth reconfigurable UWB RFID tag with on-chip antenna. In *Loughborough Antennas and Propagation Conference*, pp. 1–3, Loughborough, United Kingdom, November 2011. IEEE. (Also available as TD-11-01035).

[GXH+15] M. Gan, Z. Xu, M. Hofer, G. Steinboeck, and T. Zemen," A sub-band divided ray tracing algorithm using the DPS subspace in UWB indoor scenarios", in IEEE 81st Vehicular Technology Conference (VTC-Spring), Glasgow, Scotland, May 2015. (Also available as TD(15)12022).

[GXH+15] M. Gan, Z. Xu, M. Hofer, G. Steinbock, and T. Zemen. A Sub-band divided ray tracing algorithm using the DPS subspace in UWB indoor scenarios. In *IC1004*, Dublin, Ireland, 2015 (TD(15)12022).

[GYGR14] D. González G, H. Yanikomeroglu, M. Garcia-Lozano, and S. R. Boqué. A novel multiobjective framework for cell switch-off in dense cellular networks. *COST IC1004 8th*, pp. 1–8, February 2014.

[GZ11] A. A. Glazunov and J. Zhang. Some examples of uncorrelated antenna radiation patterns for MIO applications. In *PIERS Proceedings, Marrakesh, MOROCCO*, pp. 598–602, March 2011. (Also available as TD-11-01026).

[GZAK13a] K. Guan, Z. Zhong, B. Ai, and T. Kürner. Deterministic propagation modeling for the realistic high-speed railway environment. In *Proceedings of IEEE VTC'13*, pp. 1–5. IEEE, 2013.

[GZAK13b] K. Guan, Z. Zhong, B. Ai, and T. Kürner. Semi-deterministic path-loss modeling for viaduct and cutting scenarios of high-speed railway. *IEEE Antennas and Wireless Propagation Letters*, 12:789–792, 2013.

[GZAK14] K. Guan, Z. Zhong, B. Ai, and T. Kürner. Propagation prediction for composite scenarios of dense semi-closed obstacles in high-speed railway. In *General Assembly and Scientific Symposium (URSI GASS), 2014 XXXIth URSI*, pp. 1–4, 2014.

[GZT+15] X. Gao, M. Zhu, F. Tufvesson, F. Rusek, and O. Edfors. Extension of the COST 2100 channel model for massive MIMO. In COST IC1004, January 2015.

H

[HA14] M. N. Hasan and K. Anwar. Capacity bound for uncoordinated transmissions on multi-way relay networks. Technical Report TD-11-004, Krakow, Poland, September 2014.

[Had14] Y. Haddad. Topology of Wireless Networks. Technical Report TD-14-10027, Aalborg, Denmark, May 2014.

[Haf90] C. Hafner. *The Generalized Multipole Technique for Computational Electrodynamics.* London: Artech, 1990.

[Han88] J. E. Hansen, editor. *Spherical near-field antenna measurements.* IEE Electromagnetic waves series 26, London, UK, 1988.

[Har93] R. F. Harrington. *Field Computation by Moment Methods.* Wiley-IEEE Press, NJ, 1993.

[Hat80] M. Hata. Empirical formula for propagation loss in land mobile radio services. *IEEE Transaction on Vehicular Technology*, 29(3):317–325, 1980.

[Hay05] S. Haykin. Cognitive radio: brain-empowered wireless communications. *IEEE Journal on Selected Areas in Communications*, 23(2):201–220, 2005.

[HBB11] Z. Hasan, H. Boostanimehr, and V. K Bhargava. Green cellular networks: a survey, some research issues and challenges. *IEEE on Communications Surveys and Tutorials*, 13(4):524–540, 2011.

[HBK+15] M. A. Hein, C. Bornkessel, W. Kotterman, C. Schneider, and K. Rajeshand, F. Wollenschläger, R. S. Thomä, G. Del Galdo, and M. Landmann. Emulation of virtual radio environments for realistic end-to-end testing for intelligent traffic systems, in *IEEE MTT-S International Conference on Microwaves for Intelligent Mobility (ICMIM),* Heidelberg, Germany, 2015. IEEE Xplore digital library, New York, NY: IEEE. http://dx.doi.org/10.1109/ICMIM.2015.7117934 (Also available as TD(15)12037).

[HCB+14] S. Hur, Y. Chang, S. Baek, Y.-J. Lee, and J. Park, mmwave propagation models based on 3D ray-tracing in urban environments. Technical report, TD(14)10054, Aalborg, Denmark, May 2014.

[HCKP14] S. Hur, Y. Cho, T. Kim, and J. Park. Millimeter-wave Channel modeling based on measurements in In-building, campus and Urban Environments at 28 GHz. Technical Report TD(14)11029, Krakow, Poland, September 2014.

[HCL+14] S. Hur, Y. Cho, K. Lee, J. Ko, and J. Park. Millimeter-wave Channel modeling based on Measurements in In-building and Campus Environments at 28 GHz. Technical Report TD(14)10053, Aalborg, Denmark, May 2014.

[HDM+15] S. Häfner, D. Dupleich, R. Müller, C. Schneider, J. Luo, E. Schulz, X. Lu, C. Cao, T. Wang, and R. S. Thomä. Characterisation of channel measurements at 70ghz in indoor femtocells. In *COST IC1004, Dublin, Ireland,* 2015.

[HGW13] K. Haneda, C. Gustafson, and S. Wyne. 60 GHz spatial transmission: multiplexing or beamforming? IEEE Transactions on Antennas and Propagation, 61:5735–5743. (Also available as TD(13)07023).

[HHG11] Z. H. Hu, P. S. Hall, and P. Gardner. Reconfigurable dipole-chassis antennas for small terminal MIMO applications. Electronics Letters, 47(17):953–955, 2011.

[HHS12] M. D. Huang, M. H. A. J. Herben, and P. F. M. Smulders. Causes of discrepancies between measurements and EM-simulations of millimeter-wave antennas, IEEE Antennas and Propagation Magazine, Vol. 55, pp. 139–149, December 2013. (Also available as TD-12-05011).

[Hil09] D. A. Hill. *Electromagnetic fields in cavities: deterministic and statistical theories.* IEEE, Wiley, Piscataway, NJ, 2009

[HJK+14] K. Haneda, J. Järveläinen, A. Karttunen, M. Kyrö, and J. Putkonen. Indoor short-range radio propagation measurements at 60 and 70 GHz. In *Proceedings of 8th European Conference Antennas and Propagation* (EuCAP2014), pp. 1–5, Den Haag, the Netherlands, April 2014. (Also available as TD(14)09063).

[HK14] L. Haring and C. Kisters. Signaling-assisted map-based modulation classification in adaptive MIMO OFDM systems. In *2014 IEEE 80th Vehicular Technology Conference (VTC Fall)*, pp. 1–5, September 2014.

[HKD11] R. Hu, M. Kieffer, and P. Duhamel. Protocol-assisted channel decoding. Technical Report TD-01-037, Lund, Sweden, June 2011.

[HKD+13] K. Haneda, A. Khatun, M. Dashti, T. A. Laitinen, V. Kolmonen, J. Takada, and P. Vainikainen. Measurement-based analysis of spatial degrees of freedom in multipath propagation channels. *IEEE Transactions on Antennas and Propagation*, 61(2):890–900, 2013. (Also available as TD (13)06056).

[HKGW13] K. Haneda, A. Khatun, C. Gustafson, and S. Wyne. Spatial degrees-of-freedom of 60 GHz multiple-antenna channels. In *77th Vehicular Technical Conference (VTC 2013-Spring)*, pp. 1–5, June 2013. (Also available as TD(13)06059).

[HM71] R. F. Harrington and J. R. Mautz. Computation of characteristic modes for conducting bodies. *IEEE Transactions on Antennas Propagation*, 19(5):629–639, Sep 1971.

[HM72] R. F. Harrington and J. R. Mautz. Control of radar scattering by reactive loading. *IEEE Transactions on Antennas Propagation*, 20(4):446–454, 1972.

[HMHT13] F. Harrysson, J. Medbo, T. Hult, and F. Tufvesson. Experimental investigation of the directional outdoor-to-in-car propagation channel. *IEEE Transactions on Vehicular Technology*, 62(6):2532–2543, 2013. (Also available as TD(12)05022).

[HMS+10] L. Hanlen, D. Miniutti, D. Smith, D. Rodda, and B. Gilbert. Co-channel interference in body area networks with indoor *measurements* at 2.4 GHz: distance-to-interferer is a poor estimate of received interference power. *International Journal of Wireless Information Networks, Springer*, 17 (3–4):113–125, 2010.

[HRK14] S. Hahn, D. M. Rose, and T. Küner. Mobility load balancing – a case study: simplified vs. realistic scenarios. Technical Report TD(14)10-030, Aalborg, Denmark, May 2014.

[HRK+14] R. He, O. *Renaudin*, V.-M. Kolmonen, K. Haneda, Z. Zhong, B. Ai, and C. Oestges. Statistical characterization of dynamic multi-path components for vehicle-to-vehicle radio channels. In *Proceedings of 81st Vehicular*

Technology Conference (VTC2015-Spring), Glasgow, Scotland, May 2015. (Also available as TD(14)11007).

[HRK+15] R. He, O. Renaudin, V.-M. Kolmonen, K. Haneda, Z. Zhong, B. Ai, and C. Oestges. Characterization of quasi-stationarity *regions* for vehicle-to-vehicle radio channels. *IEEE Transactions on Antennas and Propagation*, 63(5):2237–2251, 2015. (Also available as TD(14)10004).

[HRM12] K. Haneda, A. Richter and A. F. Molisch, "Modeling the frequency dependence of ultrawide-band spatio-temporal radio channels," IEEE Transactions on Antennas and Propagation, vol. 60, no. 6, pp. 2940–2950, June 2012. (Also available as TD(12)03035).

[HRSK15] S. Hahn, D. M. Rose, J. Sulak, and T. Kürner. Impact of Realistic Pedestrian Mobility Modelling in the Context of Mobile Network Simulation Scenarios. IEEE 81th Vehicular Technology Conference (VTC2015-Spring), Glasgow, Scotland, 11–14 May 2015, Technical Report TD(15)12046.

[HS12a] M. Hekrdla and J. Sykora. Lattice-constellation indexing for wireless network coding 2-way relaying with modulo-sum relay decoding. Technical Report TD05057, 2012.

[HS12b] M. Hekrdla and J. Sykora. Suppression of relative-fading by diversity reception in wireless network coding 2-way relaying. Technical Report TD04033, 2012.

[HS12c] T. Hynek and J. Sykora. Wireless network coding initialization procedure through multi-source automatic modulation classification in random connectivity networks. Technical Report TD05041, 2012.

[HS13] T. Hynek and J. Sykora. Potential games for distributed coordination of wireless network coding based cloud. Technical Report TD07031, 2013.

[HS14] M. Hekrdla and J. Sykora. Design of constellations for adaptive minimum-cardinality PLNC 2-way relaying utilising the symmetry of hexagonal lattice. Technical Report TD09025, 2014.

[HT13] S. Häfner and R. S. Thomä. Estimation of radio channel parameters in case of an unknown transmitter. Technical Report TD-07-035, Ilmenau, Germany, May 2013.

[HTK06] K. Haneda, J. I. Takada, and T. Kobayashi. Cluster properties investigated from a series of ultrawideband double directional propagation measurements in home environments. *IEEE Transactions on Antennas Propagation*, 54(12):3778–3788, 2006.

[HTKD11] K. Haneda, K. Takizawa, A. Khatun, and M. Dashti. On attainable spatial diversity gain in multipath channels. In *IC1004*, Lisbon, Portugal, October 2011 (TD(11)02059).

[HTM+14] B. Hanssens, E. Tanghe, L. Martens, W. Joseph, and C. Oestges. Measurement-based analysis of doppler characteristics for ultrawideband radio channels in an office environment. In *IC1004*, Ferrara, Italy, February 2014 (TD(14)09011).

[HWL99] R. Hoppe, G. Wolfle, and F. M. Landstorfer. Measurement of building penetration loss and propagation models for radio transmission into buildings. IEEE 50th Vehicular Technology Conference (VTC Fall 1999), 4:2298–2302, 1999.

[HWZJ12] Y. Huang, W. Wang, X. Zhang, and J. Jiang. Analysis and design of energy efficient traffic transmission scheme based on user convergence behaviour in wireless system. In *2012 IEEE 23rd International Symposium on Personal Indoor and Mobile Radio Communications (PIMRC)*, pp. 815–819, September 2012.

[HXS12] S. Haykin, Y. Xue, and P. Setoodeh. Cognitive radar: Step toward bridging the gap between neuroscience and engineering. *Proceedings of the IEEE*, 100(11):3102–3130, 2012.

[HXZ15] M. Hofer, Z. Xu, and T. Zemen. On the optimum number of hypotheses for adaptive reduced-rank subspace selection. In *COST IC1004 12th MC & Scientific* Meeting, number TD(15)12079, Dublin, Irland, January 2015.

[HYKY03] L. Hanzo, L.-L. Yang, E-L. Kuan, and K. Yen. *Single and multi-carrier DS-CDMA: Multi-user detection, space-time spreading, synchronization, networking and standards.* Wiley-IEEE Press, New Jersey, 2003.

[HZ14] M. Hofer and T. Zemen. Iterative non-stationary channel estimation for LTE downlink communications. In *Proceedings of ICC 2014 – IEEE International Conference on Communication Workshops*, pp. 26–31, June 2014. (Also available as TD(14)10013).

[HZA+13a] R. He, Z. Zhong, B. Ai, J. Ding, Y. Yang, and A. F. Molisch. Short-term fading behavior in highspeed railway cutting scenario: measurements, analysis, and statistical models. *IEEE Transactions on Antennas and Propagation*, 61(4):2209–2222, 2013.

[HZA+13b] R. He, Z. Zhong, B. Ai, G. Wang, J. Ding, and A. F. Molisch. Measurements and analysis of propagation channels in high-speed railway viaducts. *IEEE Transactions on Wireless Communications*, 12(2):794–805, 2013.

[HZA+14] R. He, Z. Zhong, B. Ai, J. Ding, W. Jiang, H. Zhang, and X. Li. A standardized path loss model for the gsm-railway based high-speed railway communication systems. *Proceedings of IEEE VTC14*, pp. 1–5, 2014.

[HZAD11a] R. He, Z. Zhong, B. Ai, and J. Ding. An empirical path loss model and fading analysis for high-speed railway viaduct scenarios. *IEEE Antennas and Wireless Propagation Letters*, 10:808–812, 2011.

[HZAD11b] R. He, Z. Zhong, B. Ai, and J. Ding. Propagation measurements and analysis for high-speed railway cutting scenario. *Electronics letters*, 47(21):1167–1168, 2011.

[HZAM13] X. He, X. Zhou, K. Anwar, and T. Matsumoto. Joint estimation of observation-error-probability and iterative decoding in wireless sensor networks. Technical Report TD-06-085, Malaga, Spain, February 2013.

[HZAG15] R. He, Z. Zhong, B. Ai, and K. Guan. Reducing the cost of the high-speed railway communications: from propagation channel view. *IEEE Transactions on Intelligent Transportation Systems* 2015 (to be published).

[HZAO14] R. He, Z. Zhong, B. Ai, and C. Oestges. A heuristic cross-correlation model of shadow fading in high-speed railway environments. In *Proceedings of General Assembly and Scientific Symposium (URSI GASS)*, pp. 1–4, 2014.

[HZAO15] R. He, Z. Zhong, B. Ai, and C. Oestges. Shadow fading correlation in high-speed railway environments. *IEEE Transactions on Vehicular Technology*, 2015 (to be published).

[HZAZ14] R. He, Z. Zhong, B. Ai, and B. Zhang. Measurement-based auto-correlation model of shadow fading for the highspeed railways in urban, suburban, and rural environments. In *Proceedings of IEEE Antennas and Propagation Society International Symposium (APSURSI)*, pp. 949–950, 2014.

[HZM14] X. He, X. Zhou, and T. Matsumoto. Performance Analysis of a Binary CEO Problem. Technical Report TD-09-038, Ferrara, Italy, February 2014.

I

[IDPD13] M. Ilić-Delibašić and M. Pejanović-Djurisić. Improving energy efficiency of relay systems using dual-polarized antenna. *Journal of Green Engineering*, 3(2):167–179, 2013.

[IEEE06] IEEE standard for information technology – telecommunications and information exchange between systems – local and metropolitan area networks-specific requirements part 15.4: Wireless medium access control (MAC) and physical layer (PHY) specifications for low-rate wireless personal area networks (WPANs). *IEEE Std 802.15.4 2006 (Revision of IEEE Std 802.15.4-2003)*, pp. 1–305, 2006.

[IEEE11] IEEE 802.11 Std. IEEE Local and metropolitan area networks – Specific requirements Part 11: Wireless LAN Medium Access Control (MAC) and Physical Layer (PHY) specifications, 2011.

[IEEE12a] IEEE standard for information technology–telecommunications and information exchange between systems–local and metropolitan area networks-specific requirements-part 11: Wireless LAN medium access control (MAC) and physical layer (PHY) specifications amendment 3: Enhancements for very high throughput in the 60 GHz band. *IEEE Std 802.11ad-2012*, pp. 1–628, December 2012.

[IEEE12b] IEEE standard for local and metropolitan area networks – part 15.6: Wireless body area networks. *IEEE Std 802. 15.6-2012*, pp. 1–271, February 2012.

[IF14] M. Ignatenko and D. S. Filipovic. Application of characteristic mode analysis to hf low profile vehicular antennas. In *IEEE Antennas and Propagation Society International Symposium (APSURSI)*, pp. 850–851, July 2014.

[IH98] J. How and D. Hatzinakos. Analytic alpha-stable noise modeling in a Poisson field of interferers or scatterers. *IEEE Transactions Signal Processing*, 46(6):1601–1611, 1998.

[IKA+11] J. Ilvonen, O. Kivekas, A. A. H. Azremi, R. Valkonen, J. Holopainen, and P. Vainikainen. Isolation improvement method for mobile terminal antennas at lower UHF band. In *5th European Conference on Antennas and Propagation (EUCAP)*, pp. 1238–1242, April 2011.

[IKR15] Institute of Communication Networks and Computer Engineering. IKR simulation library. http://www.ikr.uni-stuttgart.de/Content/IKRSimLib/ 2015.

[IMT-2020_2015] IMT-2020. (5G) promotion group, 5G concept white paper. February 2015.

[IMU13] Intel, Motorola Mobility, and Aalborg University. MIMO AAS spatially filtered channel model simulations. Technical Report R4-137059, 3GPP TSG-RAN WG4 no. 69, San Francisco, USA, 2013. In parts also available as TD(14)10061.

[Iof14] A. Ioffe. MIMO OTA FoM Recommendation. Technical Report MOSG140304R1, CTIA MOSG, 2014. (Also available as TD(14)10061).

[ITU12] Propagation data and prediction methods required for the design of terrestrial line-of-sight systems. Technical Report ITU Recommendations pp. 530–514, 2012.

[ITU2012] Final Acts of the World Radio communication Conference (online), 2012. Available at: http://www.itu.int/pub/R-ACT-WRC.9-2012/en

[ITU2015] Final Acts of the World Radio communication Conference (online), 2015. Available at: http://www.itu.int/pub/R-ACT-WRC.11-2015/en

[ITU-R 2015] ITU-R. Workplan, timeline, process and deliverables for the future development of IMT. International Telecommunication Union Radiocommunications, ITU-R, 2015.

[IV14] E. Tanghe W. Joseph P. Laly B. Hanssens-M. Liénard-L. Martens I. Vin, D. Gaillot. Polarisation properties of specular and dense multipath component in a large industrial hall. In *COST IC1004 9th Scientific Meeting*, pp. 1–8, Ferrara, Italy, February 2014.

[IY02] M. F. Iskander and Z. Yun. Propagation prediction models for wireless communication systems. *IEEE Transaction Microwave Theory Technology*, 50(2):662–673, 2002.

[IY14] A. Ioffe and B. Yanakiev. MPAC test zone size simulations based on throughput. Technical Report TD(14)10062, 10th COST IC1004 Scientific Meeting, Aalborg, Denmark, May 2014.

J

[Jak14] M. L. Jakobsen. Parameter estimation in suzuki's delta-k model using interference tools for renewal point processes. Technical Report TD-011-054, Krakow, Poland, September 2014.

[JCK+11] M. Jain, J. I. Choi, T. Kim, D. Bharadia, S. Seth, K. Srinivasan, P. Levis, S. Katti, and P. Sinha. Practical, real-time, full duplex wireless. In *17th Annual International Conference on Mobile Computing and Networking*, pp. 301–312. ACM, 2011.

[Jen11] P. Jensen. Temperature impact on drive test throughput measurements. Technical Report TD(11)02069, 2nd COST IC1004 Scientific Meeting, Lisbon, Portugal, Oct 2011.

[JGLR14] M. Joud, M. Garćia-Lozano, and S. Ruiz. On the use of c-/u-plane split and comp to enhance small cells with moderate speed users. Technical Report TD(14)10063, Aalborg, Denmark, May 2014.

[JH14] J. Järveläinen and K. Haneda. Sixty gigahertz indoor radio wave propagation prediction based on full scattering model. *Radio Science*, 49:293–305, 2014.

[JHK+11] J. Järveläinen, K. Haneda, V.-M. Kolmonen, J. Poutanen, and P. Vainikainen. Propagation Channel Models for Indoor Localization. In *IC1004*, Lisbon, Portugal, October 2011 (TD(11)02049).

[JHK+12] J. Järveläinen, K. Haneda, M. Kyrö, V.-M. Kolmonen, J. Takada, and H. Hagiwara. 60 GHz radio wave propagation prediction in a hospital environment using an accurate room structural model. In *Proceeding of 2014 Loughborough Conference of Antennas Propagation (LAPC 2012)*, pp. 1–4, November 2012. (Also available as TD(12)05049).

[Jin12a] Y. Jing. Analysis of Reference Antenna performance over different 2D elevations. Technical Report TD(12)05068, 5th COST IC1004 Scientific Meeting, Bristol, UK, September 2012.

[Jin12b] Y. Jing. Evaluating self interference using UE reports. Technical Report TD(12)05069, 5th COST IC1004 Scientific Meeting, Bristol, UK, September 2012.

[JKR13] Y. Jing, H. Kong, and M. Rumney. Interference impact on OTA performance. Technical Report TD(TD(13)06076, 6th COST IC1004 Scientific Meeting, Malaga, Spain, January 2013.

[JLR02] L. Juan-Llacer and J. L. Rodriguez. A utd-po solution for diffraction of plane waves by an array pf perfectly conducting wedges. *IEEE Transaction on Antennas Propagation*, 50(9):1207–1211, 2002.

[Joh96] D. H. Johnson. Optimal linear detectors for additive noisechannels. *IEEE Transactions on Signal Processing*, 44(12):3079–3084, 1996.

[JPF14] Lishuai Jing, T. Pedersen, and B. H. Fleury. Direct ranging in multi-path channels using ofdm pilot signals. In *Proc. SPAWC 2014 – Sig. Proc. Advances in Wireless Commun.*, pp. 150–154, June 2014. (Also available as TD(14)10017).

[JPF14a] M. L. Jakobsen, T. Pedersen, and B. H. Fleury. Analysis of stochastic radio channels with temporal birth-death dynamics: A marked spatial point process perspective. *IEEE Transactions on Antennas Propagation*, 62(7):3761–3775, 2014. (Also available as TD(08)048).

[JPF14b] M. L. Jakobsen, T. Pedersen, and B. H. Fleury. Simulation of birth-death dynamics in time-variant stochastic radio channels. In *Proceedings of 2014 International Zurich Seminar on Communication (IZS2014),* pp. 124–127, 2014.

[JPGLL] J. Garcia-Pardo, J. Rodriguez, J. Pascual-Garcia, M. Martinez-Inglés, and L. Juan-Llacer. Using tuned diffuse *scattering* parameters in ray tracing channel modeling. In IC1004 12th MC and Scientific Meeting, Dublin, 28–30 January 2015.

[JW04] M. A. Jensen and J. W. Wallace. A review of antennas and propagation for MIMO wireless communications. *IEEE Transactions on Antennas and Propagation*, 52(11):2810–2824, 2004.

[JZG+14] S. Jovanoska, R. Zetik, F. Govaers, W. Koch, and R. Thomä. Localization of multiple persons using UWB in indoor scenarios. In *11th COST IC1004 Management Committee Meeting*, Krakow, Poland, September 2014 (TD(14)11014).

[JZK12] Y. Jing, X. Zhao, and H. Kong. Analysis of MIMO OTA performance using 2D and 3D channel models. Technical Report TD(12)04036, 4th COST IC1004 Scientific Meeting, Lyon, France, May 2012.

K

[K11] M. Käske. Impact of spectral windowing on the estimation of dmc. Technical Report TD-02-053, Lisbon, Portugal, 2011.

[Kak14] C. G. Kakoyiannis. Robust, electrically small, circular inverted-f antenna for energy-efficient wireless microsensors. *IET Microwaves, Antennas Propagation*, 8(13):1047–1056, October 2014.

[KAPT09] T. Koike-Akino, P. Popovski, and V. Tarokh. Optimized constellations for two-way wireless relaying with physical network coding. *IEEE J. Sel. Areas Commun.*, 27(5):773–787, June 2009.

[KAT13] Y. Kuang, K. Astrom, and F. Tufvesson. Single antenna anchor-free UWB positioning based on multipath propagation. In *Communications (ICC), 2013 IEEE International Conference on*, pp. 5814–5818, June 2013. (Also available in TD(13)08063).

[KAY10] M. M. Khan, A. Alomainy, and H. Yang. Off-body radio channel characterisation using ultra wideband wireless tags. In *Proceeding of International Conference on Body Sensor Networks (BSN2010)*, pp. 80–83, Singapore, Singapore, 2010.

[KB15] I. Kavanagh and C. Brennan. Path loss modelling for indoor environments using an integrated equation formulation. In *COST IC1004 12th Management Committee and Scientific Meeting*, February 2015.

[KBA+14] LKK11F. Kaltenberger, A. Byiringiro, G. Arvanitakis, R. Ghaddab, D. Nussbaum, R. Knopp, et al. Broadband wireless channel measurements for high speed Trains. Technical Report TD(14) 11017, September. 2014.

[KBA+15] F. Kaltenberger, A. Byiringiro, G. Arvanitakis, R. Ghaddab, D. Nussbaum, R. Knopp, M. Bernineau, Y. Cocheril, H. Philippe, and E. Simon. Broadband wireless channel measurements for high speed trains. In *Proceedings of IEEE ICC'15*, 2015.

[KC12a] C. G. Kakoyiannis and P. Constantinou. One-off wideband efficiency measurements of electrically small and large antennas inside fixed-geometry wheeler cavities. Technical Report TD-12-03065, Barcelona, Spain, 2012.

[KC12b] C. G. Kakoyiannis and P. Constantinou. Electrically small, circular inverted-f antenna based on a robust, minimal design space. In *European Microwave Conference (EuMC)*, pp. 404–407, October 2012.

[KC14a] S. Khatibi and L. M. Correia. Modelling virtual radio resource management with traffic offloading support. Technical Report TD-14-11026, COST IC1004, Krakow, Poland, September 2014.

[KC14b] S. Khatibi and L. M. Correia. Virtualisation of radio resources – next step in radio access virtualisation. In *Proc. PIMRC 2014 – IEEE 25th International Sympouism on Personal Indoor and Mobile Radio Communication* Washington, USA, September 2014. (Also available as TD(14)09044).

[KC15] S. Khatibi and L. M. Correia. The effect of channel quality on virtual radio resource management. Technical Report TD-15-12034, COST IC1004, Dublin, Ireland, January 2015.

[KCD+14] P. Kntor, L. Csurgai-Horvth, A. Drozdy, P. Horvth, and J. Bit. Rain attenuation modelling for performance evaluation of microwave 5G mesh networks. Technical Report TD(14) 11047, September 2014.

[KCO+12] P.-S. Kildal, Xiaoming Chen, C. Orlenius, M. Franzen, and C. S. L. Patane. Characterization of reverberation chambers for OTA measurements of wireless devices: Physical formulations of channel matrix and new uncertainty formula. *IEEE Transactions on Antennas and Propagation*, 60(8):3875–3891, 2012.

[KDB14] P. Kantor, A. Drozdy, and J. Bito. Effects of rain fading in ad-hoc microwave 5G mesh networks. Technical Report TD(14) 10082, May 2014.

[Ke13] K. Guan, Z. Zhong, B. Ai, and T. Kurner. Semi-Deterministic Modeling for Propagation of High-Speed Railway. Technical Report TD(13) 06018, February 2013.

[Kha13] A. K. Khandani. Two-way (true full-duplex) wireless. In *13th Canadian Workshop on Information Theory (CWIT)*, pp. 33–38. IEEE, 2013.

[KH12] P. Kyösti and L. Hentilä. Criteria for physical dimensions of MIMO OTA multi-probe test setup. In *2012 6th European Conference on Antennas and Propagation (EUCAP)*, pp. 2055–2059, March 2012.

[KHJP15] A. Karttunen, K. Haneda, J. Järveläinen, and J. Putkonen. Polarisation characteristics of propagation paths in indoor 70 GHz channels. In *Proceeding of 9th European Conference on Antennas Propagation (EuCAP2015)*, Lisbon, Portugal, April 2015. (Also available as TD(14)11031).

[KHM02] K. Kuroe, N. Hashimoto, and T. Mori. On energy function for complex-valued neural networks and its applications. In *9th International Conference on Neural Information Processing*, pp. 1079–1083, November 2002.

[KHNK11] P. Kyoösti, L. Hentilä, M. Nurkkala, and J. Kallankari. Impact of various channel models on LTE mobile terminal performance in MIMO OTA test environment.

Technical Report TD(11)01042, 1st COST IC1004 Scientific Meeting, Lund, Sweden, June 2011.

[KHS+12] M. Kyrö, K. Haneda, J. Simola, K.-I. Takizawa, H. Hagiwara, and P. Vainikainen. Statistical channel models for 60 GHz radio propagation in hospital environments. *IEEE Trans. Antennas Propagation*, 60(3):1569–1577, March 2012. (Also available as TD(11)02037).

[KHT+13] T. Kumpuniemi, M. Hämäläinen, T. Tuovinen, K. Y. Yazdandoost, and J. Iinatti. Generic small scale channel model for on-body UWB WBAN communications. In *Proceeding of 2nd Ultra Wideband for Body Area Networking Workshop (UWBAN) – Co-located with the 8th International Conference on Body Area Networks (BodyNets)*, Vol. 1, pp. 570–574, October 2013. (Also available as TD(13) 08011 with title Measurement-Based Small Scale On-Body UWB WBAN Channel Model).

[KHT+14] T. Kumpuniemi, M. Hämäläinen, T. Tuovinen, K. Yekeh Yazdandoost, and J. Iinatti. Radio channel modelling for pseudo-dynamic WBAN on-body UWB links. In *Proceeding of 8th IEEE International Symposium on Medical Information and Communication Technology (ISMICT2014)*, Vol. 1, pp. 1–5, April 2014. (Also available as TD(14)09016 with title Pseudo-Dynamic UWB Radio Channel Modelling for WBAN Communications).

[KHYI14] T. Kumpuniemi, M. Hämäläinen, K. Y. Yazdandoost, and J. Iinatti. Human body size and shape effect on UWB on-body WBAN radio channels – preliminary results. In *Proceeding of 3rd Ultra Wideband for Body Area Networking Workshop (UWBAN) – Co-located with the 8th International Conference on Body Area Networks (BodyNets*, Vol. 1, pp. 1–7, October 2014. (Also available as TD(14)11008 with title Comparison of UWB On-Body WBAN Radio Channels Between Various Test Persons – Preliminary Results).

[KHYI15] T. Kumpuniemi, M. Hämäläinen, K. Y. Yazdandoost, and J. Iinatti. Dynamic on-body UWB radio channel modelling. In *Proceeding of 9th IEEE International Symposium on Medical Information and Communication Techniques (ISMICT2015)*, Vol. 1, p. 1–5, March 2015. (Also available as TD(15)12051 with title Dynamic on-body UWB radio channel measurements).

[Kil13] P.-S. Kildal. Rethinking the wireless channel for OTA testing and network optimization by including user statistic: RIMP, pure-LOS, throughput and detection probability. In *2013 International Symposium on Antennas and Propagation (ISAP2013)*, Nanjing, China, October 2013.

[KJ13] V. Kafedziski and T. Javornik. Frequency–space interference alignment in multi-cell MIMO OFDM downlink systems. In *IEEE 77th Vehicular Technology Conference (VTC Spring), 2013*, pp. 1–5, June 2013.

[KJN12] P. Kyösti, T. Jämsä, and J. P. Nuutinen. Channel modelling for multiprobe over-the-air MIMO testing. *International Journal of Antennas and Propagation*, 2012.

[KJVM11] D. Kurup, W. Joseph, G. Vermeeren, and L. Martens. Path loss and SAR at specific in-body locations in heterogeneous human model. Technical Report TD-11-01027, Lisbon, Portugal, October 2011.

[KJZ11a] H. Kong, Y. Jing, and X. Zhao. Experimental validation on handset antenna pattern measurement accuracy. Technical Report TD(11)01060, 1st COST IC1004 Scientific Meeting, Lund, Sweden, June 2011.

[KJZ11b] H. Kong, Y. Jing, and X. Zhao. Preliminary LTE MIMO OTA test results using two-stage method. Technical Report TD(11)01062, 1st COST IC1004 Scientific Meeting, Lund, Sweden, June 2011.

[KK04] K. Kaemarungsi and P. Krishnamurthy. Properties of indoor received signal strength for WLAN location fingerprinting. In *MobiQuitous*, pp. 14–23. IEEE Computer Society, 2004.

[KK12] J. Kosciow and P. Kulakowski. Ray-tracing analysis of an indoor passive localization system. In *3rd COST IC1004 Management Committee Meeting*, Barcelona, Spain, February 2012 (TD(12)03066).

[KK13a] P. Kyösti and P. Kemppainen. Radio channel measurements in live LTE networks. Technical Report TD(13)07053, 7th COST IC1004 Scientific Meeting, Ilmenau, Germany, May 2013.

[KK13b] P. Kyösti and A. Khatun. Probe configurations for 3D MIMO over-the-air testing. In *2013 7th European Conference on Antennas and Propagation (EuCAP)*, pp. 1421–1425, April 2013.

[KK14] J. Kmiecik and P. Kulakowski. Passive indoor tracking with (ultra) wideband systems. In *11th COST IC1004*

Management Committee Meeting, Krakow, Poland, September 2014 (TD(14)11020).

[KKCT12] M. Kim, Y. Konishi, Y. Chang, and J. Takada. Measurement results of indoor wideband mimo channels at 11 ghz. In *IC1004*, Bristol, UK, September 2012 (TD(12)05029).

[KKJ14] P. Kyösti, P. Kemppainen, and T. Jämsä. Radio channel measurements in live LTE networks for MIMO over-the-air emulation. In *8th European Conference on Antennas and Propagation (EuCAP2014)*, The Hague, Netherlands, April 2014. (Also available as TD(13)08046).

[KKK11] S. Karimkashi, A. A. Kishk, and D. Kajfez. Antenna array optimisation using dipole models for MIMO applications. *IEEE Transactions on Antennas Propagation*, 59(8):3112–3116, 2011.

[KKP08] A. Kalis, A. G. Kanatas, and C. B. Papadias. A novel approach to MIMO transmission using a single RF front end. *IEEE Journal of Selection Areas Communication*, 26(6):972–980, August 2008. (Also available as TD(02)071).

[KL12] W. Kotterman and M. Landmann. Use of effective aperture distribution function (EADF) for simply and fast characterising, compressing, and embedding antenna patterns. Technical Report TD-12-03029, Barcelona, Spain, February 2012.

[KLHT11] W. A. Th. Kotterman, M. Landmann, A. Heuberger, and R. S. Thomä. New laboratory for over-the-air testing and wave field synthesis. In *XXX General Assembly and Scientific Symposium of the International Union of Radio Science*, Istanbul, Turkey, 2011. URSI (Also available as TD(11)01023).

[KLMA13] K. Witrisal, Erik Leitinger, Paul Meissner, and Daniel Arnitz. Cognitive radar for the localization of RFID transponders in dense multipath environments. In *IEEE Radar Conference*, Ottawa, Canada, April 2013. (Also available as TD(13)08022).

[KLV$^+$03] K. Kalliola, H. Laitinen, P. Vainikainen, M. Toeltsch, J. Laurila, and E. Bonek. 3-D double-directional radio channel characterisation for urban macrocellular applications. *IEEE Transactions on Antennas and Propagation*, 51:3122–3133, 2003.

[KM02] T. Kürner and A. Meier. Prediction of Outdoor and Outdoor-to-Indoor Coverage in Urban Areas at 1.8 GHz.

IEEE Journal on Selected Areas in Communications, 20(3):496–506, April 2002.

[KM96] G. Kechriotes and E. S. Manolakos. Hopfield neural network implementation of the optimal CDMA multiuser detector. *IEEE Transactions on Neural Networks*, 7(1):131–141, 1996.

[KMH+08] P. Kyösti, J. Meinilä, L. Hentilä, X. Zhao, T. Jämsä, C. Schneider, M. Narandzić, M. Milojević, A. Hong, J. Ylitalo, V.-M. Holappa, M. Alatossava, R. Bultitude, Y. de Jong, and T. Rautiainen. Winner II Channel Models. Technical report, IST-WINNER, 2008.

[KML+15] J. Kmiecik, P. Meissner, E. Leitinger, K. Witrisal, and P. Kulakowski. Experimental validation of passive indoor localization and tracking using UWB systems. In *12th COST IC1004 Management Committee Meeting*, Dublin, Ireland, January 2015 (TD(15)12084).

[KMM11] B. Krasniqi, C. Mehlfürer, and C. F. Mecklenbräuker. Bandwidth Re-allocation Depending on Large-Scale Path-Loss for Two Users in Partial Frequency Reuse Cellular Networks. Technical Report TD-11-01009, Lund, Sweden, 2011.

[Kot12] W. Kotterman. Increasing the volume of test zones in anechoic chamber MIMO over-the-air test set-ups. In *2012 International Symposium on Antennas and Propagation (ISAP12)*, pp. 786–789, Nagoya, Japan, October 2012. (Also available as TD(12)05023).

[Kot13] W. Kotterman. Influence of complex amplitude errors on the quality of synthesised wave fields. Technical Report TD(13)07054, 7th COST IC1004 Scientific Meeting, Ilmenau, Germany, May 2013.

[KPCB15] I. Kavanagh, X. V. Pham, M. Condon, and C. Brennan. A method of moments based indoor propagation model. In *Proceedings 9th European Conference on Antennas and Propagation, EuCAP 2015*, Lisbon, Portugal, April 2015.

[KPD12] E. Kocan M. Pejanovic-Djurisic, T. Javornik, BER Performance Enhancement in OFDM AF Fixed Gain Relay Systems, in Proc. of IEEE conf. EUROCON 2013, pp. 502–507, Zagreb, Croatia, July 2013. (Also available as TD(12)05021).

[KPSL14] O. Koymen, A. Partyka, S. Subramanian, and J. Li. Indoor mm-wave channel measurements: Comparative study of 2.9 ghz and 29 ghz. *COST IC1004, TD(13)11066, Krakow, Poland*, 2014.

[KRG+11] P. Kulakowski, F. Royo-Sanchez, R. Galindo-Moreno, and L. Orozco-Barbosa. Can indoor RSS localization with 802.15.4 sensors be viable? In *2nd COST IC1004 Management Committee* Meeting, Lisbon, Portugal, October 2011 (TD(11)02068).

[KRO13] P. Kulakowski, F. Royo-Sanchez, and L. Orozco-Barbosa. Indoor RSS localization: measurements in static and mobile scenarios. In *7th COST IC1004 Management Committee Meeting*, Ilmenau, Germany, May 2013 (TD(13)07048).

[KS10] T. Kürner and M. Schack. 3D-Ray-Tracing embedded into an integrated simulator for car-to-x-communication. *URSI International Symposium on Electromagnetic Theory (EMTS)*, pp. 880–882, August 2010.

[KSLDG14] W. Kotterman, C. Schirmer, M. Landmann, and G. Del Galdo. On arranging dual-polarised antennas in 3D wave field synthesis. In *8th European Conference on Antennas and Propagation (EuCAP)*, pp. 3406–3410, The Hague, The Netherlands, April 2014. (Also available as TD(13)08062).

[KSST14] M. Käske, G. Sommerkorn, C. Schneider, and R. S. Thomä. On the reliability of measurement based channel synthesis from a mimo performance evaluation point of view. In *2014 8th European Conference on Antennas and Propagation (EuCAP)*, pp. 2541–2545, April 2014.

[KT14] P. Komulainen. and A. Tolli. Coordinated beamforming for cellular underlay device-to-device communication. Technical Report TD-14-11021, Krakow, Poland, September 2014.

[KTH+13] T. Kumpuniemi, T. Tuovinen, M. Hämäläinen, K. Yekeh Yazdandoost, R. Vuohtoniemi, and J. Iinatti. Measurement-based on-body path loss modelling for UWB WBAN communications. In *Proceeding of 7th IEEE International Symposium on Medical Information and Communication Technology (ISMICT2013)*, Vol. 1, pp. 233–237, March 2013. (Also available as TD(13)07002 with title On-Body Path Loss Modelling Based On UWB WBAN Measurements).

[Kul12] P. Kulakowski. On NLOS conditions in wireless localization. In *5th COST IC1004 Management Committee Meeting*, Bristol, UK, September 2012 (TD(12)05056).

[Kut00] H. Kuttruff. *Room Acoustics, 4th ed.* London: Taylor & Francis, 2000.

[Kür99] T. Kürner. A run-time efficient 3D propagation model for urban areas including vegetation and terrain effects. *49th IEEE Vehicular Technology Conference*, 1:782–786, July 1999.

[Kwo06] D. H. Kwon. Effect of antenna gain and group delay variations on pulse-preserving capabilities of ultrawideband antennas. *IEEE Transactions on Antennas Propagation*, 54:2208–2215, 2006.

[KWPW14] W. Keusgen, R. J. Weiler, M. Peter, and M. Wisotzki. Propagation measurements and simulations for millimeter-wave mobile access in a busy urban environment. In *9th International Conference on Infrared, Millimeter, and Terahertz Waves, IRMMW-THz 2014*, Tucson, USA, September 2014.

[KyK13] J. Kyröläinen and P. Kyösti. Statistics of phase difference between orthogonal polarization components in propagation channel, 2013 September. Ghent, Belgium, TD-08-051.

[Kyo07] P. Kyosti et al. WINNER II channel models, D1.1.2. https://www.ist-winner.org/WINNER2-Deliverables/D1.1 2v1.1.pdf, 2007.

[Kyo12] P. Kyösti. Impact of random initialization of channel model implementations on fading statistics. Technical Report TD(12)03064, 3rd COST IC1004 Scientific Meeting, Barcelona, Spain, February 2012.

[KYRC07] K. Kim, G. M. Yeo, B. H. Ryu, and K. Chang. Interference analysis and subchannel allocation schemes in tri-sectored ofdma systems. In *Vehicular Technology Conference, 2007. VTC-2007 Fall. 2007 IEEE 66th*, pp. 1857–1861, September 2007.

[KZM12] X. Kang, R. Zhang, and M. Motani. Price-based resource allocation for spectrum-sharing femtocell networks: a stackelberg game approach. *IEEE Journal on Selected Areas in Communications*, 30(3):538–549, 2012.

L

[LA12] B. K. Lau and J. B. Andersen. Simple and efficient decoupling of compact arrays with parasitic scatterers. *IEEE Transactions on Antennas Propagation*, 60(2):464–472, 2012.

[LAC+13] Y. Li, B. Ai, X. Cheng, S. Lin, and Z. Zhong. A TDL based Non-WSSUS vehicle-to-vehicle channel model. *International Journal of Antennas and Propagation*, 2013 (Article ID 103461):8, 2013. (Also available as TD(13)08024).

[LAKM06] B. K. Lau, J. B. Andersen, G. Kristensson, and A. F. Molisch. Impact of matching network on bandwidth of compact antenna arrays. *IEEE Transactions on Antennas Propagation*, 54(11):3225–3238, 2006.

[Lau11] B. K. Lau. Multiple antenna terminals. In MIMO: from theory to implementation, C. Oestges, A. Sibille, and A. Zanella, eds. Academic Press, San Diego, CA, pp. 267–298, 2011.

[Lax51] M. Lax. Multiple scattering of waves. *Reviews of Modern Physics*, 23(4):287–310, 1951.

[LB11] T. Zemen and Laura Bernado. Time-varying radio channel parameters: Characterization in vehicular channels. In *COST IC1004 2nd MC & Scientific Meeting*, number TD(11)02022, 2011.

[LB14] L. Laughlin and M. Beach. Performance variation in electrical balance duplexers due to user influence on antenna. Technical Report TD-14-09067, Ferrara, Italy, February 2014.

[LBC12] J. Llorca, M. Barbi, and N. Cardona. Hardware emulation of on-body channels. A proposal of IC1004 measurement-based reference channel models for BAN. Technical Report TD-12-03031, Barcelona, Spain, February 2012.

[LFMW14] E. Leitinger, M. Froehle, P. Meissner, and K. Witrisal. Multipath-Assisted Maximum-Likelihood Indoor Positioning using UWB Signals. In *Proceeding of ICC 2014 – IEEE International Conference on Communication Workshops; Workshop on Advances in Network Localization and Tracking*, pp. 170–175, June 2014. (Also available as TD(14)09054).

[LFP15] I. C. Llorente, W. Fan, and G. F. Pedersen. Emulation of ray tracing models in multi-probe anechoic chamber setups. Technical Report TD(15)12054, 12th COST IC1004 Scientific Meeting, Dublin, Ireland, January 2015.

[LGAJ13] A. Ligata, H. Gacanin, F. Adachi, and T. Javornik. On performance of analogue network coding in the presence of phase noise. Technical Report TD-06-027, Malaga, Spain, February 2013.

[LGJ14] A. Ligata, H. Gacanin, and T. Javornik. On performance of MIMO-OFDM/TDM using MMSE-FDE with nonlinear HPA in a multipath fading channel. *IEICE Transactions on Communications*, 97(9):1947–1957, 2014. (Also available as TD(12)03037).

[LGOR11] M. A. Lema, M. Garc'ia-Lozano, J. Olmos, and S. Ruiz. Uplink system and link level simulation of 3GPP-LTE rel. 8. Technical Report TD(11)01-040, Lund, Sweden, June 2011.

[LGP+13] M. Landmann, M. Grossmann, N. Phatak, C. Schneider, R. Thomä, and G. Del Galdo. Performance analysis of channel model simplifications for MIMO OTA LTE UE testing. In *Antennas and Propagation (EuCAP), 2013 7th European Conference on*, pp. 1856–1860, April 2013. (Also available as TD(13)06057).

[LGRO+11] M. A. Lema, M. Garcia-Lozano, S. Ruiz, J. Olmos, and D. Perez Diaz de Cerio. LTE UL power control evaluation in a system level simulator for synthetic and realistic scenarios. Technical Report TD(11)02-028, Lisbon, Portugal, October 2011.

[LGR14] M. Lema, M. Garcia-Lozano, and S. Ruiz. Sounding Reference Signals Management for CSI Improvement in LTE Uplink with Carrier Aggregation. Technical Report TD-14-11023, Krakow, Poland, September 2014.

[LGRO11] M. Lema, M. Garcia-Lozano, S. Ruiz, and J. Olmos. LTE Power Control Evaluation in a System Level Simulator for Synthetic and Realistic Scenarios. Technical Report TD-11-02028, Lisbon, Portugal, October 2011.

[LGRO13] M. Lema, M. Garcia-Lozano, S. Ruiz, and J. Olmos. Improved Scheduling Decisions for LTE-A Uplink Based on MPR Information. Technical Report TD-13-06070, Málaga, Spain, February 2013.

[LGS+13] R. Litjens, F. Gunnarsson, B. Sayrac, K. Spaey, C. Willcock, A. Eisenblätter, B. González Rodriguez, and T. Kürner. Self-management for unified heterogeneous radio access networks. Technical Report TD(13)06022, Malaga, Spain, February 2013.

[LH12] C.-H. Lee and M. Haenggi. Interference and outage in Poisson cognitive networks. *IEEE Transactions on Wireless Communications*, 11(4):1392–1401, 2012.

[LHB+14] L. Laughlin, T. *Hawkins*, M. Beach, K. Morris, and J. Haine. Temporal variation in electrical balance duplexer isolation due to device motion. Technical Report TD-14-10055, Aalborg, Danmark, May 2014.

[LHOV09] A. Lea, P. Hui, J. Ollikainen, and R. G. Vaughan. Propagation between on-body antennas. *IEEE Transactions on, Antennas and Propagation*, 57(11):3619–3627, 2009.

[Lin99] J. Lindner. MC-CDMA in the context of general multiuser/multisubchannel transmission methods. *European Transactions on Telecommunications (ETT)*, 10(4):351–367, 1999.

[LK12] Y. Lostanlen and T. Kürner. Ray Tracing Modeling, pp. 271–291. Wiley, Hoboken, NJ, 2012.

[LKGC11] S. H. Lim, Y.-H. Kim, A. El Gamal, and S.-Y. Chung. Noisy network coding. *IEEE Trans. Inf. Theory*, 57(5):3132–3152, May 2011.

[LKK11] I. Latif, F. Kaltenberger, and R. Knopp. Link abstraction for multi-user MIMO in LTE using interference-aware receiver. WCNC 2012, IEEE Wireless Communications and Networking Conference, April 1–4, 2012, Paris, France. (Also available as (TD(11)02044)).

[LKKO12] I. Latif, F. Kaltenberger, R. Knopp, and J. Olmos. Link abstraction for variable bandwidth with incremental redundancy HARQ in LTE. WIMEE 2013, 9th International Workshop on Wireless Network Measurements, In conjunction with 11th International Symposium on Modeling and Optimization in Mobile, Ad Hoc, and Wireless Networks (WiOpt 2013), May 13–17, 2013, Tsukuba Science City, Japan; (Also available as (TD(12)05060).

[LKT12] M. Landmann, M. Käske, and R. S. Thomä. Impact of incomplete and inaccurate data models on high resolution parameter estimation in multidimensional channel sounding. *IEEE Transactions on Antennas and Propagation*, 60(2):557–573, 2012.

[LL15] H. Li and B. K. Lau. Efficient evaluation of specific absorption rate (SAR) for MIMO terminals. *Electronics Letters,* 50(22):1561–1562, October 2014. (Also available as TD-15-12061).

[LLH11] H. Li, B. K. Lau, and S. He. Angle and polarization diversity in compact dual-antenna terminals with chassis excitation. In *URSI General Assembly (URSI GA'2011),* Istanbul, Turkey, August, 2011. (Also available as TD-11-02052).

[LLLH13] H. Li, X. Lin, B. K. Lau, and S. He. Equivalent circuit based calculation of signal correlation in lossy antenna arrays. *IEEE Transactions on Antennas and Propagation,* 61(10): 5214–5222, October 2013. (Also available as TD-13-06013).

[LLM14a] G. Lasser, D. Löschenbrand, and C. Mecklenbräuker. Near field scans of tyre mounted dipoles using a separate phase reference antenna. In *2014 IEEE-APC Topical Conference on Antennas and Propagation in Wireless Communications (APWC)*, Palm Beach, Aruba, August 2014. Selene Srl.

[LLM14b] G. Lasser, D. Löschenbrand, and C. F. Mecklenbräuker. Update on distorted measured antenna patterns. Technical Report TD-14-10033, Aalborg, Denmark, 2014.

[LLM15] D. Löschenbrand, G. Lasser, and C. F. Mecklenbräuker. Reflection suppression in anechoic chambers for spherical near-field antenna measurements using modal filtering. Technical Report TD-15-12058, Dublin, Ireland, 2015.

[LLV$^+$12a] M. Luo, N. Lebedev, G. Villemaud, G. de la Roche, J. Zhang, and J. M. Gorce. On predicting large scale fading characteristics with a finite difference method. Technical Report TD(12)03023, Barcelona, Spain, February 2012.

[LLV$^+$12b] M. Luo, N. Lebedev, G. Villemaud, G. de la Roche, J. Zhang, and J. M. Gorce. On predicting large scale fading characteristics with the mr-fdpf method. In *Proceedings of 6th European Conference on Antennas and Propagation, EuCAP 2012*, Prague, Czech Republic, April 2012.

[LLYH12] H. Li, B. K. Lau, Z. Ying, and S. He. Decoupling of multiple antennas in terminals with chassis excitation using polarization diversity, angle diversity and current control. *IEEE Transactions on Antennas and Propagation,* 60(12):5947–5957, December 2012. (Also available as TD-12-05043).

[LM14] G. Lasser and C. F. Mecklenbräuker. Distortions of measured antenna patterns caused by dielectric support structures. Technical Report TD-14-09048, Ferrara, Italy, 2014.

[LMCL15] H. Li, R. Ma, J. Chountalas, and B. K. Lau. Characteristic mode based pattern reconfigurable antenna for mobile handset. In *9th European Conference on Antennas and Propagation (EuCAP)*, Lisbon, Portugal, April 2015. (Also available as TD-15-12059).

[LMFW14] E. Leitinger, P. Meissner, M. Froehle, and K. Witrisal. Performance bounds for multipath-assisted indoor localization on backscatter channels. In *IEEE RadarCon 2014*, 2014. (Also available as TD(13)07012).

[LMG+15] P. H. Lehne, K. Mahmood, A. A. Glazunov, P. Grønsund, and P.-S. Kildal. Measuring user-induced randomness to evaluate smart phone performance in real environments. In *9th European Conference on Antennas and Propagation (EuCAP2015)*, Lisbon, Portugal, April 2015. (In parts, also available as TD(14)09026 and TD(14)11035).

[LML13] H. Li, Z. Miers, and B. K. Lau. Generating multiple characteristic modes below 1 GHz in small terminals for MIMO antenna design. In *2013 Antennas and Propagation Society International Symposium (APSURSI)*, Orlando, USA, July 2013. (Also available as TD-13-08041).

[LMLW15] E. Leitinger, P Meissner, M. Lafer, and K. Witrisal. Simultaneous localisation and mapping using multipath channel information. In *IEEE ICC 2015 Workshop on Advances in Network Localization and Navigation (ANLN)*, June 2015. (Also partly available as TD(15)12068).

[LMPM15] G. Lasser, L. W. Mayer, Z. Popović, and C. F. Mecklenbräuker. Fss shield for a switched beam antenna. In *COST IC1004 12th MC & Scientific Meeting*, number TD(15)12073, Dublin, Ireland, 2015.

[LMV14] V. Degli-Espposti L. Minghini, R. D'Errico, and E. Vitucci. Electromagnetic simulation and measurement of *diffuse* scattering from building walls. In *IC1004 9th MC and Scientific Meeting*, Ferrara, 2014.

[LMW13] Henghui Lu, S. Mazuelas, and M. Z. Win. Ranging likelihood for wideband wireless localisation. In *Communications (ICC), 2013 IEEE International Conference on*, pp. 5804–5808, June 2013.

[LN10] J. Lee and S. Nam. Fundamental aspects of near-field coupling small antennas for wireless power transfer. *IEEE Transactions on Antennas Propagation*, 58(11):3442–3449, 2010.

[LPQ+12] L. Liu, J. Poutanen, F. Quitin, K. Haneda, F. Tufvesson, P. D. Doncker, P. Vainikainen, and C. Oestges. The COST2100 MIMO channel model. *IEEE Wireless Communications*, 19(6):92–99, 2012.

[LRG+12] M. Lema, S. Ramon-Ferran, M. Garcia-Lozano, S. Ruiz, and J. Olmos. Admission Control in LTE Uplink Systems with Sounding Reference Signals based Channel State Indicator. Technical Report TD-12-03040, Barcelona, Spain, February 2012.

[LRYG14] C. Ling, S. Ruiz-Boqué, X. Ying, and M. Garcia-Lozano; "Optimal power allocation and relay selection in spectrum sharing cooperative networks"; General Assembly and Scientific Symposium (URSI GASS), 2014 XXXIth URSI, vol., no., pp. 1, 4, 16–23 Aug. 2014. (Also available as TD(14)10012).

[LTCS10] S. Y. Lien, C. C. Tseng, K. C. Chen, and C. W. Su. Cognitive radio resource management for qos guarantees in autonomous femtocell networks. In *IEEE International Conference on Communications (ICC)*, pp. 1–6, May 2010.

[LTDL14] H. Li, A. Tsiaras, B. Derat, and B. K. Lau. Analysis of SAR on flat phantom for different multi-antenna mobile terminals. In *8th European Conference on Antennas and Propagation (EuCAP'2014),* The Hague, The Netherlands, April 2014. (Also available as TD-14-09043).

[LTEM14] E. G. Larsson, F. Tufvesson, O. Edfors, and T. L. Marzetta. Massive MIMO for next generation wireless systems. *IEEE Communication Magazine*, 52(2):186–195, February 2014.

[LTY14] L. Tian, V. Degli-Esposti, E. M. Vitucci, X. Yin, F. Mani, S. X. Lu, "Semi-Deterministic Modeling of Diffuse Scattering Component Based on Propagation Graph Theory", In *Proceedings of the 25th IEEE annual International Symposium on Personal, Indoor and Mobile Radio Communications* (PIMRC'14), Washington DC, USA, 2–5 September 2014.

[LTZB13] R. Litjens, Y. Toh, H. Zang, and O. Blume. Assessment of the energy efficiency enhancement of future mobile networks. *COST IC1004 8th Scientific Meeting*, pp. 1–7, September 2013.

[Lue13] P. M. Luengo. Sensitivity Analysis and Self-Optimization of LTE Intra-Frequency Handover. Technical Report TD-13-06045, Málaga, Spain, February 2013.

[LXX+11] G. Ye Li, Z. Xu, C. Xiong, C. Yang, S. Zhang, Y. Chen, and S. Xu. Energy-efficient wireless communications: tutorial, survey, and open issues. *IEEE Wireless Communications*, 18(6):28–35, 2011.

[LZB+15] L. Laughlin, C. Zhang, M. Beach, K. Morris, and J. Haine. A widely tunable full duplex transceiver combining electrical balance isolation and analog cancellation. In *2015 IEEE Vehicular Technology Conference (VTC Spring)*, May 2015. (Also available as TD-15-12023).

[LZN+13] R. Litjens, H. Zhang, I. Noppen, L. Yu, E. Karipidis, and K. Bönier. System-level assessment of non-orthogonal spectrum sharing via transmit beamforming. *COST IC1004 6th MCM, Malaga, Spain*, (TD(13)06003), 2013.

M

[MAAU14] M. Mhedhbi, N. Amiot, S. Avrillon, and B. Uguen. Indoor on-body channel ray tracing and motion capture based simulation. Technical Report TD-14-10058, Aalborg, Denmark, May 2014.

[Mal09] A. Maltsev et al. Channel models for 60Hz WLAN systems. IEEE 802.11-09/0334r3, July 2009.

[Mal+10] A. Maltsev et al. Channel models for 60 GHz wlan systems. Technical Report doc. 802.11-09/0334r8, 2010.

[Maltsev] A. Maltsev, V. Erceg, and E. Perahia. Channel models for 60 GHz WLAN systems. Document IEEE 802.11-09/0334r8, 2010.

[MAB15] J. Medbo, H. Asplund, and J.-E. Berg. 60 GHz channel directional characterization using extreme size virtual array antenna. Technical report, TD(15)12055, Dublin, Ireland, January 2015.

[Mar10] T. L. Marzetta. Noncooperative cellular wireless with unlimited number of base station antennas. *IEEE Transactions on Wireless Communications*, 9(11):3590–3600, 2010.

[Mar14] Marta Fernandez, Iratxe Landa, Amaia Arrinda, Ruben Torre, and Manuel M. Velez, "Measurements of impulsive noise from specific sources in medium wave band"; IEEE Antennas and wireless propagation letters, July 2014, vol.13, pp. 1263 – 1266. (Also available as TD(14) 10016).

[Mas11] C. Masouros. Correlation rotation linear precoding for MIMO broadcast communications. *IEEE Transactions on Signal Processing*, 59(1):252–262, 2011.

[MB13] M. Mhedhbi, and S. Avrillon and B. Bernard. Modeling of UWB antenna disrupted by human phantom in spherical harmonic space. Technical Report TD(13)08001, Ghent, Belgium, September 2013.

[MB14] M. Moulu and A. Burr. Relay Beamforming with statistical channel state information. Technical Report TD-10-068, Aalborg, Denmark, May 2014.

[MBH13] H. Ma, T. W. C. Brown, and A. T. S. Ho. UWB near field detection of egg quantity in a fridge: A preliminary study. In *2013 Loughborough Antennas and Propagation Conference (LAPC)*, pp. 246–249, November 2013. (Also available as TD-14-10043).

[MBH+14] J. Medbo, K. Borner, K. Haneda, V. Hovinen, T. Imai, J. Jarvelainen, T. Jamsa, A. Karttunen, K. Kusume, J. Kyrolainen, P. Kyosti, J. Meinila, V. Nurmela, L. Raschkowski, A. Roivainen, and J. Ylitalo. Channel Modelling for the Fifth Generation Mobile Communications. In *Antennas and Propagation (EuCAP), 2014 8th European Conference on*, pp. 219–223, April 2014.

[MBV12] F. Martelli, Ch. Buratti, and R. Verdone. A Mathematical Model for Interference-limited Wireless Access Networks based on IEEE 802.15.4. Technical Report TD-12-04050, Lyon, France, May 2012.

[MBZV14] S. Mijovic, C. Buratti, A. Zanella, and R. Verdone. A cooperative beamforming technique for body area networks. In *Proceeding of International conference on selected topics in Mobile and Wireless Networking, (MoWNet2014)*, Rome, Italy, September 2014. (Also available as TD(14)09037).

[MC06] C. Medeiros and A. M. Castela. Reconfigurable antennas for multi-services (in portuges). Technical report, IST-TUL, Lisbon, Portugal, September 2006.

[MC12] M. Mackowiak and L. M. Correia. Towards a radio channel model for off-body communications in a multipath environment. In *Proceeding of 18th European Wireless Conference (EW2012)*, pp. 1–7, Poznan, Poland, April 2012. (Also available as TD(12)03009).

[MC13] M. Mackowiak and L. M. Correia. A statistical model for the influence of body dynamics on the gain pattern of wearable antennas in off-body radio channels. *Wireless Personal Communications, Springer*, 73(3):381–399, 2013. (Also available as TD(11)02017).

[MC14] M. Mackowiak and L. M. Correia. Statistical path loss model for dynamic off-body channels. In *Proceeding of PIMRC 2014 – IEEE 25th International Symposium on Personal, Indoor and Mobile Radio Communication*, pp. x–x, Washington, DC, USA, September 2014. (Also available as TD(14)09012).

[MCA12] A. Mahmood, M. Chitre, and M. A. Armand. PSK communication with passband additive symmetric α-stable noise. *IEEE Transactions on Communication*, 60(10):2990–3000, 2012.

[MCIŠ+11] C. Mehlführer, J. Colom Ikuno, M. Šimko, S. Schwarz, M. Wrulich, and M. Rupp. The Vienna LTE simulators-enabling reproducibility in wireless communications research. *EURASIP Journal on Advances in Signal Processing*, 2011(1), 2011.

[MD14] M. Matis, J. Dobos, and L. Papaj. Hybrid MANET DSR-DTN routing protocol. Technical Report TD-14-10020, Aalborg, Denmark, May 2014.

[MDDEV14] L. Minghini, R. D'Errico, V. Degli-Esposti, and E. M. Vitucci. Electromagnetic simulation and measurement of diffuse scattering from building walls. In *Proceedings of the 8th European Conference on Antennas and Propagation* (EuCAP 2014), The Hague, The Netherlands, 6–11 April 2014.

[MDPRD12] M. Maman, A. Di Paolo, R. Rosini, and R. D'Errico. Dynamic management of cooperative communications in Body Area Networks. In *Proceeding of Future Network Mobile Summit, (FutureNetw2012)*, Berlin, Germany, July 2012. (Also available as TD(11)02071).

[MDS+14] R. Müller, D. A. Dupleich, C. Schneider, R. Herrmann, J. Luo, and R. S. Thomä. Ultra-wideband millimetre-wave measurements at 70 GHz in an outdoor scenario for future cellular networks. Technical report, TD(14)11050, Krakow, Poland, September 2014.

[Med15] J. Medbo. 60 GHz channel directional characterization using extreme sise virtual antenna array. *COST IC1004, Dublin, Ireland*, 2015.

[METIS] METIS. Web site http://www.metis2020.com/

[MF11] P. Marsch, and G. P. Fettweis. *Coordinated MultiPoint in Mobile Communications*. Cambridge University Press, Cambridge, 2011.

[MFGC15] G. Martinez-Pinzon, M. Fuentes, C. Garcia-Pardo, and N. Cardona. Spectrum sharing for lte and dtt. study case: Indoor lte-a femtocell in dvb-t2 service area. Technical Report TD(15)12071, Action COST IC1004, Cooperative Radio Communications for Green Smart Environments, Dublin, Ireland, January 2015.

[MFP03] A. F. Molisch, J. R. Foerster, and M. Pendergrass. Channel models for ultrawideband personal area networks. *IEEE Transactions on Wireless Communications*, 10(6):14–21, 2003.

[MGCG13] H. B. Maad, A. Goupil, L. Clavier, and G. Gelle. Clipping demapper for ldpc decoding in impulsive channel. *IEEE Communication Letters*, 17(5):968–971, 2013.

[MGG+12] M. T. Martinez-Ingles, C. Garcia-Pardo, J. P. Garcia, J. Molina-Garcia-Pardo, J. Rodriguez, J. Reig, and L. Juan-Llacer. Initial experimental characterization of the millimeter-wave radio channel. In *Antennas and Propagation (EUCAP), 2012 6th European Conference on*, pp. 1118–1120, March 2012. (Also available as TD(12)03004).

[MGI06] J. Takada, M. Ghoraishi, and T. Imai. Identification of scattering objects in microcell urban mobile propagation channel. *IEEE Journal Antennas and Propagation*, 54(11):3473–3480, 2006.

[MGM+13] P. Meissner, M. Gan, F. Mani, E. Leitinger, M. Froehle, C. Oestges, T. Zemen, and K. Witrisal. On the Use of Ray Tracing for Performance Prediction of UWB Indoor Localisation Systems. In *IEEE ICC Workshop on Advances in Network Localization and Navigation (ANLN)*, Budapest, Hungary, 2013. (Also available as TD(13)07010).

[MGWW10] S. Marano, W. M. Gifford, H. Wymeersch, and M. Z. Win. Nlos identification and mitigation for localization based on UWB experimental data. *IEEE Journal on Selected Areas in Communications*, 28(7):1026–1035, 2010.

[MH13] A. Mahmud and K. A. Hamdi. Hybrid femtocell resource allocation strategy in fractional frequency reuse. In *2013 IEEE Wireless Communications and Networking Conference (WCNC)*, pp. 2283–2288, April 2013.

[MHS+08] R. Y. Mesleh, H. Haas, S. Sinanovic, C. W. Ahn, and S. Yun. Spatial modulation. *IEEE* Transactions *on Vehicle Technology*, 57(4):2228–2241, 2008.

[Mid77] D. Middleton. Statistical-physical models of electromagnetic interference. *IEEE Transactions on Electromagnetic Compatibility*, EMC-19(3):106–127, 1977.

[Min10] Minesok Kim, Yohei Konishi, and Jun-ichi Takada. Effect of IQ imbalance and phase noise in fully parallel MIMO channel sounder. Technical Report TD(11) 02005, Oct., 2010.

[Min15] Minesok Kim, Karma Wangchuk, Shigenobu Sasaki, Kazuhiko Fukawa, Jun-ichi Takada. Development of low cost mm-wave radio channel sounder and phase noise calibration. Technical Report TD(15) 12036, Jan., 2015.

[MIRMGP+12] M.-T. Martínez-Inglés, J.-V. Rodríguez, J.-M. Molina-Garcia-Pardo, J. Pascual-García, and L. Juan-Llácer, "Comparison of a UTD-PO Formulation for Multiple-Plateau Diffraction With Measurements at 62 GHz", IEEE Transactions on Antennas and Propagation, vol. 61, no. 2, 2013, pp. 1000–1003. DOI 10.1109/TAP.2012.2224836

[MiW14] MiWEBA Deliverable FP7-ICT 368721/D5.1, Channel modeling and characterization, June 2014.

[MiWEBA] MiWEBA. Channel modelling and characterization, ICT-608637-MiWEBA/D5.1, June 2014, available online.

[MKH+14] R. Müller, M. Käske, S. Häfner, P. Rauschenbach, G. Sommerkorn, C. Schneider, F. Wollenschläger, and R. S. Thomä, "Design of a circular antenna array for MIMO channel sounding application at 2.53 GHz," in 8th European Conference on Antennas and Propagation (EuCAP), 2014, pp. 239–243. The Hague, Netherlands, 6–11 April 2014, Piscataway, NJ: IEEE. http://dx.doi.org/10.1109/EuCAP.2014.6901734 (Also available as TD(14)10024).

[MKT14] G. Sommerkorn, M. Käske, C. Schneider, and R. Thomä. Vehicle-to-vehicle double-directional MIMO channel sounding campaign for the RESCUE project. In *COST IC1004 10th MC & Scientific Meeting*, number TD(14)11046, 2014.

[ML12] M. Mackowiak and Correia L. M. Correlation analysis of off-body radio channels in a street envirnment. In *Proceeding of PIMRC 2012 – IEEE 23th International Symposium on Personal, Indoor and Mobile Radio Communication*, pp. 1769–1773, Sydney, Australia, September 2012. (Also available as TD(12)05016).

[ML13] M. Mackowiak and L. M. Correia L. M. MIMO capacity performance of off-body radio channels in a street environment. In *Proceeding of PIMRC 2013 – IEEE 24th International Symposium on Personal, Indoor and Mobile Radio Communication*, pp. x–x, London, UK, September 2013. (Also available as TD(13)08003).

[ML14] Z. Miers and B. K. Lau. Characteristic mode tracking with farfield patterns. Technical Report TD-14-11048, Krakow, Poland, September 2014.

[MLFW13] P. Meissner, E. Leitinger, M. Froehle, and K. Witrisal. Accurate and robust indoor localisation systems using ultra-wideband signals. In *European Navigation Conference (ENC)*, Vienna, Austria, 2013. (Also available as TD(13)0711).

[Mli11] F. Mlinarsky. MIMO OTA test methodology consideration for small anechoic chambers. Technical Report TD(11)01003, 1st COST IC1004 Scientific Meeting, Lund, Sweden, June 2011.

[MLL13] Z. Miers, H. Li, and B. K. Lau. Design of multi-antenna feeding for MIMO terminals based on characteristic modes. In *IEEE Antennas and Propagation Society International Symposium (APSURSI),* Orlando, USA, Jul. 7–13, 2013. (Also available as TD-13-08041).

[MLL14] Z. Miers, H. Li, and B. K. Lau. Design of bezel antennas for multiband MIMO terminals using characteristic modes. In *8th European Conference on Antennas and Propagation (EuCAP'2014),* pp. 2556–2560, The Hague, The Netherlands, April, 2014. (Also available as TD-14-10075).

[MLLW13] P. Meissner, E. Leitinger, M. Lafer, and K. Witrisal. Measure MINT UWB database, 2013. Publicly available database of UWB indoor channel measurements (www.spsc.tugraz.at/tools/UWBmeasurements).

[MLLW14] P. Meissner, E. Leitinger, M. Lafer, and K. Witrisal. Real-time demonstration of multipath-assisted indoor navigation and tracking (MINT). In *2014 IEEE International Conference on, Communications Workshops (ICC)*, pp. 144–149, June 2014. (Also available as TD(14)10003).

[MLW14] P. Meissner, E. Leitinger, and K. Witrisal. UWB for robust indoor tracking: Weighting of multipath components for efficient estimation. *IEEE Wireless Communication Letter*, 3(5):501–504, 2014 (An extended version is available as TD(15)12067).

[MLW15] P. Meissner, E. Leitinger, and K. Witrisal. Robust and Accurate Indoor Localization using Channel Information. In *COST Action IC1004 Scientific Meeting*, Dublin, Ireland, 2015 (TD(15)12067).

[MM99] J. Mitola and G. Q. Jr. Maguire, Cognitive radio: making software radios more personal. *IEEE Personal Communications*, 6(4):13–18, 1999.

[MME13] R. Brem, M. Mocker, and T. Eibert. Ray tracing with improved accuracy for virtual drive simulation. In *IC1004 6th MC and Scientific Meeting*, Malaga, 2013.

[MMIJL12] M.-T. Martinez-Ingles, D. P. Gaillot, J. Pascual-García, J. M. Molina-Garcia-Pardo, M. Liénard and J.-V. Rodríguez, "Deterministic and Experimental Indoor mmW Channel Modeling", IEEE Antennas and Propagation Letters, vol. 13, pp. 1047–1050. December 2014.

[MMP+14] M.-T. Martinez-Ingles, J.-M. Molina-Garcia-Pardo, J. Pascual-García, J.-V. Rodríguez and L. Juan-Llácer, "Experimental Comparison between Centimeter-and Millimeter-Wave Ultra-Wideband Radio Channels", Radio Sci., 49, 450–458, doi:10.1002/2014RS005439. 2014.

[MMS+10] A. Maltsev, R. Maslennikov, A. Sevastyanov, A. Lomayev, and A. Khoryaev. Statistical channel model for 60 GHz WLAN systems in conference room environment. In *4th European Conference on Antennas and Propagation (EuCAP 2010)*, Barcelona, pp. 1–5, 2010.

[MMS14] T. Mazloum, F. Mani, *and* A. Sibille. A disc of scatterers based radio channel model for secure key generation. In *Proceeding of IEEE 8th European Conference on Antennas and Propagation (EuCAP2014)*, pp. 1290–1294, April 2014.

[MN12] J. Medbo and A. Nilsson. Leaky Coaxial Cable MIMO Performance in an Indoor Office Environment. In *IEEE 23rd International Symposium on Personal Indoor and Mobile Radio Communications (PIMRC), 2012*, pp. 2061–2066, September 2012.

[MnL13] P. Muñoz Luengo. Sensitivity Analysis and Self-Optimization of LTE Intra-Frequency Handover. Technical Report TD-13-06045, Málaga, Spain, February 2013.

[MNRK14] A. Moäller, J. Nuckelt, D. M. Rose, and T. Kürner. Physical layer performance comparison of LTE and IEEE 802.11p for vehicular communication in an urban NLOS scenario. In *Proceedings VTC 2014 Fall – IEEE 80th Vehicular Technology Conference*, pp. 1–5, September 2014. (Also available as TD(14)10010).

[MO11] M. Maman and L. Ouvry. BATMAC: An adaptive TDMA MAC for body area networks performed with a spacetime dependent channel model. In *Proceeding of IEEE International Symposium on Medical Information and Communication Technology (ISMICT 2011)*, pp. 1–5, Montreux, Switzerland, March 2011.

[MO12] F. Mani and C. Oestges. A ray based method to evaluate scattering by vegetation elements. In *IC1004 4th MC and Scientific Meeting*, Lyon, May 2012, Technical Report TD(12)04018.

[MOC12a] M. Machkowiak, C. Oliveira, and L. M. Correia. A Comparison of Phantom Models for On-Body Communications", in Proc. of PIMRC'2012 – 23rd IEEE Symposium on Personal, Indoor, Mobile and Radio Communications, Sydney, Australia, Sep. 2012. (Also available as TD(12)04038).

[MOC12b] M. Machkowiak, C. Oliveira, and L. M. Correia. Radiation Pattern of Wearable Antennas: A Statistical Analysis of the Influence of the Human Body. Springer Int. Journal of Wireless Information Networks, July 2012. (Also available as TD(12)03010).

[Mor99] S. P. Morgan. Prediction of Indoor Wireless Coverage by Leaky Coaxial Cable using Ray Tracing. *Vehicular Technology, IEEE Transactions on*, 48(6):2005–2014, 1999.

[Mosl4] M. Mostafa. *Equalization and decoding: A continuous-time dynamical approach.* PhD thesis, University of Ulm, Institute of Communications Engineering, Ulm, Germany, 2014.

[Moz12] P. Mozola. The Compatibility Analyses between DVB-H and DVB-T2. Technical Report TD(12)03049, Action COST IC1004, Cooperative Radio Communications for Green Smart Environments, Barcelona, Spain, February 2012.

[Moz14] P. Mozola. The Compatibility Analyses Between Broadcast Systems. Technical Report TD(14)09072, Action COST IC1004, Cooperative Radio Communications for Green Smart Environments, Ferrara, Italy, February 2014.

[MPDD+13] T. Mavridis, L. Petrillo, P. De Doncker, J. Sarrazin, D. Lautru, and A. Benlarbi-Dela. A 60 GHz indoor off-body channel implementation. In *2013 IEEE Antennas and Propagation Society International Symposium (APSURSI)*, pp. 1786–1787, July 2013. (Also available as TD(13) 08025 with title A 60 GHz Indoor Off-Body Channel Implementation).

[MPK+14] A. Maltsev, A. Pudeyev, I. Karls, I. Bolotin, G. Morozov, W. Keusgen, R. J. Weiler, M. Peter, M. Danchenko, and A. Kuznetsov. Quasi-deterministic Approach to mmWave Channel Modeling in a Non-stationary Environment, IEEE

Globecom; Workshop on Emerging Technologies for 5G Wireless Cellular Networks, Austin, TX, USA, Dec. 2014. (Also available as TD(14)11065).

[MPM55] D. A. McNamara, C. W. I. Pistorius, and J. A. G. Malherbe. *Introduction to the Uniform Geometrical Theory of Diffraction*. Artech House, Norwood, MA, 1955.

[MPR+13] M.-T. Martinez-Ingles, J. Pascual-Garcia, J. Rodriguez, J. Molina-Garcia-Pardo, L. Juan-Llacer, D. P. Gaillot, M. Lienard, and P. Degauque. Indoor radio channel characterization at 60 ghz. In *Antennas and Propagation (EuCAP), 2013 7th European Conference on*, pp. 2796–2799, April 2013. (Also available as TD(12)05050).

[MPS+14a] T. Mavridis, L. Petrillo, J. Sarrazin, D. Lautru, A. Benlarbi-Delai, and P. De Doncker. Theoretical and experimental investigation of a 60 GHz off-body propagation model. *IEEE Translation on Antennas and Propagation*, 62(1):393–402, 2014.

[MPS+14b] T. Mavridis, L. Petrillo, J. Sarrazin, D. Lautru, A. Benlarbi-Delai, and P. De Doncker. Theoretical and experimental investigation of a 60-GHz off-body propagation model. *IEEE Translation Antennas Propagation*, 62(1):393–402, 2014.

[MQO11] F. Mani, F. Quitin, and C. Oestges. Directional spreads of dense multipath components: a ray-tracing approach. Technical Report TD(11)02003, Lisbon, Portugal, October 2011.

[MQO12] F. Mani, F. Quitin, and C. Oestges. Directional spreads of dense multipath components in indoor environments: experimental validation of a ray-tracing approach. *IEEE Transactions on Antennas and Propagation*, 60(7): 3389–3396, 2012.

[MRDC13] M. Mackowiak, R. Rosini, R. D'Errico, and L. M. Correia. Comparing off-body dynamic channel model with realtime measurements. In *Proceeding of 7th International Symposium on Medical Information and Communication Technology (ISMICT 2013)*, pp. 121–125, Tokyo, Japan, March 2013. (Also available as TD(13)06006).

[MRH09] D. Matolak, K. Remley, and C. Holloway. Outdoor-to-indoor channel measurements and models. Technical report, CTIA, September 2009.

[MS11] Z. Mhanna and A. Sibille. Statistical modeling of the power gain pattern of a random set of parameterized planar dipoles. Technical Report TD-11-01051, Lund, Sweden, June 2011.

[MS14] Z. Mhanna and A. Sibille. Monostatic vs. bistatic backscattering gain for UWB RFID real time location systems. In *COST Action IC1004 Scientific Meeting*, Krakow, Poland, 2014 (TD(14)11059).

[MSLM13] T. F. Eibert, M. S. L. Mocker, and R. Brem. Ray tracing with improved accuracy for virtual drive simulation. In *COST IC1004 6th MC & Scientific Meeting*, number TD(13)06023, 2013.

[MSM12] S. De Moor, D. Schuurman, and L. De Marez. Digimeter report 5: Adoption and usage of Media & ICT in Flanders – Wave 5. August–September 2012.

[MSM+13] M.-T. Martinez-Ingles, C. Sanchis-Borras, J.-M. Molina-Garcia-Pardo, J.-V. Rodriguez, and L. Juan-Llacer. Experimental evaluation of an indoor mimo-ofdm system at 60 ghz based on the ieee802.15.3c standard. *Antennas and Wireless Propagation Letters, IEEE*, 12:1562–1565, 2013. (Also available as TD(13)08004).

[MST11] Z. Meifang, A. Singh, and F. Tufvesson. Measurement based ray launching foe analyses of outdoor propagation. Technical Report TD(11)03022, Barcelona, Spain, February 2011.

[MT13] F. Kaltenberger, C. Oestges, M. Gan, F. Mani, and T. Zemen. A ray tracing algorithm using the discrete prolate spheroidal subspace. In *IC1004 7th MC and Scientific Meeting*, 2013.

[MTL10] M. Mostafa, W. G. Teich, and J. Lindner. A modified discrete recurrent neural network as vector detector. In *Circuits and Systems (APCCAS), 2010 IEEE Asia Pacific Conference on*, pp. 620–623, Kuala Lumpur, Malaysia, December 2010.

[MTLlla] M. Mostafa, W. G. Teich, and J. Lindner. Analog iterative decoding based on recurrent neural networks. Technical Report TD-01-011, Lund, Sweden, June 2011.

[MTLllb] M. Mostafa, W. G. Teich, and J. Lindner. Global vs. local stability for recurrent neural networks as vector equalizer. In *2011 5th International Conference on Signal Processing and Communication Systems (ICSPCS)*, pp. 1–5, Honolulu, HI, December 2011. (Also available as TD(03)021).

[MTL12] M. Mostafa, W. G. Teich, and J. Lindner. Iterative methods: a continuous-time dynamical approach. Technical Report TD-05-042, Bristol, UK, September 2012.

[MTL14a] M. Mostafa, W. G. Teich, and J. Lindner. Approximation of activation functions for vector equalization based on recurrent neural networks. In *2014 8th International Symposium on Turbo Codes and Iterative Information Processing (ISTC)*, pp. 52–56, Bremen, Germany, August 2014. (Also available as TD(12)028).

[MTL14b] M. Mostafa, W. G. Teich, and J. Lindner. Comparison of belief propagation and iterative threshold decoding based dynamical systems. IEEE International Symposium on Information Theory ISIT, Istanbul, Turkey, 2013. (Also available as TD(14)09010).

[MTL14c] M. Mostafa, W. G. Teich, and J. Lindner. Local stability analysis of discrete-time, continuous-state, complex-valued recurrent neural networks with inner state feedback. *IEEE Transactions on Neural Networks and Learning Systems*, 25(4):830–836, 2014.

[MTT+13] M. Marinova, E. Tanghe, A. Thielens, L. Vallozzi, G. Vetmeeren, and J. Wout. Diversity performance of off-body UWB-MIMO. Technical Report TD(13)08032, Ghent, Belgium, September 2013.

[MV12] F. Martelli and R. Verdone. Coexistence issues for wireless body area networks at 2.45 GHz. In *Proceeding of 18th European Wireless Conference European Wireless (EW 2012)*, pp. 1–6, Poznan, Poland, April 2012.

[MVK13a] K. Maliatsos, P. N. Vasileiou, and A. G. Kanatas. Channel estimation and link level evaluation of adaptive beamspace MIMO systems. In *2013 3rd International Conference on Wireless Communications, Vehicular Technology, Information Theory and Aerospace Electronic Systems (VI-TAE)*, pp. 1–5, June 2013. (Also available as TD(13)07021).

[MVK13b] K. Maliatsos, P. N. Vasileiou, and A. G. Kanatas. V-BLAST reception for beamspace MIMO systems with limited feedback. In *2013 IEEE 24th International Symposium on Personal Indoor and Mobile Radio Communications (PIMRC)*, pp. 1034–1039, September 2013. (Also available as TD(13)07022).

[MW12] P. Meissner and K. Witrisal. Analysis of Position-Related Information in Measured UWB Indoor Channels. In *6th European Conference on Antennas and Propagation*

(EuCAP), Prague, Czech Repuplic, 2012. (Also available as TD(12)03001).

[MZK13] N. Mataga, R. Zentner, and A. Katalini´c. Ray entity based postprocessing of ray-tracing data for continuous modeling of radio channel, Radio science, vol.49 (2014), no.3; pp. 217–230. (Also available as TD(13)08073).

N

[NA13] J. Navarro, R. Andres. A new method for spectrum monitoring networks design. In *2013 IEEE Antennas and Propagation Society International Symposium (APSURSI)*, 2013.

[NAH15] A. Narbudowicz, M. J. Ammann, and D. Heberling. Compact reconfigurable antennas for MIMO. In *9th European Conference on Antennas and Propagation (EuCAP)*, Lisbon, Portugal, April 2015. (Also available as TD-15-12025).

[NAT+13] J. Nuckelt, T. Abbas, F. Tufvesson, C. Mecklenbräuker, L. Bernado, and T. Kürner. "Comparison of ray tracing and channel-sounder measurements for vehicular communications." In Vehicular Technology Conference (VTC Spring), 2013 IEEE 77th. June 2013. (Also available as TD-05-031).

[Nav13] A. Navarro. A discussion about measurement-based ray-tracing models calibration in urban environments. Technical Report TD(13)06035, Malaga, Spain, February 2013.

[NAV+14] A. Navarro, A. Arteaga, L. Vargas, J. Aguilar, and M. Arciniegas. Spectrum monitoring system and benchmarking of mobile networks using open software radios simones. In *2014 IEEE Latin-America Conference on Communications (LATINCOM)*, 2014.

[NAVA12] A. Navarro, A. Arteaga, L. Vargas, and J. Aguilar. Spectrum monitoring. *ITU News*, 2012.

[NBS14] K. K. Nagalapur, F. Brännström, and E. G. Ström. On channel estimation for 802.11p in highly time-varying vehicular channels. In *Proceedings ICC 2014 – IEEE International Conference on* Communication, pp. 5659–5664, June 2014. (Also available as TD(14)10008).

[NC11] N. Chahat, M. Zhadobov, R. Sauleau, and K. Ito. A compact UWB antenna for on-body communication. *IEEE Translation on Antenna and Propagation*, 59(4):1123–1131, 2011.

[NCC+14] E. C. Neira, U. Carlberg, J. Carlsson, K. Karlsson, and E. G. Ström. Evaluation of V2X antenna performance using a multipath simulation tool. In *8th European Conference on Antennas and Propagation (EuCAP)*, pp. 2534–2538, The Hague, The Netherlands, April 2014. (Also available as TD(14)10039).

[NEGJ11] N. Farah, E. Peiker-Feil, W. G. Teich, and J. Lindner Comparison of CDD and MC-*CAFS* for the MIMO-OFDM DL in LTE. In *Proceedings of 6th International OFDM-Workshop (InOWo'11)*, Vol. 1, pp. 81–85, August 2011. (Also available as TD(12)04046).

[NFP13] J. Ø. Nielsen, W. Fan, and G. F. Pedersen. Characterization of interference for over the air terminal testing. Technical Report TD(13)07041, 7th COST IC1004 Scientific Meeting, Ilmenau, Germany, May 2013.

[NFSS03] V. Nikolopoulos, M. Fiacco, S. Stavrou, and S. R. Saunders. Narrowband fading analysis of indoor distributed antenna systems. *Antennas and Wireless Propagation Letters, IEEE*, 2(1):89–92, 2003.

[NG11] B. Nazer and M. Gastpar. Compute-and-forward: Harnessing interference through structured codes. *IEEE Trans. Inf. Theory*, 57(10):6463–6486, Oct. 2011.

[NGMN] NGMN. 5G White Paper, Next Generation Mobile Networks (NGMN) Alliance, 17 February 2015.

[NGMN13] NGMN, Suggestions on Potential Solutions to C-RAN by NGMN Alliance, Technical report, January 2013.

[NHA+14] M. G. Nilsson, P. Hallbjörner, N. Arabäck, B. Bergqvist, T. Abbas, and F. Tufvesson. "Measurement uncertainty, channel simulation, and disturbance characterization of an over-the-air multiprobe setup for cars at 5.9 GHz." IEEE Transactions on Industrial Electronics, vol. 62, no. 12, pp. 7859–7869, Dec 2015. (Also available as TD-10-029).

[NIC12] NICTA. Body area network radio channel data. Online, December 2012. http://tinyurl.com/jwp55gs

[Nik12] Niklas Jalden. Channel extrapolation based on wideband MIMO channel measurements. Technical Report TD(12) 03020, February, 2012.

[NKS11] M. Narandzic, M. Käske, and G. Sommerkorn and S. Jackel, and C. Schneider and R. S. Thomä. Variation of estimated Large-scale MIMO channel properties between repeated measurements. In *2011 IEEE* 73rd *Vehicular Technology Conference (VTC Spring),* 2011. doi: 10.1109/ VETECS.2011.5956676, 1550–2252

[NL14] A. Navarro and S. Londoño. A proposal for a 3D reference scenario for ray tracing systems. Technical Report TD(14)10-093, Aalborg, Denmark, May 2014.

[NLM13] H. Ngo, E. G. Larsson, and T. L. Marzetta. Energy and spectral efficiency of very large multiuser MIMO systems. *IEEE Transactions on Wireless Communications*, 61(4):1436–1449, 2013.

[NM65] J. A. Nelder and R. Mead. A simplex method for function minimisation. *Computer Journal*, 7:308–313, 1965.

[NO15] K. Kitao, N. Omaki, T. Imai, and Y. Okumura. Impact of building shape in the intersection on ray tracing accuracy for non line-of-site (nlos) scenario in street cell environment. In *IC1004 12th MC and Scientific Meeting*, Dublin, 2015.

[NPGP13] A. Navarro, A. Pachon, and U. Garcia-Palomares. A novel approach to optimal resource allocation in hetnets based on the cinr requested by users. Temporary document, IC1004 COST action, May 2013.

[NPHaBT13] M. Nilsson, N. Arabäck, P. Hallbjörner, B. Bergqvist, and F. Tufvesson. Multipath propagation simulator for V2X communication tests on cars. In *7th European Conference on Antennas and Propagation*, pp. 1342–1346, Gothenburg, Sweden, 2013. In parts also available as TD(14)10029.

[NRJK13] J. Nuckelt, D. M. Rose, T. Jansen, and T. Kürner. On the use of OpenStreetMap data for V2X channel modeling in urban scenarios. In *Proceedings 7th European Conference on Antennas and Propagation (EuCAP)*, pp. 3984–3988, April 2013. (Also available as TD(13)07005).

[NSK15] M. Narandzic, C. Schneider, and W. Kotterman, and Thomä, R. S. Required number of propagation scenarios for acceptable reproduction of spectral efficiency distribution in (heterogeneous) network simulations. In *Proceedings of 2015 9th European Conference on Antennas and Propagation (EuCAP),* 2015. (Also available as TD-11-022).

[NSK11] J. Nuckelt, M. Schack, and T. Kürner. Performance evaluation of wiener filter designs for channel estimation in vehicular environments. In *Proceedings of VTC 2011 Fall – IEEE 74th Vehicular Technology Conference*, pp. 1–5, September 2011. (Also available as TD(11)02031).

[NS95] C. L. Nikias and M. Shao. *Signal Processing with Alpha-Stable Distributions and Applications.* John Wiley and Sons, Inc., New Jersy, 1995.

[NVA+15] M. Nilsson, D. Vlastaras, T. Abbas, B. Bergqvist, and F. Tufvesson. On multilink shadowing effects in measured V2V channels on highway. In *Proceedings 9th European Conference on Antennas and Propagation (EuCAP)*, 2015.

O

[Obr13] N. Obriot. Impact of random phases on SCME channel model ergodicity. Technical Report TD(13)08059, 8th COST IC1004 Scientific Meeting, Gent, Belgium, September 2013.

[OC04] Y. Okano and K. Cho. Monopole antenna array arrangement for card-type mobile terminal. In *IEEE Radio and Wireless Conference*, pp. 415–418, September 2004.

[OC13] N. Obaid and A. Czylwik. The impact of deploying pico base stations on capacity and energy efficiency of heterogeneous cellular networks. Temporary document, IC1004 COST action, September 2013.

[OC13a] C. Oliveira and L. M. Correia. Perspectives for the use of MIMO in Dynamic Body Area Networks. in Procceedings of EuCAP'2013 – 7th European Conf. on Antennas and Propagation, Gothenburg, Sweden, Apr. 2013. (Also available as TD(13)06014).

[OC13b] C. Oliveira and L. M. Correia. Perspectives for the use of mimo in dynamic body area networks. In *Antennas and Propagation (EuCAP), 2013 7th European Conference on*, pp. 482–486, April 2013. (Also available as TD(13)06014).

[OC13c] C. Oliveira and L. M. Correia. Signal correlation and power imbalance in dynamic on-body communications. In *Vehicular Technology Conference (VTC Spring), 2013 IEEE 77th*, pp. 1–5, June 2013. (Also available as TD(12)0515).

[OC13d] C. Oliveira and L. M. Correia. Signal correlation and power imbalance in dynamic on-body communications. In *Vehicular Technology Conference (VTC Spring), 2013 IEEE 77th*, pp. 1–5, June 2013. (Also available as TD(12)05015).

[OCD+12] C. Oestges, N. Czink, P. De Doncker, V. Degli-Esposti, K. Haneda, W. Joseph, M. Liénard, L. Liu, J. Molina-García-Pardo, M. Narandžić, J. Poutanen, F. Quitin, and E. Tanghe. *Pervasive Mobile and Ambient Wireless Communications – COST Action 2100*, chapter Radio Channel Modelling for 4G Networks, pp. 67–148. Springer, 2012.

[Oes14] C. Oestges. Experimental validation of ray-tracing at 12 and 30 GHz. Technical report, TD(14)11002, Krakow, Poland, September 2014.

[Ofc09] Digital dividend: cognitive access. Statement on licence-exempting cognitive devices using interleaved spectrum. Technical report, Ofcom, July 2009.

[Ofc10] Implementing geolocation. Technical report, Ofcom, November 2010.

[OFGZ13] J. Olmos, R. Ferrús, and H. Galeana-Zapién. A new cell selection algorithm for mobile networks with limited backhaul capacity. Technical Report TD(13)08064, Ghent, Belgium, September 2013.

[OKI09] H. Okamoto, K. Kitao, and S. Ichitsubo. Outdoor-to-indoor propagation loss prediction in 800-MHz to 8-GHz band for an urban area. *IEEE Transactions on Vehicular Technology*, 58(3): 1059–1067, 2009

[OM12] C. Oliveira, M. Mackowiak, and L. M. Correia. Correlation analysis in on-body communications. In *Antennas and Propagation (EUCAP), 2012 6th European Conference on*, pp. 3383–3387, March 2012. (Also available as TD(11)02021).

[OSM+14] Giuseppe Oliveri, Mohamad Mostafa, Werner G. Teich, Jürgen Lindner, Hermann Schumacher, "Advanced Low-Power High-Speed Nonlinear Signal Processing: An Analog VLSI Example", *Proceedings WInnComm-Europe 2015*, Erlangen, 05–09. October 2015, pp. 48–56. (Also available as TD(14)10037).

[OSRL12] J. Olmos, A. Serra, S. Ruiz, and I. Latif. On the Use of Mutual Information at Bit Level for Accurate Link Abstraction in LTE with Incremental Redundancy H-ARQ. *COST IC1004 5th MCM, Bristol, UK*, (TD(12)05046), 2012.

[OZW10] H. Osman, H. Zhu, and J. Wang. Downlink distributed antenna systems in indoor high building femtocell environments. In *IEEE 21st International Symposium on Personal Indoor and Mobile Radio Communications (PIMRC)*, pp. 1016–1020, September 2010.

P

[PAC12] S. Piersanti, L. A. Annoni, and D. Cassioli. Millimeter waves channel measurements and path loss models. In *IEEE International Conference on Communications (ICC)*, pp. 4552–4556, 2012.

[PD14] P. O Pasquero and R. D'Errico. Angle of arrival characterization in UWB indoor off-body channel. Technical Report TD(14)09052, Ferrara, Italy, February 2014.

[PDC12] J. Papaj, L. Dobos, and A. Cizmar. An overview of security in Opportunistic networks. Technical Report TD-12-05004, Bristol, United Kindom, September 2012.

[PDC13] J. Papaj, L. Dobos, and A. Cizmar. Performance analysis of the enhanced DSR routing protocol for the temporary disconnected MANET. Technical Report TD-13-07042, Illmenau, Germany, May 2013.

[PDID13] M. Pejanović-Djurišić and M. Ilic-Delibašić. A Solution for Improving Energy Efficiency of Relay Systems. *COST IC1004 7th MCM, Ilmenau, Germany*, (TD(13)07027), 2013.

[PDP14] J. Papaj, L. Dobos, and R. Palitefka. Trust based candidate node selection in hybrid MANET – DTN. Technical report, May 2014.

[PDRR+12a] D. Perez-Daz de Cerio, S. Ruiz, J. Rosell-Ferrer, J. Ramos-Castro, and J. M. Colome. Help4Mood: Testing and achievements of the first wireless sensor network prototype. Technical Report TD-12-03033, Barcelona, Spain, February 2012.

[PDRR+12b] D. Perez-Daz de Cerio, S. Ruiz, J. Rosell-Ferrer, J. Ramos-Castro, and J. M. Colome. Mobile ubiquitous system for distance health monitoring and preliminary medical diagnostic using GSM network. Technical Report TD-12-03033, Barcelona, Spain, February 2012.

[Peh11] J. M. Peha. Cellular systems and rotating radar using the same spectrum. In *Proceeding of ISART, Boulder, CO, USA*, July 2011.

[PEK+14] A. Posselt, L. Ekiz, O. Klemp, B. Geck, and C. F. Mecklenbräuker. System level evaluation for vehicular MIMO antennas in simulated and measured channels. In *8th European Conference on Antennas and Propagation (EuCAP)*, pp. 3487–3490, The Hague, The Netherlands, April 2014. (Also available as TD(14)09029).

[PF07] T. Pedersen and B. Fleury. Radio channel modelling using stochastic propagation graphs. In *Proceeding of ICC 2007 – IEEE International Conference Communication*, Glasgow, Scotland, June 2007.

[PF11a] C. L. Patane and M. Franzen. MIMO LTE Round Robin: Analysis. Technical Report TD(11)01030, 1st COST IC1004 Scientific Meeting, Lund, Sweden, June 2011.

[PF11b] C. L. Patane and M. Franzen. MIMO LTE Round Robin: Results, Accuracy and Repeatability. Technical Report TD(11)01029, 1st COST IC1004 Scientific Meeting, Lund, Sweden, June 2011.

[PF14] E. Peiker-Feil. *Increasing the bandwidth efficiency of OFDM-MFSK*. PhD thesis, University of Ulm, Institute of Communications Engineering, Ulm, Germany, 2014.

[PFE+14] A. Posselt, A. Friedrich, L. Ekiz, O. Klemp, and B. Geck. System-Level assessment of volumetric 3D vehicular MIMO antenna based on measurement. In *International Conference on Connected Vehicles & Expo (ICCVE)*, Vienna, Austria, November 2014. (Also available as TD(14)11011).

[PFMJ12] M. Porcius, C. Fortuna, M. Mohorcic, and T. Javornik. TopoSWiM: A Tool for Topology Design and Control in Large Scale Wireless Mesh Networks. Technical Report TD-12-04037, Lyon, France, May 2012.

[PFP12] M. Pelosi, O. Franek, and G. Pedersen. The effect of the user's body on high-Q and low-Q planar inverted F antennas for LTE frequencies. Technical Report TD-12-03041, Barcelona, Spain, February 2012.

[PFTL13] E. Peiker-Feil, W. G. Teich, and J. Lindner. Extending the bandwidth efficiency of OFDM-MSFSK. Technical Report TD-06-046, Malaga, Spain, February 2013.

[PFWTL12] E. Peiker-Feil, M. Wetz, W. G. Teich, and J. Lindner. OFDM-MSFSK as a special case of noncoherent communication based on subspaces. Technical Report TD-05-017, Bristol, UK, September 2012.

[Pie15] Pierre Laly, Davy P. Gaillot, Emmeric Tanghe, Eric Simon, Martine Lienard, Wout Jospeh and Luc Martens. MIMOSA: A real time MIMO channel sounder based on highly flexible software architecture. Technical Report TD(15) 01021, January, 2015.

[PJA+14] D. Plets, W. Joseph, S. Aerts, K. Vanhecke, G. Vermeeren, and L. Martens. Prediction and comparison of downlink electric-field and uplink localised SAR values for realistic indoor wireless planning. *Radiation Protection Dosimetry*, 2014. (Also available as TD(13)08026).

[PJV+12] D. Plets, W. Joseph, K. Vanhecke, E. Tanghe, and L. Martens. Coverage prediction and optimisation algorithms for indoor environments. *EURASIP Journal of Wireless Communication and Networks*, 2012.

[PK11] Z. Pi and F. Khan. An introduction to millimeter-wave mobile broadband systems. *IEEE M COM*, (6):101–107, 2011.

[PK13a] V. Peng and T. Kürner. Iterative Power Control Algorithm for Maximum Data Rate Sum in Cellular Networks with Distributed Implementation. Technical Report TD-13-07006, Ilmenau, Germany, May 2013.

[PK13b] V. Petrini and H. R. Karimi. Tv white spaces data bases: Algorithms for the calculation of maximum permitted radiated power levels and application to a real secenario. Technical Report TD(13)06019, Action COST IC1004, Cooperative Radio Communications for Green Smart Environments, Ilmenau, Germany, February 2013.

[PKHAM13] B. Peng, T. Kürner, J. Häfen, and A. Ayadi-Miessen. Application of Soft Frequency Reuse for the Realistic Irregular and Sectorised Cells in Cellular Mobile Networks. Technical Report TD-13-07007, Ilmenau, Germany, May 2013.

[PLK+12] D. Parveg, T. Laitinen, A. Khatun, V.-M. Kolmonen, and P. Vainikainen, "Calibration Procedure for 2-D MIMO Over-The-Air Multi-Probe Test System," 6th European Conference on Antennas and Propagation (EuCAP 2012), Prague, Czech Republic, pp. 1594–1598, March 2012. (Also available as TD(11)02060).

[PMB13] V. Petrini, M. Missiroli, and M. Barbiroli. A C/I based approach to setting the maximum eirp levels for database-assisted wsds. Technical Report TD(14)09070, Action COST IC1004, Cooperative Radio Communications for Green Smart Environments, Ferrara, Italy, February 2013.

[PMK+11] P. Paschalidis, K. Mahler, A. Kortke, M. Wisotzki, M. Peter, and W. Keusgen. 2×2 MIMO measurements of the wideband car-to-car channel at 5.7 GHz on urban street intersections. In *Proceedings IEEE Vehicular Technology Conference (VTC Fall)*, pp. 1–5, Sept 2011. (Also available as TD(11)02036).

[PMK+12] P. Paschalidis, K. Mahler, A. Kortke, M. Peter, and W. Keusgen. Statistical evaluation of multipath component lifetime in the car-to-car channel at urban street intersections based on geometrical tracking. In *Proceedings IEEE Vehicular Technology Conference (VTC Spring)*, pp. 1–5, May 2012. (Also available as TD(13)06054).

[PMOR12] R. Protzmann, K. Mahler, K. Oltmann, and I. Radusch. Extending the V2X simulation environment VSim-RTI with advanced communication models. In *12th International Conference on ITS Telecommunications (ITST)*, pp. 683–688, Taipei, Taiwan, November 2012. (Also available as TD(14)09071).

[PNSR14] S. K. Pal, T. A. Nugraha, S. Shams, and A. Rahman. Resource allocation strategy using optimal power control for mitigating two-tier interference in heterogeneous networks. In *IEEE Wireless Communications and Networking Conference Workshops (WCNCW)*, pp. 104–109, April 2014.

[PR12] R. Pirkl and K. A. Remley. MIMO channel capacity in 2-D and 3-D isotropic environments. *International Journal of Antennas and Propagation*, 2012 (Special Issue on MIMO OTA), 2012.

[PRDO14] O. P. Pasquero, R. Rosini, R. D'Errico, and C. Oestges. On-body channel correlation in various walking scenarios. In *Antennas and Propagation (EuCAP), 2014 8th European Conference on*, pp. 840–844, April 2014. (Also available as TD(13)07001).

[PS12] P. Prochazka and J. Sykora. Block-structure based extended layered design of network coded modulation for arbitrary individual rates using hierarchical decode & forward strategy. Technical Report TD05054, 2012.

[PS13] P. Prochazka and J. Sykora. Generic FG-SPA update rules design of linear canonical message representation. Technical Report TD07036, 2013.

[PSH$^+$11] J. Poutanen, J. Salmi, K. Haneda, V.-M. Kolmonen, and P. Vainikainen. Angular and Shadowing Characteristics of Dense Multipath Components in Indoor Radio Channels. *IEEE Transactions on Antennas and Propagation*, 59(1): 1–9, 2011.

[PT14] G. Pedersen and A. Tatomirescu. Antenna performance labelling of mobile phones in EU+ performance of today's phones. Technical Report TD-14-10077, Aalborg, Denmark, May 2014.

[PT12] S. Payami and F. Tufvesson. Channel measurements and analysis for very-large array systems at 2.6 GHz. In *Proceedings of the EuCAP 2012 – 6th European Conference in Antennas and Propagation*, pp. 433–437, March 2012.

[PVTL11] V. Plicanic, I. Vasilev, R. Tian, and B. K. Lau. Capacity maximisation of handheld MIMO terminal with adaptive matching in indoor environment. *Electronics Letters*, 47(16):900–901, 2011. (Also available as TD-11-02054).

[PW10a] N. Patwari and J. Wilson. Rf sensor networks for device-free localization: Measurements, models, and algorithms. *Proceedings of the IEEE*, 98(11):1961–1973, 2010.

[PW10b] M. Patel and Jianfeng Wang. Applications, challenges, and prospective in emerging body area networking technologies. *Wireless Communications, IEEE*, 17(1):80–88, 2010.

[PW13] M. Gan, F. Mani, E. Leitinger, M. Fröhle, C. Oestges, T. Zemen, P. Meissner, and K. Witrisal. On the use of ray tracing for performance prediction of UWB indoor localization systems. In *IC1004 7th MC and Scientific Meeting*, Ilmenau, 2013.

[PWK+08] P. Paschalidis, M. Wisotzki, A. Kortke, W. Keusgen, and M. Peter. A wideband channel sounder for car-to-car radio channel measurements at 5.7 GHz and results for an urban scenario. In *Proceedings VTC 2008 Fall – IEEE 68th Vehicular Technology Conference*, Calgary, Canada, 2008.

Q

[QOHD10] F. Quitin, C. Oestges, F. Horlin, and P. De Doncker. Diffuse Multi-path Component Characterization for Indoor MIMO Channels. In *European Conference on Antennas and Propagation*, pp. 1–5, Barcelona, ES, April 2010.

R

[RALRT+11] J. M. Ruiz-Avilés, S. Luna-Ram'irez, M. Toril, F. Ruiz, I. de la Bandera-Cascales, and P. Munoz-Luengo. Analysis of load sharing techniques in enterprise LTE femtocells. In *Wireless Advanced (WiAd)*, 2011, pp. 195–200, June 2011. (Also available as TD(11)01-012).

[Raschowski] L. Raschowski, P. Kyösti, K. Kusume, and T. Jämsä, Eds. METIS Channel Models, ICT-317669-METIS/D1.4, available online: https://www.metis2020.com/documents/deliverables.

[RBHK15] D. M. Rose, J. Baumgarten, S. Hahn, and T. Kürner. SiMoNe -simulator for mobile networks: system-level simulations in the context of realistic scenarios. In *5th International Workshop on Self-Organizing Networks (IWSON)*, May 2015.

[RBO12] M. Redzic, C. Brennan, and N. E. O'Connor. Advances in indoor location based on signal strength and image data. In *3rd COST IC1004 Management Committee Meeting*, Barcelona, Spain, February 2012 (TD(12)03056).

[RBRH14] D. Reed, R. Borsato, and A. Rodriguez-Herrera. Antenna evaluation with spatial channel models: How standardized channel models are utilized for over-the-air testing. In *2014 8th European Conference on Antennas and Propagation (EuCAP)*, pp. 3483–3487, April 2014.

[RBSK70] G. T. Ruck, D. E. Barrick, W. D. Stuart, and C. K. Krichbaum. *Radar Cross Section Handbook.* Ed. Plenom Press, New York, NY, 1970.

[RD12a] R. Rosini and R. D'Errico. Comparing on-body dynamic channels for two antenna designs. In *Antennas and Propagation Conference (LAPC), 2012 Loughborough*, pp. 1–4, Nov 2012. (Also available as TD(12)05081 with title Dynamic On-Body Channel Characterization: a Comparison for Two Antennas).

[RD12b] R. Rosini and R. D'Errico. Comparing on-body dynamic channels for two antenna designs. In *Proceeding of 8th Loughborough Antennas Propagation Conference (LAPC2009)*, pp. 1–4, Loughborough, UK, 2012.

[RD12c] R. Rosini and R. D'Errico. Off-body channel modeling at 2.45 GHz for two different antennas. In *Proceeding of 6th European Conference on Antennas and Propagation (EuCAP2012)*, Prague, Czech Republic, March 2012.

[RD13] R. Rosini and R. D'Errico. Space-time correlation for on-to-off body channels at 2.45 GHz. In *Proceeding of 7th European Conference on Antennas and Propagation (EuCAP 2013)*, pp. 3529–3533, Gothenburg, Sweden, April 2013. (Also available as TD(13)06061).

[RDV12] R. Rosini, R. D'Errico, and R. Verdone. Body-to-body communications: a measurements-based channel model at 2.45 GHz. In *Proceeding of PIMRC 2012 – IEEE 23th International Symposium on Personal, Indoor and Mobile Radio Communication*, pp. 1763–1768, Sydney, Astralia, September 2012. (Also available as TD(12)04026).

[RE12] C. Rowell and Lam Y. E. Mobile-phone *antenna* design. *IEEE Antennas and Propagation Magazine*, 54(4):14–34, August 2012.

[Rei11] A. Reichman. Scenarios for design and performance evaluation in body environment. Technical Report TD-11-02075, Lisbon, Portugal, October 2011.

[Rei12] A. Reichman. IEEE Standard for Local and Metropolitan Area Networks. IEEE 802.15.6-2012 Part 15.6: Wireless Body Area networks. Technical report, 2012.

[RFMF10] F. Richter, A. J. Fehske, P. Marsch, and G. P. Fettweis. Traffic demand and energy efficiency in heterogeneous cellular mobile radio networks. In *IEEE 71st Vehicular Technology Conference (VTC 2010-Spring)*, pp. 1–6, May 2010.

[RHV$^+$12] D. Robalo, O. Holland, F. J. Velez, J. Oliveira, and A. Aghvami. Energy Saving in the Optimization of the Planning of Fixed WiMAX with Relays in Hilly Terrains: Impact of Sleep Modes and Cell zooming, (invited paper) in Proc. of ISWCS 2012 – The Ninth International Symposium on Wireless Communication Systems (invited Session on Green Wireless Communications), Paris, France, Aug. 2012. (Also avalibale as TD-12-05025).

[Ric05] A. Richter. *Estimation of Radio Channel Parameters: Models and Algorithms.* PhD thesis, Technische Universität Ilmenau, Fakultät für Elektrotechnik und Informationstechnik, Ilmenau, DE, 2005.

[RJ12] R. J. C. Bultitude, T. Smith, D. Cule, and H. Zhu. Comparison of quasi-simultaneous outdoor-to-indoor propagation loss and delay dispersion measurements at 150, 450, and 700 MHz. Technical Report TD(12) 04056, May 2012.

[RJHK13] D. M. Rose, T. Jansen, S. Hahn, and T. Kürner. Impact of realistic indoor mobility modelling in the context of propagation modelling on the user and network experience. In *Proceedings of the 7th European Conference on Antennas and Propagation (EuCAP 2013)*, Gothenbürg, Sweden, April 2013, Technical Report TD(13) 06011, Feb. 2013.

[RJK11] D. M. Rose, T. Jansen, and T. Kürner. Indoor to outdoor propagation measuring and modeling of femto cells in LTE networks at 800 and 2600 MHz. *IEEE GLOBECOM Workshops*, pp. 203–207, Houston, Texas, December 5–9, 2011.

[RJK12] D. M. Rose, T. Jansen, and T. Kurner. Indoor to outdoor propagation – measuring and modeling of femto cells in LTE networks at 800 and 2600 MHz. Technical Report TD(12) 04027, May 2012.

[RJSJ14] T. Rappaport, G. R. MacCartney Jr., M. K. Samimi, and T. Jämsä. Millimeter-wave channel modeling based on measurements in in-building and campus environments at 28 GHz. Technical Report TD(14)10072, Aalborg, Denmark, May 2014.

[RJT+13] D. M. Rose, T. Jansen, U. Türke, T. Werthmann, and T. Kürner. The IC1004 Urban Hannover Scenario 3D Tg3c channel modeling sub-committee final report patterns. *European Cooperation in the Field of Scientific and Technical Research*, COST IC1004 TD(13)08054, Ghent, Belgium, 2013.

[RK12] D. M. Rose and T. Kürner. Outdoor-to-indoor propagation accurate measuring and modeling of indoor environments at 900 and 1800 MHz. In *Proceedings of the 6th European Conference on Antennas and Propagation (EuCAP 2012)*, pp. 1440–1444, Prague, Czech Republic, March 26–30, 2012, Technical Report TD(12) 04028.

[RK13] K. Remley and B. Kaslon. 2 GHz building-penetration channel measurements. Technical report, CTIA, March 2013.

[RK14] D. Rose and T. Kürner. An analytical 3d ray-launching method using arbitrary polygonal shapes for wireless propagation prediction. In *IC1004 10th MC and Scientific Meeting*, Aalborg, 2014.

[RKJZ15] M. Rumney, H. Kong, Y. Jing, and X. Zhao. Advances in antenna pattern-based MIMO OTA test methods. In *9th European Conference on Antennas and Propagation*, Lisbon, Portugal, 2015. (Also available as TD(15)12064).

[RLTR12] J. M. Ruiz-Avilés, S. Luna-Ram'irez, M. Toril, and F. Ruiz. Traffic steering by self-tuning controllers in enterprise ltefemtocells. *EURASIP Journal on Wireless Communications and Networking*, 2012(1), 2012. (Also available as TD(12)03-013).

[RMM+] Rosini R., F. Martelli, M. Maman, R. Verdone, and R. D'Errico. Technical report.

[RMM+12] R. Rosini, F. Martelli, M. Maman, R. D'Errico, C. Buratti, and R. Verdone. On-body area networks: from channel measurements to MAC layer performance evaluation.

In *Proceeding of 18th European Wireless Conference (EW2012)*, Poznan, Poland, April 2012. (Also available as TD(12)03060).

[RO14] O. Renaudin and C. Oestges. UCL-Aalto experimental channel characterization for vehicle-to-vehicle scenarios. In *COST IC1004 9th MC & Scientific Meeting*, number TD(14)09014, 2014.

[ROA] ROADSAFE project. Available: https://portal.ftw.at/projects/roadsafe

[Ros14] R. Rosini. *From Radio Channel Modeling to a System Level Perspective in Body-centric Communications.* PhD thesis, University of Bologna, Bologna, Italy, May 2014.

[RPJZ11] L. Reichardt, J. Pontes, G. Jereczek, and T. Zwick. Capacity maximizing MIMO antenna design for car-to-car communication. In *International Workshop on Antenna Technology (iWAT)*, pp. 243–246, Hong Kong, March 2011. (Also available as TD(12)05035).

[RPL+13] F. Rusek, D. Persson, B. Lau, E. G. Larsson, T. L. Marzetta, OveEdfors, and F. Tufvesson. Scaling up MIMO: opportunities and challenges with very large arrays. *IEEE Signal Processing Magazine*, 30(1):40–60, January 2013.

[RPW14] A. Reichman, M. Priesler, and S. Wayer. Distributed Network Synchronization. Technical Report TD-13-09005, Ferrara, Italy, February 2014.

[RPWD13] A. Reichman, M. Priesler, S. Wayer, and A Danilin. Distributed Network Synchronization. Technical Report TD-13-06039, Málaga, Spain, February 2013.

[RSM+13] T. S. Rappaport, S. Sun, R. Mayzus, H. Zhao, Y. Azar, K. Wang, G. N. Wong, J. K. Schulz, M. Samimi, and F. Gutierrez. Millimeter wave mobile communications for 5G cellular: it will work! *IEEE Access*, 1:335–349, 2013.

[RSP+14] W. Roh, J-Y. Seol, J. Park, B. Lee, J. Lee, Y. Kim, J. Cho, K. Cheun, and F. Aryanfar. Millimeter-wave beamforming as an enabling technology for 5G cellular communications: Theoretical feasibility and prototype results. *IEEE Communication Magazine*, 52(2):106–113, 2014.

[RSSZ12] L. Reichardt, Y. L. Sit, T. Schipper, and T. Zwick. Using the car-to-car communication standard simultaneously for radar sensing and communication. In *COST IC1004 5th MC & Scientific Meeting*, number TD(12)05036, Bristol, UK, September 2012.

[RSZ+13] T. S. Rappaport, S. Sun, H. Zhao, Y. Azar, K. Wang, G. N. Wong, J. K. Schulz, M. Samimi, and F. Gutierrez. Millimeter wave mobile communication for 5G cellular: It will work! *IEEE Access*, 1:335–349, 2013.

[RV13] D. Robalo and F. J. Velez. Wireless networks services and applications parameters characterization and model for the mapping between the quality of service and experience for multimedia applications. Technical Report TD(13)08069, Ghent, Belgium, September 2013.

[RV14] D. Robalo and F. J. Velez. Enhanced Multi-Band Scheduling for Carrier Aggregation in LTEAdvanced Scenarios. Technical Report TD-14-11006, Krakow, Poland, September 2014.

[RVD14] R. Rosini, R. Verdone, and R. D'Errico. Body-to-body indoor channel modelling at 2.45 GHz. *IEEE Transactions Antennas Propagation*, 62(11):5807–5819, 2014. (Also available as TD(13)08055).

[RvDHH13] Ad C. F. Reniers, A. R. van Dommeie, M. D. Huang, and M. H. A. J. Herben. Disturbing effects of microwave probe on mm-wave antenna pattern measurements", Proceedings of the 8th European Conference on Antennas and Propagation (EuCAP2014), The Hague, Netherlands, 6–11 April 2014. (also available as TD-13-06008).

[RvDHH14a] Ad C. F. Renier, A. R. van Dommele, M. D. Huang, and M. H. A. J. Herben. Analyzing the disturbing effects of microwave probe on mm-wave antenna pattern measurements. Technical Report TD-14-09001, Ferrara, Italy, 2014.

[RvDHH14b] A. C. F. Reniers, A. R. van Dommele, M. D. Huang, and M. H. A. J. Herben. Disturbing effects of microwave probe on mm-wave antenna pattern measurements. In *8th European Conference on Antennas and Propagation (EuCAP)*, pp. 161–164, 2014.

[RVPP15] D. Robalo, F. J. Velez, R. R. Paulo, and G. Piro. Extending the LTE-sim simulator with multi-band scheduling algorithms for carrier aggregation in lte-advanced scenarios. Technical Report TD(15)12082, Dublin, Ireland, January 2015.

[RZD12] A. Katalinic, R. Zentner, and T. Delac. Diffraction in urban scenarios: where is the (virtual) source? In *IC1004 5th MC and Scientific Meeting*, Bristol, 2012.

S

[Sab92] W. C. Sabine. *Collected papers on acoustics.* Cambridge: Harvard University Press, 1992.

[SAG14] C. Sun, X. An, and L. Guo. Multi-probe anechoic chamber MIMO-OTA approach and future work. Technical Report TD(14)11003, 11th COST IC1004 Scientific Meeting, Krakow, Poland, September 2014.

[Sal14] S. Salous. Wideband measurements in the 60 GHz band. *COST IC1004, Aalborg, Denmark*, 2014.

[Sam12] W. Sami. How can mobile and broadcasting networks use adjacent bands? Technical Report 2011 Q1, European Broadacasting Union, EBU, 2012.

[San13] Sana Salous, Yuriy Nechaye, Costas Constantinou and Adnan Cheema. On body network measurements in the 60 GHz band using a VNA and a custom designed channel sounder. Technical Report TD(13) 08042, September, 2013.

[San15] Sana Salous, Mustafa Abdallah and Yuteng Gao. Channel sounder in the 29 GHz band for 5G systems. Technical Report TD(15) 12090, January, 2015.

[SAR09] S. Sadr, A. Anpalagan, and K. Raahemifar. Radio resource allocation algorithms for the downlink of multiuser ofdm communication systems. *IEEE Communications Surveys Tutorials*, 11(3):92–106, rd 2009.

[Say10] K. Sayrafian, W.-B. Yang, J. Hagedorn, J. Terrill, and K. Y. Yazdandoost. Channel models for medical implant communication. *International Journal of Wireless Information Networks*, 17(3–4):105–112, 2010.

[SB11] J. Sykora and A. Burr. Layered design of hierarchical exclusive codebook and its capacity regions for HDF strategy in parametric wireless 2-WRC. *IEEE Transactions on Vehicular Technology*, 60(7):3241–3252, 2011.

[SB12] J. Sykora and A. Burr. Design and rate regions of network coded modulation for random channel class in WNC with HDF relaying strategy. Technical Report TD04030, 2012.

[SB13a] J. Sykora and A. Burr. Advances in wireless network coding — the future of cloud communications. In *Proceedings of IEEE International Symposium on Personal, Indoor and Mobile Radio Communications (PIMRC)*, pp. 1–190, London, UK, September 2013 (tutorial).

[SB13b] J. Sykora and A. Burr. Network coded modulation for random channel class in WNC with HDF relaying strategy. *IEEE Communication Letters*, 17(5):818–821, 2013.

[SBB12] U. Schilcher, C. Bettstetter, and G. Brandner. Temporal correlation of interference in wireless networks with rayleigh block fading. *IEEE Transactions on Mobile Computing*, 11(12):2109–2120, 2012.

[Sch12a] Rohde & Schwarz. Extending the two-channel method for additional implementation of channel models. Technical report, 3GPP, 3GPP TSG-RAN WG4 R4-123274, Prague, Czech Republic, May 2012. Parts also available as TD(13)07047.

[Sch12b] Rohde and Schwarz. Results from CTIA's IL/IT test campaign using the two-channel method. Technical report, 3GPP, 3GPP TSG-RAN WG4 R4-125747, Santa Rosa, USA, October 2012. Parts also available as TD(13)07047.

[SCO13a] S. K. Sharma, S. Chatzinotas, and B. Ottersten. Interference alignment for spectral coexistence of heterogeneous networks. *EURASIP Journal On Wireless Communications and Networking*, 46:1–14, 2013.

[SCO13b] N. K. Sharma, S. Chatzinotas, and B. Ottersten. Frequency packing for cognitive dual satellite systems. In *8th Meeting of the Management Committee of COST IC1004 – Cooperative Radio Communications for Green Smart Environments*, 2013.

[SCO14] S. K. Sharma, S. Chatzinotas, and B. Ottersten, "A Hybrid Cognitive Transceiver Architecture: Sensing-Throughput Tradeoff", in Proc. Int. Conf. CROWNCOM, Oulu, Finland, June 2014, pp. 143–149. (Also available as TD(14)09065).

[SCO15] S. K. Sharma, S. Chatzinotas, and B. Ottersten, "Cooperative Spectrum Sensing for Heterogeneous Sensor Networks Using Multiple Decision Statistics", Int. Conf. CROWNCOM 2015, submitted, (Also available as IC1004 TD(14)10049).

[SCR14] S. Salous, A. Cheema, and X. Raimundo. Radio channel propagation measurements and spectrum sensing using an agile chirp sounder. Technical Report TD(14) 09032, February 2014.

[SCS13] S. Subramani, W. H. Chin, and M. Sooriyabandara. Performance analysis of ieee 802.15.6 mac layer access modes. Technical Report TD(13)06030, Malaga, Spain, February 2013.

[SDNH14] S. Salous, V. Degliesposti, M. Nekovee, and S. Hur. Millimeter-wave Propagation Characterization and Modelling Towards 5G Systems. 2015 EuCAP. (Also available as TD(14)10091).

[SF11] W. L. Schroeder and Y. Feng. A critical review of MIMO OTA test concepts – Lessons learned from actual measurements. In *XXX General Assembly and Scientific Symposium of the International Union of Radio Science*, Istanbul, Turkey, August 2011.

[SFC11] T. Snow, C. Fulton, and W. J. Chappell. Transmit–receive duplexing using digital beamforming system to cancel self-interference. *IEEE Transactions on Microwave Theory and Techniques*, 59(12):3494–3503, 2011.

[SFF12] J. Andersen, S. Fu, G. Pedersen, and O. Franek. Accuracy of 3d ray tracing by comparing electromagnetic propagation mechanisms. In *3rd MC and Scientific Meeting*, Barcelona, 2012.

[SFR+14] I. Szini, M. Foegelle, D. Reed, T. Brown, and G. F. Pedersen. On antenna polarization discrimination, validating MIMO OTA test methodologies. *IEEE Antennas and Wireless Propagation Letters*, 13:265–268, 2014.

[SH14] J. Sykora and J. Hejtmanek. Joint and recursive hierarchical interference cancellation with successive compute & forward decoding. Technical Report TD10040, 2014.

[SHJ+09] D. Smith, L. Hanlen, Z. Jian, D. Miniutti, D. Rodda, and B. Gilbert. Characterization of the dynamic narrowband on-body to off-body area channel. In *Proceeding of ICC 2009 – IEEE International Conference on Communication*, pp. 1–6, Dresden, Germany, 2009.

[SHK13] J. Sachs, R. Herrmann, and M. Kmec. Time and range accuracy of short-range ultra-wideband pseudo-noise radar. *Applied Radio Electronics*, 12(1):105–113, 2013.

[SHP14a] S. Baek, Y. Lee, S. Hur, Y. Chang, and J. Park. mmwave channel modeling based on 3d ray-tracing in urban environments. In *IC1004 11th MC and Scientific Meeting*, Krakow, 2014.

[SHP14b] S. Baek, Y. Lee, S. Hur, Y. Chang, and J. Park. mmwave propagation models based on 3d ray-tracing in urban environments. In *IC1004 10th MC and Scientific Meeting*, Aalborg, 2014.

[SHPTG13] J. Sánchez-Heredia, D. Pugachov, Bolin T., and A. M. González. Antenna effect on LTE terminals exposed to realistic fading conditions. Technical Report TD-13-08030, Ghent, Belgium, September 2013.

[Sib13] A. Sibille. Statistical modeling of antennas in the context of radio channel modeling. URSI-GASS, Beijing, China, August, 2014. (Also available as TD(06)083).

[Sib14-LAPC] A. Sibille. Towards standardised statistical models for wireless communications terminals?, 2012 Loughborough Antennas and Propagation Conference (LAPC), November 2014.

[Sib15] A. Sibille. Joint antenna-channel aspects in MIMO terminals performance evaluation. Technical Report TD-15-12027, Dublin, Ireland, January 2015.

[SIH+14] C. Schneider, M. Ibraheam, S. Häfner, M. Käske, M. Hein, and R. S. Thomä. On the reliability of multipath cluster estimation in realistic channel data sets. In *2014 8th European Conference on Antennas and Propagation (EuCAP)*, pp. 449–453, April 2014. (Also available as TD (14)09085).

[SIO14] K. Saito, T. lmai, and Y. Okumura. 2GHz Band MIMO Channel Properties in Urban Small Cell Scenario in Crowded Area. Technical Report TD(14) 10083, May 2014.

[SJ11] J. Sykora and E. A. Jorswieck. Optimized beam-forming and achievable rates of network coded modulation with HDF strategy in 2-source 2-relay network. Technical Report TD01007, 2011.

[SKL+13] R. K. Sharma, W. Kotterman, M. H. Landmann, C. Schirmer, C. Schneider, F. Wollenschläger, G. Del Galdo, M. A. Hein, and R. S. Thomä. Over-the-air testing of cognitive radio nodes in a virtual electromagnetic environment. *International Journal of Antennas and Propagation*, 2013, 2013 (Available as part of TD(13)07039).

[Sko13] O. Skoblikov. Mobility-Based Authentication in Wireless Ad-hoc Networks. Technical Report TD-13-07055, Illmenau, Germany, May 2013.

[SKSO13] S. Chatzinotas, S. K. Sharma, and B. Ottersten. Eigenvalue based blind spectrum sensing techniques. In *7th Meeting of the Management Committee of COST IC1004 – Cooperative Radio Communications for Green Smart Environments*, 2013.

[SKSO14a] S. Chatzinotas, S. K. Sharma, and B. Ottersten. Cooperative spectrum sensing for heterogeneous sensor networks. In *10th Meeting of the Management Committee of COST*

IC1004 – Cooperative Radio Communications for Green Smart Environments, 2014.

[SKSO14b] S. Chatzinotas, S. K. Sharma, and B. Ottersten. Performance analysis of a hybrid transceiver for cognitive radio. In *9th Meeting of the Management Committee of COST IC1004 – Cooperative Radio Communications for Green Smart Environments*, 2014.

[SKST12] G. Sommerkorn, M. Käske, C. Schneider, and R. Thomä. MIMO channel sounding in an urban macro cell with an extreme elevated base station. Technical Report TD(12) 03030, Feb. 2012.

[SKST13] G. Sommerkorn, M. Käske, C. Schneider, and R. S. Thomä. On the reliability of channel synthesis based on realistic multipath parameter from a mimo performance evaluation point of view. Technical Report TD-06-044, Malaga, Spain, February 2013.

[Sla13] S. J. Ambroziak and R. J. Katulski. Empirical propagation model for the container terminal environment. Technical Report TD(13) 08002, September 2013.

[SLC97] J. M. Song, C. C. Lu, and W. C. Chew. Mlfma for electromagnetic scattering from large complex objects. *IEEE Transactions on Antennas Propagation*, 45(10):1488–1493, 1997.

[SLR15] A. Skårbratt, C. P. Lötbäck, and R. Rehammar. Reverberation chamber channel models and MIMO measurements. Technical Report TD(15)12009, 12th COST IC1004 Scientific Meeting, Dublin, Ireland, January 2015.

[SM12] M. Shemshaki and C. F. Mecklenbräuker. Empirical pathloss models applicable for LTE systems in wireless vehicular communications. In *COST IC1004 5th MC & Scientific Meeting*, number TD(12)05008, 2012.

[SM14a] M. Shemshaki and C. F. Mecklenbräuker. Antenna selection algorithm with improved channel predictor for vehicular environment. In *IEEE 6th International Symposium on Wireless Vehicular Communications (WiVeC)*, pp. 1–5, September 2014. (Also available as TD(14)09074).

[SM14b] M. Shemshaki and C. F. Mecklenbräuker. Antenna selection scheme with channel prediction for IEEE 802.11p. In *COST IC1004 10th MC & Scientific Meeting*, number TD(14)10059, Aalborg, Denmark, May 2014.

[SM14c] V. Shivaldova and C. F. Mecklenbräuker. Quantization-based complexity reduction for range-dependent modified

Gilbert model. In *IEEE 8th Sensor Array and Multichannel Signal Processing Workshop (SAM)*, pp. 345–348, A Coruna, Spain, June 2014. (Also available as TD(14)10014).

[SMM03] M. Sabattini, E. Masry, and L. B. Milstein. A non-gaussian approach to the performance analysis of UWB TH-BPPM systems. In *IEEE Conference on Ultra Wideband Systems and Technologies*, pp. 52–55, November 2003.

[SMS+12a] I. S. Comsa, M. Aydin, S. Zhang, P. Kuonen, and J. Wagen. A Dynamic Q-Learning-Based Scheduler Technique for LTE-Advanced Technologies Using Neural Networks. Technical Report TD-12-05026, Bristol, United Kingdom, September 2012.

[SMS+12b] I. S. Comsa, M. Aydin, S. Zhang, P. Kuonen, and J. Wagen. Multi Objective Resource Scheduling using LTE-A Simulator. Technical Report TD-12-03003, Barcelona, Spain, February 2012.

[SMS+12c] I. S. Comsa, M. Aydin, S. Zhang, P. Kuonen, and J. Wagen. Multi Objective Resource Scheduling using LTE-A Simulator. Technical Report TD-12-04047, Lyon, France, May 2012.

[SMS+14] G. Sommerkorn, K. Martinand, C. Schneider, S. Hafner and R. Thomä. "Full 3D MIMO channel sounding and characterization in an urban macro cell", in *XXXIth URSI general assembly and scientific symposium:* (URSI GASS 2014); Beijing, China, 16–23 August, 2014", Piscataway, NJ: IEEE. http://dx.doi.org/10.1109/URSIGASS.2014.6929298 (Also available as TD(14)10042).

[SMW12] Y. Shen, S. Mazuelas, and M. Z. Win. Network navigation: Theory and interpretation. *Selected Areas in Communications, IEEE Journal on*, 30(9):1823–1834, 2012.

[SNB13] G. Song, A. Nix, and M. Beach. Interference analysis with relay deployment. Technical Report TD-08-014, Gent, Belgium, September 2013.

[SNCC13] S. Salous, Y. Nechayev, C. Constantinou, and A. Cheema. On body network measurements in the 60 GHz band using a VNA and a custom designed channel sounder. Technical Report TD-13-08042, Ghent, Belgium, September 2013.

[SNG+11] M. Schack, J. Nuckelt, R. Geise, L. Thiele, and T. Kürner. Comparison of path loss measurements and predictions at urban crossroads for C2C communications. In *Proceedings of the 5th European Conference on Antennas and*

Propagation (EUCAP), pp. 2896–2900, April 2011. (Also available as TD(11)01036).

[SO14] C. S. Sharma, and B. Ottersten. Cooperative spectrum sensing for heterogeneous sensor networks. Technical Report TD-14-10049, Aalborg, Denmark, May 2014.

[SO15] C. S. Sharma, and B. Ottersten. Applications of matching theory in cooperative and cognitive radio networks. Technical Report TD-15-12047, Dublin, Ireland, January 2015.

[Sou92] E. S. Sousa. Performance of a spread spectrum packet radio network link in a Poisson field of interferers. *IEEE Transactions on Information Theory*, 38(6):1743–1754, 1992.

[SPC11] R. Saruthirathanaworakun, J. M. Peha, and L. M. Correia. Opportunistic primary-secondary spectrum sharing with a rotating radar. In *COST IC1004 – Cooperative Radio Communications for Green Smart Environments TD(11) 02014*, October 2011.

[SPDBF12] I. Szini, G. F. Pedersen, S. C. Del Barrio, and M. D. Foegelle. LTE radiated data throughput measurements, adopting MIMO 2×2 reference antennas. In *2012 IEEE Vehicular Technology Conference (VTC Fall)*, pp. 1–5, September 2012.

[SPSF12] I. Szini, G. F. Pedersen, A. Scannavini, and L. J. Foged. MIMO 2×2 reference antennas concept. In *Antennas and Propagation (EUCAP), 2012 6th European Conference on*, pp. 1540–1543, March 2012.

[SPTI13] I. Szini, G. Pedersen, A. Tatomirescu, and A. Ioffe. MIMO 2×2 absolute data throughput concept. In *2013 7th European Conference on Antennas and Propagation (EuCAP)*, pp. 299–302, April 2013.

[SPF$^+$12] G. Steinböck, T. Pedersen, B. H. Fleury, W. Wang, and R. Raulefs. In-room reverberant multi-link channels: Preliminary investigations. In *COST IC1004 3th Scientific Meeting*, pp. 1–7, Barcelona, Spain, February 2012.

[SPF$^+$13a] G. Steinböck, T. Pedersen, B. H. Fleury, W. Wang, and R. Raulefs. Experimental validation of the reverberation effect in room electromagnetics. In *COST IC1004 8th Scientific Meeting*, pp. 1–10, Ghent, BE, September 2013.

[SPF$^+$13b] G. Steinbock, T. Pedersen, B. H. Fleury, Wei Wang, and R. Raulefs. Distance dependent model for the delay power spectrum of in-room radio channels. *Antennas and*

Propagation, IEEE Transactions on, 61(8):4327–4340, 2013.

[SPF15] G. Steinböck, T. Pedersen, and B. H. Fleury. Estimation of the human absorption cross section via reverberation models. In *COST IC1004 12th Scientific Meeting*, pp. 1–5, Dublin, Ireland, January 2015.

[SPG+14] G. Steinboeck, T. Pedersen, M. Gan, T. Zemen, P. Meissner, Leitinger E., and K. Witrisal. Preliminary hybrid model for reverberant indoor radio channels. In *IC1004*, Aalborg, Denmark, May 2014 (TD(14)10056).

[SPP+11] V. Shivaldova, A. Paier, T. Paulin, G. Maier, P. Fuxjäger, A. Alonso, D. Smely, B. Rainer, and C. F. Mecklenbräuker. Overview of an IEEE 802.11p PHY performance measurement campaign 2011. In *COST IC1004 2nd MC & Scientific Meeting*, number TD(11)02056, Lisbon, Portugal, 2011.

[SRC14a] S. Salous, X. Raimundo, and A. Cheema. Small cell wideband measurements in the 60 GHz band. Technical report, TD(14)11024, Krakow, Poland, September 2014.

[SRC14b] S. Salous, X. Raimundo, and A. Cheema. Wideband measurements in the 60 GHz band. Technical report, TD(14)10066, Aalborg, Denmark, May 2014.

[SRL02] C. Savarese, J. M. Rabaey, and K. Langendoen. Robust positioning algorithms for distributed ad-hoc wireless sensor networks. In *Proceedings of the General Track of the Annual Conference on USENIX Annual Technical Conference*, ATEC '02, pp. 317–327, Berkeley, CA, USA, 2002 (USENIX Association).

[SRL+12] Y. L. Sit, L. Reichardt, R. Liu, H. Liu, and T. Zwick. Maximum capacity antenna design for an indoor mimo-UWB communication system. In *10th International Symposium on Antennas, Propagation EM Theory (ISAPE)*, pp. 73–76, October 2012. (Also available as TD-12-05033).

[SRL15] A. Skårbratt, R. Rehammar, and C. P. Lötbäck. LTE adaptive mode and 802.11p OTA measurements in reverberation chamber. Technical Report TD(15)12010, 12th COST IC1004 Scientific Meeting, Dublin, Ireland, January 2015.

[SRR87] A. A. M. Saleh, A. J. Rustako, and R. Roman. Distributed Antennas for Indoor Radio Communications. *IEEE Transactions on Communications*, 35(12):1245–1251, 1987.

[SS10] J. Sommer and J. Scharf. IKR simulation library. In K. Wehrle, M. Günes, and J. Gross (eds), *Modeling and Tools for Network Simulation*, pp. 61–68, Springer, Berlin, 2010.

[SS11] M. Sonkki and E. Salonen. Wideband dual polarized Vivaldi antenna for MIMO-OTA measurement system. Technical Report TD(11)02051, 2nd COST IC1004 Scientific Meeting, Lisbon, Portugal, October 2011.

[SSBS12] W. Sun, E. G. Ström, F. Brännström, and D. Sen. Longterm clock synchronization in wireless sensor networks with arbitrary delay distributions. In *Proceedings of Globecom 2012 – IEEE Global Telecommunications Conference*, December 2012. (Also available as TD(12)05038).

[SSEH+15] M. Sonkki, D. Sanchez-Escuderos, V. Hovinen, E. T. Salonen, and P. Ferrando. Wideband dual-polarized *cross*-shaped vivaldi antenna. *Antennas and Propagation, IEEE Transactions on*, 2015 (In parts also available as TD(11)02051 and TD(12)03058).

[Str13] E. G. Ström. On 20 MHz channel spacing for V2X communication based on 802.11 OFDM. In *39th Annual Conference of the IEEE Industrial Electronics Society (IECON)*, pp. 6891–6896, November 2013. (Also available as TD(13)08006).

[ST94] G. Samorodnitsky and M. S. Taqqu. *Stable Non-Gaussian Random Processes: Stochastic Models with Infinite Variance.* Chapman and Hall, London, 1994.

[ST13a] D. Schulz and R. S. Thomä. Polarimetric quaternion effective aperture distribution function. Technical Report TD-07-028, Ilmenau, Germany, May 2013.

[ST13b] S. Skoblikov and R. S. Thomä. Influence of fixed elevation angles in high resolution parameter estimation using a linear antenna array. Technical Report TD-07-030, Ilmenau, Germany, May 2013.

[Ste15] Stephan Häfner, Diego A. Dupleich, Robert Müller, Christian Schneider, Jian Luo, Egon Schulz, Xiaofeng Lu, Chang Cao, Tianxiang Wang, and Reiner S. Thomä. Characterization of channel measurements at 70 GHz in indoor femtocells. Technical Report TD(15) 12043, January, 2015.

[STFvG13] W. L. Schroeder, A. Tankielun, Y. Feng, and C. von Gagern. Characterization of UE antenna systems by means the two-antenna MIMO OTA measurement method. Technical Report TD(13)06072, 6th COST IC1004 Scientific Meeting, Malaga, Spain, February 2013.

[Stu12] S. Salous, S. Feeney, X. Raimundo, and A. Cheema (2015). Wideband MIMO channel sounder for radio measurements in the 60 GHz band, *IEEE transactions on Wireless Communications*. doi: 10.1109/TWC.2015.2511006

[SV12] Ch. Sergiou and V. Vassiliou. Source-Based Routing Trees for Efficient Congestion Control in Wireless Sensor Networks. Technical Report TD-12-04051, Lyon, France, May 2012.

[SWM13] V. Shivaldova, A. Winkelbauer, and C. F. Mecklenbräuker. Vehicular link performance: From real-world experiments to reliability models and performance analysis. *IEEE Vehicular Technology Magazine*, 8(4):35–44, December 2013. (Also available as TD(13)08077).

[SWM14a] V. Shivaldova, A. Winkelbauer, and C. F Mecklenbräuker. Signal-to-noise ratio modeling for vehicle-to-infrastructure communications. In *Proceedings IEEE 6th International Symposium on Wireless Vehicular Communications (WiVeC)*, pp. 1–5. IEEE, 2014. (Also available as TD(14)11016).

[SWM14b] V. Shivaldova, A. Winkelbauer, and C. F. Mecklenbräuker. Signal-to-noise ratio modeling for vehicle-to-infrastructure communications. In IEEE 6th International Symposium on Wireless Vehicular Communications (WiVeC), Vancouver, Canada, September 2014. (Also available as TD(14)09049).

[SWPR13] B. Soret, H. Wang, K. I. Pedersen, and C. Rosa. Multicell cooperation for lte-advanced heterogeneous network scenarios. *IEEE Wireless Communications*, 20(1):27–34, 2013.

[SYP14] I. Szini, B. Yanakiev, and G. F. Pedersen. MIMO reference antennas performance in anisotropic channel environments. *IEEE Transactions on Antennas and Propagation*, 62(6):3270–3280, 2014.

[Szi11a] I. Szini. Simplified SCME MIMO OTA test method. Technical Report TD(11)01006, 1st COST IC1004 Scientific Meeting, Lund, Sweden, June 2011.

[Szi11b] I. Szini. Simplified SCME test method. Technical Report TD(11)02010, 2nd COST IC1004 Scientific Meeting, Lisbon, Portugal, October 2011.

[Szi14a] I. Szini. Anechoic chamber multi cluster boundary array, an empiric analysis of test volume limitations. Technical Report TD(14)10050, 10th COST IC1004 Scientific Meeting, Aalborg, Denmark, May 2014.

[Szi14b] I. Szini. Measurement campaign on AC MC boundary array, comparing B13 devices in different mechanical modes and test methodologies. Technical Report

TD(14)10086, 10th COST IC1004 Scientific Meeting, Aalborg, Denmark, May 2014.

T

[TAP12] A. Tatomirescu, O. N. Alrabadi, and G. F. Pedersen. Tx-Rx isolation exploiting tunable balanced – unbalanced antennas architecture. In *2012 Loughborough Antennas and Propagation Conference (LAPC)*, pp. 1–4, November 2012.

[TAP13] A. Tatomirescu, O. Alrabadi, and G. F. Pedersen. Optimal placement of MIMO antenna pairs with different quality factors in smart-phone platforms. In *7th European Conference on Antennas and Propagation (EuCAP)*, pp. 732–735, April 2013. (Also available as TD-12-05053).

[TBP13] A. Tatomirescu, E. Buskgaard, and G. F. Pedersen. Compact MIMO antenna for multi-band operation. Technical Report TD-13-08074, Ghent, Belgium, September 2013.

[TC11] V. Tralli and A. Conti. Energy efficiency of relay-assisted communications with interference. In *Proceedings of the 4th International Symposium on Applied Sciences in Biomedical and Communication Technologies, ISABEL 2011, Barcelona, Spain*, page Art. No. 182, October 26–29, 2011.

[TC12] V. Tralli and A. Conti. Energy efficiency of relay-assisted communications with interference. Technical Report TD-03-062, Barcelona, Spain, February 2012.

[TC13] T. Tao and A. Czylwik. Beamforming design and relay selection for multiple MIMOAF relay systems with limited feedback. Technical Report TD-06-025, Malaga, Spain, February 2013.

[TCK12] S. Torrico, K. L. Chee, and T. Kurner. A propagation prediction model in vegetated residential environments a simplified analytical approach. Technical Report TD(12) 03014, February 2012.

[TCZ12] J. D. Ser, T. Chen, M. Matinmikko, and J. Zhang. Energy efficient spectrum access in cognitive wireless access networks. In *5th Meeting of the Management Committee of COST IC1004 – Cooperative Radio Communications for Green Smart Environments*, 2012.

[TDEV+14] L. Tian, V. Degli-Esposti, E. M. Vitucci, X. Yin, F. Mani, and S. X. Lu. Deterministic modeling of diffuse scattering component based on propagation graph theory. In *Proceedings of the 25th IEEE annual International Symposium on Personal, Indoor and Mobile Radio Communications* (PIMRC '14), Washington DC, USA, 2–5 September 2014.

[TdPE02] S. Thoen, L. Van der Perre, and M. Engels. Modelling the channel time-variance for fixed wireless communications. *IEEE Communications Letters*, 6(8):331–333, 2002.

[TGJ+12] E. Tanghe, D. P. Gaillot, W. Joseph, M. Liénard, P. Degauque, and L. Martens. Robustness of high-resolution channel parameter estimators in presence of dense multipath components. *Electronics Letters*, 48(2):130–132, 2012. (Also available as TD (12)04013).

[THR+00] R. S. Thomä, D. Hampicke, A. Richter, G. Sommerkorn, A. Schneider, U. Trautwein, and W. Wirnitzer. Identification of time-variant directional mobile radio channels. *IEEE Instrumentation and Measurement Society*, 49(2):357–364, 2000.

[Tia] Y. Tian. *Evaluation of Diversity and Adaptive Antenna Techniques on a Small-Cell LTE Base-Station Prototype.* PhD thesis, University of Bristol. MSc. Thesis.

[TJT12] T. Werthmann, T. Jansen, and U. Tuerke. Theic 1004 urban hannover scenario – 3d pathloss predictions and realistic traffic and mobility patterns. In *IC1004 5th MC and Scientific Meeting*, Bristol, 2012.

[TKH+12] K. Takizawa, M. Kyrö, K. Haneda, H. Hagiwara, and P. Vainikainen. Performance evaluation of 60 GHz radio systems in hospital environments. In *IEEE International Conference on Communication (ICC 2012)*, pp. 3291–3295, June 2012. (Also available as TD(11)02055).

[TKT14] R. Tsuji, M. Kim, and J. Takada. Visualization of the propagation channels and interpretation of the mechanisms in macro cellular environments at 11 GHz. Technical Report TD(14)09045, Ferrara, Italy, February 2014.

[TL12] R. Tian and B. K. Lau. Effective degree-of-freedom of a compact six-port MIMO antenna. In *2012 IEEE Antennas and Propagation Society International Symposium (APSURSI),* Chicago, USA, July 2012. (Also available as TD-12-03050).

[TLU13] S. A. Torrico, R. H. Lang, and C. Utku. Radiative transport theory vs. monte-carlo simulations: Propagation loss predictions for mobile-to-mobile communications in a trunk dominated park environment. Technical Report TD(13)08013, Ghent, Belgium, September 2013.

[TLY11] R. Tian, B. K. Lau, and Z. Ying. Multiplexing efficiency of MIMO antennas. *IEEE Antennas Wireless Propagation Letter*, 10:183–186, 2011. (Also available as TD-11-01025).

[TLY12] R. Tian, B. K. Lau, and Z. Ying. Multiplexing efficiency of MIMO antennas with user effects. In *2012 IEEE Antennas and Propagation Society International Symposium (APSURSI),* Chicago, USA, July 2012. (Also available as TD-12-03051).

[TMS09] R. Tandra, S. M. Mishra, and A. Sahai. What is a spectrum hole and what does it take to recognise one? *Proceedings of the IEEE*, 97(5):824–848, 2009.

[TN95] G. A. Tsihrintzis and C. L. Nikias. Performance of optimum and suboptimum receivers in the presence of impulsive noise modelled as an α-stable process. *IEEE Transaction Commununication*, 43:904–914, 1995.

[TNB14a] Y. Tian, A. Nix, and M. Beach. 4G Femtocell LTE Base-Station with Diversity and Adaptive Antenna. In *10th International Conference on Wireless Communications, Networking and Mobile Computing (WiCOM)*, pp. 1–7, September 2014.

[TNB14b] Y. Tian, A. Nix, and M. Beach. Improve the degree of freedom with mixed CSIT in distributed cooperative relay system. Technical Report TD-10-006, Aalborg, Denmark, May 2014.

[TP13] A. Tatomirescu and G. Pedersen. Body-loss for popular thin smart phones. Technical Report TD-13-06049, Malaga, Spain, February 2013.

[TP14] A. Tatomirescu and G. F. Pedersen. Reconfigurable dual-band PCB antenna. Technical Report TD-14-10081, Aalborg, Denmark, May 2014.

[TPFP11] A. Tatomirescu, M. Pelosi, O. Franek, and G. F. Pedersen. Port isolation methods for compact MIMO antennas. Technical Report TD-11-02045, Lisbon, Portugal, October 2011.

[TPFP12] A. Tatomirescu, M. Pelosi, O. Franek, and G. Pedersen. The user's body effect on decoupling network for compact MIMO hansets. Technical Report TD-12-03071, Barcelona, Spain, February 2012.

[TPK+11] A. Tatomirescu, M. Pelosi, M. B. Knudsen, O. Franek, and G. F. Pedersen. Port isolation method for MIMO antenna in small terminals for next generation mobile networks.

In *2011 IEEE Vehicular Technology Conference (VTC Fall)*, pp. 1–5, 2011.

[Tuf14] F. Tufvesson. Propagation channel models for next-generation wireless communications systems, COST IC1004 TD(14)10071.

[TvG14] A. Tankielun and C. von Gagern. On spatial characteristics of the decomposition method in MIMO OTA testing. In *European Conference on Antennas and Propagation (EuCAP 2014)*, The Hague, The Netherlands, April 2014.

[TW12 W. G. Teich and P. Wallner. Soft iterative interference cancellation with SOR for digital transmission schemes based on multiple sets of orthogonal spreading codes. In *2012 6th International Conference on Signal Processing and Communication Systems (ICSPCS)*, pp. 1–5, Gold Coast, QLD, December 2012. (Also available as TD(06)066).

[TZY12] L. Tian, X. Zhou, and X. Yin. Experimental investigation on joint spatial and polarisation cross-correlation of propagation channel in indoor distributed antenna systems. In *IC1004*, Lyon, France, May 2012 (TD(12)04016).

U

[UE13] Z. Utkovski and T. Eftimov. Non-coherent two-way relaying with amplify-and-forward. Technical Report TD-07-032, Ilmenau, Germany, May 2013.

[UF12] S. Uppoor and M. Fiore. Urban road traffic modeling and impact on vehicular networking. In *COST IC1004 4th MC & Scientific Meeting*, number TD(12)04009, Lyon, France, May 2012.

[UP12] Z. Utkovski and P. Popovski. Non-coherent and semi-coherent denoise-and-forward schemes for two-way relaying. Technical Report TD03026, 2012.

[UPS14] T. Uricar, P. Prochazka, and J. Sykora. Hierarchical interference cancellation in BPSK 3-source multiple access channel with wireless physical layer network coding. Technical Report TD09068, 2014.

[US12a] T. Uricar and J. Sykora. Hierarchical network code mapper design for adaptive relaying in butterfly network. Technical Report TD03048, 2012.

[US12b] T. Uricar and J. Sykora. Non-uniform 2-slot constellations: Design algorithm and 2-way relay channel performance. Technical Report TD04041, 2012.

[USQH13] T. Uricar, J. Sykora, B. Qian, and W. Ho Mow. Wireless (physical layer) network coding in 5-node butterfly network: superposition coding approach. Technical Report TD06026, 2013.

[UTFB14] S. Uppoor, O. Trullols-Cruces, M. Fiore, and J. M. Barcelo-Ordinas. Generation and analysis of a large-scale urban vehicular mobility dataset. *IEEE Transactions on Mobile Computing*, 13(5):1061–1075, May 2014.

[UVPD13] U. Urosevic, Z. Veljovic, and M. Pejanovic-Djurisic. A solution for increasing energy efficiency and code rate of simple cooperative relaying. Technical Report TD-08-031, Gent, Belgium, September 2013.

V

[VA90] M. K. Varansi and B. Aazhang. Multistage detection in asynchronous code-division multiple-access communications. *IEEE Transactions on Communication*, 38(4):509–519, 1990.

[VAN+14] D. Vlastaras, T. Abbas, M. Nilsson, R. Whiton, M. Olback, and F. Tufvesson. Impact of a truck as an obstacle on vehicle-to-vehicle communications in rural and highway scenarios. In *Proceedings 6th International Symposium on Wireless Vehicular Communications.* IEEE, 2014. (Also available as TD(14)10019).

[Vas84] K. S. Vastola. Threshold detection in narrow-band non-gaussian noise. *IEEE Transactions on Communication*, 32(2): 134–139, 1984.

[VBM+14] J. Vychodil, J. Blumenstein, T. Mikulasek, A. Prokes, and V. Derbek. Measurement of in-vehicle channel feasibility of ranging in UWB and MMW band. In *COST IC1004 11th MC & Scientific Meeting.*

[VBB12] M. Varga, M. A. Badiu, and V. Bota. Mean mutual information based link adaptation algorithm for cooperative relaying in wireless systems. Technical Report TD-04-034, Lyon, France, May 2012.

[VDEM+13] E. M. Vitucci, V. Degli-Esposti, F. Mani, and C. Oestges. Analysis and modeling of the polarisation characteristics of diffuse scattering in indoor and outdoor radio propagation. In *Proceedings of 21st International Conference on Applied Electromagnetics and Communications* (ICECom 2013), Dubrovnik, Croatia, October 14–16, 2013.

[VDEMO13] E. M. Vitucci, V. Degli-Esposti, F. Mani, and C. Oestges. Analysis and modeling of the polarisation characteristics of diffuse scattering in indoor and outdoor radio propagation. In *Proceedings of 21st International Conference on Applied Electromagnetics and Communications* (ICECom 2013), Dubrovnik, Croatia, October 14–16, 2013.

[Ver12] R. Verdone, and A. Zanella, Eds. Pervasive Mobile and Ambient Wireless Communications – COST 2100. Springer Verlag, London, 2012.

[VFDE14] E. M. Vitucci, F. Fuschini, and V. Degli-Esposti. Ray tracing simulation of the radio channel time- and angle-dispersion in large indoor environments. In *Proceedings of the 8th European Conference on Antennas and Propagation* (EuCAP 2014), The Hague, The Netherlands, 6–11 April 2014.

[VFL13] I. Vasilev, E. Foroozanfard, and B. K. Lau. Adaptive impedance matching performance of MIMO terminals with different bandwidth and isolation properties in realistic user scenarios. In *7th European Conference on Antennas and Propagation (EuCAP)*, pp. 2590–2594, Gothenburg, Sweden, April 2013. (Also available as TD-13-06069).

[VHK$^+$13] U.-T. Virk, K. Haneda, V.-M. Kolmonen, J. F. Wagen and P. Vainikainen, "Full-wave characterization of indoor office environment for accurate coverage analysis," International Conference on Electromagnetics in Advanced Applications (ICEAA'13), Turin, Italy, September 2013. (Also available as TD(13)08071).

[VHK$^+$14] U. T. Virk, K. Haneda, V.-M. Kolmonen, P. Vainikainen, and Y. Kaipainen. Characterization of vehicle penetration loss at wireless communication frequencies. In *Proceedings 8th European Conference on Antennas and Propagation (EuCAP)*, pp. 234–238, April 2014. (Also available as TD(13)06080).

[VHW14] U.-T. Virk, K. Haneda and J.-F. Wagen. Dense multipath components add-on for COST 2100 channel model. In *Proceedings of 9th European Conference on Antennas and Propagation* (EuCAP 2015), Lisbon, Portugal, April 2015.

[VLDB14a] E. Van Lil and J. W. De Bleser. On the accuracy of MoM solutions for RCS and propagation computations. In *URSI General Assembly and Scientific Symposium Proceedings*, Beijing, China, August 2014.

[VLDB14b] E. Van Lil, J. De Bleser, "On the Accuracy of MoM Solutions for RCS and Propagation Computations", URSI GASS, pp. p. 1–p. 5, Beijing, China, 16–23 August 2014. (Also available as TD(14)09058).

[VM13] A. B. Vallejo-Mora. A Sensitivity Analysis of Uplink Power Control Parameters in LTE Networks. Technical Report TD-13-06079, Málaga, Spain, February 2013.

[VMDE+11] E. M. Vitucci, F. Mani, V. Degli-Esposti, and C. Oestges. "Polarimetric Properties of Diffuse Scattering from Building Walls: Experimental Parameterization of a Ray-Tracing Model", IEEE Transactions on Antennas and Propagation, Vol. 60 No. 6, pp. 2961–2969, June 2012.

[VMTK13] P. N. Vasileiou, K. Maliatsos, E. D. Thomatos, and A. G. Kanatas. Reconfigurable orthonormal basis patterns using ESPAR antennas. *IEEE Antennas Wireless Propagation Letters*, 12:448–451, 2013. (Also available as TD-12-05013).

[VO13] E. Vinogradov and C. Oestges. Modelling time-variant fast fading statistics in indoor peer-to-peer scenarios. In *IC1004*, Ghent, Belgium, September 2013 (TD(13)08019).

[VPB14] M. Varga, D. G. Popescu, and V. Bota. Hermite polynomials based determination of the mean mutual information per coded bit for cooperative H-ARQ. Technical Report TD-10-047, Aalborg, Denmark, May 2014.

[VPL13] I. Vasilev, V. Plicanic, and B. K. Lau. On user effect compensation of MIMO terminals with adaptive impedance matching. In *2013 IEEE Antennas and Propagation Society International Symposium (APSURSI)*, pp. 174–175, Orlando, USA, July 2013. (Also available as TD-13-07033).

[VPTL13] I. Vasilev, V. Plicanic, R. Tian, and B. K. Lau. Measured adaptive matching performance of a MIMO terminal with user effects. *IEEE Antennas Wireless Propagation Letter*, 12:1720–1723, 2013. (Also available as TD-14-09050).

[VRLO+09] S. van Roy, L. Liu, C. Oestges, and P. De Doncker. An ultra-wideband SAGE algorithm for body area networks. In *Proceeding of International Conference on Electromagnetics in Advanced Applications (ICEAA 2009)*, pp. 584–587, September 2009.

[VRQL+13] S. Van Roy, F. Quitin, L. Liu, C. Oestges, F. Horlin, J. Dricot, and P. De Doncker. Dynamic channel modeling for multi-sensor body area networks. IEEE Transactions on *Antennas and Propagation*, 61(4):2200–2208, 2013. (Also available as TD(12)04001 with title Dynamic channel modeling and validation for multisensor body area networks).

[VTF+14] E. M. Vitucci, L. *Tarlazzi*, F. Fuschini, P. Faccin, and V. Degli-Esposti. Interleaved MIMO-DAS for Indoor Radio Coverage: Concept and Performance Assessment. *IEEE Transaction on Antennas and Propagation*, 62(6):3299–3308, 2014.

[VTK10] C. Votis, G. Tatsis, and P. Kostarakis. Envelope correlation parameter measurements in a MIMO antenna array configuration. *International Journal of Communication, Network, and System Science*, 3(4):350–354, 2010.

[VTNH14] H. V. Vu, N. H. Tran, T. V. Nguyen, and S. I. Hariharan. Estimating shannon and constrained capacities of Bernoulli–Gaussian impulsive noise channels in Rayleigh fading. *IEEE Transactions Communication*, 62(6):1845–1856, 2014.

[VVH13] P. Vainikainen, U. T. Virk, and K. Haneda. Signal flow graph based framework for physical stochastic radiowave propagation models – example: Indoor propagation modeling for mobile communications and navigation. Technical Report TD-08-070, Ghent, Belgium, September 2013.

[VVW13] U. T. Virk, V. Viikari, and J. F. Wagen. Full wave time domain propagation modelling for radio system design. Technical Report TD(13)07057, Ilmenau, Germany, May 2013.

[VVMGSH09] J. F. Valenzuela-Valdes, A. M. Martinez-Gonzalez, and D. A. Sanchez-Hernandez. Diversity gain and MIMO capacity for nonisotropic environments using a reverberation chamber. *IEEE Antennas Wireless Propagation Letters*, 8:112–115, 2009.

[VWMG12] D. Vargas, A. Winkelbauer, G. Matz, and D. Gomez-Barquero. Low complexity iterative MIMO receivers for DVB-NGH with MMSE demapping and log-likelihood ration quantization. Technical Report TD-03-061, Barcelona, Spain, February 2012.

[VZ12] R. Verdone and A. Zanella. *Pervasive Mobile and Ambient Wireless Communications: COST Action 2100. Springer-Verlag London Ltd.* 2012 ISSN 1860–4862.

W

[WAM+] W. Thompson, A. Younis, M. Beach, H. Haas, J. McGeehan, P. Grant, P. Chambers, Z. Chen, and C. X. Wang. Technical report.

[WAM14] K. Wu, K. Anwar and T. Matsumoto, "BICM-ID-based IDMA: Convergence and Rate Region Analyses", IEICE Trans. on Communications, Vol. E97-B, No. 7, July, 2014. (Also available as TD(14)09024).

[Wan13] S. Wang. Periodic spectrum management in cognitive relay networks. In *6th Meeting of the Management Committee of COST IC1004 – Cooperative Radio Communications for Green Smart Environments*, 2013.

[Wan14] S. Wang. Dynamic body channel measurement in different scenarios. Technical Report TD-14-11053, Krakow, Poland, September 2014.

[WB10] C. Wright and S. Basuki. Utilizing a channel emulator with a reverberation chamber to create the optimal MIMO OTA test methodology. In *Mobile Congress (GMC), 2010 Global*, pp. 1–5, October 2010.

[WB14a] S. Wang and M. Z. Bocus. Dynamic body channel measurement and key parameters exploitation in a periodic walking scenario. Technical Report TD-14-11055, Krakow, Poland, September 2014.

[WB14b] Y. Wang and A. Burr. Physical-layer network coding via low density lattice codes. 2014 European Conference on Networks and Communications (EuCNC), vol., no., pp.1, 5, 23–26 June 2014. (Also available as TD(14)10067).

[WBN04] M. Webb, M. Beach, and A. Nix. Capacity limits of MIMO channels with co-channel interference. In *2004 IEEE 59th* Vehicular *Technology Conference, 2004. VTC 2004-Spring*, Vol. 2, pp. 703–707, 2004. (Also available as TD(12)05047).

[WBN+09] Yu Wang, I. B. Bonev, J. O. Nielsen, I. Z. Kovacs, and G. F. Pedersen. Characterization of the indoor multiantenna body-to-body radio channel. *IEEE Transaction Antennas Propagation*, 57(4):972–979, 2009.

[WCY+12] X. Wu, Z. Cao, F. Yang, Y. Wang, Y. Li, J. He, and Y. Gao. MIMO OTA measurement system analysis proposals for a specific laboratory. Technical Report TD(12)05009, 5th COST IC1004 Scientific Meeting, Bristol, UK, September 2012.

[WCY+13] X. Wu, Z. Cao, F. Yang, Y. Ma, C. Zhang, L. Jin, and Z. Zhang. Proposal and verification of single cluster solution for MIMO OTA in CTIA IL/IT test. Technical Report TD(13)08028, 8th COST IC1004 Scientific Meeting, Gent, Belgium, September 2013.

[WCF13] S. Wang, F. Cao, and Z. Fan. Periodic spectrum management in cognitive relay networks. Technical Report TD-13-06031, Malaga, Spain, February 2013.

[WD92] C. J. C. H. Watkins and P. Dayan. Q-learning. Machine *Learning*, 2(3–4):279–292, 1992.

[Wet 11] M. Wetz. *Transmission methods for wireless multi carrier systems in time-varying environments.* PhD thesis, University of Ulm, Institute of Information Technology, Ulm, Germany, 2011.

[WiS] WiSpry. http://www.wispry.com/products-capacitors.php.

[WKBM11] T. Werthmann, M. Kaschub, C. Blankenhorn, and C. M. Mueller. Approaches for evaluating the application performance of future mobile networks. Technical Report TD(11)01038, Lund, Sweden, June 2011.

[WM12] K. Witrisal and P. Meissner. Performance bounds for multipath-aided indoor navigation and tracking (MINT). In *International Conference on Communications (ICC)*, Ottawa, Canada, October 2012. (Also available as TD(11)02001).

[WMGW12] H. Wymeersch, S. Marano, W. M. Gifford, and M. Z. Win. A machine learning approach to ranging error mitigation for UWB localization. *IEEE Transactions on Communications*, 60(6):1719–1728, 2012.

[WMJ+] W. Thompson, M. Beach, J. McGeehan, A. Younis, H. Haas, P. Grant, P. Chambers, Z. Chen, and C. X. Wang and M. Di Renzo. Technical report.

[WML+14] K. Witrisal, P. Meissner, E. Leitinger, T. Zemen, M. Gan, C. Oestges, B. Fleury, G. Steinboeck, T. Pedersen, and M. Jakobsen. UWB channel Modelling for indoor localization. In *IC1004*, Ferrara, Italy, February 2014 (TD(14)09053).

[WPKW15] Richard J. Weiler, Michael Peter, Wilhelm Keusgen, and MikeWizotski. Millimeter-Wave Channel Sounding of Outdoor Ground Reflections", IEEE Radio Wireless Symposium, San Diego, USA, Jan. 2015. (Also available as TD(15)12049).

[WPS09] M. Z. Win, P. C. Pinto, and L. A. Shepp. A mathematical theory of network interference and its applications. *Proceedings of the IEEE*, 97(2):205–230, 2009.

[WR12] C. Wei and Y. Roblin. Desensitization of ban antenna and ban channel model. Technical Report TD(12)04053, Lyon, France, May 2012.

[WR13] C. Wei, and Y. Roblin. Compact UWB fractal slot ANT for wearable applications. Technical Report TD(13)08067, Ghent, Belgium, September 2013.

[WVH15] U.-T. Virk, J.-F. Wagen and K. Haneda. Simulating specular reflections for point cloud geometrical database of the environment. In *Proceedings of 2015 Loughborough Antennas and Propagation Conference,* Loughborough, UK, November 2015.

X

[XB92] H. H. Xia and H. L. Bertoni. Diffraction of cylindrical and plane waves by an array of absorbing half-screens. *IEEE Transactions on Antennas Propagation*, 40(2):170–177, 1992.

[XB12] N. Xie and A. Burr. Distributed Cooperative Spatial Multiplexing with Slepian Wolf Code, Vehicular Technology Conference (VTC Spring), 2013 IEEE 77th, vol., no., pp. 1, 5, 2–5 June 2013. (Also available as TD(12)05064).

[XZG14] W. Xing, Y. Zhang, and M. Guizani. Evaluation of diversity and adaptive antenna techniques on a small-cell lte base-*station* prototype. Temporary document, IC1004 COST action, May 2014.

[XPKC12] X. Yin, J. Park, M. Kim, and H. Chung. Stochastic geometrical modelling of small-scale fading cross-correlation for cooperative relay communications. Technical Report TD-12-03068, Barcelona, Spain, February 2012.

[XS12] R. Xavier and S. Salous. Bit error rate comparison of the tapped delay line channel model measured data in a WiMAX IEEE 802.16-d simulator. Technical Report TD(12)05082, Bristol, UK, September 2012.

[XZ13] Z. Xu and T. Zemen. Grassmannian delay-tolerant limited feedback for interference alignment. In *2013 Asilomar Conference on Signals, Systems and Computers*, pp. 230–235, November 2013.

[XZ14] Z. Xu and T. Zemen. Time-variant channel prediction for interference alignment with limited feedback. In *IEEE International Conference on Communications Workshops (ICC), 2011* pp. 653–658, June 2014.

Y

[YCP+15] X. Yan, L. Clavier, G. W. Peters, N. Azzaoui, F. Septier, and I. Nevatn. Skew-t copula for dependence modelling of impulsive (α-stable) interference. In *International Conference on Communications*, 2015.

[YM12] K. Y. Yazdandoost and R. Miura. Creeping wave antenna for body area network communication. Technical Report TD(11)02002, Lisbon, Portugal, October 2012.

[YM13] K. Y. Yazdandoost and R. Miura. Antenna polarization mismatch in ban communications. Technical Report TD(11)02002, Ghent, Belgium, September 2013.

[YNCP12] B. Yanakiev, J. Ø. Nielsen, M. Christensen, and G. F. Pedersen. Long-range channel measurements on small terminal antennas using optics. In *IEEE Transactions on Instrumentation and Measurement*, 61:2749–2758, October 2012.

[Yon07] S.-K. Yong. Tg3c channel modeling sub-committee final report. Technical report, IEEE Tech. Rep. 802.15-07-0584-01-003c, March 2007.

[Yoh12] Yohei Konishi, Minseok Kim, Yuyuan Chang, Jun-ichi Takada. Development of versatile MIMO channel sounder for double directional and multi-link channel characteriation at 11 GHz. Technical Report TD(12) 05051, September, 2012.

[YØCP10] B. Yanakiev, J. Ø. Nielsen, M. Christensen, and G. Fr. Pedersen. Optical measurements on small terminal antennas. Technical Report TD-11-01048, Lund, Sweden, 2010.

[YPKC12] X. Yin, J.-J. Park, M.-D. Kim, and H.-K. Chung. Stochastic geometrical modelling of SSF cross-correlation for cooperative relay communications. Technical Report TD(12) 03068, February, 2012.

[YTB13] Y. Tian, A. Nix, and M. Beach. Evaluation of diversity and adaptive antenna techniques on a small-cell lte base-station prototype. Temporary document, IC1004 COST action, September 2013.

[YWZ14] X. Yin, S. Wang, and N. Zhang. Cluster-of-Scatterer-based Stochastic Channel Modelling for Massive MIMO Scenarios. In *COST IC1004 10th Management Committee and Scientific Meeting*, May 2014.

[YY06] K. Y. Yazdandoost, and R. Kohno. UWB antenna for wireless body area network. In *Proceeding of Asia Pacific Microwave Conference (APMC2006)*, Yokohama, Japan, December 2006.

[YY09] K. Y. Yazdandoost, and R. Kohno. Body implanted medical device communications. *IEICE Transactions on Communication*, E92-B(2):410–417, 2009.

[YY10] K. Y. Yazdandoost, and K. Sayrafian-Pour. Channel model for body area network (ban)- part 15.6: Wireless body area net Sayrafian-Pour works. *IEEEP802.15-08-0780-12-0006*, November 2010.

[YY11a] K. Y. Yazdandoost. Antenna for over body surface communication. In *Proceeding of IEEE Asia Pacific Microwave Conference (APMC2011)*, Melbourne, Australia, December 2011.

[YY11b] K. Y. Yazdandoost and R. Miura. Creeping wave antenna for body area network communication. Technical Report TD(11)02002, Lisbon, Portugal, October 2011.

[YY11c] K. Y. Yazdandoost and R. Miura. In-body channel modelling. Technical Report TD(11)01008, Lund, Sweden, June 2011.

[YY12] K. Y. Yazdandoost. UWB antenna for implanted applications. In *Proceeding of European Microwave Week (EuMW2012)*, Amsterdam, Netherlands, 2012.

[YY13a] K. Y. Yazdandoost. UWB loop antenna for in-body wireless body area network. In *Proceeding of European Conference on antenna and propagation (EuCAP2013)*, Gothenburg, Sweden, April 2013.

[YY13b] K. Y. Yazdandoost and R. Miura. UWB loop antenna for in-body wireless body area network. Technical Report TD(13)07009, Ilmenau, Germany, May 2013.

[YY14a] K. Y. Yazdandoost and R. Miura. Miniaturized UWB antenna for brain-machine-interface. Technical Report TD(14)11002, Krakow, Poland, September 2014.

[YY14b] K. Y. Yazdandoost and R. Miura. UWB antenna and propagation for wireless endoscopy. Technical Report TD(14)10002, Aalborg, Denmark, May 2014.

[YYC$^+$13] X. Yin, H. Yongyu, L. Cen, T. Li, and Z. Zhimeng. A preliminary study on anisotropic characteristics of propagation channels for Tx-Rx polarizations. In *Proceeding of WCNC 2013 – IEEE Wireless Communication and Networking Conference*, pp. 3455–3459, April 2013.

[YYM12a] K. Y. Yazdandoost and R. Miura. Body implanted UWB antenna. Technical Report TD(12)03005, Barcelona, Spain, February 2012.

[YYM12b] K. Y. Yazdandoost and R. Miura. BAN channel mode with multi-scenario applications. Technical Report TD-12-05001, Bristol, UK, September 2012.

[YYSP10] K. Y. Yazdandoost and K. Sayrafian-Pour. Channel mode for Body Area Network. IEEE P802.15 Working Group for Wireless Personal Area Networks, IEEEP802.15-08-0780-12-0006. Technical report, 2010.

[YZX+13] R. Yu, Y. Zhang, S. Xie, Y. Liu, L. Song, and M. Guizani. Secondary users cooperation in cognitive radio networks: Balancing sensing accuracy and efficiency. Technical Report TD-13-06010, Malaga, Spain, February 2013.

Z

[ZB12] K. Zhu and A. Burr. "Iterative non-coherent detection of serially-concatenated codes with differential modulation," Wireless Communications and Networking Conference (WCNC), 2013 IEEE, vol., no., pp. 3969, 3973, 7–10 April 2013. (Also available as TD(12)05062).

[ZBD14] N. Zogovic, D. Bajić, and G. Dimić. A real ARQ Scheme for Improving Decision Space Resolution. In *Proceedings of TELFOR 2014, Belgrade, Serbia*, pp. 328–331, November 25–27, 2014.

[ZBMP13] A. Zanella, A. Bazzi, B. M. Masini, and G. Pasolini. Optimal transmission policies for energy harvesting nodes with partial information of energy arrivals. *PIMRC 2013, London, UK*, pp. 954–959, 8–11 September 2013.

[ZCAM12] X. Zhou, M. Cheng, K. Anwar, and T. Matsumoto. Distributed joint source-channel coding for relay systems exploiting spatial and temporal correlations. Technical Report TD-03-072, Barcelona, Spain, February 2012.

[ZCSE04] K. I. Ziri-Castro, W. G. Scanlon, and N. E. Evans. Indoor radio channel characterization and modeling for a 5.2 GHz bodyworn receiver. *IEEE Antennas Wireless Propagation Letter*, 3(1):219–222, 2004.

[ZDB14] N. Zogović, G. Dimić, and D. Bajić. A "Raised-Fractional-Power" wireless transmitter power consumption model. *29th International Conference on Microelectronics, MIEL 2014, Belgrade, Serbia*, pp. 401–404, May 12–14, 2014.

[Zetll] P. Zetterberg. Experimental investigation of TDDn reciprocity-based zero-forcing transmit precoding.

EURASIP Journal on Advances in Signal Processing, 2011:5, 2011. (Also available as TD (12) 05039).

[ZH14] R. Zentner and A. Hrovat. Aoa, aod, delay and doppler shift calculation within ray entity for arbitrary sloped single diffraction case. In *IC1004 9th MC and Scientific Meeting*, Ferrara, 2014.

[Zha13a] Y. Zhang. Secondary users cooperation in cognitive radio networks: Balancing sensing accuracy and efficiency. In *6th Meeting of the Management Committee of COST IC1004 – Cooperative Radio Communications for Green Smart Environments*, 2013.

[Zha13b] Y. Zhang. Sensing-performance tradeoff in cognitive radio enabled smart grid. In *6th Meeting of the Management Committee of COST IC1004 – Cooperative Radio Communications for Green Smart Environments*, 2013.

[Zha14] H. Zhang. Separation of User and Control Planes in Mobile Radio Networks. Technical Report TD-14-11010, Krakow, Poland, September 2014.

[ZHKT14] M. Zhu, K. Haneda, V. M. Kolmonen, and F. Tufvesson. Parameter Based and Physical Clusters for Channel Modeling in Sub-urban and Urban Scenarios. IEEE Transactions on Antennas and Propagation, in press.

[Zhu] M. Zhu. Geometry-based radio channel characterization and modeling: parameterization, implementation and validation. 2014, Lund University, Sweden. http://lup.lub.lu.se/record/4587340/file/4587341.pdf

[ZK13] R. Zentner and A. Katalinic. Ray tracing interpolation for continuous modelling of double directional radio channel. Proceedings of EUROCON 2013, 1–4 July 2013, Zagreb, Croatia, IEEE, pp. 212–217. (Also available as TD(13)06052).

[ZKD12] R. Zentner, A. Katalinić, and Delac¢. Diffraction in urban scenarios: Where is the (virtual) source? Technical Report TD(12)05028, Bristol, UK, September 2012.

[ZLSH11] S. Zhang, B. K. Lau, A. Sunesson, and S. He. Closely located dual PIFAs with T-slot induced high isolation for MIMO terminals. In *2011 IEEE Antennas and Propagation Society International Symposium (APSURSI)*, vol. 1, pp. 2205–2207, Spokane, USA, July 2011. (Also available as TD-11-01063).

[ZLSH12] S. Zhang, B. K. Lau, A. Sunesson, and S. He. Compact UWB MIMO antenna for USB dongles with angle and polarization diversity. In *International Symposium on Antennas and Propagation (ISAP'2012),* Nagoya, Japan, October 2012. (Also available as TD-12-04040).

[ZM12] P. Zetterberg and N. N. Moghadam. An experimental investigation of SIMO, MIMO, interference-alignment (IA) and coordinated multi-point (CoMP). In *19th International Conference on Systems, Signals and Image Processing (IWSSIP), 2012*, pp. 211–216, April 2012.

[ZMD12] R. Zentner, A. K. Mucalo, and T. Delac. Diffraction in urban scenarios: where is the (virtual) source? Technical Report TD(12) 05028, September 2012.

[ZST12] M. Zhu, A. Singh, and F. Tufvesson. Measurement based ray launching for analysis of outdoor propagation. March 2012. (Also available as TDXXX).

[ZT12] M. Zhu and F. Tufvesson. A study on the relation between parameter based cluster and physical clusters. Technical Report TD-005-066, Bristol, UK, September 2012.

[ZVK+15] Meifang Zhu, Joao Vieira, Yubin Kuang, Andreas F. Molisch, and Fredrik Tufvesson. Tracking and positioning using phase information of multi-path components from measured radio channels. In *Proceeding of ICC 2015 –* IEEE *International Conference on Communication Workshops; Workshop on Advances in Network Localization and Tracking*, submitted for publication 2015. (Also available as TD(14)11037).

[ZY10] L. Zhu and K. L. Yeung. Optimization of resource allocation for the downlink of multiuser miso-ofdm systems. In *IEEE 17th International Conference on Telecommunications (ICT)*, pp. 266–271, April 2010.

[ZYL+14] N. Zhang, X. Yin, S. X. Lu, M. Du, and X. Cai. Measurement-based angular characterization for 72 ghz propagation channels in indoor environments. In *Globecom Workshops (GC Wkshps), 2014*, pp. 370–376, December 2014. (Also available as TD(14) 11032).

Index

About the Author

Prof. Dr. Narcis Cardona

Chairman of EU COST IC1004,
Coordinator of COST IRACON and ARCO5G
Steering Committee Member of METIS and METIS2
Vice-Director of iTEAM Research Institute
Director of Master degree in Mobile Communications
Universitat Politecnica de Valencia
Camino de Vera S/N
46022 Valencia | Spain
Tel: +34 963 879 580 | Fax: +34 963 879 583
ncardona@iteam.upv.es | www.iteam.es

Telecommunications Engineer, Universitat Politecnica de Catalunya, Spain, 1990

PhD in Telecommunications Engineering, Universitat Politecnicade Valencia, Spain, 1995

Professional Biography

MsC (1990), PhD (1995), Prof. (2001). Since October 1990 he is with the Communications Department of the Polytechnic University of Valencia (UPV), currently Full Professor on Signal Theory and Communications. Prof. Cardona is in head of the Mobile Communications Group at the UPV, with 30 researchers including assistant professors & research fellows. Additionally he is the Director of the Mobile Communications Master Degree (since 2006) and Vice-Director of the Research Institute of Telecommunications and Multimedia Applications (iTEAM) since 2004. Prof. Cardona has led National research projects and has participated to European projects, Networks of Excellence and other research forums in FP6, FP7 and H2020, always in Mobile Communications aspects. At European scale he has been Vice-Chairman of COST273 Action (2003–2006), Chairman of the EU Action COST IC1004 (2011–2015), coordinator of the National Network of

Excellence ARCO5G since January 2015, Vice-Chairman of the COST Action IRACON from March 2016, member of the Steering Board of METIS (7FP; 2011–2015), METIS2 (H2020 5GPPP; 2015–2017), and WIBEC (H2020 ITN; 2016–2019). His current research topics are Radio wave propagation, Planning and Optimisation of Mobile Access Networks, Digital Multimedia Broadcasting, Dynamic Spectrum Management and Wireless Body Environment Communications.